Springer Proceedings in Mathematics & Statistics

Volume 25

For further volumes:
http://www.springer.com/series/10533

Springer Proceedings in Mathematics & Statistics

This book series features volumes composed of select contributions from workshops and conferences in all areas of current research in mathematics and statistics, including OR and optimization. In addition to an overall evaluation of the interest, scientific quality, and timeliness of each proposal at the hands of the publisher, individual contributions are all refereed to the high quality standards of leading journals in the field. Thus, this series provides the research community with well-edited, authoritative reports on developments in the most exciting areas of mathematical and statistical research today.

Dmitriy Bilyk • Laura De Carli
Alexander Petukhov • Alexander M. Stokolos
Brett D. Wick
Editors

Recent Advances in Harmonic Analysis and Applications

In Honor of Konstantin Oskolkov

Editors
Dmitriy Bilyk
Department of Mathematics
University of South Carolina
Columbia, South Carolina
USA

Laura De Carli
Department of Mathematics
Florida International University
University Park, Miami, Florida
USA

Alexander Petukhov
Department of Mathematics
University of Georgia
Athens, Georgia
USA

Alexander M. Stokolos
Department of Mathematics
Georgia Southern University
Statesboro, Georgia
USA

Brett D. Wick
Department of Mathematics
Georgia Institute of Technology
Atlanta, Georgia
USA

ISSN 2194-1009 ISSN 2194-1017 (electronic)
ISBN 978-1-4899-9346-5 ISBN 978-1-4614-4565-4 (eBook)
DOI 10.1007/978-1-4614-4565-4
Springer New York Heidelberg Dordrecht London

Mathematics Subject Classification (2010): 42-XX, 35-XX, 11Zxx, 5T60, 8Wxx

Printed on acid-free paper

Springer is part of Springer Science+Business Media (www.springer.com)

Konstantin Oskolkov
Drawn by Nikolay Oskolkov

Preface

This volume is dedicated to Konstantin Oskolkov's 65th birthday and is a celebration of his contributions to mathematical analysis. It grew out of the AMS Sectional Meeting held at Georgia Southern University, March 11–13, 2011. Many of the chapters appearing in this volume are close to Kostya's broad mathematical interests.

The editors wish to thank all those who made this volume possible. Special thanks go to all the authors who have contributed chapters to this volume. We also want to acknowledge the time and the hard work of the referees.

Columbia, SC, USA	Dmitriy Bilyk
Miami, FL, USA	Laura DeCarli
Athens, GA, USA	Alexander Petukhov
Statesboro, GA, USA	Alexander M. Stokolos
Atlanta, GA, USA	Brett D. Wick

Preface

Contents

Part I
Konstantin Oskolkov

On the Scientific Work of Konstantin Ilyich Oskolkov

Dmitriy Bilyk, Laura De Carli, Alexander Petukhov, Alexander M. Stokolos, and Brett D. Wick

Abstract This chapter is a brief account of the life and the scientific work of K.I. Oskolkov.

Konstantin Ilyich Oskolkov, or Kostya for his friends and colleagues, was born in Moscow on February 17, 1946. Kostya's father, Ilya Nikolayevich, worked as an engineer at the Research Institute of Cinema and Photography. His mother, Maria Konstantinovna, was a distinguished pediatric cardiology surgeon. Since Maria's father was a priest, during Stalin's purges, her parents had to hide away, and for a long time, she grew up without them and was forced to hide her background. Kostya's paternal grandfather, Nikolay Innokent'evich Oskolkov, was a famous engineer who built bridges, dams, and subways across all of Russia and USSR. At the age of 25, he directed the reconstruction of the famous Borodinsky bridge in Moscow, giving the bridge the look that it still has today. Nikolay Innokent'evich's

D. Bilyk
University of South Carolina, Columbia, SC 29208, USA
e-mail: bilyk@math.sc.edu

L. De Carli
Florida International University, Miami, FL 33199, USA
e-mail: decarlil@fiu.edu

A.M. Stokolos (✉)
Georgia Southern University, Statesboro, GA 30460, USA
e-mail: astokolos@georgiasouthern.edu

A. Petukhov
University of Georgia, Athens, GA 30602, USA
e-mail: petukhov@math.uga.edu

B.D. Wick
Georgia Institute of Technology, Atlanta, GA 30308–0332, USA
e-mail: wick@math.gatech.edu

D. Bilyk et al. (eds.), *Recent Advances in Harmonic Analysis and Applications*, Springer Proceedings in Mathematics & Statistics 25, DOI 10.1007/978-1-4614-4565-4_1,

wife, Anna Vladimirovna Speer, came from the lineage of Karl von Knorre, a famous astronomer, a student of V.Ya. Struve, and the founder and director of the Nikolaev branch of the Pulkovo observatory.

The early 1970s was a time of scientific bloom in the USSR. Physicists, engineers, and mathematicians were honored members of the society—newspaper articles, movies, and TV shows were created about them. It was during this time that Kostya's academic career began. In 1969, Kostya graduated with distinction from the Moscow Institute of Physics and Technology, one of the leading institutions of Soviet higher education specializing in science and technology, with a major in applied mathematics. One of Kostya's professors was Sergey Alexandrovich Telyakovskii, who encouraged Kostya to start graduate school at the Steklov Mathematical Institute of the Academy of Sciences of USSR under his supervision. In 1972, Kostya received the degree of Candidate of Sciences (the equivalent of Ph.D.), and then in 1979 at the same institute, he defended the dissertation for the degree of Doctor of Sciences (Dr. Hab.), a nationally recognized scientific degree which was exceptionally hard to achieve.

The beginning of Kostya's scientific work coincided with a revolutionary period of breakthrough results in multidimensional harmonic analysis. In 1971, Ch. Fefferman [72] proved the duality of the real Hardy space H^1 and BMO. In that same year, Fefferman [71] constructed an example of a continuous function on the two-dimensional torus whose rectangular Fourier series diverges almost everywhere. In 1972 L. Carleson and P. Sjölin [70] found the sharp region of L^p-convergence of two-dimensional Bochner-Riesz averages. In 1972 Fefferman [73] disproved a long-standing "disc multiplier" conjecture by showing that the spherical sums of multidimensional Fourier series converge in the L^p norm only in the trivial case $p = 2$.

In the 1970s, the Function Theory seminar at Moscow State University was led by D.E. Menshov and P.L. Ulyanov. During that time, an extremely talented group of mathematicians working in harmonic analysis, approximatively of Kostya's age, was active in Moscow. Notable names include S.V. Bochkarev, B.S. Kashin, E.M. Nikishin, and A.M. Olevskii. It was in this academic environment that Kostya began his career. His research activity was also greatly influenced by such well-known Soviet mathematicians as members of the Academy of Sciences S.M. Nikol'skii and L.S. Pontryagin, as well as his Ph.D. advisor S.A. Telyakovskii.

Between 1972 and 1991, Kostya worked at the Steklov Institute. Together with Boris Kashin, they led a seminar. The atmosphere of this seminar was extremely welcoming and informal. Both supervisors always tried to encourage the speakers and provide suggestions on how they could improve the results or the presentation (which was not very typical in the Russian academia). He also worked at the Department of Computational Mathematics and Cybernetics of Moscow State University, where he taught one of the main courses on Optimal Control.

Much of Kostya's time and effort was invested into the collaboration between the Academy of Sciences of USSR and Hungary. In particular, for a long time, he was an editor of the journal "Analysis Mathematica."

Kostya extensively traveled to different cities and towns of the Soviet Union, where he lectured on various topics, served as an opponent in dissertation defenses, and chaired the State Examination Committee. In the former USSR, where much of the scientific activity and potential was concentrated in big centers like Moscow or Leningrad, such visits greatly enriched the mathematical life of other cities. In particular, Kostya often visited Odessa. Numerous mathematicians from Odessa have been inspired by their communication with Kostya. The papers of V. Kolyada, V. Krotov, A. Korenovsky, P. Oswald, and A. Stokolos in the present volume attest to this fact.

At that time Kostya was one of few members of the Steklov Institute who spoke English and German fluently. Because of that, he was constantly involved in receiving frequent foreign visitors to the institute, which he always did with great pleasure. In particular, he often spoke with L. Carleson, who visited the institute on several occasions.

The work of L. Carleson profoundly influenced Kostya's mathematical research. From the start of his scientific career, Kostya was very enthusiastic about Carleson's theorem, which establishes the a.e. convergence of Fourier series of L^2 functions (1966). The original proof was so complicated that soon after its publication there appeared more detailed proofs in several books (e.g., Mozzochi [86]; Jørsboe and Mejlbro [78]), as well as an alternative proof by Fefferman [74]. Lecturing in various parts of the Soviet Union, Kostya often stressed the importance of this proof and attracted attention on this theorem in which he saw great potential for future research. His predictions came true when in the mid-1990s, M. Lacey and C. Thiele (as well as other authors later on) further developed the techniques used in the proof of Carleson's theorem and successfully applied them to problems in multilinear harmonic analysis. In particular, they provided a short proof of Carleson's theorem based on their method of time-frequency analysis of combinatorial model sums [85].

We now highlight some of Kostya's contributions to mathematics. We choose to violate the chronological order and start with the topic, which we find most interesting and influential (although this choice inevitably reflects the personal tastes of the authors). The focus of our exposition is on the results in the area of harmonic analysis. The subsequent articles by M. Chakhkiev, V. Kolyada, V. Maiorov, and V. Temlyakov give a snapshot of Oskolkov's contribution in the areas of Approximation Theory and Optimal Control.

Kostya's research activity was to a great extent inspired and motivated by his participation in the seminar of Luzin and Men'shov at Moscow State University. For a long time, this seminar was supervised by P.L. Ul'yanov. As a student of N.K. Bari, P.L. Ul'yanov was deeply interested in the finest features of convergence of Fourier series, in particular the problem of finding *spectra of uniform convergence*.

Let us turn to rigorous definitions. Let $\mathscr{K} = \{k_n\}$ be a sequence of pairwise distinct integers. Denote by $\mathscr{C}(\mathscr{K})$ the subspace of continuous 1-periodic functions with uniform norm, whose Fourier spectrum is contained in $\mathscr{K}$, i.e.,

$$\mathscr{C}(\mathscr{K}) = \left\{ f(t) : f(t+1) = f(t) \in \mathscr{C}, \hat{f}_k = \int_0^1 f(t) e^{-2\pi i k t} dt = 0, k \notin \mathscr{K} \right\}.$$

Denote

$$S_N f(t) = \sum_{n=0}^{N} \hat{f}_k e^{-2\pi i k_n t}, \qquad L_N(\mathscr{K}) = \sup_{0 \neq f \in \mathscr{C}(\mathscr{K})} \frac{\|S_N f\|}{\|f\|}.$$

The sequence $\mathscr{K}$ is called a *spectrum of uniform convergence* if for any function f in $\mathscr{C}(\mathscr{K})$, the sequence $S_N(f)$ converges to $f(t)$ uniformly in t as $N \to \infty$. The boundedness of the sequence L_N suffices to deduce that $\mathscr{K}$ is a spectrum of uniform convergence; however, the main difficulty lies precisely in obtaining good bounds on L_N in terms of the spectrum $\mathscr{K}$.

The classical result of du Bois-Reymond on the existence of a continuous function whose Fourier series diverges at one point shows that the sequence of all integers is not a spectrum of uniform convergence, while all lacunary sequences are spectra of uniform convergence. For a long time, it was not known whether the sequence n^2 (or more general polynomial sequences) is a spectrum of uniform convergence. This problem was repeatedly mentioned by P.L. Ulyanov, in particular, in his 1965 survey [94]. In his remarkable publication [34] Kostya gave a negative answer to this question. His proof is very transparent, elegant, short, and inspiring and led to a series of outstanding results.

We shall briefly outline Kostya's approach. If one denotes

$$h_N(P) = \sum_{1 \leq |n| \leq N} \frac{e^{2\pi i P(n)}}{n},$$

it is then evident that

$$|h_N(P)| \leq \sum_{1 \leq |n| \leq N} \frac{1}{n} \sim 2 \log N \to \infty.$$

This is a trivial bound of h_N. At the same time, any nontrivial estimate of the type $|h_N(P)| \leq (\log N)^{1-\varepsilon}$ for all polynomials of degree r would easily imply the bound $L_N \geq (\log N)^{\varepsilon}$, and the growth of the Lebesgue constants would then disprove the uniform convergence. Therefore, the question reduces to improving the trivial bounds for the trigonometric sums, which is far from being simple.

Kostya has demonstrated that no power sequence and, more generally, no polynomial sequence can be a spectrum of uniform convergence. In addition, a remarkable lower bound $L_N > a_r (\log N)^{\varepsilon_r}$ for the Lebesgue constants of polynomial spectra has been established. Here $\varepsilon_r = 2^{-r+1}$, the constant a_r is positive and depends only on the degree of the polynomial defining the spectrum, *but not on the polynomial itself.*

Kostya's ingenious insight consisted of applying the method of trigonometric sums to the solution of this problem. His main observation was that the sequence h_N is nothing but the Hilbert transform of the sequence $\{e^{2\pi i P(n)}\}$ and the algebraically regular nature of this sequence allows one to obtain a substantially improved result. For instance, when $r = 1$ and $P(x) = \alpha x$, the following canonical relations hold

$$h(P) \equiv \sum_{n \neq 0} \frac{e^{2\pi i \alpha n}}{n} = 2i \sum_{n=1}^{\infty} \frac{\sin(2\pi i \alpha n)}{n} = 2\pi i \left(\frac{1}{2} - \{\alpha\}\right),$$

where $\{\alpha\}$ is the fractional part of the number α and $\alpha \notin \mathbb{Z}$. Moreover, the supremum of the partial sums is nicely bounded by

$$\sup_{N,\alpha} \left| 2i \sum_{n=1}^{N} \frac{\sin(2\pi i \alpha n)}{n} \right| < \infty, \tag{1}$$

as opposed to the aforementioned logarithmic bound, which can be interpreted as boundedness in two parameters: the upper limof the partial sums and all polynomials of the first degree.

On one hand, this estimate demonstrates the applicability of the method of trigonometric sums; on the other hand, it shows the type of bound one may expect to obtain by using this method for polynomials of higher degrees.

Consequently, Kostya managed to improve the trivial bound and to deduce the estimate $L_N > a_r(\log N)^{\varepsilon_r}$ with some constant a_r depending on r from the bound

$$|h_N(P)| \leq c_r(\log N)^{1-\varepsilon_r}, \tag{2}$$

where P is a polynomial of degree r with real coefficients and $\varepsilon_r = 2^{1-r}$.

The method employed in [34] to prove Eq. (2) is elegant and essentially elementary. It is roughly as follows: by squaring out the quantity $|h_N(P)|$, one obtains a double sum

$$|h_N(P)|^2 = \sum_{1 \leq |n|,\, |m| \leq N} \sum \frac{e^{2\pi i(P(n)-P(m))}}{nm}.$$

Introducing the summation index $\nu = n - m$ and invoking elementary estimates, one obtains a relation of the type

$$|h_N(P)|^2 \leq \sum_{1 \leq |\nu| \leq N} \frac{|h_N(P_\nu)|}{\nu} + 1,$$

where $P_\nu(x) = P(x+\nu) - P(x)$, $(\nu = \pm 1, \pm 2, \dots)$. Since for each ν the polynomial $P_\nu(x)$ has degree strictly less than r, the proof may be completed by induction on r.

Notice that if $r = 1$, inequality (2) turns into Eq. (1). Kostya and his coauthor and friend G.I. Arkhipov came up with the brilliant idea that Eq. (2) can be substantially improved; in fact, the logarithmic growth of Eq. (2) may be replaced with boundedness, as in Eq. (1), for polynomials P of arbitrary degree, not just of degree $r = 1$. The proof is not simple and requires heavy machinery like the Hardy-Littlewood-Vinogradov circle method for trigonometric sums. The following remarkable theorem was proved in [36]:

Theorem A (G.I. Arkhipov and K.I. Oskolkov, 1987). *Let $\mathscr{P}_r$ be the class of algebraic polynomials P of degree r with real coefficients. Then*

$$\sup_N \sup_{\{P\in\mathscr{P}_r\}} \left| \sum_{1\le|n|\le N} \frac{e^{2\pi i P(n)}}{n} \right| \equiv g_r < \infty$$

and for every $P \in \mathscr{P}_r$, the sequence of symmetric partial sums convergences and the sum is bounded uniformly in $\mathscr{P}_r$.

Of course, this stronger bound brought forth new results that did not take long to appear. The first application was obtained for the discrete Radon transform. Namely, let $P \in \mathscr{P}_r$ and define

$$Tf(x) = \sum_{j\neq 0} \frac{f(x-P(j))}{j}.$$

Then

$$\widehat{Tf}(n) = \hat{f}(n) \sum_{j\neq 0} \frac{e^{2\pi i n P(j)}}{j},$$

therefore

$$|\widehat{Tf}(n)| \le |\hat{f}(n)| \sup_N \sup_{\{Q\in\mathscr{P}_r\}} \left| \sum_{1\le|j|\le N} \frac{e^{2\pi i Q(j)}}{j} \right| \le g_r |\hat{f}(n)|$$

and

$$T : L^2 \to L^2.$$

In 1990 E.M. Stein and S. Wainger [92] independently proved the boundedness of the discrete Radon transform in the range $3/2 < p < 3$. A. Ionescu and S. Wainger [77] subsequently extended the result to all $1 < p < \infty$. See [84] for a good source of information about the current state of the subject.

Later, Kostya found a new and unexpected method of proof for Theorem A by interplacing the theory of trigonometric sums with PDEs. His key observation was that formal differentiation of the trigonometric sum

$$h(t,x) := (\text{p. v.}) \sum_{|n|\in\mathbb{N}} \frac{e^{\pi i(n^2 t + 2nx)}}{2\pi i n}$$

yields the solution of the Cauchy initial value problem for the Schrödinger equation of a free particle with the initial data $1/2 - \{x\}$

$$\frac{\partial \psi}{\partial t} = \frac{1}{2\pi i} \frac{\partial^2 \psi}{\partial x^2}, \qquad \psi(t,x)\,|_{t=0} = 1/2 - \{x\}.$$

However, one has to make rigorous sense of this formalism, which is highly non-trivial. For instance, the series $\vartheta(t,x) := \sum_{n\in\mathbb{Z}} e^{\pi i(n^2t+2nx)}$, which arises naturally, is not summable by any regular methods for irrational values of t as observed by G.H. Hardy and J.E. Littlewood, see [75].

Using the Green function $\Gamma(t,x) = \sqrt{\frac{i}{t}}e^{-\frac{\pi i x^2}{t}}$ and the Poisson summation formula, Kostya established the following identity, which must be understood in the sense of distributions.

$$\vartheta(t,x) = \Gamma(t,x)\vartheta\left(-\frac{1}{t},-\frac{x}{t}\right).$$

This might be viewed as a generalization of the well-known reciprocity of truncated Gauss sums, see [75, p. 22]:

$$\sum_{n=1}^{q} e^{\frac{\pi i n^2 p}{q}} = \sqrt{\frac{iq}{p}}\sum_{m=1}^{p} e^{-\frac{\pi i m^2 q}{p}}.$$

From this identity, Kostya derives the existence and global boundedness for the discrete oscillatory Hilbert transforms with polynomial phase $h(t,x)$, i.e., a particular case of Theorem A for the polynomials of second degree. The case of higher-degree polynomials, for example, cubic, requires the analysis of linearized periodic KdV equation. The general case was considered in the remarkable paper [41].

The success achieved by Kostya in the study of the Schrödinger equation of a free particle with the periodic initial data has been developed even further. Z. Ciesielski suggested that Kostya tries to use Jacobi's elliptic ϑ-function as a periodic initial data. This function has lots of internal symmetries, and the problem sounded quite promising.

Formally, the problem is the following:

$$\frac{\partial\psi}{\partial t} = \frac{1}{2\pi i}\frac{\partial^2\psi}{\partial x^2}, \qquad \psi(t,x)\,|_{t=0} = \vartheta_\varepsilon(x) = c(\varepsilon)\sum_{m\in\mathbb{Z}} e^{-\frac{\pi(x-m)^2}{\varepsilon}}.$$

Here, ε is a small positive parameter which tends to 0 and $c(\varepsilon)$ a positive factor, normalizing the data in the space $L^2(\mathbb{T})$, i.e., on the period.

D. Dix, Kostya's colleague from the University of South Carolina, conducted a series of computer experiments (unpublished) and plotted the 3D graph of the density function $\rho = \rho(\theta_\varepsilon,t,x) = |\psi(\theta_\varepsilon,t,x)|^2, (t,x) \in \mathbb{R}^2$, for $\varepsilon = 0.01$. The result was astonishing, see Fig. 1.

Instead of expected chaos, the picture turned out to be well structured. First, the graphs represent a rugged mountain landscape, and second, the landscape is not a completely random combination of "peaks and trenches." In particular, it is criss-crossed by a rather well-organized set of deep rectilinear canyons, or "the valleys of shadows." The solutions exhibit deep self-similarity features, and complete rational Gauss sums play the role of scaling factors. Effects of such nature are labeled in the modern physics literature as quantum carpets.

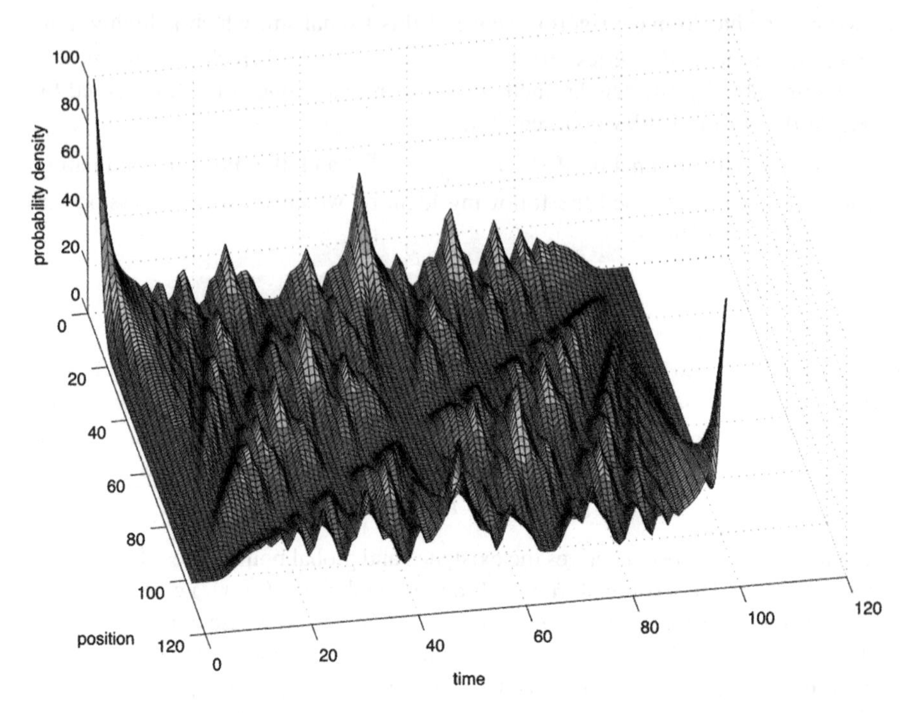

Fig. 1 The Schrödinger landscape

Moreover, Kostya showed that semiorganized and semi-chaotic features, exhibited by the bivariate Schrödinger densities $|\psi(t,x)|^2$, also occur for a wide class of $\sqrt{\delta}$-type initial data where $\delta = \delta(x)$ denotes the periodic Dirac's delta-function. By definition, $\sqrt{\delta}$ is a family of regular periodic initial data $\{f_\varepsilon(x)\}_{\varepsilon>0}$ such that in the distributional sense, $|f_\varepsilon|^2 \to \delta$ for $\varepsilon \to 0$.

These phenomena were mathematically justified by Kostya using the expansions of densities $|\psi_\varepsilon|^2$ into ridge-series (infinite sums of planar waves) consisting of Wigner's functions and by analyzing the distribution of zeros of bivariate Gauss sums.

Figure 2 below demonstrates Bohm trajectories—the curves on which the solution ψ conserves the initial value of the phase, i.e., remains real valued and positive.

Figure 2 looks like a typical quantum carpet from the Talbot effect. The Talbot effect phenomenon, discovered in 1836 by W.H.F. Talbot [93], the British inventor of photography, consists of multi-scaled recovery (revival) of the periodic "initial signal" on the grating plane. It occurs on an observation screen positioned parallel to the original plane, at the distances that are rational multiples of the so-called Talbot distance. At the bottom of the figure, the light can be seen diffracting through a grating, and this exact pattern is reproduced at the top of the picture, one Talbot

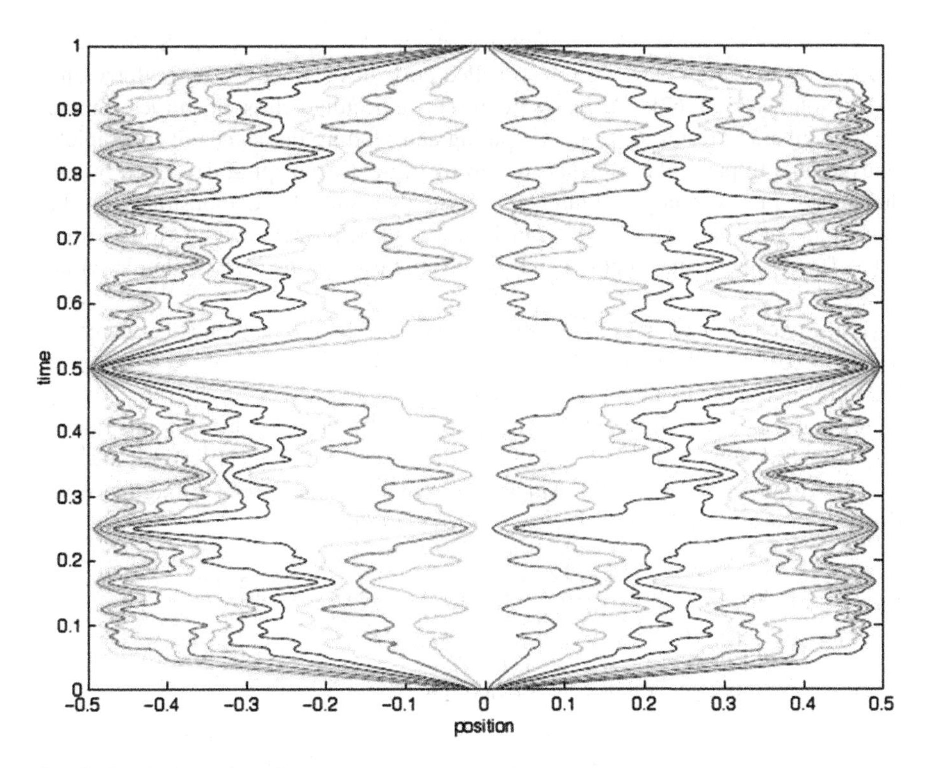

Fig. 2 The Bohm trajectories

length away from the grating. Halfway down, one sees the image shifted to the side, and at regular fractions of the Talbot length, the sub-images are clearly seen. A careful examination of Fig. 2 reveals the aforementioned features in this picture.

Kostya suggested the model that explains the Talbot effect mathematically [56]. He established the bridges between the following equations describing the Talbot effect:

$$\text{Wave} \mapsto \text{Helmholtz with small parameter} \mapsto \text{Schrödinger}$$

Subsequently, several theorems concerning the Talbot effect were proved by him, explaining the phenomenon of "the valleys of shadows"—the rectilinear domains of extremely low light intensity in Fig. 1.

In particular, it was discovered that there are surprisingly wide and very interesting relations of his results on Vingradov series with many concepts in mathematics, such as the Fresnel integral, continued fractions, Weyl exponential sums, Carleson's theorem on trigonometric Fourier series of L^2 functions, the Riemann ζ-function, and shifted truncated Gauss sums—in other words, deep connections exist between the objects of analytic number theory and partial differential equations of Schrödinger type with periodic initial data.

Kostya has explored the complexity features of solutions to the Schrödinger equation, which are related to the so-called curlicues studied by V.M. Berry and J. H. Goldberg [64–67]. Curlicues represent a peculiar class of curves on the complex plane $\mathbb{C}$ resulting from computing and plotting the values of incomplete Gauss sums. In particular, the metric entropy of the Cornu spiral described by the incomplete Fresnel integral equals 4/3. Kostya's result [44] demonstrates a very remarkable fact that, although the Cauchy initial value problem with periodic initial value $f(x)$ is linear, the solutions may be chaotic even in the case of simple initial data.

These phenomena were enthusiastically received by the mathematical community. In 2010, P. Olver published a paper [88] in the American Mathematical Monthly attempting to attract the attention of young researchers to the subject.

Kostya also took a different direction of research related to the aforementioned trigonometric sums in [51, 52, 55, 58]. In particular, in [55] he found an answer to S.D. Chowla's problem [68], which had been open since 1931. Along the way, Kostya characterized the convergence sets for the series

$$S(t) \sim \sum_{(n,m)\in\mathbb{N}^2} \frac{\sin 2\pi nmt}{nm}, \quad C(t) \sim \sum_{(n,m)\in\mathbb{N}^2} \frac{\cos 2\pi nmt}{nm},$$

as well as for more general double series of the type

$$E(\lambda,t,x,y) \sim \sum_{(n,m)\in\mathbb{N}^2} \lambda_{n,m} \frac{e^{2\pi i(nmt+nx+my)}}{nm},$$

where λ is a bounded "slowly oscillating" multiplier, satisfying, say, the Paley condition, t,x,y-independent real variables. Such series naturally arise in the study of the discrepancy of the distribution of the sequence of fractional parts $\{nt\}$ (mod 1) and Wigner's functions arising from the Schrödinger density $|\psi|^2$.

We now turn our attention to some of Kostya's earlier results, which highlight his versatile contributions to harmonic analysis and approximation theory.

In 1973, E.M. Nikishin and M. Babuh [87] demonstrated that one could construct a function of two variables whose rectangular Fourier series diverges almost everywhere (the existence of such functions was proved by Fefferman [71] in 1971) with modulus of continuity $\omega_C(f,\delta) = O\left(\log\frac{1}{\delta}\right)^{-1}$. One year later, Kostya [15] proved that this estimate is close to being sufficient. If $f \in C(\mathbb{T}^2)$ and

$$\omega_C(f,\delta) = o\left(\log\frac{1}{\delta}\log\log\log\frac{1}{\delta}\right)^{-1},$$

then the rectangular Fourier sums converge a.e.; the exact condition is still an open question. Kostya's proof used very delicate estimates of the majorant of the Fourier series of a bounded function of one variable due to R. Hunt. In addition, Kostya suggested a remarkable method for expressing the information about the smoothness of a function in terms of a certain extremal sequence which we shall discuss later. Thus, even Kostya's earliest results are elegant and complete, although very technical and far from trivial.

A natural counterpart of Carleson's theorem is Kolmogorov's example [79] of an L^1 function whose Fourier series diverges almost everywhere. Finding the optimal integrability class in S.V. Kolmogorov's theorem is an important open question. The first step in this direction was made in 1966 by V.I. Prohorenko [89]. The best result known today was obtained by Konyagin [82] in 1998. In his paper S.V. Konyagin wrote, "The author expresses his sincere thanks to K. I. Oskolkov for a very fruitful scientific discussion during his (the author's) visit to the University of South Carolina, which stimulated the results of the present paper."

One of Kostya's earliest research interests was the quest for a.e. analogues of estimates written in terms of norms. We shall take the liberty of drawing a parallel to the Diophantine approximation. The classical Dirichlet–Hurwitz estimate

$$\left|x - \frac{p}{q}\right| \leq \frac{1}{q^2\sqrt{5}}$$

holds for all real x and for infinitely many values of p and q with $(p,q) = 1$. Moreover, for some values of x (such as the "golden ratio" $\frac{\sqrt{5}-1}{2}$), the constant $\sqrt{5}$ cannot be increased. At the same time, as shown by A. Khinchin for almost all x, the order of approximation can be greatly improved. For example, for almost all x, there exist infinitely many p, q with $(p,q) = 1$ such that

$$\left|x - \frac{p}{q}\right| \leq \frac{1}{q^2 \log q}.$$

More generally, instead of $\log q$, one can use any increasing function $\varphi(q)$, where the series $\sum \frac{1}{q\varphi(q)}$ diverges. The divergence condition is sharp, which easily follows from the Borel–Cantelli lemma. Therefore, the Dirichlet–Hurwitz estimate can be improved by a logarithmic factor almost everywhere.

In the same spirit, Kostya improved Lebesgue's result on the approximation of continuous functions with the partial sums of Fourier series. Uniform estimates may be substantially strengthened in the a.e. sense. More precisely, Lebesgue's theorem [83] implies that if $f \in \mathrm{Lip}_\alpha$, $0 < \alpha < 1$, then the following uniform estimate of the rate of approximation is valid:

$$|f(x) - S_n f(x)| \leq C \frac{\log n}{n^\alpha},$$

and there is a function $f \in \mathrm{Lip}_\alpha$ such that

$$\limsup_{n \to \infty} \frac{n^\alpha}{\log n} |f(0) - S_n f(0)| > 0.$$

In [20], using the exponential estimates on the majorants of the Fourier sums of a bounded function due to Hunt [76], Kostya showed that for almost all $x \in \mathbb{T}$, where $\mathbb{T} = [0, 2\pi)$, the estimate can be improved to

$$|f(x) - S_n f(x)| \le C_x \frac{\log\log n}{n^\alpha},$$

and there is a function $f \in \mathrm{Lip}_\alpha$ such that for almost all $x \in \mathbb{T}$

$$\limsup_{n\to\infty} \frac{n^\alpha}{\log\log n} |f(x) - S_n f(x)| > 0.$$

We would like to mention that the parallel with the Diophantine approximation is more than just formal. In his later works, Kostya used continued fractions, the main tool of Diophantine approximation, to obtain convergence theorems for trigonometric series. See, for example, [51, 52, 55, 58].

The proof of the aforementioned metric version of Lebesgue's theorem was based on a remarkable sequence δ_k, defined for a modulus of continuity $\omega(\delta)$ by the following rule:

$$\delta_0 = 1,\ \delta_{k+1} = \min\left\{\delta :\ \max\left(\frac{\omega(\delta)}{\omega(\delta_k)};\ \frac{\delta\omega(\delta_k)}{\delta_k\omega(\delta)}\right) \le \frac{1}{2}\right\},\ k = 0, 1, \ldots.$$

One can view this sequence as a discrete K-functional. Namely, it is well known that the modulus of continuity $\omega(\delta)$ controls the rate of convergence while the ratio $\omega(\delta)/\delta$ controls the growth of the derivative of a smooth approximation process when $\delta \to 0$. So, the δ_k system controls both, which is similar to the idea of the K-functional.

The idea of such partitions was already in the air, probably since the work of S.B. Stechkin [91] in the early 1950s. Simultaneous partition of a modulus of continuity $\omega(\delta)$ and the function $\delta/\omega(\delta)$ apparently was first used by V.A. Andrienko [63]. As in the work of Stechkin, Andrienko used such partitions to construct counterexamples.

Kostya however was the first who wrote this sequence explicitly and employed it to obtain positive results. Amazingly, this sequence turns out to be very useful in the description of phenomena that are either close to or seemingly far from the rate of a.e. approximations. For instance, the classical Bari-Stechkin-Zygmund condition on the modulus of continuity just means that δ_k/δ_{k+1} is bounded. Later on, this method was widely used by many authors, see, for example, [80, 81].

Another example of application of δ_k sequence is the a.e. form of a Jackson-type theorems from constructive approximation theory. Namely, let $f \in L^p(\mathbb{T})$, $1 \le p < \infty$; let $\omega_p(f, \delta)$ denote the L^p-modulus of continuity of a function f; and let $S_\delta(f)(x) = \delta^{-1} \int_x^{x+\delta} f(y)dy$. Then

$$\|f - S_\delta(f)\|_p \le C_p \omega_p(f, \delta).$$

In [20] Kostya suggested an a.e. version of the above theorem. Let $\omega(t)/t$, $w(t)$ and $\omega(t)/w(t)$ be increasing, and assume also that

$$\sum_{k=0}^{\infty}\left(\frac{\omega(\delta_k)}{w(\delta_k)}\right)^p<\infty. \tag{3}$$

If $\omega_p(f,\delta)=O(\omega(\delta))$, then

$$f(x)-S_\delta(f)(x)=O_x(w(\delta)) \qquad \text{a.e. on } \mathbb{T}.$$

If (1) diverges, then there is a function f such that $\omega_p(f,\delta)=O(\omega(\delta))$ and

$$\limsup_{\delta\to 0+}\frac{f(x)-S_\delta(f)(x)}{w(\delta)}=\infty \qquad \text{a.e. on } \mathbb{T}.$$

Further applications of the sequence δ_k include a quantitative characterization of the Luzin C-property. By Luzin's theorem, an integrable function is continuous if restricted to a proper subset of the domain whose complement has arbitrarily small measure. It is then natural to ask the following: if the function has some smoothness in the integral metric, what can be concluded about the uniform smoothness of this restriction?

Kostya [23, 31] suggested the following sharp statement: let $\omega(\delta)$ be a modulus of continuity, and let f be such that $\omega_p(f,\delta)\le\omega(\delta)$. Let another modulus of continuity $w(\delta)$ be as above [see Eq. (3)]. Then for some measurable function $C(t)\in L^{p,\infty}$

$$|f(x)-f(y)|\le(C(x)+C(y))\,w(|x-y|).$$

The convergence of the series in Eq. (1) is a sharp condition. Since any $L^{p,\infty}$ function is bounded modulo a proper set of arbitrary small measure, the above inequality provides the quantitative version of the Luzin C-property.

Later, that property was generalized to functions in H^p, $0<p\le 1$ and in L^p, $p\ge 0$ by A. Solyanik [90]. Also V. G. Krotov and his collaborators have studied the C-property in more general settings (see his paper in this volume).

Kostya's interest in the convergence of Fourier series lead him to consider the question of the best approximation of a continuous function f with trigonometric polynomials. This problem has a long history and tradition, especially in the Russian school. Here Kostya again used a combination of deep and simple ideas and obtained optimal results.

To be specific, let f be a continuous periodic function with Fourier sums $S_n(f)$, and let $E_n(f)=E_n$ be the best approximation of f by trigonometric polynomials of order n. Classic estimates due to Lebesgue state that

$$\|f-S_n(f)\|\le(L_n+1)E_n(f),$$

where L_n are Lebesgue constants. From this inequality it follows that

$$\|f-S_n(f)\|\le C(\log n)E_n(f).$$

This inequality is sharp in many function classes defined in terms of a slowly decreasing majorant of best approximations. But the inequality is not sharp if the best approximations decrease quickly.

The following estimate was proved by Kostya in [17] :

$$\|f - S_n(f)\| \leq C \sum_{k=n}^{2n} \frac{E_k(f)}{n-k+1}.$$

Here, C is an absolute constant, and $\|\cdot\|$ is a norm in the space of continuous functions. This estimate sharpens Lebesgue's classical inequality for fast decreasing E_k. The sharpness of this estimate is proved for an arbitrary class of functions having a given majorant of best approximation. Kostya also investigated the sharpness of the corresponding estimate for the rate of almost everywhere convergence of Fourier series. See the note by V. Kolyada in this volume.

When f is continuous with no extra regularity assumptions, the partial Fourier sums may not provide a good approximation of f. In a paper with D. Offin [6], Kostya constructed a simple and explicit orthonormal trigonometric polynomial basis in the space of continuous periodic functions by simply periodizing a well-known wavelet on the real line. They obtained trigonometric polynomials whose degrees have optimal order of growth if their indices are powers of 2. Also, Fourier sums with respect to this polynomial basis have almost best approximation properties.

More recently, Kostya wrote an interesting series of papers on the approximation of multivariate functions. He became interested in the *ridge approximation* (approximation by finite linear combination of planar waves) and the algorithms used to generate such approximations. His interest in these problems was motivated by the connections between the ridge approximation and optimal quadrature formulas for trigonometric polynomials, which are discussed in [43]. In this chapter Kostya also studied the best ridge approximation of L^2 radial functions in the unit ball of $\mathbb{R}^2$ and showed that the orthogonal projections on the set of algebraic polynomials of degree k are linear and optimal with respect to degree n ridge approximation. The proof of this result uses, in particular, the inverse Radon transform and Fourier-Chebyshev analysis.

References

Scientific Articles by K.I. Oskolkov

1. Andreev, A., Berdyshev, V.I., Bojanov, B., Kashin, B.S., Konyagin, S.V., Nikol'skii, S.M., Oskolkov, K.I., Petrushev, P., Sendov, B., Telyakovskii, S.A., Temlyakov, V.N.: In memory of Sergei Borisovich Stechkin [1920–1995], East J. Approx. **2**, 131–133 (1996).
2. Ciesielski, Z. Approximation by algebraic polynomials on simplexes. (Russian) Translated from the English by K. I. Oskolkov. Uspekhi Mat. Nauk, **4**(244), 212–214 (1985).

3. DeVore, R.A., Oskolkov, K.I., Petrushev, P.P.: Approximation by feed-forward neural networks, Ann. Numer. Math. **4**, 261–287 (1997). The heritage of P.L. Chebyshev: A Festschrift in honor of the 70th birthday of T.J. Rivlin. MR 97i:41043
4. DeVore, R.A., Oskolkov, K.I., Petrushev, P.P.: Approximation by feed-forward neural networks. [J] Ann. Numer. Math. **4**(1–4), 261–287 (1997). ISSN 1021–2655
5. Maiorov, V.E., Oskolkov, K., Temlyakov, V.N.: In: Bojanov, B. (ed.) Gridge approximation and Radon compass, Approximation Theory: a Volume-Dedicated to Blagovest Sendov, pp. 284–309, DARBA, Sofia (2002)
6. Offin, D., Oskolkov, K.I.: A note on orthonormal polynomial bases and wavelets. Constr. Approx. **9**, 319–325 (1993). MR 94f:42047
7. Oskolkov, K.I.: Convergence of a trigonometric series to a function of bounded variation. Mat. Zametki **8**, 47–58 (1970) (Russian)
8. Oskolkov, K.I., Steckin, S.B., Teljakovskii, S.A.: Petr Vasil'evic Galkin, Mat. Zametki **10**, 597–600 (1971) MR 44 No 6436 (Russian)
9. Oskolkov, K.I.: The norm of a certain polynomial operator. Sibirsk Mat. Z. **12**, 1151–1157 (1971) MR 45 No 4021 (Russian)
10. Oskolkov, K.I., Teljakovskii, S.A.: On the estimates of P. L. Ul'janov for integral moduli of continuity. Izv. Akad. Nauk Armjan. SSR Ser. Mat. **6**, 406–411 (1971). MR 45 No 8782 (Russian)
11. Oskolkov, K.I.: The sharpness of the Lebesgue estimate for the approximation of functions with prescribed modulus of continuity by Fourier sumes, Trudy Mat. Inst. Steklov. **112**, 337–345 (1971), 389, Collection of articles dedicated to Academician Ivan Matveevic Vinogradov on his 80th birthday, I. MR 49 No No 970 (Russian)
12. Oskolkov, K.I.: Generalized variation, the Banach indicatrix and the uniform convergence of Fourier series, Mat. Zametki **12**, 313–324 (1972). MR 47 No 5507 (Russian)
13. Oskolkov, K.I.: Subsequences of Fourier sums of functions with a prescribed modulus of continuity. Mat. Sb. (N.S.) **88**(130), 447–469 (1972). MR 48 No 11874 (Russian)
14. Oskolkov, K.I.: Fourier sums for the Banach indicatrix. Mat. Zametki **15**, 527–532 (1974). MR 50 No 10177 (Russian)
15. Oskolkov, K.I.: Estimation of the rate of approximation of a continuous function and its conjugate by Fourier sums on a set of full measure. Izv. Akad. Nauk SSSR Ser. Mat. **38**, 1393–1407 (1974). MR 50 No 10663 (Russian)
16. Oskolkov, K.I.: An estimate for the approximation of continuous functions by sequences of Fourier sums, Trudy Mat. Inst. Steklov. **134**, 240–253 (1975), 410, Theory of functions and its applications (collection of articles dedicated to Sergei Mikhailovich Nikolskii on the occasion of his 70th birthday). MR 53 No 6203 (Russian)
17. Oskolkov, K.I.: Lebesgue's inequality in the uniform metric and on a set of full measure. Mat. Zametki **18**, 515–526 (1975). MR 54 No 833 (Russian)
18. Oskolkov, K.I.: On strong summability of Fourier series and differentiability of functions. Anal. Math. **2**, 41–47 (1976). MR 53 No 6210 (English, with Russian summary)
19. Oskolkov, K.I.: The uniform modulus of continuity of summable functions on sets of positive measure. Dokl. Akad. Nauk SSSR **229**, 304–306 (1976). MR 57 No 9917 (Russian)
20. Oskolkov, K.I.: Approximation properties of integrable functions on sets of full measure. Mat. Sb. (N.S.) **103**(145), 563–589 (1977), 631. MR 57 No 13343 (Russian)
21. Oskolkov, K.I.: Sequences of norms of Fourier sums of bounded functions, Trudy Mat. Inst. Steklov. **142**, 129–142 (1977), 210, Analytic number theory, mathematical analysis and their applications (dedicated to I.M. Vinogradov on his 85th birthday).
22. Oskolkov, K.I.: Polygonal approximation of functions of two variables, Mat. Sb. (N.S.) **107**(149), 601–612 (1978), 639. MR 81j:41020 (Russian)
23. Oskolkov, K.I.: Quantitative estimates of N.N. Luzin's C-property for classes of integrable functions, Approximation Theory (Papers, VIth Semester, Stefan Banach Internat. Math. Center, Warsaw, 1975), Banach Center Publ. 4 PWN, Warsaw (1979), 185–196 MR 81a:26003

24. Oskolkov, K.I.: Optimality of a quadrature formula with equidistant nodes on classes of periodic functions. Dokl. Akad. Nauk SSSR **249**, 49–52 (1979). MR 81b:41077 (Russian)
25. Oskolkov, K.I.: Lebesgue's inequality in the mean. Mat. Zametki **25**, 551–555 (1979), 636. MR 81c:42005 (Russian)
26. Oskolkov, K.I.: The upper bound of the norms of orthogonal projections onto subspaces of polygonals, Approximation Theory (Papers, VIth Semester, Sefan Banach Internat. Math. Center, Warsaw, 1975), Banach Center Publ., 4, PWN, Warsaw (1979), 177–183. MR 82e:41013
27. Oskolkov, K.I.: Approximate properties of classes of periodic functions. Mat. Zametki **27**, 651–666 (1980). MR 81j:42011 (Russian)
28. Oskolkov, K.I.: Partial sums of the Taylor series of a bounded analytic function. Trudy Mat. Inst. Steklov. **157**, 153–160 (1981), 236, Number Theory, mathematical analysis and their applications. MR 83c:300004 (Russian)
29. Oskolkov, K.I.: On optimal quadrature formulas on certain classes of periodic functions. Appl. Math. Optim. **8**, 245–263 (1982). MR 83h:41032
30. Oskolkov, K.I.: On exponential polynomials of the least Lp-norm, Constructive Function Theory '81 (Varna, 1981), Publ. House Bulgar. Acad. Sci., Sofia (1983), 464–467. MR 85a:41022
31. Oskolkov, K.I.: Luzin's C-property for a conjugate function. Trudy Mat. Inst. Steklov. **164**, 124–135 (1983). Orthogonal series and approximations of functions. MR 86e:42019 (Russian)
32. Oskolkov, K.I.: Strong summability of Fourier series Trudy Mat. Inst. Steklov. **172**, 280–290, 355 (1985) Studies in the theory of functions of several real variables and the approximation of functions. MR 87a:42021 (Russian)
33. Oskolkov, K.I.: A subsequence of Fourier sums of integrable functions. Trudy Mat. Inst. Steklov **167**, 239–260, 278 (1985) Current problems in mathematics. Mathematical analysis, algebra, topology. MR 87i:42008 (Russian)
34. Oskolkov, K.I.: Spectra of uniform convergence. Dokl. Akad. Nauk. SSSR **288**, 54–58 (1986). MR 88e:42012 (Russian)
35. Oskolkov, K.I.: Inequalities of the "large size" type and applicatiojns to problems of trigonometric approximation. Anal. Math. **12**, 143–166 (1986). MR 88i:42002 (English, with Russian summary)
36. Arkhipov, G.I., Oskolkov, K.I.: A special trigonometric series and its applications. Mat. Sb. (N.S.) **134**(176), 147–157, 287 (1987). MR 89a:42010 (Russian)
37. Oskolkov, K.I.: Continuous functions with polynomial spectra, Investigations in the theory of the approximation of functions (Russian). Akad. Nauk SSSR Bashkir. Filial Otdel Fiz. Mat., Ufa, , 187–200 (1987). MR 90b:42013 (Russian)
38. Oskolkov, K.I.: Properties of a class of I.M. Vinogradov series. Dokl. Akad. Nauk SSSR **300**, 803–807 (1988). MR 89f:11117 (Russian)
39. Oskolkov, K.I.: I.M. Vinogradov series and integrals and their applications, Trudy Mat. Inst. Steklov. **190**, 186–221 (1989), Translated in Proc. Steklov Math. 1992, no. 1, 193–229; Theory of functions (Russian) (Amberd, 1987). MR 90g:11112 (Russian)
40. Oskolkov, K.I.: On functional properties of incomplete Gaussian sums. Canad. J. Math. **43**, 182–212 (1991). MR 92e:11083
41. Oskolkov, K.I.: I.M. Vinogradov series in the Cauchy problem for Schrödinger-type equations. Trudy Mat. Inst. Steklov. **200**, 265–288 (1991). MR 93b:11104 (Russian)
42. Oskolkov, K.I.: A class of I.M. Vinogradov's series and its applications in harmonic analysis, Progress in Approximation Theory (Tampa, FL, 1990), Springer Ser. Comput. Math. **19**, 353–402. Springer, New York (1992). MR 94m:42016
43. Oskolkov, K.I.: Ridge approximation, Fourier-Chebyshev analysis, and optimal quadrature formulas. Tr. Mat. Inst. Steklov **219**, 269–285 (1997). MR 99j:41036 (Russian)
44. Oskolkov, K.I.: Schrödinger equation and oscillatory Hilbert transforms of second degree. J. Fourier Anal. Appl. **4**, 341–356 (1998). MR 99j:42004

45. Oskolkov, K.I.: Ridge approximation and the Kolmogorov-Nikolskii problem. Dokl. Akad. Nauk **368**, 445–448 (1999). MR 2001b:41024 (Russian)
46. Oskolkov, K.I.: Linear and nonlinear methods for ridge approximation, Metric theory of functions and related problems in analysis (Russian), Izd. Nauchno-Issled. Aktuarno-Finans. Tsentra (AFTs), Moscow, 165–195 (1999). MR 2001i:41039 (Russian, with Russian summary)
47. Oskolkov, K.I.: Ridge approximations and the Kolmogorov-Nikol'skij problem. [J] Dokl. Math. **60**(2), 209–212 (1999); translation from Dokl. Akad. Nauk, Ross. Akad. Nauk **368**(4), 445–448 (1999). ISSN 1064–5624; ISSN 1531–8362
48. Oskolkov, K.I.: On representations of algebraic polynomials by superpositions of plane waves. [J] Serdica Math. J. **28**(4), 379–390 (2002). ISSN 0204–4110
49. Maiorov, V.E., Oskolkov, K.I., Temlyakov, V.N.: Gridge approximation and Radon compass. [A] Bojanov, B.D. (ed.) Approximation theory. A volume dedicated to Blagovest Sendov. Sofia: DARBA. 284–309 (2002). ISBN 954-90126-5-4/hbk
50. Oskolkov, K.: On representations of algebraic polynomials by superpositions of plane waves. Serdica Math. J. **28**, 379–390 (2002). Dedicated to the memory of Vassil Popov on the occasion of his 60th birthday
51. Oskolkov, K.I.: Continued fractions and the convergence of a double trigonometric series. [J] East J. Approx. **9**(3), 375–383 (2003). ISSN 1310–6236
52. Oskolkov, K.: Continued fractions and the convergence of a double trigonometric series. East J. Approx. **9**, 375–383 (2003)
53. Oskolkov, K.: On a result of Telyakovskii and multiple Hilbert transforms with polynomial phases. Mat. Zametki **74**, 242–256 (2003)
54. Oskolkov, K.: Schrödinger equation and oscillatory Hilbert transforms of second degree. J. Fourier Anal. Appl. **4**, 341–356 (1998)
55. Oskolkov, K.I.: The series $\sum \frac{e^{2\pi i n m x}}{mn}$ and Chowla's problem. Proc. Steklov Inst. Math. **248**, 197–215 (2005)
56. Oskolkov, K.I.: The Schrödinger density and the Talbot effect. [A] Figiel, T. (ed.) et al., Approximation and probability. Papers of the conference held on the occasion of the 70th anniversary of Prof. Zbigniew Ciesielski, Bedlewo, Poland, September 20–24, 2004. Warsaw: Polish Academy of Sciences, Institute of Mathematics. Banach Center Publications 72, 189–219 (2006).
57. Oskolkov, K.I.: Linear and nonlinear methods of relief approximation. (English. Russian original) J. Math. Sci., New York **155**(1), 129–152 (2008); translation from Sovrem. Mat., Fundam. Napravl. **25**, 126–148 (2007). ISSN 1072–3374; ISSN 1573–8795
58. Oskolkov, K.I.; Chakhkiev, M.A.: On Riemann "nondifferentiable" function and Schrödinger equation. Proc. Steklov Inst. Math. **269**, 186–196 (2010); translation from Trudy Mat. Inst. Steklova **269**, 193–203 (2010).
59. Tandori, K. Systems of signs. (Russian) Translated from the German by K. I. Oskolkov. International conference on current problems in algebra and analysis (Moscow-Leningrad, 1984). Uspekhi Mat. Nauk, **4**(244), 105–108 (1985).

Books and Articles Translated or Edited by K.I.Oskolkov

60. Brensted, A.: Vvedenie v teoriyu vypuklykh mnogogrannikov, "Mir", Moscow, 1988,Translated from the English by K.I. Oskolkov; Translation edited and with a preface by B.S. Kashin. (Russian)
61. Sendov, B., Popov, V.: Usrednennye moduli gladkosti, "Mir", Moscow, 1988, Translated from the Bulgarian and with a preface by Yu. A. Kuznetsov and K.I. Oskolkov. (Russian)
62. Vinogradov, I.M., Karacuba, A.A., Oskolkov, K.I., Parsin, A.N.: Trudy mezhdunarodnoi konferenctsii po teorii chisel (Moskva, 14–18 sentyabrya 1971 g.), Izdat. "Nauka", Moscow, 1973, With an introductory address by M.V. Keldys; Trudy Mat. Inst. Steklov. 132 (1973). (Russian)

List of other references

63. Andrienko, V.A.: On imbeddings of certain classes of functions. Izv. Akad. Nauk SSSR, Ser. Mat. **31**, 1311–1326 (1967).
64. Berry, M.V., Goldberg, J.H.: Renormalisation of curlicues. Nonlinearity **1**(1), 1–26 (1988)
65. Berry, M.V.: Quantum fractals in boxes. J. Physics A: Math. Gen **29**, 6617–6629 (1996)
66. Berry, M.V., Klein, S.: Integer, fractional and fractal Talbot effects. J. Mod. Optics **43**, 2139–2164 (1996)
67. Berry, M., Marzoli, I., Schleich, W.: Quantum carpets, carpets of light. Physics World, June 2001, pp. 1–6.
68. Chowla, S.D.: Some problems of diophantine approximation (I). Mathematische Zeitschrift **33**, 544–563(1931)
69. Carleson, L.: On convergence and growth of partial sums of Fourier series. Acta Math. **116**, 135–157 (1966)
70. Carleson, L., Sjölin, P.: Oscillatory integrals and a multiplier problem for the disc. Stud. Math. **44**, 287–299 (1972)
71. Fefferman, C.: On the divergence of multiple Fourier series. Bull. Amer. Math. Soc. **77**, 191–195 (1971)
72. Fefferman, C.: Characterizations of bounded mean oscillation. [J] Bull. Am. Math. Soc. **77**, 587–588 (1971)
73. Fefferman, C.: The multiplier problem for the ball. Ann. Math. **94**(2), 330–336 (1971)
74. Fefferman, C.: Pointwise convergence of Fourier series. Ann. of Math. **98**(2), 551–571 (1973)
75. Hardy, G.H.: Collected papers of G.H. Hardy, 1 . Clarendon Press, Oxford (1966)
76. Hunt, R.A.: On the convergence of Fourier series, Orthogonal expansions and their continuous analogues. Proc. Conf. S. 111. Univ., Edwardsville, 1967. SIU Press, Carbondale, Illinois, (1968)
77. Ionescu, A., Wainger, S.: L^p boundedness of discrete singular Radon transforms. [J] J. Am. Math. Soc. **19**(2), 357–383 (2006)
78. Jørsboe, O., Mejlbro, L.: The Carleson-Hunt theorem on Fourier series. In: Lecture Notes in Mathematics, vol. 911. Springer, Berlin (1982)
79. Kolmogorov, A.: Une série de FourierLebesgue divergente presque partout. Fundamenta math. **4**, 324–328 (1923)
80. Kolyada, V.I.: Estimates of maximal functions connected with local smoothness. Sov. Math., Dokl. **35**, 345–348 (1987)
81. Kolyada, V.I.: Estimates of maximal functions measuring local smoothness. Anal. Math. **25**(4), 277–300 (1999)
82. Konyagin, S.V.: On everywhere divergence of trigonometric Fourier series. Sb. Math. **191**(1), 97–120 (2000)
83. Lebesgue, H.: Sur la représentation trigonometrique approchée des fonctions satisfaisant à une condition de Lipschitz. Bull. Soc. Math. France **38**, 184–210 (1910)
84. Lillian Pierce, B.: Discrete fractional Radon transforms and quadratic forms. Duke Math. J. **161**(1), 69–106 (2012)
85. Lacey, M. and Thiele, Ch.: A proof of boundedness of the Carleson operator, Math. Res. Lett. **7**(4), 361–370 (2000)
86. Mozzochi, C.J.: On the pointwise convergence of Fourier series. In: Lecture Notes in Mathematics, Vol. 199. Springer, Berlin (1971)
87. Nikishin, E.M., Babuh, M.: On convergence of double Fourier series of continuous functions (Russian), Sib. Math. Zh. **11**(6), 1189–1199 (1973)
88. Olver, P.J.: Dispersive quantization. Amer. Math. Monthly **117**, 599–610 (2010)
89. Prohorenko, V.I.: Divergent Fourier series. USSR Sb. **4**, 167 (1968)
90. Soljanik, A.A.: Approximation of Functions from Hardy Classes, Ph.D. thesis, Odessa (1986)
91. Stechkin, S.B.: On absolute convergence of Fourier series. Izv. AN SSSR, Ser. matem. **17**, 87–98 (1953)

92. Stein, E.M., Wainger, S.: Discrete analogies of singular Radon transforms. Bull. Amer. Math. Soc **23**, 537–534 (1990)
93. Talbot, W.H.F.: Facts relating to optical sciences. No. IV, Philosophical Magazine **9**(56), 401–407 (1836)
94. Ulyanov, P.L.: Some problems in the theory of orthogonal and biorthogonal series. Izv. Akad. Nauk Azerb. SSR, Ser. Fiz.-Tekhn. Mat. Nauk **6**, 11–13 (1965)
95. Vinogradov, I.M.: Method of Trigonometrical Sums in the Theory of Numbers. Dover, Mineola, NY (2004)

K. I. Oskolkov

Magomed A. Chakhkiev

Abstract This note presents several ideas and results of K. I. Oskolkov from a short period in the early 1980s. A that time, Professor Oskolkov worked as a researcher in the Laboratory of Real-Variable Function Theory at the Moscow Department of the Steklov Institute and, concurrently, supervised Ph.D. students and lectured at the Moscow State University.

I will only touch upon the scientific and pedagogical activities of Konstantin Ilyich Oskolkov in the very beginning of the 80s in the last century, since at that time I studied in graduate school at the Faculty of Computational Mathematics and Cybernetics of the Lomonosov Moscow State University and Konstantin Ilyich was my research advisor.

Once a week at the V.A. Steklov Mathematical Institute we participated in the graduate student seminar directed by professors V.I. Blagodatskikh and K.I. Oskolkov. The meetings were always laid-back and casual: someone would make tea, someone could make a friendly joke about the speaker even if it was one of the seminar's supervisors. Nevertheless, eight out of nine graduate students successfully defended candidate's dissertations (analog of Ph.D.), five went on to also defend doctoral dissertations, and S.M. Aseev was elected a corresponding member of the Russian Academy of Sciences.

At that time Konstantin Ilyich got interested in the problem of A.N. Kolmogorov and S.M. Nikol'skiĭ about the optimal quadrature formula of the type

$$\int_a^b f(x)\,dx = \sum_{k=1}^{n} \lambda_k f(x_k)$$

M.A. Chakhkiev (✉)
Russian State Social University, Moscow, Russia
e-mail: chahkiev_magomed@mail.ru

D. Bilyk et al. (eds.), *Recent Advances in Harmonic Analysis and Applications*, Springer Proceedings in Mathematics & Statistics 25, DOI 10.1007/978-1-4614-4565-4_2,

on function classes, which was somewhat aside from his mainstream research interests. V.P. Motornyĭ [1], $p = \infty$, and A.A. Zhensykbaev [2], $1 \leq p < \infty$, proved the optimality of equidistribution on the classes W_p^r of periodic functions $x(t)$ with the restriction $\|x^{(r)}(t)\|_p \leq 1$. In light of these results it seemed natural that for convex, centrally symmetric, shift-invariant classes of periodic functions, the optimal quadrature formulas should be given by equidistributed nodes. However, K. I. Oskolkov provided a surprising example in which equidistributed nodes were not optimal. This was achieved by the class of periodic functions with the restriction

$$\|x''(t) + \Omega^2 x(t)\|_\infty \leq 1,$$

where Ω is a real number. It turned out that for certain integer $n = n(\Omega)$ the optimal quadrature formula has non-equidistributed nodes [4]. Besides in [3], the optimality of equidistributed nodes was proved for classes of periodic functions with the condition

$$\left\| P\left(\frac{d}{dt}x(t)\right)\right\|_\infty \leq 1,$$

where $P(z) = z^2 + az + b$ is a polynomial whose both roots are real.

As a topic for my dissertation, K.I. Oskolkov proposed to investigate the optimal quadrature formulas for the classes $W_p(P_r)$ of periodic functions $x(t)$ with the restriction

$$\left\| P_r\left(\frac{d}{dt}x(t)\right)\right\|_p \leq 1,$$

where $P_r(z)$ is a polynomial of degree r with real coefficients. Here again for all $r = 1, 2, 3, \ldots$ and $1 \leq p < \infty$ it turned out that equidistributed nodes are optimal as soon as all roots of the polynomial $P_r(z)$ are real [6].

Various extremal problems of approximation theory, in particular the problem about the Kolmogorov width, on the classes $W_p(P_r)$ introduced by K.I. Oskolkov have been considered by numerous authors.

Oskolkov classes $W_p(P_r)$ present additional interest since they can be used to approximate other function classes well known in function theory. As shown in [5], one can choose polynomials $P_r(z)$, $r \to \infty$ with real roots so that in the limit, one obtains the classes, known in approximation theory as A_p^h, of periodic functions $x(t)$ which can be analytically extended to the strip $\{t + i\eta : -h < \eta < h\}$, such that for all fixed η, $|\eta| < h$:

$$\int_0^1 |\mathrm{Re}\, x(t + i\eta)|^q dt \leq 1.$$

Therefore, the results proved for the classes $W_p(P_r)$, in particular, the problems of optimal quadratures, problems about widths, and others can be transferred also to these classes.

In the same period Oskolkov, with great success, continued to work in the trigonometric theories. In [7] he studies the questions related to the spectra of uniform convergence—a circle of problems in general posed by P.L. Ul'yanov. Before this work it was not known whether the sequences of the type $\{n^2\}$, $\{n^3\}$, ... are spectra of uniform convergence. It was shown in [7] that no power sequence or, more generally, no polynomial sequence can be a spectrum of uniform convergence.

The works [8, 9] study trigonometric series and integrals with a real algebraic polynomials in the power of the imaginary exponential. Such series and integrals, which K.I. Oskolkov called Vinogradov series and integrals, due to the diversity of their functional properties present great interest as an object of the theory of trigonometric series. K.I. Oskolkov has also obtained nontrivial applications of these series and integrals in the study of the properties of the solutions of the Cauchy problem for the Schröedinger equation with periodic and nonperiodic initial data. These results have been highly regarded by the president of the International Mathematical Union, L. Carleson. His private letter to Oskolkov contained the following words: "I am impressed by your beautiful results ..." and an invitation to visit Sweden for a few months in order to give a series of lectures. However, even though all expenses were covered by the Swedish side, K.I. Oskolkov was not allowed by the Soviet officials to leave the country. This is perhaps because in his previous trip to the USA, he behaved, as he usually did, very freely and independently.

References

1. Motornyĭ, V.P.: The best quadrature formula of the form $\sum_{k=1}^{n} p_k f(x_k)$ for certain classes of periodic differentiable functions. (Russian) Izv. Akad. Nauk SSSR Ser. Mat. **38** (1974), 583614
2. Zhensykbaev, A.A.: Monosplines of minimal norm and optimal quadrature formulas. (Russian) Uspekhi Mat. Nauk **36** (1981), no. 4(220), 107159
3. Oskolkov, K.I.: Optimality of a quadrature formula with equidistant nodes on classes of periodic functions. (Russian) Dokl. Akad. Nauk SSSR **249**(1) (1979), 4952
4. Oskolkov, K.I.: On optimal quadrature formulae on certain classes of periodic functions. Appl. Math. Optim. **8**(3) (1982), 245263.
5. Oskolkov, K.I.: On exponential polynomials of the least L_p-norm. Constructive function theory '81 (Varna, 1981), 464467, Publ. House Bulgar. Acad. Sci., Sofia, 1983
6. Chakhkiev, M.A.: Linear differential operators with real spectrum, and optimal quadrature formulas. (Russian) Izv. Akad. Nauk SSSR Ser. Mat. **48**(5) (1984), 10781108
7. Oskolkov, K.I. Spectra of uniform convergence. (Russian) Dokl. Akad. Nauk SSSR **288**(1) (1986), 5458
8. Oskolkov, K.I.: (2-AOS) Properties of a class of I. M.Vinogradov series. (Russian) Dokl. Akad. Nauk SSSR **300**(4) (1988), 803807; translation in Soviet Math. Dokl. **37**(3) (1988), 737741
9. Oskolkov, K.I.: I. M.Vinogradov series and integrals and their applications. (Russian) TrudyMat. Inst. Steklov. **190** (1989), 186221; translation in Proc. Steklov Inst. Math. (1992), no. 1, 193229

My First Meetings with Konstantin Oskolkov

Viktor I. Kolyada

Abstract This note tells about our first meetings with Konstatin Oskolkov. We discuss also optimal estimates of the rate of convergence of Fourier series obtained by Oskolkov in 1975

I remember that warm evening at the end of May 1977, when I was in the Odessa airport waiting for the arrival of Kostya from Moscow. Kostya was invited for one month to give a course of lectures at the Department of Mathematical Analysis of the Odessa University. Formally I was not acquainted with him, but I knew him by sight (from a conference in Armenia in 1975).

Two hours later we walked along the Primorski boulevard, went down the famous Potemkin stairs, and came to the port. We stood at a table in a small bar in the open air, drinking white Georgian wine and talking. It was the beginning of our friendship, which we keep till now, although a great part of the past 35 years we lived very far from each other.

Kostya gave a very interesting course devoted to maximal functions, Calderón-Zygmund decomposition, and applications. This course was extremely useful for students as well as for all the members of our group in Analysis. During that month we discussed a lot of different questions with Kostya. I was very interested in Kostya's work [2] in which he found an estimate of the deviation of the partial sums of a Fourier series of a function in terms of its best approximations. This estimate was sharp for an *arbitrary* rate of decay of the sequence of best approximations. Let me discuss this remarkable result in detail.

For a 2π-periodic function f, denote by $S_n(f,x)$ the nth partial sum of its Fourier series. Also, let $E_n(f)$ be the best approximation of the function f by trigonometric

V.I. Kolyada (✉)
Department of Mathematics, Karlstad University, Universitetsgatan 1,
651 88 Karlstad, Sweden
e-mail: viktor.kolyada@kau.se

D. Bilyk et al. (eds.), *Recent Advances in Harmonic Analysis and Applications*, Springer Proceedings in Mathematics & Statistics 25, DOI 10.1007/978-1-4614-4565-4_3,

polynomials of order n in the space C of continuous functions with the uniform norm. In 1910 Lebesgue proved the following inequality:

$$||f - S_n(f)||_C \leq cE_n(f)\log(n+1). \tag{1}$$

This inequality is sharp if the sequence $E_n(f)$ decreases slowly. However, it is rough in the case of rapid decay of $E_n(f)$. For example, if $E_n(f) = O(2^{-n})$, then also $||f - S_n(f)||_C = O(2^{-n})$, and the logarithm in Eq. (1) becomes superfluous. Thus, one faces the problem of finding an estimate which takes into account the properties of the sequence $E_n(f)$ in a better way. Such an estimate was obtained by Kostya in [2]. Namely, he proved the following theorem.

Theorem 1. *For any function $f \in C$,*

$$||f - S_n(f)||_C \leq c\sum_{k=n}^{2n} \frac{E_k(f)}{k-n+1} \quad (n = 0, 1, \ldots). \tag{2}$$

It is clear that Eq. (2) implies Eq. (1). It is easy to see that in the case $E_n(f) = O(2^{-n})$, the right-hand side of Eq. (2) is $O(2^{-n})$, which agrees with the observation made above.

Moreover, Oskolkov proved that estimate (2) is sharp. Let $\mathscr{H}$ be the set of all sequences $\varepsilon = \{\varepsilon_n\}$ of nonnegative numbers converging monotonically to zero as $n \to \infty$. Denote by $C(\varepsilon)$ the class of all functions $f \in C$ such that $E_n(f) \leq \varepsilon_n$ $(n = 0, 1, \ldots)$.

Theorem 2. *There exist positive absolute constants c and c' such that for any sequence $\varepsilon \in \mathscr{H}$*

$$c'\sum_{k=n}^{2n} \frac{\varepsilon_k}{k-n+1} \leq \sup_{f \in C(\varepsilon)} ||f - S_n(f)||_C \leq c\sum_{k=n}^{2n} \frac{\varepsilon_k}{k-n+1}. \tag{3}$$

Thus, inequality (2) is sharp for any order of decay of $E_n(f)$. Similar results are also valid in the space L.

In the proof of inequality (2) Oskolkov used a partition of the sequence $E_n(f)$ in a geometric progression. That is, he considered the sequence of natural numbers n_k such that

$$E_{n_{k+1}}(f) \leq \frac{1}{2}E_{n_k}(f). \tag{4}$$

Afterwards Stechkin [3] extended the results of Kostya's work [2] to the approximation of periodic functions by de la Vallée Poussin sums.

We also discussed one result of mine [1] with Kostya. I proved estimates of the best approximations of a periodic function in L^q in terms of its best approximations in L^p $(p < q)$ which were sharp for any order of the decay of the sequence $E_n(f)_p$. In the proof, I also used partitions (4). I remember that Kostya appreciated this result very much.

One month passed quickly, and the day came to say good-bye to Kostya. Eleonora Storozhenko, Kostya, and I were sitting on the open terrace of the restaurant "Ukraine" in the very center of Odessa. In a few hours Storozhenko and I had a flight to Baku (Azerbaijan), to a conference in the Theory of Functions. Kostya did not plan to go to that conference; he had to stay in Odessa for another couple of days and then leave for Moscow. We ordered a bottle of champagne and proposed a toast to Kostya. He said that it saddened him very much to part with us. "What is the problem?!", said we, "Let's go with us!"

In the late evening of the same day we landed in the Baku airport. Kostya was with us.

References

1. Kolyada, V.I.: Imbedding theorems and inequalities in various metrics for best approximations. Mat. Sbornik **102(144)**(2), 195–215 (1977); English transl. in Math. USSR Sb. **31**, 171–189 (1977)
2. Oskolkov, K.I.: Lebesgue's inequality in a uniform metric and on a set of full measure. Mat. Zametki **18**(4), 515–526 (1975); English transl. in Math. Notes **18**, 895–902 (1975)
3. Stechkin, S.B.: On the approximation of periodic functions by de la Vallée Poussin sums. Anal. Math. **4**, 61–74 (1978)

Meetings with Kostya Oskolkov

Veniamin G. Krotov

Abstract This brief note contains the author's memories of Konstantin Oskolkov, meetings with him, and his influence on the Odessa function theory group.

I have met Konstantin Oskolkov at my very first mathematical conference. It was in May of 1975 in the mountains of Armenia (Aghveran). Victor Kolyada (my friend and collaborator at the Odessa University) and I soon became close to Kostya. We were amazed by his sincerity and friendliness, and one couldn't help but notice his broad erudition. He spoke good English and German, loved literature, and wrote outstanding poetry himself.

Later I realized that Kostya is a real sportsman in his heart. This refers to all sides of his activities. He always strives to perform any job as perfectly as possible even though such an approach requires a lot of additional intellectual, physical, and moral effort.

Since then we met on a regular basis. It happened at mathematical conferences in various cities of the former USSR (Yerevan, Tbilisi, Baku, Kiev, Saratov, Kazan', Kemerovo, Irkutsk, etc.) and abroad (Warsaw, Gdansk, Budapest, Varna). We also often met in Moscow, where we frequently came in order to give talks at the famous seminars at the Moscow State University or attend to other professional activities.

I recall that when we were traveling to a conference in Kemerovo (dedicated to N.N. Luzin's 100th birthday), we were forced to stay in Moscow overnight before a long flight to Kemerovo. The whole Odessa function theory team spent the night at Kostya's apartment.

Kostya visited us in Odessa on numerous occasions. He gave series of lectures for our students, served as an official opponent at dissertation defenses or as a chair of

V.G. Krotov (✉)
Belarusian State University, Nesavisimosti av., 4, MINSK, 220030, Belarus
e-mail: krotov@bsu.by

D. Bilyk et al. (eds.), *Recent Advances in Harmonic Analysis and Applications*, Springer Proceedings in Mathematics & Statistics 25, DOI 10.1007/978-1-4614-4565-4_4,

the state examination committee at the department of mechanics and mathematics. We were always eagerly awaiting Kostya's next visit and enjoying his company.

When Oskolkov turned 40, he was in Odessa, and in his honor, Yuriy Mikhaylovich Shmandin (E.A. Storozhenko's husband, who unfortunately died recently) wrote the following poem à la V.V. Mayakovskiy, the famous Russian futurist poet:

За Костю, сорок нынче которому,
Одесситы с улыбкой радостной,
Поднимают бокалы унций по сорок
Сорокоградусной.

Сорок - капля (без кривотолков),
Жизни маленький осколок - миг.
Побольше Косте таких осколков,
Миру - таких Осколковых.

It is, of course, difficult to translate into English, but we shall make a humble attempt and give a rough approximation:

To Kostya, who's turning forty now,
Odessa's every son and daughter
Raises a glass with forty ounces
Of firewater.
In the lives of each of us forty
Is only a splinter, a fragment.
Many such fragments to Kostya!
More Oskolkovs – to the planet!

(There is some wordplay here: the Russian word 'oskolok' means a fragment or a splinter.)

Kostya never distinguished between colleagues of his own age (such as Victor and myself) and younger mathematicians from our research group (Alex Stokolos, Lesha Solyanik, Yura Kryakin, Tolik Korenovskii). They all became his friends. This, of course, also refers to the more senior mathematicians of the Odessa team—our leader Eleanora Alexandrovna Storozhenko and Vitaliy Andrienko.

Moreover, he had tremendous influence on each and every one of us, in particular mathematically. Many of us worked on problems directly related to Oskolkov's research.

For example, at the conference in Aghveran, I first heard Kostya discuss new results on the problem of Petr Lavrentyevich Ul'yanov about the quantitative estimates of Luzin's C-property before they were even published. After some time Victor Kolyada continued these investigations, and in the recent years I have also joined this field together with my students Iya Ivanishko and Misha Prokhorovich (my survey in the present volume will in particular describe this area of research). Kostya always tried to support us with useful advice, with unobtrusive yet precise and thoughtful comments, with his interest. It should be noted however that this never stood in the way of him being very principled when it came to mathematical rigor.

I wish Kostya Oskolkov strong health and new great mathematical achievements. And I wish myself many new meetings with this wonderful person.

Konstantin Ilyich Oskolkov, Friend and Colleague

Michael T. Lacey

Abstract We make some personal and professional reflections on Konstantin Oskolkov.

1 Mathematical Interests

Konstantin Ilyich Oskolkov, Kostya to all who have known him, and I met sometime after I moved to Atlanta, in 1996, joined by our mutual interest in the subtle properties of the convergence of Fourier series. He, through his remarkable work with Arkipov, on the subtle convergence properties of arithmetical sums like

$$\sum_{n \neq 0} \frac{e^{2\pi i n^2 x}}{n}.$$

These results, extended by the equally remarkable work of Bourgain, revealed intimate connections between these sums and questions in ergodic theory, even those related to the pointwise convergence of Fourier series which was my point of connection, through for instance the proof with Christoph Thiele, of the famous theorem of Lennart Carleson. Only ten pages and three lemmas, it certainly permitted a broader audience to learn the proof.

On this point, Kostya has a favorite story, and also one of mine. He once asked Lennart Carleson himself, would it not be that one day someone would find a truly elementary proof of his great theorem. The response was "Oh no, Kostya, some

M.T. Lacey (✉)
School of Mathematics, Georgia Institute of Technology, Atlanta, GA 30332, USA
e-mail: lacey@math.gatech.edu

D. Bilyk et al. (eds.), *Recent Advances in Harmonic Analysis and Applications*, Springer Proceedings in Mathematics & Statistics 25, DOI 10.1007/978-1-4614-4565-4_5,

things are just hard." This encapsulates two fine parts of Kostya, as well as a nice nugget about Lennart Carleson: A fine taste in intricate and beautiful mathematics and an eye for the details that fill out a person.

2 Armenia, 2001

In the fall of 2001, Kostya convinced me to attend one of series of conferences in Yerevan, Armenia. On a brilliant fall weekend, I flew from Atlanta, over New York City, to Paris for a weekend, before arriving in Yerevan.

I was driven to Nor Amberd, the now disused conference center of the Armenian Academy of Sciences, in a spectacular setting on a mountainside, with a striking landscape of rock, valleys, and monasteries as far as one can see. I could hear Kostya in a nearby room, raising toasts as I went to sleep.

In the evening of the arrival day, a Tuesday, the local CNN station was showing on endless loop, and Russian voice-over, the infamous images of the twin towers collapsing into dust. It was a lesson to me in tragedy, of the terrorist attack, as seen from the eyes of the Russian participants and those of the Armenian people, both of whom have been through far greater tragedies.

Konstantin served as a guide to me at that conference, explaining who was who and their history. Memorably, Sergey Mikhailovich Nikol'skii attended that conference, as did Petr Lavrent'evich Ul'yanov, Przemysław Wojtaszczyk a friend of my family, Jean Pierre Kahane, and a young mathematician with a notable new result on the convergence of Fourier series, Nikolai Yur'evich Antonov. Of course the local talents also attended, including Aleksandr Andraniki Talalyan, Norair Arakelian, Artur Sahakian, and Grigor Karagulyan, who would become my friend and collaborator.

Kostya himself not only translated my talk to the conference but served as a wonderful guide to people and culture. A particular gift was his warm friendship with everyone, and the ready eye for the details that shape the person, both the hard, and the soft spots of his friends.

3 Friendship and Mushrooms

Kostya's long history of mathematics in Russia has been the source of many interesting conversations we have had over the years. Notably, he recently marked the 100th anniversary of his father, in Moscow, pointing to the deep personal connection he has to the intellectual atmosphere that prevailed in Steklov Institute. I remember once being interested in an announcement of one of Kostya's many friends. Next time, I saw Kostya, I described the result and my interest in it. Could the result be true? His response was to the effect that his friend was a proud person, and if he had proved such a good result, then he would surely brag about it to Kostya.

I have gained much insight into a range of people that otherwise I would only have known by the name. These stories and insights have been very generously shared over hours of conversation, with good food, especially the mushroom soup, plenty of wine, and even more toasts.

I extend my warmest regards and best wishes to Konstantin Ilyich Oskolkov, on this fine tribute to his mathematics, professionalism, and very memorable friendship.

The Activity of K. I. Oskolkov in Nonlinear Approximation of Functions

Vitaly Maiorov

Abstract The paper represents the review of activity of K.I. Oskolkov in nonlinear approximation of functions and optimal distribution of quadratures.

In the recent times, the problem of approximation by a plane wave (ridge functions) attracted the attention of many researchers. Different aspects of this problem are natural components of many theoretical and applied topics, for example, Radon mapping, tomography, equations of mathematical physics and geometry. This direction includes different variants of approximation by neural networks.

An essential contribution was made by K. I. Oskolkov in the area of nonlinear approximation by superpositions of plane wave functions. One of the directions in this field is approximation by linear combination of ridge functions or plane waves that is of functions of the form $f(a \cdot x)$, where a is a vector from $\mathbb{R}^d$ and $a \cdot x$ is the inner product of vectors a and x. The problem of approximation by ridge functions is divided into approximation by functions $f(a \cdot x)$ with free profile function f and fixed profile f (neural networks).

Constructive approximation by ridge functions, i.e., estimates of approximations depending on number of ridge functions, occupies an important place in this line of investigations. Oskolkov [5–8] developed the relevant harmonic analysis methods and applied them in approximation by ridge functions with free and fixed profiles. In particular, Oskolkov showed that in the two-dimensional case, the orders of approximation of radial functions by linear combinations of n ridge functions and by the space of algebraic polynomials of degree n coincide. In this connection, first, he obtained an estimate for the optimal number of ridge functions with free profile necessary for the best approximation. Later, these results were used for the best approximation of the Sobolev function classes $W_p^{r,d}$ (Konovalov, Leviatan,

V. Maiorov (✉)
Department of Mathematics, Technion, Haifa 32000, Israel
e-mail: maiorov@tx.technion.ac.il

D. Bilyk et al. (eds.), *Recent Advances in Harmonic Analysis and Applications*, Springer Proceedings in Mathematics & Statistics 25, DOI 10.1007/978-1-4614-4565-4_6,

Maiorov). The final result related to the best approximation of the Sobolev classes in the general case ($d \geq 2$) was obtained by Maiorov. In [1] Oskolkov, together with DeVore and Petrushev, obtained effective estimates for the approximation of function classes by ridge functions with fixed profile, using the methods of harmonic analysis.

Approximation by ridge functions with free profile was discussed by Donoho and Johnstone, who stated the conjecture that uniformly distributed direction vectors a are optimal for approximation of radial and harmonic functions. Oskolkov showed [5] that this conjecture is true for radial functions; however, for harmonic functions, uniformly distributed vectors a are nonoptimal. For the construction of optimal vectors a in the problem of approximation of harmonic functions Oskolkov developed a new method of collapsed waves for the case when some waves may be infinitely close to each other.

One of the important applied aspects of approximation by plane waves is the construction of greedy algorithm, where the dictionary is taken to be the set of all possible ridge functions. In [2] the greedy algorithm which realizes optimal estimates of approximation was constructed in many function classes, including classes of smooth functions and functions with the given form.

The problem of Kolmogorov–Nikol'skii about optimal distribution of quadratures in integration has a long history. Essential efforts were concentrated around the conjecture of uniform distribution of knots (Motornyi, Zhensykbaev): uniform knots and the formula of rectangles

$$\frac{1}{N}\sum_{j=1}^{N} f\left(\frac{2\pi}{N}\right)$$

are optimal for the recovery of integrals $\int_0^{2\pi} f(\vartheta)\mu(d\vartheta)$ on all periodic classes on the segment $[0,2\pi]$. For $p \in [1,\infty)$ and natural $n \geq 4$ the correctness of the conjecture was stated (Motornyi, Zhensykbaev). For big indexes of smoothness r one of difficulties was to prove that the knots are not collapsed.

In order to find the limits of this conjecture Oskolkov considered [3, 4] the following modification of periodic Sobolev classes:

$$\left\|P\left(\frac{d}{d\vartheta}\right) f(\vartheta), L_2([0,2\pi])\right\| \leq 1,$$

where $P(\frac{d}{d\vartheta})$ is a fixed differential operator. He found that the answer depends crucially on the spectrum of this operator. In particular, for the operator $P = \frac{d^2}{d\vartheta^2} + \omega^2$ (the oscillation differential operator), the conjecture of uniform distribution does not hold.

References

1. DeVore, R.A., Oskolkov, K.I., Petrushev, P.P.: Approximation by feed-forward neural networks. Ann. Numer. Math. **4**, 261–287 (1997), The heritage of P.L. Chebyshev: A Festschrift in honor of the 70th birthday of T.J. Rivlin. MR 97i:41043
2. Maiorov, V.E., Oskolkov, K.I., and Temlyakov, V.N.: Gridge approximation and Radon compass. In: Bojanov, B. (ed.) Approximation Theory: A Volume-Dedicated to Blagovest Sendov, pp. 284–309. DARBA, Sofia (2002)
3. Oskolkov, K.I.: On optimality of quadrature formula with uniform knots on classes of periodical functios. Dokl. AN SSSR **249**(1), 49–51 (1979)
4. Oskolkov, K.I.: On optimal quadrature formula on certain classes of periodic functions. Appl. Math. Optim. **8**, 245–263 (1982)
5. Oskolkov, K.I.: Ridge approximation, Fourier-Chebyshev analysis, and optimal quadrature formulas. Tr. Mat. Inst. Steklov **219**, 269–285 (1997), MR 99j:41036 (Russian)
6. Oskolkov, K.I.: Linear and nonlinear methods for ridge approximation, Metric theory of functions and related problems in analysis (Russian), Izd. Nauchno-Issled. Aktuarno-Finans. Tsentra (AFTs), Moscow, 165–195 (1999) MR 2001i:41039 (Russian, with Russian summary)
7. Oskolkov, K.I.: Ridge approximation and the Kolmogorov-Nikolskii problem. Dokl. Akad. Nauk **368**, 445–448 (1999), MR 2001b:41024 (Russian)
8. Oskolkov, K.I.: On representations of algebraic polynomials by superpositions of plane waves. Serdica Math. J. **28**, 379–390 (2002), Dedicated to the memory of Vassil Popov on the occasion of his 60th birthday

How Young We Were

Vladimir Temlyakov

Abstract I met Kostya in 1971 in the Steklov Institute of Mathematics. Those days he was a graduate student and I was an undergraduate student of Professor Sergei Aleksandrovich Telyakovskii. Later, we both defended our PhD dissertations under Telyakovskii's auspices. Thus, we are "scientific brothers" with Kostya. In the beginning of the 1970s trigonometric series and approximation of periodic functions by trigonometric polynomials was a "hot topic." The Russian translation of the fundamental book of A. Zygmund on trigonometric series was published in 1965. This and strong traditions of Russian school in the theory of functions and in approximation theory attracted many young mathematicians including Kostya and me to that topic. These activities were concentrated around seminars run by D.E. Men'shov and P.L. Ul'yanov in Moscow State University and by S.B. Stechkin in the Steklov Institute of Mathematics. We had a chance to interact on an everyday basis not only between ourselves but also with great mathematicians working in that area. Let me discuss in detail only one of Kostya's results from [2] which will help to understand a wonderful mathematical atmosphere in Moscow of those days.

Let C be a space of 2π-periodic continuous functions $f(x)$ with the norm $\|f\| := \max_x |f(x)|$. Denote by $\omega(f,\delta)$ the modulus of continuity of a function $f \in C$. Let $s_n(f)$ be the nth partial sum of the Fourier series of a function $f \in C$. For a given modulus of continuity $\omega(\delta)$ define the following class of functions:

$$H_\omega := \{f \in C : \omega(f,\delta) \le \omega(\delta), \delta \in [0,\pi]\}.$$

V. Temlyakov (✉)
University of South Carolina, Columbia, SC 29208, USA
e-mail: temlyak@math.sc.edu

D. Bilyk et al. (eds.), *Recent Advances in Harmonic Analysis and Applications*, Springer Proceedings in Mathematics & Statistics 25, DOI 10.1007/978-1-4614-4565-4_7,

Consider the quantities

$$S_n(\omega) := \sup_{f \in H_\omega} \|f - s_n(f)\|, \qquad n = 0, 1, 2, \ldots.$$

It is clear that for all $f \in H_\omega$ and for all n we have

$$\|f - s_n(f)\| \leq S_n(\omega). \tag{1}$$

In paper [2] Kostya studied the question on sharpness of bound (1) for individual functions from H_ω when $n \to \infty$. It is a classical and difficult problem with an interesting history [2]. It is by itself a sufficient motivation for an attempt to solve the problem. There was one more less formal motivation too. Kostya and I were in the Department of Theory of Functions at the Steklov Institute of Mathematics chaired by academician Sergei Mikhailovich Nikol'skii. We had a privilege to talk to Sergei Mikhailovich very often and to enjoy his stories. He told us that Sergei Natanovich Bernstein discussed the role of function classes and approximation of individual functions in constructive approximation in the opening session of his seminar in Approximation Theory (Moscow, Spring 1945). Let me deviate a little from the mainstream to make this story mathematically clear.

The approximation of functions by algebraic polynomials was studied in parallel with that for trigonometric polynomials. De la Vallée-Poussin proved in 1908 that for best approximations of the function $|x|$ in the uniform norm on $[-1,1]$ by algebraic polynomials of degree n, the following upper estimate holds:

$$e_n(|x|) \leq C/n.$$

He raised the question of the possibility of an improvement of this estimate in the sense of order. Bernstein (1912) proved that this order estimate is sharp. Moreover, he then established the asymptotic behavior of the sequence $\{e_n(|x|)\}$:

$$e_n(|x|) = \mu/n + o(1/n), \qquad \mu = 0.282 \mp 0.004.$$

Denote by $E_n(f)$ the best approximation of a function $f \in C$ by trigonometric polynomials of order n in the uniform norm $\|\cdot\|$ defined above. Bernstein's general attitude to the role of studying the sequences of $E_n(F) := \sup_{f \in F} E_n(f)$ for a given function class F was skeptical. One of his arguments was that the sequence $\{E_n(F)\}$ may not reflect the behavior of $\{E_n(f)\}$ for any individual $f \in F$, because usually the extremal function that realizes $\sup_{f \in F} E_n(f)$ depends on n. He formulated a problem of studying

$$\sup_{f \in F} \limsup_{n \to \infty} \frac{E_n(f)}{E_n(F)} \quad \text{and} \quad \sup_{f \in F} \liminf_{n \to \infty} \frac{E_n(f)}{E_n(F)}$$

and their analogs for approximation by algebraic polynomials for some function classes. In particular, he thought that the function $|x|$ is an extremal function in the sense of the above quantities in the class $Lip_1 1$ for approximation by algebraic polynomials in the uniform norm. However, it turned out not to be the case. Nikol'skii [1] proved in 1946 that for W^r classes, there is a function $f \in W^r$ such that

$$\limsup_{n\to\infty} E_n(f)/E_n(W^r) = 1,$$

where W^r, $r \in \mathbb{N}$, is the class of r times continuously differentiable periodic functions such that $\|f^{(r)}\| \le 1$.

Clearly, the problem of asymptotic behavior of sequences $\{E_n(f)\}$ of best approximations is different from the problem of asymptotic behavior of sequences $\{\|f - s_n(f)\|\}$. However, these problems are very close in spirit.

In [2] Kostya studied in detail the following quantity:

$$S_-(\omega) := \sup_{f\in H_\omega} \liminf_{n\to\infty} \frac{\|f - s_n(f)\|}{S_n(\omega)}.$$

It is clear that $S_-(\omega) \le 1$. Kostya proved that for any nontrivial modulus of continuity $\omega(\delta)$, we have $S_-(\omega) = 1/2$. It is a very interesting and difficult result. In particular, it supports Bernstein's skepticism. However, it was proved in [3,4] that for the class W^r, there exists a function $f \in W^r$ such that

$$\lim_{n\to\infty} E_n(f)/E_n(W^r) = 1.$$

These results discovered an interesting phenomenon that asymptotic behavior of sequences $\{E_n(f)/E_n(H_\omega)\}$ and $\{\|f - s_n(f)\|/S_n(\omega)\}$ is different.

The discussed above paper [2] illustrates very well Kostya's attitude to mathematics. He takes up difficult problems from classical areas and gives complete solutions to them.

We worked with Kostya at the Steklov Institute of Mathematics till the beginning of 90th. I moved to the University of South Carolina in 1992 and managed to convince Ron DeVore and other colleagues that Kostya would be a great addition to our group at the USC. Kostya was hired by the USC in 1993 and worked there till his retirement in 2009.

Very often writing an anniversary note like this one turns into deifying a person celebrating an anniversary. I know Kostya too well to deify him. He is not only a mathematician he is a human being too. There were a lot of episodes where I saw that clearly. I better stop here and do not go into details. I wish Kostya many happy and productive years ahead!

References

1. Nikol'skii, S.M.: On interpolation and best approximation of differentiable periodic functions by trigonometric polynomials. Izv. Akad. Nauk SSSR Ser. Mat. **10**, 393–410 (1946)
2. Oskolkov, K.I.: An estimate in the approximation of continuous functions by subsequences of Fourier sums. Proc. Steklov Inst. Math. **134**, 273–288 (1975)
3. Temlyakov, V.N.: On the asymptotic behavior of best approximations of continuous functions. Soviet Math. Dokl. **17**, 739–743 (1976)
4. Temlyakov, V.N.: Asymptotic behavior of best approximations of continuous functions. Math USSR Izvestija **11**, 551–569 (1977)

Part II
Contributed Papers

A Survey of Multidimensional Generalizations of Cantor's Uniqueness Theorem for Trigonometric Series

J. Marshall Ash

Abstract Georg Cantor's pointwise uniqueness theorem for one dimensional trigonometric series says that if, for each x in $[0,2\pi)$, $\sum c_n e^{inx} = 0$, then all $c_n = 0$. In dimension d, $d \geq 2$, we begin by assuming that for each x in $[0,2\pi)^d$, $\sum c_n e^{inx} = 0$ where $n = (n_1, \ldots, n_d)$ and $nx = n_1 x_1 + \cdots + n_d x_d$. It is quite natural to group together all terms whose indices differ only by signs. But here there are still several different natural interpretations of the infinite multiple sum, and, correspondingly, several different potential generalizations of Cantor's Theorem. For example, in two dimensions, two natural methods of convergence are circular convergence and square convergence. In the former case, the generalization is true, and this has been known since 1971. In the latter case, this is still an open question. In this historical survey, I will discuss these two cases as well as the cases of iterated convergence, unrestricted rectangular convergence, restricted rectangular convergence, and simplex convergence.

1 Introduction

The idea of this chapter is to provide an overview and an organization of other surveys I have authored or coauthored on uniqueness for multiple trigonometric series.

Let $\{d_n\}_{-\infty<n<\infty}$ be a sequence of complex numbers and let $x \in \mathbb{T}^1 = [0,2\pi)$. Suppose a function has a representation of the form

$$\sum d_n e^{inx} = \lim_{N\to\infty} d_0 + \sum_{n=1}^{N} \left(d_{-n} e^{-inx} + d_n e^{inx}\right).$$

J.M. Ash (✉)
DePaul University, Chicago, IL 60614-3504, USA
e-mail: mash@math.depaul.edu

D. Bilyk et al. (eds.), *Recent Advances in Harmonic Analysis and Applications*, Springer Proceedings in Mathematics & Statistics 25, DOI 10.1007/978-1-4614-4565-4_8,

To see why it is natural to combine the nth and $-n$th terms, suppose that a_n and b_n are real and that $d_n = (a_n + ib_n)/2$ and d_{-n} are complex conjugates. Since $e^{i\theta} = \cos\theta + i\sin\theta$, $d_n e^{inx} + d_{-n}e^{-inx}$ is immediately computed to be $a_n \cos nx + b_n \sin nx$, the "natural" nth term of a real-valued trigonometric series. Is this representation unique? In other words, if $\sum d_n e^{inx} = \sum d'_n e^{inx}$ for every x, does it necessarily follow that $d_n = d'_n$ for every n? Subtract and set $c_n = d_n - d'_n$ to get a cleaner formulation.

(U) Let $\sum c_n e^{inx} = 0$ for every $x \in \mathbb{T}^1$. Does this imply that $c_n = 0$ for every n?

In 1870, Georg Cantor showed that the answer to question (U) is "yes."

Theorem C. *Let* $\sum c_n e^{inx} = 0$ *for every* $x \in \mathbb{T}^1$. *Then* $c_n = 0$ *for every* n.

In all dimensions we will always combine terms whose indices differ only by signs. This reduction in dimension 1 leads to the definite meaning of $\sum d_n e^{inx}$ given above. When $d \geq 2$, the meaning of $\sum d_n e^{inx}$ is not yet definite, so there are many variants of question (U).

First, for each $n \in Z^{+d} = \{0,1,2,\dots\}^d$, we write $\sum_{n\in Z^d} C_n = \sum_{n\in Z^{+d}} T_n$, where $T_n = \sum_{\{\nu:\text{each } \nu_i = n_i \text{ or } -n_i\}} C_\nu$. For example, when $d = 2$, $T_{3,4} = C_{3,4} + C_{-3,4} + C_{3,-4} + C_{-3,-4}$ and $T_{5,0} = C_{5,0} + C_{-5,0}$. This reduction still leaves many possible ways of interpreting the multiple sum. Here are six very natural ones. For simplicity, each will only be described in dimension 2, and we will write (n_1, n_2) as (ℓ, m) to avoid indices.

Square convergence:

$$Sq \sum_{n\in Z^{+2}} T_n = \lim_{N\to\infty} \left(\sum_{\ell=0}^{N} \sum_{m=0}^{N} T_{(\ell,m)} \right).$$

The Nth partial sum contains all terms with indices in the square with lower left corner $(0,0)$ and upper right corner (N,N).

Spherical convergence:

$$Sp \sum_{n\in Z^{+2}} T_n = \lim_{N\to\infty} \sum_{k=0}^{N} \left(\sum_{\{(\ell,m):\ell^2+m^2\leq k\}} T_{(\ell,m)} \right).$$

The Nth partial sum contains all terms with indices in the intersection of the disk of radius $\sqrt{N}$ and the first quadrant.

One way iterated convergence:

$$It \sum_{n\in Z^{+2}} T_n = \lim_{N\to\infty} \sum_{k=0}^{N} \left(\lim_{J\to\infty} \sum_{j=0}^{J} T_{(j,k)} \right).$$

The terms with indices of height 0 are summed yielding a first intermediate number, then the terms with indices of height 1 are summed yielding a second intermediate number, and so on, producing a one-dimensional one way sequence of intermediate numbers. Finally all the numbers of that sequence are added together. In dimension d, there are $d!$ distinct versions of one way iterated convergence, but they are all very similar, and it will be enough for us to pick any one of them.

Unrestricted rectangular convergence:

$$UR \sum_{n \in Z^{+2}} T_n = \lim_{\min\{M,N\} \to \infty} \sum_{j=0}^{M} \sum_{k=0}^{N} T_{(j,k)}.$$

Restricted rectangular convergence:

$$RR \sum_{n \in Z^{+2}} T_n = t \text{ if for every } E \geq 1, \text{ no matter how large,}$$

$$\lim_{\substack{\min\{M,N\} \to \infty \\ 1/E \leq M/N \leq E}} \sum_{j=0}^{M} \sum_{k=0}^{N} T_{(j,k)} = t.$$

Simplex convergence:

$$Sm \sum_{n \in Z^{+2}} T_n = \lim_{N \to \infty} \sum_{k=0}^{N} \left(\sum_{m=0}^{k} T_{(m,k-m)} \right).$$

We discuss six generalizations of Cantor's Theorem C. They are:

Theorem 1 (Iterated). *Fix any $d \geq 2$. Let $It \sum c_n e^{inx} = 0$ for every $x \in \mathbb{T}^d$. Then $c_n = 0$ for every $n \in \mathbb{Z}^d$.*

Theorem 2 (Unrestricted Rectangular). *Fix any $d \geq 2$. Let $UR \sum c_n e^{inx} = 0$ for every $x \in \mathbb{T}^d$. Then $c_n = 0$ for every $n \in \mathbb{Z}^d$.*

Theorem 3 (Spherical). *Fix any $d \geq 2$. Let $Sp \sum c_n e^{inx} = 0$ for every $x \in \mathbb{T}^d$. Then $c_n = 0$ for every $n \in \mathbb{Z}^d$.*

Question 4 (Restricted Rectangular). Let $RR \sum c_n e^{inx} = 0$ for every $x \in \mathbb{T}^2$. Does this imply that $c_n = 0$ for every $n \in \mathbb{Z}^2$?

Question 5 (Square). Let $Sq \sum c_n e^{inx} = 0$ for every $x \in \mathbb{T}^2$. Does this imply that $c_n = 0$ for every $n \in \mathbb{Z}^2$?

Question 6 (Simplex). Let $Si \sum c_n e^{inx} = 0$ for every $x \in \mathbb{T}^2$. Does this imply that $c_n = 0$ for every $n \in \mathbb{Z}^2$?

Many additional questions can be asked. For a lot of one-dimensional generalizations of Theorem C, see Chap. 9 of [15]. We will later need to mention one higher dimensional extension of Theorem 3 which involves replacing the assumption of spherical convergence: $Sp \sum c_n e^{inx} = 0$, by the weaker assumption of spherical Abel summability: $Sp \sum c_n e^{inx} r^{\|n\|}$ exists for all positive $r < 1$, where $\|n\| = \sqrt{n_1^2 + \cdots + n_d^2}$, and $\lim_{r \to 1^-} Sp \sum c_n e^{inx} r^{\|n\|} = 0$.

2 History of the Three Theorems

Our discussion here will be informed by drawing comparisons with the steps of Cantor's original proof. Here are the four major steps of his proof:

1. Establish the Cantor-Lebesgue theorem that $|c_n| + |c_{-n}| \to 0$.
2. Show that the Riemann function $F(x) = c_0 \frac{x^2}{2} + \sum_{n \neq 0} \frac{c_n}{(in)^2} e^{inx}$ is continuous.
3. Establish the consistency of Riemann summability, i.e, show that the Schwarz second derivative D^2 defined by

$$D^2 F(x) = \lim_{h \to 0} \frac{F(x+h) - 2F(x) + F(x-h)}{h^2} \tag{1}$$

   satisfies

$$D^2 F(x) = \lim_{h \to 0} c_0 + \sum_{n \neq 0} c_n e^{inx} \left(\frac{\sin n\frac{h}{2}}{n\frac{h}{2}} \right)^2 = 0.$$

4. Prove Schwarz's theorem that continuous functions with identically zero Schwarz second derivative are of the form $ax + b$.

The theorem about iterated convergence has a direct simple inductive proof, the starting point being Cantor's theorem.

The unrestricted rectangular theorem was given an erroneous proof in 1919. The false proof was given for $d = 2$. The idea was to copy the steps of Cantor's proof very directly. There was defined the natural analogue of the Riemann function, namely the following termwise fourth integral of the original series,

$$F(x,y) := c_{00} \frac{x^2 y^2}{4} + \sum_{n \in \mathbb{Z}, n \neq 0} c_{0n} \frac{x^2}{2} \frac{e^{iny}}{(in)^2} + \sum_{m \in \mathbb{Z}, m \neq 0} c_{m0} \frac{e^{imx}}{(im)^2} \frac{y^2}{2}$$
$$+ \sum_{m \in \mathbb{Z}, n \in \mathbb{Z}, mn \neq 0} c_{mn} \frac{e^{imx}}{(im)^2} \frac{e^{iny}}{(in)^2}.$$

Differentiating formally (but without any justification) shows that

$$\frac{\partial^4 F}{\partial^2 x \partial^2 y} \text{``} = \text{''} \sum c_{mn} e^{imx} e^{iny} = 0.$$

Next the author showed that F is continuous and does have generalized fourth derivative $D_{xxyy} F(x,y) = 0$ for every point (x,y), where

$$D_{xxyy} := \lim_{\substack{h,k\to 0\\ hk\neq 0}} \frac{\begin{array}{lll} +1F(x-h,y+k) & -2F(x,y+k) & +1F(x+h,y+k) \\ -2F(x-h,y) & +4F(x,y) & -2F(x+h,y) \\ +1F(x-h,y-k) & -2F(x,y-k) & +1F(x+h.y-k) \end{array}}{h^2k^2}.$$

Finally the truth of the following "analog" of Schwarz's theorem was assumed without proof.

Conjecture 1. If $F(x,y)$ is continuous and if for all (x,y)

$$D_{xxyy}F(x,y) = 0,$$

then F behaves as if F were C^4 and satisfied $\frac{\partial^4 F}{\partial^2 x \partial^2 y} = 0$.

This was assumed to be correct and easily extendable to all dimensions.

Since it appeared that the unrestricted rectangular case had been resolved, nothing happened in that area for over 50 years. But when we looked at this around 1970, we could find no proof for the conjecture and felt strongly that one was needed.

The next area to receive attention was that of circular uniqueness. In 1957 Victor Shapiro proved a powerful d-dimensional theorem. Shapiro worked in a somewhat more general context, also considering questions of summability. He did not prove Theorem 3 because his proof required an extra assumption on the coefficient size [13]. For $m \in \mathbb{Z}^d$, let $|m|$ denote $\sqrt{m_1^2 + \cdots + m_d^2}$. A corollary of one of Shapiro's results was this:

Corollary 1. *If $Cr\sum c_n e^{inx} = 0$ for all $x \in \mathbb{T}^d$ and if*

$$\lim_{r\to\infty} \frac{1}{r} \sum_{r-1<|m|\leq r} |c_m| = 0, \tag{2}$$

then all $c_n = 0$.

The coefficient size assumption (2) is natural in dimension 2: since there are $O(r)$ lattice points being summed over in condition (2), this assumption asserts that the c_m tends to zero "on the average" as $|m| \to \infty$. But the assumption becomes much stronger as the dimension increases; specifically in dimension d there are $O\left(r^{d-1}\right)$ terms in the sum, so that the coefficients are required to be decaying like $o\left(r^{2-d}\right)$ on the average. Fourteen years later, in 1971, Roger Cooke found this generalization to the Cantor–Lebesgue theorem for dimension $d = 2$ [11].

Theorem 4 (Cooke). *Let $d = 2$. If $\{c_m\}$ is a doubly indexed set of complex numbers such that*

$$\sum_{|m|=r} c_m e^{imx}$$

tends to zero for almost all x, then

$$\sqrt{\sum_{|m|=r} |c_m|^2} \text{ tends to } 0 \text{ as } r \to \infty. \tag{3}$$

From the definition of spherical convergence it is clear that spherical convergence at x to 0 (or to any other finite value for that matter) implies that the hypothesis of Cooke's theorem holds at x. Now it is a very easy calculation that when $d = 2$, the conclusion of Cooke's theorem implies the validity of condition (2) and thus the unconditional spherical uniqueness theorem in dimension $d = 2$ [4], p. 42. But an unconditional proof of Theorem 3 seemed well out of reach.

In the early 1970s, the pendulum swung back to the unrestricted rectangular convergence uniqueness question. Just at the time of Cooke's work, Grant Welland and I looked at the 1919 paper with the gap mentioned above [1]. We were unable to fill the gap, but we did discover that when a series converges UR almost everywhere, "most" coefficients tend to zero, while all coefficients are bounded. The first fact is easy, but the second required a clever idea that we found in the unpublished thesis of Paul Cohen [9]. From this control of the coefficient size, it follows that Shapiro's condition (2) holds in dimension 2. So the UR uniqueness theorem for two dimensions would follow immediately from Corollary 1 if everywhere UR convergence implies everywhere Sp convergence. It does not. However, by a lucky stroke of fate, everywhere UR convergence *does* imply everywhere spherical Abel summability, and it turns out that the hypotheses of the quite general theorem of Victor Shapiro which yielded Corollary 1 above are satisfied. Thus Theorem 2 was proved in two dimensions.

The first precursor to a higher dimensional theory came in 1976, when Connes extended the Cantor Lebesgue result of Cooke, whose proof was exceedingly two dimensional, to all dimensions [10]. At this point, we knew that if we wanted to prove Theorem 3, we could use the fact that

$$\lim_{k\to\infty} \sum_{\{n:n_1^2+\cdots+n_d^2=k\}} |c_n|^2 = 0$$

without having to add any further hypothesis.

From the middle seventies until the early nineties was a period of hibernation.

In the early 1990s, attention turned to Theorem 2. First of all, Cris Freiling and Dan Rinne showed me the function

$$F(x,y) := (x+y)|x+y|$$

which satisfies the hypothesis of Conjecture 1 above. This analogue of Schwarz's theorem had been stated as fact in 1919. But F does not have the expected form of

$$a(y)x+b(y)+c(x)y+d(x),$$

with $a(y)$ and $c(x)$ being C^4. So the proposed analogue of Schwarz's theorem is false! We went on to give a proof for all dimensions of Theorem 2 [2]. The proof involved replacing the simple but false Conjecture 1 by adding more hypotheses. Rename D_{xxyy} as $S_{(1,1)}$. Further define

$$S_{(1,0)} := \lim_{\substack{h,k\to 0\\ hk\neq 0}} \frac{\begin{array}{lll} +1F(x-h,y+2k) & -2F(x,y+2k) & +1F(x+h,y+2k)\\ -2F(x-h,y+k) & +4F(x,y+k) & -2F(x+h,y+k)\\ +1F(x-h,y) & -2F(x,y) & +1F(x+h.y)\end{array}}{h^2},$$

$$S_{(0,1)} := \lim_{\substack{h,k\to 0\\ hk\neq 0}} \frac{\begin{array}{lll} +1F(x,y+k) & -2F(x+h,y+k) & +1F(x+2h,y+k)\\ -2F(x,y) & +4F(x+h,y) & -2F(x+2h,y)\\ +1F(x,y-k) & -2F(x+h,y-k) & +1F(x+2h.y-k)\end{array}}{k^2},$$

$$S_{(0,0)} := \lim_{\substack{h,k\to 0\\ hk\neq 0}} \frac{\begin{array}{lll} +1F(x,y+2k) & -2F(x+h,y+2k) & +1F(x+2h,y+2k)\\ -2F(x,y+k) & +4F(x+h,y+k) & -2F(x+2h,y+k)\\ +1F(x,y) & -2F(x+h,y) & +1F(x+2h.y).\end{array}}{1}$$

What is true is that all four of these are zero everywhere. Notice that the square of the step size of the second difference appears in the denominator exactly when the difference is symmetric in that direction. To motivate these definitions, note that the 1-dimensional function $A(x) = ce^{inx} + de^{-inx}$ has a second symmetric difference $-4A(x)\sin^2\frac{mh}{2}$, whereas its second forward difference is $-4A(x+h)\sin^2\frac{mh}{2}$. The symmetric differences are so nice that they can overcome the damage done to the quotient by the step size squared term in the denominator; the forward differences are not as nice, but they do not have corresponding denominator terms fighting against their movement toward zero. However, the proof of the corollary still remained quite difficult. We developed a complicated covering technique to get the job done. Probably the best way to understand the technique of the proof, is to see it applied to the much, much simpler one-dimensional case. There it gives a proof by means of covering of Schwarz's original theorem. This proof is much longer and more involved than Schwarz's original, short, and beautiful proof. It has the virtue of extending to our higher dimensional situation, and also it avoids using the maximum principle [3]. I will only give a small one-dimensional analogue of how a covering might come into play here. Suppose you want to prove that if a function is uniformly differentiable to zero on the interval $[a,b]$, then $f(a) = f(b)$. One way would be to let $\{x_i\}$ be a very fine partition of $[a,b]$ and to begin by writing

$$F(b) - F(a) = \sum_i \frac{f(x_{i+1}) - f(x_i)}{x_{i+1} - x_i}(x_{i+1} - x_i).$$

This proof can even be fiddled with to prove that having a zero derivative implies being constant. See [3] on how to deal with second differences in a similar way.

By one of those historical coincidences that happen in mathematics from time to time, even though there had been a 15 year period of complete inactivity, just as we were proving Theorem 2, Shakro Tetunashvili was also proving it in Tbilisi, Georgia [14]. He saw our article and sent me a copy of his proof. It is important to note that his proof was published first. His proof is completely different. It is very easy to give an example of a function which converges UR while diverging iteratively. Let

$$a_{mn} = \begin{cases} (-1)^{m+n} & \text{if } m \in \{0,1\} \text{ or } n \in \{0.1\}. \\ 0 & \text{otherwise} \end{cases}$$

Here is a representation of this series where the value of a_{mn} is attached to the point (m,n):

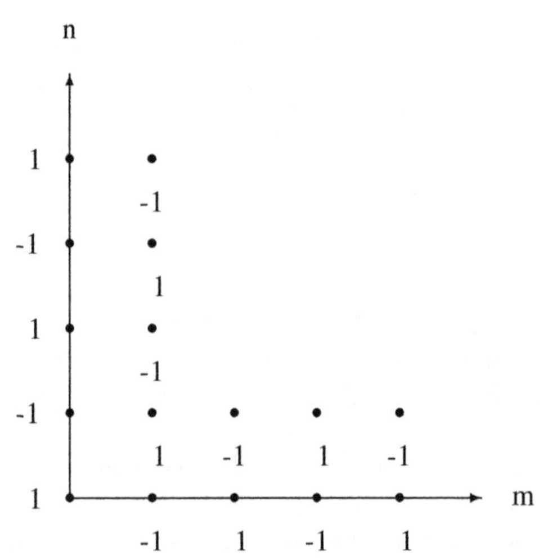

The UR limit is 0 since the rectangular partial sums S_{mn} are all 0 as soon as $\min\{m,n\} \geq 1$. Iterated convergence fails since neither $\lim_{m\to\infty} S_{m0}$ nor $\lim_{n\to\infty} S_{0n}$ exist.

Nevertheless, Tetunashvili was able to prove a lemma: UR convergence to zero everywhere implies one way iterated convergence everywhere. Applying Theorem 2 completes the proof of Theorem 2, UR uniqueness is true. Tetunashvili's Lemma uses a result from [1] that everywhere UR convergence implies that *all* partial sums are bounded which requires the aforementioned lemma from Paul Cohen's thesis. His proof also uses that Paul Cohen's lemma directly. It is much shorter than the proof of Ash, Freiling, and Rinne. A simple description of his proof can be found in [7]. An irony in this tale of two proofs is that the first, "wrong" covering proof probably would not have happened if the second "right" proof had come to our attention earlier. Only time will tell if the covering techniques we developed will eventually have useful applications elsewhere.

The last positive result is Theorem 2. This was proved in 1996 by Jean Bourgain [8]. The basic approach was already outlined in Shapiro's 1957 paper. The idea there is to follow Cantor's original proof. The analogue of the Riemann function is the formal antiLaplacian of $\sum c_n e^{inx}$, namely

$$F(x) = -\sum \frac{c_n}{\|n\|^2} e^{inx}.$$

Note that $\Delta e^{inx} = \sum_{j=1}^{d} \frac{\partial^2}{\partial x_j^2} e^{inx} = \sum_{j=1}^{d} (in_j)^2 e^{inx} = -\|n\|^2 e^{inx}$. The analogue of the one-dimensional Schwarz derivative here is this generalized Laplacian of F:

$$\lim_{\rho \to \infty} \frac{c_d}{\rho^n} \left(\frac{1}{m(B(x,\rho))} \int_{B(x,\rho)} F(t)\, dt - F(x) \right),$$

where $B(x,\rho)$ is an x-centered solid d-dimensional ball of radius ρ, and m denotes Lebesgue measure. A calculation involving Bessel functions shows that this limit is $\Delta F(x)$ when F is C^2, and a theorem of Rado asserts that if F is continuous and this generalized Laplacian is identically 0, then F is harmonic. But harmonic functions are C^∞, and hence the coefficients of F decay very rapidly. In particular, the c_n decay rapidly enough so that Shapiro's condition (2) is satisfied. Since Shapiro had shown that the assumption was enough to guarantee that the generalized Laplacian of F is 0, the proof would be complete if F could be shown to be continuous. In the one-dimensional case the continuity of the Riemann function followed very quickly from the Weierstrass M-test. Because of Conne's 1976 result, we do know that $\sum_{\{\|n\|=k\}} |c_n|^2$ tends to zero, but this is not enough to allow applying the M-test to F. Showing F to be continuous is very difficult. What Bourgain did was show F to be continuous [8].

Bourgain's proof requires numerous ideas as well as strong technique. The flow of his argument is given in [7]. The proof itself appears in at least three places. There is Bourgain's precisely but concisely written original 15-page article [8], there is a somewhat expanded 22-page version in [4], and there is a 42-page version of the proof specialized down to dimension 2 in [6].

3 The Questions

Since all three questions are completely unsolved, we will restrict our discussion to the two-dimensional situation. A square can be rotated into a 2-simplex, and a simple rotation argument shows the equivalence of the square uniqueness question and the simplex uniqueness question in two dimensions, but this equivalence is not clear in higher dimensions, since, in particular, a cube has six faces while a 3-simplex has eight faces. Since we are going to stay in two dimensions, we will naturally consider only the RR and square uniqueness questions.

There are two obvious inclusions. If a multiple numerical series converges Unrestricted Rectangularly, then it converges Restricted Rectangularly. If a multiple numerical series converges Restricted Rectangularly, then it converges Square. So

$$\text{Hypothesis of UR theorem} \implies \text{Hypothesis of RR question}$$

and

$$\text{Hypothesis of RR question} \implies \text{Hypothesis of Square question.}$$

This explains how it can be that UR uniqueness can be known while RR uniqueness remains an open question. This also suggests that in attacking the questions hoping to find affirming proofs one should try to prove the validity of RR uniqueness, while if one is thinking about a counterexample, one should try to find a counterexample to square uniqueness.

There is not much evidence either way for these questions. One reason to lean in the negative direction is the spectacular failure of the Cantor-Lebesgue theorem in this setting.

The weakest version of the one-dimensional Cantor-Lebesgue theorem says that if the sequence $\left\{c_{-n}e^{-inx} + c_n e^{inx}\right\}$ converges to zero for every x as $n \to \infty$, then

$$\text{the sequence } \{|c_{-n}| + |c_n|\} \to 0 \text{ as } n \to \infty. \tag{4}$$

The natural analogue of this for two-dimensional square convergence is to start with the assumption that the sequence with Nth term

$$\sum_{\{(m,n):\max\{|m|,|n|\}=N\}} c_{mn} e^{i(mx+ny)} \tag{5}$$

converges to zero for every (x,y) as $N \to \infty$. The conclusion should be something about the decay of the moduli of the coefficients. For $N \geq 2$, consider the sequence with Nth term

$$A_N(x,y) = e^{\frac{N}{\log\log N}} \cos^2 \frac{x}{2} \sin^{2N-2} \frac{x}{2} \cos Ny.$$

At each point (x,y) this sequence tends to zero rapidly as $N \to \infty$, for if $x = \pm\pi$, every term is identically zero, while if $|x| < \pi$ and $a = \sin^2 \frac{x}{2}$, $|A_N(x,y)| \leq$

$e^{\frac{N}{\log\log N}} a^{N-1}$, which tends very rapidly to zero since $a < 1$. Using the Euler identities to write $\sin\theta$ and $\cos\theta$ in terms of $e^{i\theta}$ and $e^{-i\theta}$ and then expanding by the binomial theorem, we see that A has the form (5), and a calculation shows that

$$c_{0N} = \frac{1}{4^N}\binom{2N}{N}\frac{1}{2N-1}e^{\frac{N}{\log\log N}}$$

which, by Stirling's formula, is on the order of $N^{-3/2}e^{\frac{N}{\log\log N}}$ and hence enormously divergent. This is an equally strong counterexample for restricted rectangular convergence, but the details are slightly messy [5]. (Because of the example I have just shown you, a Cantor-Lebesgue analogue for Square and RR convergence would have to be very, very weak. Actually, the Cantor-Lebesgue analogue here is this: if Eq. (5) tends to zero for every (x, y), the growth of the $\{c_{mn}\}$ must be less than exponential [1, Theorem 2.1]. The above example can be slightly modified to show that this is sharp.).

The example that is mentioned here also occurs in the study of spherical harmonics. The question of uniqueness is essentially open for spherical harmonics also. There is a uniqueness result that appeared 60 years ago in the PhD thesis of Walter Rudin [12]. Rudin constructs an analogue of the Riemann function, and the crux of the matter then comes down to showing that his Riemann function is continuous. He does not prove continuity, but rather restricts his result to the set of all series for which his associated Riemann function is continuous. This is a very strong auxiliary condition, but Rudin's result has never been improved. I feel that the fact that this counterexample fits both situations makes it very likely that progress on the open trigonometric questions will be strongly correlated with progress on the uniqueness question for spherical harmonics.

The first step of the one-dimensional Cantor program, namely the application of a Cantor-Lebesgue theorem to gain knowledge of some decay of the coefficients, seems out of reach because of the counterexample. However, this is not the end of the difficulties. Even the following two questions are beyond what I know how to do.

Question 4′ (Restricted Rectangular). Let $d = 2$, $RR\sum c_n e^{inx} = 0$ for every $x \in \mathbb{T}^2$, and further assume that $|c_{mn}| = o\left(\frac{1}{\sqrt{m^2+n^2}}\right)$. Does this imply that $c_n = 0$ for every $n \in \mathbb{Z}^2$?

Question 5′ (Square). Let $d = 2$, $Sq\sum c_n e^{inx} = 0$ for every $x \in \mathbb{T}^2$, and further assume that $\sum_{\{(m,n):\max\{|m|,|n|\}=k\}} |c_{mn}|^2 = o\left(\frac{1}{k\ln k}\right)$. Does this imply that $c_n = 0$ for every $n \in \mathbb{Z}^2$?

I picked the auxiliary hypotheses here pretty much at random. The idea is to keep the set of series for which the proof is valid broader than L^2. In other words, is anything at all is true about uniqueness for either of these two methods as soon as one moves out into the realm of trigonometric series that may not be Fourier series?

Here are two approaches that I have recently thought of, but have not yet tried out:

1. Try to reverse a normally irreversible hypothesis using the facts that the assumed convergence is both everywhere and to zero. For example, try to move from the RR hypothesis to the UR hypothesis, thereby using the UR uniqueness theorem as a lemma for proving RR uniqueness, or try to move from the Sq hypothesis to the RR hypothesis to prove equivalence of Questions 4 and 5. Examples of this sort of procedure can be found directly in Tetunashvili's proof of Theorem 2 and indirectly in the result of Ash and Welland that while UR convergence at a point trivially implies the finiteness of $\limsup_{\min\{m,n\}\to\infty} |S_{mn}|$ at the same point, UR convergence everywhere unexpectedly implies the boundedness of *all* partial sums at each point.
2. The Laplacian seems to be a better derivative than $\frac{\partial^4}{\partial x^2 \partial y^2}$. But the spherical generalized Laplacian used by Shapiro and Bourgain does not seem to fit the rectangular methods very well. Perhaps, a generalized Laplacian formed by integral averaging a function over the boundary of a small square and subtracting the function value at the center might work better. Another possibility is a generalized Laplacian formed by averaging a sum of function values spaced around the edge of a small square and subtracting the function value at the center.

The first seven references below can be found using links from http://condor.depaul.edu/mash/realvita.html.

References

1. Ash, J.M., Welland, G.V.: Convergence, uniqueness, and summability of multiple trigonometric series. Trans. Amer. Math. Soc. **163**, 401–436 (1972), MR **45** # 9057
2. Ash, J.M., Freiling, C., Rinne, D.: Uniqueness of rectangularly convergent trigonometric series. Ann. of Math. **137**, 145–166 (1993)
3. Ash, J.M.: A new, harder proof that continuous functions with Schwarz derivative 0 are lines. Fourier Analysis (Orono, ME, 1992) In: Bray, W.O., Milojević, P.S., Stanojević, C.V. (eds.) Lecture Notes in Pure and Appl. Math. **157**, pp. 35–46. Marcel Dekker, New York, (1994)
4. Ash, J.M., Wang, G.: A survey of uniqueness questions in multiple trigonometric series. A Conference in Harmonic Analysis and Nonlinear Differential Equations in Honor of Victor L. Shapiro. Contemp. Math. **208**, 35–71 (1997)
5. Ash, J.M., Wang, G.: One and two dimensional Cantor-Lebesgue type theorems. Trans. Amer. Math. Soc. **349**, 1663–1674 (1997)
6. Ash, J.M.: Uniqueness for Spherically Convergent Multiple Trigonometric Series. In: Anastassiou, G. (ed.) Handbook of Analytic-Computational Methods in Applied Mathematics, pp. 309–355. Chapman & Hall/CRC, Boca Raton (2000)
7. Ash, J.M.: Uniqueness for higher dimensional trigonometric series. Cubo **4**, 97–125 (2002)
8. Bourgain, J.: Spherical summation and uniqueness of multiple trigonometric series. Internat. Math. Res. Notices, **1996**(3), 93–107 (1996)
9. Cohen, P.J.: Topics in the theory of uniqueness of trigonometrical series, doctoral thesis, University of Chicago, Chicago, IL., 1958

10. Connes, B.: Sur les coefficients des séries trigonometriques convergents sphériquement. CRA Sc Paris t. Ser. A **283**, 159–161 (1976)
11. Cooke, R.L.: A Cantor-Lebesgue theorem in two dimensions. Proc. Amer. Math. Soc. **30**, 547–550 (1971), MR **43** #7847
12. Rudin, W.: Uniqueness theory for Laplace series. Trans. Amer. Math. Soc. **68**, 287–303 (1950)
13. Shapiro, V.L.: Uniqueness of multiple trigonometric series. Ann. of Math. **66**, 467–480 (1957), MR **19**, 854; 1432
14. Tetunashvili, S.: On some multiple function series and the solution of the uniqueness problem for Pringsheim convergence of multiple trigonometric series. Mat. Sb. **182**, 1158–1176 (1991) (Russian), Math. USSR Sbornik **73**, 517–534 (1992) (English). MR 93b:42022
15. Zygmund, A.: Trigonometric Series, v.2, 2nd edn. Cambridge University Press, Cambridge (1958)

The L_2 Discrepancy of Two-Dimensional Lattices

Dmitriy Bilyk, Vladimir N. Temlyakov, and Rui Yu

Dedicated to Konstantin Oskolkov, a dear friend and colleague, on the occasion of his 65th birthday.

Abstract Let α be an irrational number with bounded partial quotients of the continued fraction a_k. It is well known that *symmetrizations* of the irrational lattice $\{(\mu/N, \{\mu\alpha\})\}_{\mu=0}^{N-1}$ have optimal order of L_2 discrepancy, $\sqrt{\log N}$. The same is true for their rational approximations $\mathscr{L}_n(\alpha) = \{(\mu/q_n, \{\mu p_n/q_n\})\}_{\mu=0}^{q_n-1}$, where p_n/q_n is the nth convergent of α. However, the question whether and when the symmetrization is really necessary remained wide open.

We show that the L_2 discrepancy of the nonsymmetrized lattice $\mathscr{L}_n(\alpha)$ grows as

$$\|D(\mathscr{L}_n(\alpha), \mathbf{x})\|_2 \approx \max\left\{ \log^{\frac{1}{2}} q_n, \left| \sum_{k=0}^{n} (-1)^k a_k \right| \right\},$$

in particular, characterizing the lattices for which the L_2 discrepancy is optimal.

1 Introduction

1.1 Discrepancy

The extent of equidistribution of a finite point set can be naturally measured using the discrepancy function. Let $\mathscr{P}_N$ be a set of N points in the unit cube $[0,1]^d$ in dimension d. The discrepancy function is then defined as

---

D. Bilyk (✉) • V.N. Temlyakov • R. Yu
University of South Carolina, Columbia, SC 29208, USA
e-mail: bilyk@math.sc.edu; temlyak@math.sc.edu; yur@math.sc.edu

D. Bilyk et al. (eds.), *Recent Advances in Harmonic Analysis and Applications*, Springer Proceedings in Mathematics & Statistics 25, DOI 10.1007/978-1-4614-4565-4_9,

$$D(\mathscr{P}_N,\mathbf{x}) := \#\{\mathscr{P}_N \cap [\mathbf{0},\mathbf{x})\} - N \cdot |[\mathbf{0},\mathbf{x})|, \tag{1}$$

where $\mathbf{x} = (x_1,\dots,x_d)$, $[\mathbf{0},\mathbf{x}) = \prod_{j=1}^{d}[0,x_j)$, and $|\cdot|$ denotes the Lebesgue measure. The L_p norm of the discrepancy function, usually referred to as the L_p discrepancy, is a benchmark that one uses to evaluate the quality of a particular set of N points. The fundamental problems of the discrepancy theory are to construct sets with small L_p discrepancy and to find optimal bounds.

The main principle of the theory of irregularities of distribution states the L_p discrepancy of a finite point set cannot be too small, that is, the quantity

$$D(N,d)_p := \inf_{\mathscr{P}_N} \|D(\mathscr{P}_N,\mathbf{x})\|_p$$

must necessarily go to infinity with N when $d \geq 2$. We refer the reader to [2,6,21,23] for detailed surveys. The famous lower bounds for $D(N,d)_p$ are:

Theorem 1 (Roth [26]). *In all dimensions $d \geq 2$, we have*

$$D(N,d)_2 \geq C(d)(\log N)^{\frac{d-1}{2}}, \tag{2}$$

where $C(d)$ is a positive constant that may depend on d.

This bound has been extended to L_p discrepancy ($1 < p < \infty$) by Schmidt [29], who has also obtained a lower estimate for the L_∞ (extremal) discrepancy:

Theorem 2 (Schmidt [28]). *In dimension $d = 2$,*

$$D(N,2)_\infty \geq C \log N, \tag{3}$$

where C is a positive absolute constant.

It is well known that both bounds are sharp in the order of magnitude. While Eq. (3) is harder to prove than Eq. (2) (in fact, higher-dimensional analogs of Eq. (3) are still very far from being understood; see [3]), its sharpness had been known long before Eq. (3) has been established, [10, 22]. The example which is relevant to our discussion is the irrational lattice:

$$\mathscr{A}_N(\alpha) := \left\{ \left(\frac{\mu}{N}, \{\mu\alpha\} \right) \right\}_{\mu=0}^{N-1}, \tag{4}$$

where α is an irrational number and $\{x\}$ is the fractional part of the number x. If the partial quotients of the continued fraction of α are bounded, then the L_∞ discrepancy of this set is of the order $\log N$ (see, e.g., [20, 23]). The idea of this example goes back to Lerch [22].

It is often more convenient and effective to work with rational approximations of such lattices. For an irrational number α with the continued fraction expansion

$$\alpha = [a_0; a_1, a_2, \dots] = a_0 + \cfrac{1}{a_1 + \frac{1}{a_2 + \frac{1}{a_3 + \dots}}}, \tag{5}$$

where $a_0 \in \mathbb{Z}$, $a_k \in \mathbb{N}$, $k \geq 1$, we denote by p_n/q_n the nth order convergents of α, that is, $p_n/q_n = [a_0; a_1, \dots, a_n]$.

We consider the sets

$$\mathscr{L}_n(\alpha) := \{(\mu/q_n, \{\mu p_n/q_n\})\}_{\mu=0}^{q_n-1}, \tag{6}$$

consisting of q_n points, which approximate $\mathscr{A}_{q_n}(\alpha)$. A particular example of such sets is the popular *Fibonacci lattice*. Let $\{b_n\}_{n=0}^{\infty}$ be the sequence of Fibonacci numbers:

$$b_0 = b_1 = 1, \quad b_n = b_{n-1} + b_{n-2}, \quad \text{for} \quad n \geq 2. \tag{7}$$

The b_n-point Fibonacci set $\mathscr{F}_n \subset [0,1]^2$ is then defined as

$$\mathscr{F}_n := \{(\mu/b_n, \{\mu b_{n-1}/b_n\})\}_{\mu=0}^{b_n-1}. \tag{8}$$

Obviously, for large n, the set $\mathscr{F}_n$ is close to the irrational lattice $\mathscr{A}_N(\alpha)$ with $N = b_n$ and $\alpha = \frac{\sqrt{5}-1}{2}$, that is, the reciprocal of the *golden section*. For this value of α, we have $a_0 = 1$, $a_k = 1$ for $k \geq 1$, and $p_n = b_{n-1}$, $q_n = b_n$. Therefore, $\mathscr{F}_n = \mathscr{L}_n\big((\sqrt{5}-1)/2\big)$. It is well known (see e.g., [24]) that

$$\|D(\mathscr{F}_n, \mathbf{x})\|_\infty \leq C \log b_n \leq C' n; \tag{9}$$

hence, Fibonacci sets also have optimal L_∞ discrepancy. Similar bounds hold for more general lattices $\mathscr{L}_n(\alpha)$ whenever the sequence $\{a_k\}$ of the partial quotients of α is bounded. These results can be derived either directly or as a perturbation of the corresponding results for the irrational lattice $\mathscr{A}_N(\alpha)$.

Another standard example of a set with optimal L_∞ discrepancy is the *van der Corput set* $\mathscr{V}_n$ defined as the collection of 2^n points of the form

$$\big(0.x_1x_2\dots x_n, 0.x_nx_{n-1}\dots x_2x_1\big), \ x_k = 0 \text{ or } 1, \tag{10}$$

where the coordinates are written in binary expansion. While this set is not directly related to our discussion, we shall sometimes compare the properties of $\mathscr{V}_n$ and the lattices $\mathscr{A}_N(\alpha)$ or $\mathscr{L}_n(\alpha)$. An interesting relation between the Fibonacci and van der Corput sequences is discussed in [15].

In contrast to the L_∞ case, the sharpness of the L_2 bound (2) is harder to demonstrate. Most constructions are obtained as modifications of the classical distributions with low L_∞ discrepancy. These modifications are often necessary—for instance, it is known that the L_2 discrepancy of the van der Corput set is not optimal: it is of the order $\log N$ rather than $\sqrt{\log N}$. The first example of a set with L_2 discrepancy of the order $\sqrt{\log N}$ has been constructed by Davenport [11] in 1956 by symmetrizing the irrational lattice $\mathscr{A}_N(\alpha)$:

$$\begin{aligned}\mathscr{A}'_{\mathscr{N}}(\alpha) &:= \{(\{\mu/N\}, \{\mu\alpha\})\}_{\mu=-(N-1)}^{N-1} \\ &= \mathscr{A}_{\mathscr{N}}(\alpha) \cup \{(1-x,y) : (x,y) \in \mathscr{A}_{\mathscr{N}}(\alpha)\}.\end{aligned} \tag{11}$$

It has been shown by the authors of this chapter [5] that the same holds for the L_2 discrepancy of an analogous symmetrization $\mathscr{F}'_n$ of the Fibonacci set $\mathscr{F}_n$ (in this case, a naive perturbation argument does not work) and their method can be easily generalized to obtain the L_2 optimality of the symmetrizations of $\mathscr{L}_n(\alpha)$. Both in the case of $\mathscr{A}'_N(\alpha)$ and of $\mathscr{F}'_n$, the proofs used the Fourier series of the discrepancy function.

Davenport's work, however, has not addressed the question whether this symmetrization is really necessary; in other words, what is the L_2 discrepancy of the non-symmetrized lattices $\mathscr{A}_N(\alpha)$ or $\mathscr{L}_n(\alpha)$? In 1979, Sós and Zaremba [31] gave a partial answer to this question by proving that, when all the partial quotients of the (finite or infinite) continued fraction of α are equal, the set $\mathscr{A}_N(\alpha)$ has optimal L_2 discrepancy. In particular, their result covers the Fibonacci set $\mathscr{F}_n$ and the irrational lattice $\mathscr{A}_N((\sqrt{5}-1)/2)$ when all the partial quotients are equal to 1:

$$\|D(\mathscr{F}_n,\mathbf{x})\|_2 \asymp \|D\Big(\mathscr{A}_{b_n}\left(\left(\sqrt{5}-1\right)/2,\mathbf{x}\right)\|_2 \asymp \sqrt{\log b_n}. \tag{12}$$

It is also suggested in the same paper that perhaps the L_2 discrepancy is not optimal for some other values of α. This means that the L_2 discrepancy depends on much finer properties of α than simply the boundedness of its partial quotients.

In this chapter we continue this line of investigation. In Sect. 2, we give a Fourier-analytic proof of the fact that the nonsymmetrized Fibonacci lattice $\mathscr{F}_n$ has optimal order of magnitude of L_2 discrepancy. While this result is just a partial case of the aforementioned result of Sós and Zaremba, our proof, based on the computation of the Fourier coefficients, is much more direct and transparent. It also yields an exact formula for the L_2 discrepancy of $\mathscr{F}_n$, which opens the door to numerical experiments. In addition, this method easily generalizes and allows one to investigate other rational lattices $\mathscr{L}_n(\alpha)$.

In Sect. 3, we demonstrate how one can adapt the arguments used for the Fibonacci sets $\mathscr{F}_n$ to more general lattices. It is often the case that, when a low-discrepancy set fails to have the optimal L_2 discrepancy, the problem lies already in the Fourier coefficient of order zero: the integral $\int_{[0,1]^d} D(\mathscr{P}_N,\mathbf{x})d\mathbf{x}$; see, for example, [4, 17]. We show that this is indeed the case for the lattices $\mathscr{L}_n(\alpha)$, that is, the contribution of the other Fourier coefficients to the L_2 norm is of the order $\sqrt{\log N}$.

We also observe that the main term [the integral of the discrepancy function of $\mathscr{L}_n(\alpha)$] is closely related to the Dedekind sums, an object often arising in number theory and geometry. In particular, it allows us to show that the behavior of the integral of the discrepancy function is controlled by the growth of the alternating sums of the partial quotients of α: $\sum_{k=0}^{n}(-1)^k a_k$, which, in particular, reveals the nature of the condition that all a_k's are the same in the result of Sós and Zaremba [31]. To be more precise, we prove the following theorem:

Theorem 3. *Let $\alpha = [a_0; a_1, a_2, \ldots]$ be an irrational number with bounded partial quotients. Denote the nth convergents of α by p_n/q_n and consider the lattice $\mathscr{L}_n(\alpha)$ as defined in Eq.* (6). *Then its L_2 discrepancy satisfies*

$$\|D(\mathscr{L}_n(\alpha), \mathbf{x})\|_2 \approx \max\left\{\sqrt{\log q_n}, \left|\sum_{k=0}^{n}(-1)^k a_k\right|\right\}. \tag{13}$$

By $A \approx B$, we mean that "A is of the same order as B", that is, $A = \mathscr{O}(B)$ and $B = \mathscr{O}(A)$, as n (or N) tends to infinity (the implicit constants are independent of n or N, but may depend on the number α). Therefore, when the alternating sum of a_k's does not grow too fast, the lattices $\mathscr{L}_n(\alpha)$ have optimal order of L_2 discrepancy even *without the symmetrization*. Since $\log q_n \approx n$, this happens precisely when

$$\left|\sum_{k=0}^{n}(-1)^k a_k\right| \ll \sqrt{n}, \tag{14}$$

where $A \ll B$ means $A = \mathscr{O}(B)$. We thus characterize all the lattices $\mathscr{L}_n(\alpha)$ for which the L_2 discrepancy is minimal in the sense of order. In Sect. 4, we make some further remarks concerning the behavior of the L_2 discrepancy for certain specific values of α.

To finish the introduction, we would like to mention that constructions of sets with optimal L_2 or L_p discrepancies are an important problem in discrepancy theory and quasi-Monte Carlo methods. Following the result of Davenport [11] in dimension $d = 2$, Roth [27] constructed sets with optimal L_2 discrepancy in all dimensions. Chen [7] and Frolov [14] have constructed sets with minimal order of L_p discrepancy for $1 < p < \infty$. It is interesting to point out that in dimensions $d \geq 3$, all the known constructions until recently were probabilistic. The first deterministic examples were provided in the last decade by Chen and Skriganov [8, 9, 30].

Cubature formulas based on Fibonacci lattice have been thoroughly studied in approximation theory [34, 35]. Rational lattices $\mathscr{L}_n(\alpha)$ are obviously more practical than the irrational lattices $\mathscr{A}_N(\alpha)$; besides, cubature formulas built on $\mathscr{L}_n(\alpha)$ perform better in spaces with mixed smoothness of order $r > 2$; see [5, 18, 36].

2 The L_2 Discrepancy of the Fibonacci Set

In this section, we prove the L_2 bound for the discrepancy of the Fibonacci set, particularly giving a new proof of the result of Sós and Zaremba [31]. Recall that the discrepancy function of the Fibonacci set is

$$D(\mathscr{F}_n,\mathbf{x}) := \#\{\mathscr{F}_n \cap [\mathbf{0},\mathbf{x})\} - b_n x_1 x_2 = \sum_{\mathbf{p}=(p_1,p_2)\in\mathscr{F}_n} \chi_{[p_1,1)\times[p_2,1)}(\mathbf{x}) - b_n x_1 x_2,$$

where $\mathbf{x} = (x_1,x_2) \in [0,1]^2$. We compute the Fourier coefficients of the $D(\mathscr{F}_n,\mathbf{x})$:

$$\begin{aligned}\widehat{D}(\mathscr{F}_n,\mathbf{k}) &= \sum_{\mathbf{p}\in\mathscr{F}_n} \int_{p_1}^{1} e^{-2\pi i k_1 x_1} dx_1 \int_{p_2}^{1} e^{-2\pi i k_2 x_2} dx_2 \\ &\quad - b_n \int_0^1 x_1 e^{-2\pi i k_1 x_1} dx_1 \int_0^1 x_2 e^{-2\pi i k_2 x_2} dx_2. \end{aligned} \tag{15}$$

Note that

$$\sum_{\mu=1}^{b_n} e^{-2\pi i l \mu / b_n} = \begin{cases} b_n, & l \equiv 0 \pmod{b_n}, \\ 0, & l \not\equiv 0 \pmod{b_n}. \end{cases} \tag{16}$$

Let $L(n) := \{\mathbf{k} = (k_1,k_2) \in \mathbf{Z}^2 : k_1 + b_{n-1}k_2 \equiv 0 \pmod{b_n}\}$, then

$$\sum_{\mu=1}^{b_n} e^{-2\pi i (k_1 + b_{n-1}k_2)\mu / b_n} = \begin{cases} b_n, & (k_1,k_2) \in L(n), \\ 0, & (k_1,k_2) \notin L(n). \end{cases} \tag{17}$$

We consider different cases.

Case 1. $\mathbf{k} = (0,0)$. (*The integral of* $D(\mathscr{F}_n,\mathbf{x})$). Standard heuristics in discrepancy theory state that this case usually presents the most important complications in obtaining favorable L_2 estimates. In fact, in the case of the van der Corput set, this term is solely responsible for the L_2 discrepancy being too large; see [4, 17]. Davenport's symmetrization was created precisely to eliminate this term in the Fourier series of the discrepancy function. One can compute this term precisely in the case of the Fibonacci lattice:

Lemma 1.

$$\widehat{D}(\mathscr{F}_n,\mathbf{0}) = \begin{cases} \dfrac{3}{4}, & \textit{for } n \textit{ even}, \\ \dfrac{b_{n-1}}{6b_n} + \dfrac{7}{12}, & \textit{for } n \textit{ odd}. \end{cases} \tag{18}$$

Proof. From Eq. (15), we obtain

$$\begin{aligned}\widehat{D}(\mathscr{F}_n,\mathbf{0}) &= \sum_{\mu=0}^{b_n-1} \left(1 - \frac{\mu}{b_n}\right)\left(1 - \left\{\frac{\mu b_{n-1}}{b_n}\right\}\right) - \frac{b_n}{4} \\ &= \sum_{\mu=1}^{b_n-1} \mu/b_n \cdot \{\mu b_{n-1}/b_n\} - \frac{b_n}{4} + 1, \end{aligned} \tag{19}$$

where we have used the fact that $\sum_{\mu=0}^{b_n-1} \mu/b_n = \sum_{\mu=0}^{b_n-1} \{\mu b_{n-1}/b_n\} = \frac{b_n-1}{2}$.

We shall now connect $\widehat{D}(\mathscr{F}_n,\mathbf{0})$ to a well-known object in number theory—the Dedekind sum. The (inhomogeneous) Dedekind sum is defined as

$$\mathscr{D}(p,q) = \sum_{\mu=1}^{q-1} \rho\left(\frac{\mu}{q}\right)\rho\left(\frac{p\mu}{q}\right), \tag{20}$$

where $\rho(x) = \frac{1}{2} - \{x\}$ is the sawtooth function and p, q are positive integers. These sums have already appeared in the context of discrepancy, uniform distribution, and Fibonacci numbers, for example, [25, 37, 38]. We have the following relation:

$$\begin{aligned}
\mathscr{D}(b_{n-1},b_n) &= \sum_{\mu=1}^{b_n-1}\left(\frac{1}{2}-\frac{\mu}{b_n}\right)\left(\frac{1}{2}-\left\{\frac{\mu b_{n-1}}{b_n}\right\}\right) \\
&= \sum_{\mu=1}^{b_n-1} \mu/b_n\cdot\{\mu b_{n-1}/b_n\} - \frac{b_n}{4} + \frac{1}{4}.
\end{aligned} \tag{21}$$

We thus see from Eqs. (19) and (21) that

$$\widehat{D}(\mathscr{F}_n,\mathbf{0}) = \mathscr{D}(b_{n-1},b_n) + \frac{3}{4}. \tag{22}$$

The Dedekind sum $\mathscr{D}(p,q)$ can be computed in terms of the continued fraction expansion of p/q. The following formula holds [1, 16]:

Proposition 1. *Let n be the length of the continued fraction expansion of p/q and let p_k/q_k denote the kth convergents of p/q, $k \le n$, and a_0, a_1, …, a_n be the partial quotients. Then*

$$\mathscr{D}(p,q) = \begin{cases} \dfrac{1}{12}\left(\dfrac{p_n - q_{n-1}}{q_n} - \displaystyle\sum_{k=0}^{n}(-1)^k a_k\right), & \text{for } n \text{ even}, \\[2ex] \dfrac{1}{12}\left(\dfrac{p_n + q_{n-1}}{q_n} - \displaystyle\sum_{k=0}^{n}(-1)^k a_k\right) - \dfrac{1}{4}, & \text{for } n \text{ odd}. \end{cases} \tag{23}$$

In our case, $p = b_{n-1}$ and $q = b_n$, $p_n = q_{n-1} = b_{n-1}$, $q_n = b_n$, $a_0 = 0$, and $a_1 = a_2 = \cdots = a_n = 1$, which yields the result of Lemma 1. □

Case 2. $k_1 \neq 0, k_2 \neq 0$. In this case, Eq. (15) becomes

$$\widehat{D}(\mathscr{F}_n,\mathbf{k}) = \frac{-1}{4\pi^2 k_1 k_2}\sum_{\mathbf{p}\in\mathscr{F}_n}(1-e^{-2\pi i k_1 p_1})(1-e^{-2\pi i k_2 p_2}) + \frac{b_n}{4\pi^2 k_1 k_2}, \tag{24}$$

which together with Eqs. (16) and (17) leads to the following lemmas:

Lemma 2. *If $k_1 \neq 0$, $k_2 \neq 0$, then*

$$\widehat{D}(\mathscr{F}_n,\mathbf{k}) = \frac{b_n}{4\pi^2 k_1 k_2} \tag{25}$$

provided that at least one of k_1 and k_2 is 0 modulo b_n.

Lemma 3. *Assume $k_1 \not\equiv 0 \pmod{b_n}$ and $k_2 \not\equiv 0 \pmod{b_n}$, then*

$$\widehat{D}(\mathscr{F}_n,\mathbf{k}) = \begin{cases} \dfrac{-b_n}{4\pi^2 k_1 k_2}, & k_1 + k_2 b_{n-1} \equiv 0, \text{ i.e. } \mathbf{k} \in L(n) \\ 0, & k_1 + k_2 b_{n-1} \not\equiv 0, \text{ i.e. } \mathbf{k} \notin L(n), \end{cases} \tag{26}$$

where all congruences are taken modulo b_n.

Case 3. $k_1 = 0, k_2 \neq 0$. We have the following lemma:

Lemma 4. *If $k_1 = 0, k_2 \neq 0$,*

$$\widehat{D}(\mathscr{F}_n,\mathbf{k}) = \begin{cases} \dfrac{b_n}{4\pi i k_2}, & k_2 \equiv 0 \pmod{b_n}, \\ -\dfrac{1}{4\pi i k_2} \cdot \dfrac{e^{-2\pi i k_2 b_{n-1}/b_n} + 1}{e^{-2\pi i k_2 b_{n-1}/b_n} - 1}, & k_2 \not\equiv 0 \pmod{b_n}. \end{cases} \tag{27}$$

Proof. From Eq. (15), we obtain

$$\widehat{D}(\mathscr{F}_n,\mathbf{k}) = \frac{-1}{2\pi i k_2} \sum_{\mathbf{p} \in \mathscr{F}_n} (1 - p_1)(1 - e^{-2\pi i k_2 p_2}) + \frac{b_n}{4\pi i k_2}.$$

The case $k_2 \equiv 0$ is trivial. When $k_2 \not\equiv 0$, one is faced with the sum $\sum_{\mu=0}^{b_n-1} \frac{\mu}{b_n} e^{-\frac{2\pi i k_2 \mu b_{n-1}}{b_n}}$, which can be computed by considering the function $f(x) = \sum_{\mu=0}^{b_n-1} e^{\frac{2\pi i \mu x}{b_n}} = \frac{e^{2\pi i x} - 1}{e^{\frac{2\pi i x}{b_n}} - 1}$ and observing that the aforementioned sum equals $f'(-k_2 b_{n-1})/2\pi i$.

Case 4. $k_1 \neq 0$, $k_2 = 0$. This case is the same as the previous case due to the well-known relation $b_n^2 - b_{n+1} b_{n-1} = (-1)^n$, which implies that

$$\mathscr{F}_n = \left\{ \left(\frac{\mu}{b_n}, \left\{ \frac{\mu b_{n-1}}{b_n} \right\} \right) \right\}_{\mu=0}^{b_n-1} = \left\{ \left(\left\{ \frac{(-1)^{n-1} r b_{n-1}}{b_n} \right\}, \frac{r}{b_n} \right) \right\}_{r=0}^{b_n-1}.$$

In other words, the Fibonacci set possesses some inner symmetry: if n is odd, it is simply symmetric with respect to the diagonal $x = y$; if n is even, the symmetry involves an additional reflection about the axis $x = \frac{1}{2}$.

Lemma 5. *If* $k_1 \neq 0, k_2 = 0,$

$$\widehat{D}(\mathscr{F}_n,\mathbf{k}) = \begin{cases} \dfrac{b_n}{4\pi i k_1}, & k_1 \equiv 0 \pmod{b_n}, \\ -\dfrac{1}{4\pi i k_1}\cdot\dfrac{e^{(-1)^n 2\pi i k_1 b_{n-1}/b_n}+1}{e^{(-1)^n 2\pi i k_1 b_{n-1}/b_n}-1}, & k_1 \not\equiv 0 \pmod{b_n}. \end{cases}$$

Theorem 4. *For the Fibonacci set* $\mathscr{F}_n \subset [0,1]^2$, *we have*

$$\|D(\mathscr{F}_n,\mathbf{x})\|_2 \ll \sqrt{\log b_n}. \tag{28}$$

We first derive a formula providing the exact value of $\|D(\mathscr{F}_n,\mathbf{x})\|_2$. We start with the contribution of Lemma 3. In this case, $\widehat{D}(\mathscr{F}_n,\mathbf{k}) = -\frac{b_n}{4\pi^2 k_1 k_2}$. We make use of the well-known identity (see e.g., [32], p. 165, ex. 15):

$$\sum_{n\in\mathbb{Z}} \frac{1}{(n+x)^2} = \frac{\pi^2}{\sin^2(\pi x)}. \tag{29}$$

For $\mathbf{k}\in L(n)$, $k_i \not\equiv 0 \pmod{b_n}$, denote $k_1 + k_2 b_{n-1} = l b_n$, for $l\in\mathbb{Z}$ and $k_2 = m b_n + r$, where $m\in\mathbb{Z}$ and $r = 1,\dots,b_n - 1$. Lemma 3 implies

$$\begin{aligned}\sum_{\mathbf{k}\in L(n),k_i\neq 0} \left|\widehat{D}(\mathscr{F}_n,\mathbf{k})\right|^2 &= \frac{b_n^2}{16\pi^4} \sum_{k_2\not\equiv 0 \bmod b_n} \frac{1}{k_2^2} \sum_{l\in\mathbb{Z}} \frac{1}{b_n^2}\cdot\frac{1}{\left(l-\frac{b_{n-1}k_2}{b_n}\right)^2} \\ &= \frac{1}{16\pi^2}\sum_{r=1}^{b_n-1}\frac{1}{\sin^2\left(\frac{\pi b_{n-1} r}{b_n}\right)}\sum_{m\in\mathbb{Z}}\frac{1}{b_n^2}\cdot\frac{1}{\left(m+\frac{r}{b_n}\right)^2} \\ &= \frac{1}{16 b_n^2}\sum_{r=1}^{b_n-1}\frac{1}{\sin^2\left(\frac{\pi b_{n-1} r}{b_n}\right)\cdot\sin^2\left(\frac{\pi r}{b_n}\right)}. \end{aligned}\tag{30}$$

In the setting of Lemma 2 ($k_1, k_2 \neq 0$ and at least one of them is zero modulo b_n), inclusion-exclusion principle and the identity $\sum_{l\in\mathbb{N}}\frac{1}{l^2} = \frac{\pi^2}{6}$ yield

$$\begin{aligned}\sum_{\mathbf{k}\sim\text{Lemma 2}} \left|\widehat{D}(\mathscr{F}_n,\mathbf{k})\right|^2 &= 4\cdot\frac{b_n^2}{16\pi^4}\cdot\left(2\cdot\sum_{l,k\in\mathbb{N}}\frac{1}{l^2 b_n^2\cdot k^2} - \sum_{l_1,l_2\in\mathbb{N}}\frac{1}{b_n^4 l_1^2 l_2^2}\right) \\ &= \frac{b_n^2}{4\pi^4}\cdot\left(\frac{2\pi^4}{36 b_n^2} - \frac{\pi^4}{36 b_n^4}\right) = \frac{1}{72}\left(1-\frac{1}{2b_n^2}\right), \end{aligned}\tag{31}$$

where multiplication by 4 accounts for all possible choices of signs. We now turn to the first contribution of Lemma 4: $k_1 = 0$, $k_2 \not\equiv 0 \pmod{b_n}$:

$$\left|\widehat{D}(\mathscr{F}_n,\mathbf{k})\right|^2 = \frac{1}{16\pi^2 k_2^2}\cdot\frac{e^{-2\pi i k_2 b_{n-1}/b_n}+1}{e^{-2\pi i k_2 b_{n-1}/b_n}-1}\cdot\frac{e^{2\pi i k_2 b_{n-1}/b_n}+1}{e^{2\pi i k_2 b_{n-1}/b_n}-1}$$

$$= \frac{1}{16\pi^2 k_2^2}\cdot\frac{1+\cos\left(\frac{2\pi k_2 b_{n-1}}{b_n}\right)}{1-\cos\left(2\pi\frac{k_2 b_{n-1}}{b_n}\right)} = \frac{1}{16\pi^2 k_2^2}\cdot\frac{\cos^2\left(\frac{\pi k_2 b_{n-1}}{b_n}\right)}{\sin^2\left(\frac{\pi k_2 b_{n-1}}{b_n}\right)}. \quad (32)$$

Writing $k_2 = lb_n + r$, $l \in \mathbb{Z}$, $r = 1,\ldots,b_n-1$ and using Eq. (29), we obtain

$$\sum_{k_2\neq 0}\left|\widehat{D}(\mathscr{F}_n,(0,k_2))\right|^2 = \frac{1}{16\pi^2}\sum_{l\in\mathbb{Z}}\sum_{r=1}^{b_n-1}\frac{1}{b_n^2\cdot\left(l+\frac{r}{b_n}\right)^2}\cdot\frac{\cos^2(\pi k_2 b_{n-1}/b_n)}{\sin^2(\pi k_2 b_{n-1}/b_n)}$$

$$= \frac{1}{16b_n^2}\sum_{r=1}^{b_n-1}\frac{\cos^2\left(\frac{\pi r b_{n-1}}{b_n}\right)}{\sin^2\left(\frac{\pi r}{b_n}\right)\cdot\sin^2\left(\frac{\pi r b_{n-1}}{b_n}\right)}. \quad (33)$$

Finally, the second contribution of Lemma 4 ($k_2 \neq 0$, $k_2 \equiv 0 \pmod{b_n}$) is

$$\sum_{k_2\equiv 0,\, k_2\neq 0}\left|\widehat{D}(\mathscr{F}_n,(0,k_2))\right|^2 = \frac{b_n^2}{16\pi^2}\sum_{l\in\mathbb{Z}\setminus\{0\}}\frac{1}{b_n^2 l^2} = \frac{1}{48}. \quad (34)$$

Obviously, when $k_2 = 0$, the contributions are identical to Eqs. (33) and (34).

We are now ready to prove the main theorem and to derive the exact formula for $\|D(\mathscr{F}_n,\mathbf{x})\|_2^2$.

Proof of Theorem 4: Both $\widehat{D}(\mathscr{F}_n,\mathbf{0})$ and the contributions described in Eqs. (31) and (34) are bounded by an absolute constant. By comparing Eq. (33) to Eq. (30), we see that all the other contributions to the L_2 norm are dominated by the contribution of the terms corresponding to Lemma 3, that is, $\mathbf{k} \in L(n)$. However, dealing with these terms is a standard issue, which relies on the properties of $L(n)$. See, for example, Sect. 3 (Lemma 7) or [5, 34] for details. □

Putting together Eqs. (30), (31), (33), and (34), and the value of $\widehat{D}(\mathscr{F}_n,\mathbf{0})$ (Lemma 1), we obtain

Theorem 5. *For $n \geq 2$, we have*

$$\|D(\mathscr{F}_n,\mathbf{x})\|_2^2 = \frac{1}{16b_n^2}\sum_{r=1}^{b_n-1}\frac{1+2\cos^2\left(\frac{\pi r b_{n-1}}{b_n}\right)}{\sin^2\left(\frac{\pi r}{b_n}\right)\cdot\sin^2\left(\frac{\pi r b_{n-1}}{b_n}\right)} + \frac{89}{144} - \frac{1}{144b_n^2} \quad (35)$$

when n is even and

$$\|D(\mathscr{F}_n,\mathbf{x})\|_2^2 = \frac{1}{16b_n^2}\sum_{r=1}^{b_n-1}\frac{1+2\cos^2\left(\frac{\pi r b_{n-1}}{b_n}\right)}{\sin^2\left(\frac{\pi r}{b_n}\right)\cdot\sin^2\left(\frac{\pi r b_{n-1}}{b_n}\right)} + \frac{1}{18} - \frac{1}{144b_n^2} + \left(\frac{b_{n-1}}{6b_n} + \frac{7}{12}\right)^2$$

when n is odd.

Numerical experiments indicate that the main term in the formulae above

$$S_n = \frac{1}{16b_n^2}\sum_{r=1}^{b_n-1}\frac{1+2\cos^2\left(\frac{\pi r b_{n-1}}{b_n}\right)}{\sin^2\left(\frac{\pi b_{n-1}}{b_n}\right)\cdot\sin^2\left(\frac{\pi r b_{n-1}}{b_n}\right)} \approx 0.0224\cdot n, \tag{36}$$

which is worse than $S_n' = \frac{1}{8b_n^2}\sum_{r=1}^{b_n-1}\frac{1}{\sin^2\left(\frac{\pi b_{n-1} r}{b_n}\right)\cdot\sin^2\left(\frac{\pi r}{b_n}\right)} \approx 0.0149\cdot n$, the corresponding leading term of the analogous formula for the *symmetrized* Fibonacci set [5]. In fact, it is easy to see that $S_n = \frac{3}{2}S_n' + \mathscr{E}_n$, where the error term $\mathscr{E}_n$ converges to a finite limit as $n\to\infty$. Hence, the L_2 discrepancy of the Fibonacci set exceeds the L_2 discrepancy of its symmetrized version by about 50 % for large n.

One can show directly that the term S_n is of the order $\log b_n \approx n$, which would give a different proof of Theorem 4. A number-theoretic argument to that effect has been pointed out to us by Konstantin Oskolkov.

It is worth pointing out that the best currently known value of the constant $\lim_n \frac{\|D(\mathscr{P}_N,\mathbf{x}\|_2}{\sqrt{\log N}}$ is about 0.17907 [12]. The results of our numerical experiments (see also [5]) suggest that perhaps for the symmetrized Fibonacci lattice, this value may be slightly better, ≈ 0.176006, while for $\mathscr{F}_n$, it is about 0.264009. The largest known constant in the lower bound (2) in dimension $d=2$ is approximately 0.038925 [19].

3 General Lattices

It is fairly straightforward to extend the argument of the previous section to general lattices $\mathscr{L}_n(\alpha)$. Denote

$$L_\alpha(n) := \{\mathbf{k} : k_1 + p_n k_2 \equiv 0 \pmod{q_n}\}. \tag{37}$$

Repeating all the computations almost line by line, we obtain

Lemma 6. *When $k_1\neq 0$, $k_2\neq 0$, we have*

$$\widehat{D}(\mathscr{L}_n(\alpha),\mathbf{k}) = \begin{cases} \dfrac{q_n}{4\pi^2 k_1 k_2}, & k_1\equiv 0 \text{ or } k_2\equiv 0,\\ \dfrac{-q_n}{4\pi^2 k_1 k_2}, & k_1,k_2\not\equiv 0,\ \mathbf{k}\in L_\alpha(n),\\ 0, & k_1,k_2\not\equiv 0,\ \mathbf{k}\notin L_\alpha(n). \end{cases} \tag{38}$$

If $k_1=0, k_2\neq 0$,

$$\widehat{D}(\mathscr{L}_n(\alpha),(0,k_2)) = \begin{cases} \dfrac{q_n}{4\pi i k_2}, & k_2\equiv 0,\\ -\dfrac{1}{4\pi i k_2}\cdot\dfrac{e^{-2\pi i k_2 p_n/q_n}+1}{e^{-2\pi i k_2 p_n/q_n}-1}, & k_2\not\equiv 0. \end{cases} \tag{39}$$

If $k_1 \neq 0, k_2 = 0$,

$$\widehat{D}(\mathscr{L}_n(\alpha),(k_1,0)) = \begin{cases} \dfrac{q_n}{4\pi i k_1}, & k_1 \equiv 0, \\ -\dfrac{1}{4\pi i k_1} \cdot \dfrac{e^{(-1)^n 2\pi i k_1 q_{n-1}/q_n}+1}{e^{(-1)^n 2\pi i k_1 q_{n-1}/q_n}-1}, & k_1 \not\equiv 0. \end{cases} \tag{40}$$

Moreover, we have

$$|\widehat{D}(\mathscr{L}_n(\alpha),\mathbf{0})| = \left|\mathscr{D}(p_n,q_n)+\frac{3}{4}\right| \leq \mathscr{O}(1)+\frac{1}{12}\left|\sum_{k=0}^{n}(-1)^k a_k\right|, \tag{41}$$

where $\mathscr{D}(p,q)$ *is the Dedekind sum defined in Eq.* (20) *and the implicit constant in* $\mathscr{O}(1)$ *depends only on* α. *All the congruences above are modulo* q_n.

To pass from the case $k_2 = 0$ to $k_1 = 0$, we used the identity $p_n q_{n-1} - p_{n-1} q_n = (-1)^{n-1}$, which implies that $\mathscr{L}_n(\alpha) = \{(\{(-1)^{n-1} q_{n-1} r/q_n\}, r/q_n\}_{r=0}^{q_n-1}$. The exact formula for the L_2 discrepancy can also be derived.

Theorem 6. *We have the following relation:*

$$\|D(\mathscr{L}_n(\alpha),\mathbf{x})\|_2^2 = \frac{1}{16q_n^2}\sum_{r=1}^{q_n-1}\frac{1+2\cos^2(\pi r p_n/q_n)}{\sin^2(\pi r/q_n)\cdot\sin^2(\pi r p_n/q_n)} + \left(\mathscr{D}(p_n,q_n)+\frac{3}{4}\right)^2 + \frac{1}{18} - \frac{1}{144q_n^2}. \tag{42}$$

We are now ready to estimate the size of L_2 discrepancy of $\mathscr{L}_n(\alpha)$.

Proof of Theorem 3. Obviously, the zero-order Fourier coefficient (41) grows exactly as the alternating sum of a_k's. We shall show that the contribution of the other terms is of the order $\sqrt{\log q_n}$. One can easily see from Lemma 6 and Eq. (42) that this contribution is dominated by the input of the coefficients corresponding to $\mathbf{k} \in L_\alpha(n)$. Define the hyperbolic cross $\Gamma(M) = \{(k_1,k_2) \in \mathbb{Z}^2 : |k_1 k_2| \leq M; |k_1|,|k_2| \leq M\}$. We have the following lemma concerning the structure of the set $L_\alpha(n)$:

Lemma 7. *There exists* $\gamma > 0$ *depending only on* α, *such that for* $n > 2$,

$$\Gamma(\gamma q_n) \cap L_\alpha(n) = \mathbf{0}. \tag{43}$$

A version of this lemma restricted to $p_n = b_{n-1}$, $q_n = b_n$ is known and has been used repeatedly to obtain discrepancy estimates and errors of cubature formulas for the Fibonacci set [5, 33, 34]. Here, we prove the lemma in full generality:

Proof. Let $k_1 + p_n k_2 = l q_n$, $l \in \mathbb{Z}$. It suffices to assume $0 < |k_1|, |k_2| < q_n$. We have $k_1 k_2 = q_n k_2^2 \left(\frac{l}{k_2} - \frac{p_n}{q_n}\right)$. Denote $\Delta = \frac{l}{k_2} - \frac{p_n}{q_n}$. Since $|k_2| < q_n$ and the convergent p_n/q_n is the best approximation to α, we have $|\alpha - l/k_2| > |\alpha - p_n/q_n|$. Choose $\nu \in \mathbb{N}$ to be the smallest index such that $q_\nu > |k_2|$ and l/k_2 and p_ν/q_ν lie on the same side from α. Then $\nu \le n$, $q_{\nu-2} \le |k_2| < q_\nu$, and $|\alpha - l/k_2| > |\alpha - p_\nu/q_\nu|$. Moreover, the relation $q_\nu = a_\nu q_{\nu-1} + q_{\nu-2}$ implies that $q_\nu \le (A+1)^2 q_{\nu-2}$, where $A = \max a_k$: incidentally, here we use the fact that α has bounded partial quotients. We have $|\Delta| \ge \left|\frac{l}{k_2} - \frac{p_\nu}{q_\nu}\right|$ (if p_n/q_n is on the other side of α it is obvious; otherwise, it is closer to α since $n \ge \nu$). Therefore, since p_ν/q_ν is irreducible and $\frac{l}{k_2} \ne \frac{p_\nu}{q_\nu}$,

$$|\Delta| \ge \left|\frac{l}{k_2} - \frac{p_\nu}{q_\nu}\right| \ge \frac{1}{|k_2| q_\nu} \ge \frac{1}{(A+1)^2} \frac{1}{|k_2| q_{\nu-2}} \ge \frac{\gamma}{k_2^2}. \tag{44}$$

Hence, $|k_1 k_2| \ge \gamma q_n$ with $\gamma = 1/(A+1)^2$. □

Denoting $Z_l := \left(\Gamma(2^{l+1}\gamma b_n) \setminus \Gamma(2^l \gamma b_n)\right) \cap L_\alpha(n)$, it is now easy to deduce from Eq. (43) that $\# Z_l \ll 2^l (l+1) \log q_n$. Then one obtains using Eq. (38)

$$\sum_{\mathbf{k} \in L_\alpha(n)} |\widehat{D}(\mathcal{L}_n(\alpha), \mathbf{k})|^2 \ll \sum_{l=0}^{\infty} \sum_{\mathbf{k} \in Z_l} \frac{1}{(2^l)^2} \ll \log q_n. \tag{45}$$

Together with Roth's lower estimate (2), this finishes the proof of Theorem 3. □

4 Further Remarks

It is interesting to discuss how the L_2 discrepancy of various specific lattices $\mathcal{L}_n(\alpha)$ behaves depending on the value of α. We list only a few observations here; a more comprehensive study of the number-theoretic aspects of this question will be conducted in the subsequent work of the authors.

It is evident from Theorem 3 that, while some lattices $\mathcal{L}_n(\alpha)$ have optimal L_2 discrepancy, others do not. Set, for example, $a_{2j} = 2$ and $a_{2j+1} = 1$, in this case, the alternating sums grow as $n \approx \log q_n$. At the same time, it follows from the arguments in [5] that symmetrizations of these lattices always have asymptotically minimal L_2 discrepancy. We make some peculiar remarks:

- It is not hard to construct numbers α such that the corresponding lattices $\mathcal{L}_n(\alpha)$ would have any prescribed rate of growth of L_2 discrepancy between $\sqrt{\log N}$ and $\log N$—one just needs to build a sequence $\{a_k\}$ for which the alternating sums behave appropriately. We are not aware of any prior results of this flavor.
- However, if $\alpha = k + l\sqrt{m}$ is a quadratic irrationality, there is a certain dichotomy: the L_2 discrepancy of $\mathcal{L}_n(\alpha)$ grows either as $\sqrt{\log N}$ or as $\log N$, intermediate rates are not possible. Indeed, it is well known that the continued fractions of quadratic irrationalities are periodic. Hence, the alternating sums of a_k are either bounded by a constant (e.g., if the length of the period is odd) or grow as n (when the period is even and the alternating sum within one cycle is nonzero).

- In particular, $\mathscr{L}_n(\sqrt{2})$ has optimal L_2 discrepancy since $\sqrt{2}=[1;\overline{2}]$, while the L_2 discrepancy of $\mathscr{L}_n(\sqrt{3})$ is of the order n. In general, it would be interesting to understand which square roots $\alpha=\sqrt{m}$ have odd periods of continued fractions. One common example is $m=p^2+1$. In this case, $\sqrt{m}=[p;\overline{2p}]$, so this example was essentially covered by the Sós and Zaremba [31] result.
- We list those values of m between 1 and 250 (other than $m=p^2+1$) for which the period of $\sqrt{m}$ is odd: 13, 29, 41, 53, 58, 61, 73, 74, 85, 89, 97, 106, 109, 113, 125, 130, 137, 149, 157, 173, 181, 185, 193, 202, 218, 229, 241, and 250. For these values of m, $\mathscr{L}_n(\sqrt{m})$ has optimal L_2 discrepancy.
- It is known that for any $P\in\mathbb{N}$, there exists $m\in\mathbb{N}$ such that the length of the period of the continued fraction of $\sqrt{m}$ is P [13].
- We do not know any values of m such that the length of the period of $\sqrt{m}$ is even, but the L_2 discrepancy of $\mathscr{L}_n(\sqrt{m})$ is of the order $\sqrt{n}$ (i.e., the alternating sum of a_k over one period is zero). In the following examples, for instance, the alternating sum of the partial quotients grows as n: $m=p^2+2$, $(p+1)^2-1$, or p^2+p ($\sqrt{p^2+2}=[p;\overline{p,2p}]$, $\sqrt{(p+1)^2-1}=[p;\overline{1,p-1,1,2p}]$, and $\sqrt{p^2+p}=[p;\overline{2,2p}]$). For such m, the $\mathscr{L}_n(\sqrt{m})$ is of the order $\log N$.

Acknowledgements The research of the authors is sponsored by NSF grants DMS 1101519, DMS 0906260, and EAR 0934747.

References

1. Barkan, P.: Sur les sommes de Dedekind et les fractions continues finies, C. R. Acad. Sci. Paris Sér. A **284**, 923–926 (1977)
2. Beck, J., Chen, W.W.L.: Irregularities of Distribution, Cambridge Tracts in Mathematics 89, pp. 294. Cambridge University Press, Cambridge (1987)
3. Bilyk, D., Lacey, M., Vagharshakyan, A.: On the small ball inequality in all dimensions. J. Funct. Anal. **254**(9), 2470–2502 (2008)
4. Bilyk, D., Lacey, M., Parissis, I., Vagharshakyan, A.: Exponential squared integrability of the discrepancy function in two dimensions. Mathematika **55**, 1–27 (2009)
5. Bilyk, D., Temlyakov, V.N., Yu, R.: Fibonacci sets and symmetrization in discrepancy theory. J. Complex. **28**, 18–36 (2012)
6. Chazelle, B.: The Discrepancy Method: Randomness and complexity, pp. 463. Cambridge University Press, Cambridge (2000)
7. Chen, W.W.L.: On irregularities of distribution. Mathematika **27**, 153–170 (1980)
8. Chen, W.W.L., Skriganov, M.M.: Explicit construction in the classical mean squares problem in irregularities of point distribution. J. Reine Angew. Math. **545**, 67–95 (2002)
9. Chen, W.W.L., Skriganov, M.M.: Orthogonality and digit shifts in the classical mean square problem in irregularities of point distribution. In: Diophantine approximation, Dev. in Math. 16, pp. 141–159. Springer, Wien (2008)
10. van der Corput, J.G.: Verteilungsfunktionen I. Mitt, Proc. Akad. Wet. Amsterdam **38**, 813–821 (1935)
11. Davenport, H.: Note on irregularities of distribution. Mathematika **3**, 131–135 (1956)
12. Faure, H., Pillichshammer, F., Pirsic, G., Schmid, W.Ch.: L_2 discrepancy of generalized two-dimensional Hammersley point sets scrambled with arbitrary permutations. Acta Arith. **141**, 395–418 (2010)

13. Friesen, C.: On continied fractions of given period. Proc. of AMS **103**(1), 9–14 (1988)
14. Frolov, K.K.: Upper error bounds for quadrature formulas on function classes. Dokl. Akad. Nauk SSSR **231**, 818–821 (1976)
15. Gowers, T.: Problems related to Littlewoods conjecture. In: Gowers's Weblog. (2009). http://gowers.wordpress.com/2009/11/17/problems-related-to-littlewoods-conjecture-2/.Cited July 20, 2012
16. Hall, R.R., Huxley, M.N.: Dedekind sums and continued fractions. Acta Arithmetica LXIII. **1**, 79–90 (1993)
17. Halton, J.H., Zaremba, S.K.: The extreme and L_2 discrepancies of some plane sets. Monatsh. Math. **73**, 316–328 (1969)
18. Haselgrove, C.B.: A method for numerical integration. Math. Comp. **15**, 323–337 (1961)
19. Hinrichs, A., Markhasin, L.: On lower bounds for the L_2 discrepancy. J Complex. **27**, 127–132 (2011)
20. Hlawka, E.: The Theory of Uniform Distribution. AB Academic Publishers, Herts, UK (1984)
21. Kuipers, L., Niederreiter, H.: Uniform Distribution of Sequences. Wiley-Interscience Publication, New York (1974)
22. Lerch, M.: Question 1547. L'Intermediaire Math. **11**, 144–145 (1904)
23. Matoušek, J.: Geometric Discrepancy: An illustrated guide, Algorithms and Combinatorics 18, pp. 288. Springer, Berlin (1999)
24. Niederreiter, H.: Random Number Generation and quasi-Monte Carlo methods. SIAM, Philadelphia (1992)
25. Rocadas, L., Schoissengeier, J.: An Explicit Formula for the L_2-Discrepancy of $(n\alpha)$-Sequences. Computing **77**, 113–128 (2006)
26. Roth, K.F.: On irregularities of distribution. Mathematika **1**, 73–79 (1954)
27. Roth, K.F.: On irregularities of distribution. III, Acta Arith. **35**, 373–384 (1979)
28. Schmidt, W.M.: Irregularities of distribution. VII, Acta Arith. **21**, 45–50 (1972)
29. Schmidt, W.M.: Irregularities of distribution. X. In: Number theory and algebra, pp. 311–329. Academic Press, New York (1977)
30. Skriganov, M.M.: Harmonic analysis on totally disconnected groups and irregularities of point distributions. J. Reine Angew. Math. **600**, 25–49 (1993)
31. Sós, V.T., Zaremba, S.K.: The mean-square discrepancies of some two-dimensional lattices. Studia Sci. Math. Hungar. **14**, 255–271 (1982)
32. Stein, E.M., Shakarchi, R.: Fourier analysis. An Introduction, Princeton Lectures in Analysis 1, pp. 311. Princeton University Press, Princeton, NJ (2003)
33. Temlyakov, V.N.: Approximation of Periodic Functions: Computational Mathematics and Analysis Series. Nova Science Publishers, New York (1993)
34. Temlyakov, V.N.: Error estimates for Fibonacci quadrature formulas for classes of functions with a bounded mixed derivative. Proc. Steklov Inst. Math. **2**(200), 359–367 (1993)
35. Temlyakov, V.N.: Cubature formulas, discrepancy, and nonlinear approximation. J. Complex. **19**, 352–391 (2003)
36. Vandewoestyne, B., Cools, R.: On obtaining higher order convergence for smooth periodic functions. J. Complex. **24**, 328–340 (2008)
37. Wenpeng, Z., Yuan, Y.: On the Fibonacci Numbers and the Dedekind Sums. The Fibonacci Quarterly **38**(3), 223–226 (2000)
38. Zheng, Z.: Dedekind Sums and Uniform Distribution (mod 1). Acta Mathematica Sinica **11**, 62–67 (1995)

On Fourier Multipliers Over Tube Domains

Laura De Carli

Abstract We provide $L^p \to L^q$ estimates for a class of Fourier multipliers supported in convex cones of $\mathbf{R}^{n+1}$. In particular we consider cones whose boundary has $n > 1$ nonvanishing principal curvatures, and cones which are the convex envelope of N linearly independent half lines passing through the origin of $\mathbf{R}^{n+1}$. In some case our estimates are best possible.

1 Introduction

Let Γ be an open cone in $\mathbf{R}^{n+1}$ with vertex at 0. Define the *dual cone of* Γ as

$$\Gamma^0 = \{\, y \in \mathbf{R}^{n+1} : \langle y, \zeta \rangle > 0 \quad \forall \zeta \in \Gamma \}, \tag{1}$$

where $\langle\ ,\ \rangle$ denotes the usual scalar product in $\mathbf{R}^{n+1}$.

Γ^0 has empty interior if Γ contains straight lines. We consider the holomorphic function over the tube domain $\mathbf{R}^{n+1} + i\Gamma$

$$K_\Gamma(x+iy) = \int_{\Gamma^0} e^{2\pi i \langle x+iy,\, \zeta \rangle}\, d\zeta. \tag{2}$$

$K_\Gamma(x+iy)$ is sometimes called the *Cauchy kernel of* Γ.

Let $f \in C_0^\infty(\mathbf{R}^{n+1})$; we let

$$T_\Gamma(f) : \mathbf{R}^{n+1} + i\Gamma \to \mathbf{C},$$

$$T_\Gamma f\,(x+iy) = \int_{\Gamma^0} \hat{f}(\zeta) e^{2\pi i \langle x+iy,\, \zeta \rangle}\, d\zeta, \tag{3}$$

L. De Carli (✉)
Department of Mathematics, Florida International University, Miami, FL, USA
e-mail: decarlil@fiu.edu

D. Bilyk et al. (eds.), *Recent Advances in Harmonic Analysis and Applications*, Springer Proceedings in Mathematics & Statistics 25, DOI 10.1007/978-1-4614-4565-4_10,

where $\hat{f}(\zeta) = \int_{\mathbf{R}^{n+1}} f(x) e^{-2\pi i x \zeta} dx$ denotes the Fourier transform of f. The operator $f \to T_\Gamma(f)$ is defined in [11], and it is important in microlocal analysis to study the wave front set of distributions. See, e.g., [12].

Let $T_{\Gamma,y} f(x) = T_\Gamma f(x+iy)$. Observe that $T_\Gamma f(x+iy) = (K_{\Gamma,y} * f)(x)$, where $*$ denotes the convolution, and we have let $K_{\Gamma,y}(x) = K_\Gamma(x+iy)$.

The operator $T_{\Gamma,y}$ can also be described as follows. For $y \in \Gamma$, we let $m_y(\zeta) = e^{-2\pi\langle y,\zeta\rangle} \chi_{\Gamma^0}(\zeta)$. Then,

$$\widehat{T_{\Gamma,y} f}(\zeta) = m_y \hat{f}(\zeta) \tag{4}$$

is the Fourier multiplier operator associated with $m_y(\zeta)$.

Let $y \in \Gamma$; we wish to find optimal sets of exponents (p,q) for which $T_{\Gamma,y}$ extends to a bounded operator from $L^p(\mathbf{R}^{n+1}, \mathbf{C})$ to $L^q(\mathbf{R}^{n+1}, \mathbf{C})$. In short,

$$\left\| T_{\Gamma,y} f \right\|_q \leq C \| f \|_p, \qquad f \in C_0^\infty(\mathbf{R}^{n+1}, \mathbf{C}), \tag{5}$$

where $C = C(p,q,y,\Gamma) = \sup_{\|f\|_p = 1} \left\| T_{\Gamma,y} f \right\|_q < \infty$ is a positive constant, and we have denoted by $\| \ \|_r$ the standard norm on $L^r(\mathbf{R}^{n+1}, \mathbf{C})$.

It is not too difficult to see that if Eq. (5) holds, necessarily, $p \leq q$. We prove this in the next section, and we also estimate of the operator norm of $T_{\Gamma,y}$ in terms of $|y|$.

The main result of this chapter is also proved in Sect. 2; it generalizes a Theorem in [1] and is related to some classical restriction theorems to cones. See, e.g., [7,9].

Theorem 1.1 *Let Γ be a convex cone in $\mathbf{R}^{n+1}$. Suppose that $\partial\Gamma/\{0\}$ has $n-1$ everywhere nonvanishing principal curvatures. Then, for every (p,q) such that $(\frac{1}{p}, \frac{1}{q})$ is in the open trapezoid of vertices $(1,0)$, $(\frac{n-1}{2n}, 0)$, $(1, \frac{n+1}{2n})$, $(\frac{1}{2}, \frac{1}{2})$,*

$$\left\| T_{\Gamma,y} f \right\|_q \leq C |y|^{-(n+1)\left(\frac{1}{p} - \frac{1}{q}\right)} \| f \|_p \tag{6}$$

where C depends on p, q, and Γ but not on y.

Let us recall the definition of Gaussian curvature and principal curvatures. Let Σ be a submanifold of $\mathbf{R}^{n+1}$ of codimension 1 equipped with a smooth compactly supported measure $d\mu$. Let $J : \Sigma \to S^N$ be the usual Gauss map that takes each point on Σ to the outward unit normal of Σ at that point. We say that Σ has everywhere *nonvanishing Gaussian curvature* if the differential of the Gauss map dJ is always nonsingular. The *principal curvatures* of Σ are the eigenvalues of dJ.

We prove Theorem 1.1 using Young inequality for convolution and the following estimates for the $L^p(\mathbf{R}^{n+1}, \mathbf{C})$ norm of $K_{\Gamma,y}$.

Theorem 1.2 *Let Γ be as in Theorem 1.1; then, $K_{\Gamma,y}(x) \in L^r(\mathbf{R}^{n+1}, \mathbf{C})$ for $\frac{2n}{n+1} < r \leq \infty$, and*

$$\left\| K_{\Gamma,y} \right\|_r \leq C(\Gamma) |y|^{\frac{2n}{r} - (n+1)}. \tag{7}$$

We have proved in [1] that when $\Gamma = \{(x_0, x') : x_0 \geq |x'|\}$, the range of exponents for r cannot be improved, in the sense that if $r \leq \frac{2n}{n+1}$, for every $n \in \mathbf{N}$, there exist $y_n \in \Gamma$ for which $\| K_{\Gamma,y} \|_r > n|y_n|^{\frac{2n}{r}-(n+1)}$.

In Sect. 3 we consider cases when $m_y(\zeta) = e^{-2\pi\langle y,\zeta\rangle}\chi_{\Gamma^0}(\zeta)$ is a (p,p) Fourier multiplier for $p > 1$. In particular, we evaluate the (p,p) norm of $m_y(\zeta)$ when Γ is the convex envelope of $N \leq n+1$ linearly independent half lines passing through the origin of $\mathbf{R}^{n+1}$.

Section 2

Let $\mathscr{D}_\Gamma$ be the set of $(\frac{1}{p}, \frac{1}{q})$'s such that $T_{\Gamma,y}$ extends to a bounded operator from $L^p(\mathbf{R}^{n+1}, \mathbf{C})$ to $L^q(\mathbf{R}^{n+1}, \mathbf{C})$ for every $y \in \Gamma$. We let $C(p,q,\Gamma,y) < \infty$ be the operator norm of $T_{\Gamma,y}$. When there is no ambiguity we will write $C(\Gamma,y)$ instead of $C(p,q,\Gamma,y)$.

Let $\mathscr{Q}$ be the closed square of vertices $(\frac{1}{2}, 0)$, $(1, 0)$, $(\frac{1}{2}, \frac{1}{2})$, and $(1, \frac{1}{2})$, and let $\mathscr{T}$ be the closed triangle of vertices $(0,0)$, $(1,1)$, and $(1,0)$. Thus, $\mathscr{Q} = \{(\frac{1}{p}, \frac{1}{q}) : q \geq 2 \text{ and } p \leq 2\}$, and $\mathscr{T} = \{(\frac{1}{p}, \frac{1}{q}) : q \geq p\}$. We prove the following.

Proposition 2.1 *(i) For every cone $\Gamma \subset \mathbf{R}^n$,*

$$\mathscr{Q} \subset \mathscr{D}_\Gamma \subset \mathscr{T}.$$

(ii) $\left(\frac{1}{p}, \frac{1}{q}\right) \in \mathscr{D}_\Gamma$ if and only if $\left(\frac{1}{q'}, \frac{1}{p'}\right) \in \mathscr{D}_\Gamma$, and $C_{p,q}(\Gamma,y) = C_{q',p'}(\Gamma,y)$.
(iii) If $\left(\frac{1}{p}, \frac{1}{q}\right) \in \mathscr{D}_\Gamma$, then

$$C_{p,q}(\Gamma,y) \leq |y|^{(n+1)\left(\frac{1}{q}-\frac{1}{p}\right)} C_{p,q}\left(\Gamma, \frac{y}{|y|}\right). \tag{8}$$

Proof. To prove (iii) we use a standard scaling argument. Let $y \in \Gamma$, with $|y| = \varepsilon$. Let $f_\varepsilon(x) = f(\varepsilon^{-1}x)$. It is easy to see that

$$T_\Gamma f_\varepsilon(x+iy) = T_\Gamma f(\varepsilon^{-1}(x+iy))$$

and hence $\| T_{\Gamma,y} f_\varepsilon \|_q = \varepsilon^{\frac{n+1}{q}} \left\| T_{\Gamma,\varepsilon^{-1}y} f \right\|_q$. Moreover, $\| f_\varepsilon \|_p = \varepsilon^{\frac{n+1}{p}} \| f \|_p$. Thus,

$$\frac{\| T_{\Gamma,y} f_\varepsilon \|_q}{\| f_\varepsilon \|_p} = \varepsilon^{(n+1)\left(\frac{1}{q}-\frac{1}{p}\right)} \frac{\left\| T_{\Gamma,\varepsilon^{-1}y} f \right\|_q}{\|f\|_p} \leq \varepsilon^{(n+1)\left(\frac{1}{q}-\frac{1}{p}\right)} C(\varepsilon^{-1}y, \Gamma)$$

and so

$$C(y,\Gamma) = \sup_{f \in L^p(\mathbf{R}^{n+1})} \frac{\| T_{\Gamma,y} f_\varepsilon \|_q}{\| f_\varepsilon \|_p} \leq \varepsilon^{(n+1)\left(\frac{1}{q}-\frac{1}{p}\right)} C(\varepsilon^{-1}y, \Gamma);$$

thus,

$$C(y,\Gamma) \leq \varepsilon^{(n+1)\left(\frac{1}{q}-\frac{1}{p}\right)} C(\varepsilon^{-1}y,\Gamma),$$

from which Eq. (8) follows.

(ii) follows because $T_{\Gamma,y}$ is self-adjoint. Let us prove (i). Suppose that $q < p$, and $(\frac{1}{p},\frac{1}{q}) \in \mathscr{D}_\Gamma$. Then, the exponent of $|y|$ in Eq. (8) is positive, and so, for every function $f \in L^p(\mathbf{R}^{n+1})$ of norm 1,

$$\left(\int_{\mathbf{R}^{n+1}} |T_{\Gamma,y}f(x)|^q \, dx\right)^{\frac{1}{q}} \leq C|y|^{(n+1)\left(\frac{1}{q}-\frac{1}{p}\right)}.$$

Let $\overline{y} \in \Gamma$; by Fatou's Lemma,

$$0 = \lim_{t\to 0} \int_{\mathbf{R}^{n+1}} |T_{\Gamma,t\overline{y}}f(x)|^q \, dx \geq \int_{\mathbf{R}^{n+1}} |T_{\Gamma,0}f(x)|^q \, dx$$

which cannot be true for every f; Thus, q cannot be less that p, and $\mathscr{D}_\Gamma \subset \mathscr{T}$. We now show that $\mathscr{D}_\Gamma \supset \mathscr{Q}$. If $q \geq 2$ and $y \in \Gamma$, by the Hausdorff–Young inequality

$$\| T_{\Gamma,y}f \|_q \leq \left\| e^{-2\pi\langle y,.\rangle} \hat{f} \right\|_{L^{q'}(\Gamma^0)}, \tag{9}$$

where q' denotes the dual exponent of q. Note that $q' \leq 2$. Take $p \leq 2$. Let $\frac{1}{r} = \frac{1}{q'} - \frac{1}{p'} = \frac{1}{p} - \frac{1}{q}$. By Hölder's inequality,

$$\left\| e^{-2\pi\langle y,.\rangle} \hat{f} \right\|_{L^{q'}(\Gamma^0)} \leq \left\| e^{-2\pi\langle y,.\rangle} \right\|_{L^{r}(\Gamma^0)} \| \hat{f} \|_{p'}. \tag{10}$$

By homogeneity, $\left(\int_{\Gamma^0} e^{-2\pi r\langle y,\zeta\rangle} d\zeta\right)^{\frac{1}{r}} = C|y|^{-\frac{n+1}{r}}$, where C depends only on Γ and r. By Eq. (10), and the above observation, we obtain

$$\| T_{\Gamma,y}f \|_q \leq C|y|^{(n+1)\left(\frac{1}{q}-\frac{1}{p}\right)} \| f \|_p,$$

as required. □

Remark. We will prove in the next section that when Γ is a polygonal cone, $\mathscr{D}_\Gamma = \mathscr{T}$ We do not know whether there are cones for which $\mathscr{D}_\Gamma = \mathscr{Q}$ or not.

To prove Theorem 1.2 we shall use the following lemma, which is an easy consequence of the classical sharp estimates for the Fourier transform of the measure carried by a surface with non vanishing Gaussian curvature, that we recall here.

Lemma 2.2 *Let $S \subset \mathbf{R}^n$ be a convex domain with smooth boundary, which is such that ∂S has everywhere nonvanishing Gaussian curvature. Let χ_S be the characteristic function of S. Then,*

$$|\widehat{\chi_S}(\xi)| \leq C(1+|\xi|)^{-\frac{n+1}{2}}. \tag{11}$$

This Lemma is proved in [8] when S is the unit sphere of $\mathbf{R}^n$, but the proof can be easily generalized.

We also need the following lemma.

Lemma 2.3 *Let $\Gamma \subset \mathbf{R}^{n+1}$ be a convex cone which does not contain straight lines. If $\partial\Gamma/\{0\}$ is smooth and has $n-1$ everywhere non vanishing principal curvatures, then also $\partial\Gamma^0$ is smooth and has $n-1$ everywhere non vanishing principal curvatures.*

Proof. We can assume that $\Gamma = \{(x_0, x') : x_0 \geq \phi(x)\}$, where $\phi \in C^\infty(\mathbf{R}^n/\{0\})$ is a convex, smooth, and homogeneous function of degree 1 that vanishes only at the origin. The level set $S = \{x' \in \mathbf{R}^n : \phi(x') \leq 1\}$ is convex (and also compact), and ∂S has $n-1$ nonvanishing Gaussian curvature. Recalling the definition of Γ^0, we can see at once that $(\zeta_0, \zeta') \in \Gamma^0$ if $\zeta_0 > 0$, and

$$\min_{x' \in S}\left(1+\left\langle \frac{x'}{x_0}, \frac{\zeta'}{\zeta_0}\right\rangle\right) \geq 0 \tag{12}$$

for every $x_0 > 0$. Without loss of generality we can let $x_0 = \zeta_0 = 1$. We shall find the minimum of $F_\zeta(x') = 1 + \langle x', \zeta'\rangle$ on S. Since the gradient of $F_\zeta(x')$ vanishes in the interior of S only if $\zeta' = 0$, we shall find the critical points of $F_{\zeta'}(x')$ on the boundary of S. By the Lagrange multiplier theorem, these points are the solutions of the system

$$\zeta' = \lambda \nabla\phi(x'); \quad \phi(x') = 1. \tag{13}$$

for some $\lambda \neq 0$. By the Bolzano–Weirstrass theorem, $F_{\zeta'}$ has a minimum and a maximum on S, and hence the system (13) has two solutions, say $(x_1(\zeta'), \lambda_1(\zeta'))$ and $(x_2(\zeta'), \lambda_2(\zeta'))$. Since S is convex, necessarily, $\lambda_1(\zeta') > 0$ and $\lambda_2(\zeta') < 0$. So, the outward unit normal to $\Sigma = \{x' : \phi(x') = 1\}$ at $x_1(\zeta')$ is $\frac{\zeta'}{|\zeta'|}$, and the outward unit normal at $x_2(\zeta')$ is $-\frac{\zeta'}{|\zeta'|}$. Thus,

$$F_\zeta(x_j(\zeta')) = 1 + \lambda_j(\zeta')\langle\nabla\phi(x_j(\zeta')), x_j(\zeta')\rangle, \qquad j = 1, 2.$$

Since ϕ is homogeneous of degree 1, by Euler's homogeneity relation,

$$F_\zeta(x_j(\zeta')) = 1 + \lambda_j(\zeta')\phi(x_j(\zeta')) = 1 + \lambda_j(\zeta').$$

Then, ζ' satisfies Eq. (12) if and only if

$$1 + \lambda_j(\zeta') \geq 0, \qquad j = 1, 2. \tag{14}$$

Since $\lambda_1(\zeta') > 0$, Eq. (14) is satisfied if and only if $1+\lambda_2(\zeta') \geq 0$. We show that $\lambda_2(\zeta')$ is a homogeneous function of degree 1 and is smooth away from the origin. We will drop the subscript for simplicity of notation. If we dot multiply both sides of Eq. (13) by $x(\zeta')$, by Euler's homogeneity relation,

$$\lambda(\zeta') = \langle x'(\zeta), \zeta' \rangle. \tag{15}$$

After differentiating both sides of Eq. (15),

$$\nabla\lambda(\zeta') = x'(\zeta) + Dx'(\zeta')\zeta', \tag{16}$$

where $Dx'(\zeta')$ denotes the Jacobian matrix of $x'(\zeta')$. But $x(\zeta')$ is a critical point of ϕ on Σ, and so, using Eq. (13),

$$0 = \nabla_{\zeta'}(\phi(x(\zeta')) = \lambda^{-1}Dx(\zeta')\zeta', \tag{17}$$

which implies that $\nabla\lambda(\zeta') = x(\zeta')$. Note that $Dx(\zeta')\zeta' = 0$ implies that $x(\zeta')$ is homogeneous of degree 0. Thus, from Eq. (15),

$$\langle \nabla\lambda(\zeta'), \zeta' \rangle = \langle x(\zeta'), \zeta' \rangle = \lambda(\zeta'),$$

and we have proved that $\lambda(\zeta')$ is homogeneous of degree 1. Observe that Eq. (16) and the fact that $Dx(\zeta')\zeta' = 0$ yields $\nabla\lambda(\zeta') = x(\zeta')$.

Until now we have not used the assumption that Σ has non vanishing Gaussian curvature. This assumption is necessary to prove that λ is smooth away from the origin. We recall in fact that $x(\zeta')_{|S^{n-1}}$ is the inverse of the Gauss map of Σ, which is smooth and nonsingular by assumption. Then, $x(\zeta')$ is also smooth and nonsingular on $\mathbf{R}^n/\{0\}$ because it is homogeneous of degree 0. By Eq. (15), $\lambda(\zeta')$ is also smooth on $\mathbf{R}^n/\{0\}$. Since $\nabla\lambda(\zeta') = x(\zeta')$, the Gauss map of the set $\Sigma^0 = \{\zeta : \lambda(\zeta') = 1\}$ is $Dx(\zeta')$, which is nonsingular. This concludes the proof of the lemma. □

Remark. If $\partial\Gamma$ has points where more that $n-1$ principal curvatures vanish, then $\partial\Gamma^0$ is not necessarily smooth. Consider the cone

$$\Gamma = \{x \in \mathbf{R}^3 : x_0 \geq (x_1^4 + x_2^4)^{\frac{1}{4}}\}.$$

The curvature of $\Sigma = \{x_1^4 + x_2^4 = 1\}$ vanishes at $(0,\pm 1)$ and $(\pm 1, 0)$. We can see at once that

$$\Gamma^0 = \left\{\zeta \in \mathbf{R}^3 : \zeta_0 \geq (\zeta_1^{\frac{4}{3}} + \zeta_2^{\frac{4}{3}})^{\frac{3}{4}}\right\},$$

and $\Sigma^0 = \{\zeta_1^{\frac{4}{3}} + \zeta_2^{\frac{4}{3}} = 1\}$ is not everywhere smooth.

Proof of Theorem 1.2. Since Γ does not contain straight lines, there is not loss of generality to assume that the cone $G = \{x : x_0 \geq |x'|\}$ contains Γ. Let $y \in \Gamma$. Let $L : \mathbf{R}^{n+1} \to \mathbf{R}^{n+1}$ be a linear orthogonal transformation that maps y to $(|y|, 0, \cdots, 0)$, and leaves G unchanged (this transformation is explicit in [1]). Let $L(\Gamma) = \Gamma'$. It is easy to verify that Γ' is a convex cone and $\partial\Gamma'/\{0\}$ has everywhere $n-1$ nonvanishing principal curvatures. Moreover, $(\Gamma')^0 = L(\Gamma^0)$. In fact, $\zeta \in (\Gamma')^0$ if and only if $\langle \bar{x}, \zeta \rangle \geq 0$ for every $\bar{x} \in \Gamma'$. Since $\bar{x} = L(x)$ for some $x \in \Gamma$, we have that $\zeta \in (\Gamma')^0$ if and only if $L^*(\zeta) \in \Gamma^0$, where L^* denotes the adjoint of L. But $L^* = L^{-1}$, and so $\zeta \in L(\Gamma^0)$.

This argument shows that we can replace Γ by Γ' and assume $y = (y_0, 0, \cdots 0)$ without loss of generality. In order to simplify notation, we let $\Gamma' = \Gamma$.

In what follows we will always let $x = (x_0, x')$, with $x_0 \in \mathbf{R}$ and $x' \in \mathbf{R}^n$. We will also assume that C denotes a constant that can vary from line to line. When there is no ambiguity, we will often write $x.\zeta$ instead of $\langle x, \zeta \rangle$. With this notation,

$$
\begin{aligned}
K_\Gamma(x+iy) &= \int_{\Gamma^0} e^{2\pi i(\zeta_0(x_0+iy_0)+\langle \zeta', x' \rangle)} d\zeta' \, d\zeta_0 \\
&= \int_0^\infty \int_{\{\Gamma^0 \cap \{\zeta_0 = t\}} e^{2\pi i(t(x_0+iy_0)+\langle \zeta', x' \rangle)} d\zeta' \, dt. \qquad (18)
\end{aligned}
$$

If we make the change of variables $\zeta' \to t\zeta'$,

$$
K(x_0+iy_0, x') = \int_0^\infty t^n \int_{\Gamma^0 \cap \{\zeta_0 = 1\}} e^{2\pi i(t(x_0+iy_0)+t\langle \zeta' x' \rangle)} d\zeta' \, dt. \qquad (19)
$$

Let $S = \Gamma^0 \cap \{\zeta_0 = 1\}$. Then,

$$
K(x_0+iy_0, x') = \int_0^\infty e^{2\pi i t(x_0+iy_0)} t^n \int_S e^{2\pi i t \langle \zeta', x' \rangle} \, d\zeta' \, dt.
$$

By assumption, S is convex, and ∂S has everywhere nonvanishing Gaussian curvature. By Lemma 9,

$$
\int_S e^{2\pi i \langle \zeta', tx' \rangle} \, d\zeta' = B(tx'), \qquad (20)
$$

where $B(tx') \leq C(1+t|x'|)^{-\frac{n+1}{2}}$. Next, we observe that

$$
K(x_0+iy_0, x') = \int_0^\infty e^{2\pi i t(x_0+iy_0)} t^n B(tx') dt
$$

is the Fourier transform of $t \to \chi_{(0,\infty)} e^{-2\pi t y_0} t^n B(tx')$. By Hausdorff–Young inequality, for every $r \geq 2$,

$$
\left(\int_{-\infty}^\infty |K(x_0+iy_0, x')|^r dx_0 \right)^{\frac{1}{r}} \leq \left(\int_0^\infty \left| e^{-2\pi t y_0} t^n B(tx') \right|^{r'} dt \right)^{\frac{1}{r'}},
$$

and by Minkowski inequality,

$$||K_{y_0}||_{L^r(\mathbf{R}^{n+1})} \leq \left\| \left(\int_0^\infty \left| e^{-2\pi t y_0} t^n B(tx') \right|^{r'} dt \right)^{\frac{1}{r'}} \right\|_{L^r(\mathbf{R}^n)}$$

$$\leq \left(\int_0^\infty e^{-2\pi t y_0 r'} t^{nr'} \, ||B(t.)||_{L^r(\mathbf{R}^n)}^{r'} dt \right)^{\frac{1}{r'}}. \tag{21}$$

Recalling that $B(tx') \leq C(1+t|x'|)^{-\frac{n+1}{2}}$, we can see at once that

$$||B(t.)||_{L^r(\mathbf{R}^n)}^{r} \leq C \int_{\mathbf{R}^n} \frac{dx'}{(1+t|x'|)^{\frac{r(n+1)}{2}}} = Ct^{-n}$$

provided that $r > \frac{2n}{n+1}$. Thus, by Eq. (21),

$$||K_{y_0}||_{L^r(\mathbf{R}^{n+1})} \leq C \left(\int_0^\infty e^{-2\pi t y_0 r'} t^{n(1-r'/r)} dt \right)^{\frac{1}{r'}}$$

$$\leq C y_0^{\frac{1}{r'}\left(-1-n(1-r'/r)\right)} = C y_0^{\frac{2n}{r}-(n+1)}$$

as required. This concludes the proof of Theorem 1.2. □

Proof of Theorem 1.1 Recalling that $T_\Gamma f(x+iy) = K_\Gamma * f$, we can apply Young inequality for convolution to show that if $r > \frac{2n}{n+1}$, and $1 + \frac{1}{q} = \frac{1}{p} + \frac{1}{r}$, then $T_{\Gamma,y} : L^p(\mathbf{R}^{n+1}) \to L^q(\mathbf{R}^{n+1})$ is bounded. This imply

$$\frac{1}{q} < \frac{1}{p} - \frac{n-1}{2n}.$$

We can also say that $\mathscr{D}_\Gamma$ contains the open triangle of vertices $\left(\frac{n-1}{2n}, 0\right)$, $(1,0)$ and $\left(1, \frac{n+1}{2n}\right)$. Since Plancherel theorem implies that $\left(\frac{1}{2}, \frac{1}{2}\right) \in \mathscr{D}_\Gamma$ for every cone Γ, we can use M. Riesz interpolation theorem, to conclude that $\mathscr{D}_\Gamma$ contains the open trapezoid $\mathscr{T}$ of vertices $(1,0)$, $\left(\frac{n-1}{2n}, 0\right)$, $\left(1, \frac{n+1}{2n}\right)$,and $\left(\frac{1}{2}, \frac{1}{2}\right)$, as required. □

We do not know whether $\mathscr{D}_\Gamma = \mathscr{T}$ or not, but the next Proposition shows that $\mathscr{D}_\Gamma$ and $\mathscr{T}$ are not too far off.

Proposition 2.4 *Let Γ be as in Theorem 1.2. Then, if $\left(\frac{1}{p}, \frac{1}{q}\right) \in \mathscr{D}_\Gamma$, necessarily,*

$$\frac{1}{q} \leq \frac{n+1}{2n} \quad \textit{and} \quad \frac{1}{p} \geq \frac{n-1}{2n}.$$

Proof. Let $G \supset \Gamma^0$ be the convex envelope of a finite number of linearly independent half lines passing through the origin of $\mathbf{R}^{n+1}$. Assume that G does not contain straight lines. Let $\bar{y} \in \Gamma$ and let $\hat{f}(\zeta) = e^{-2\pi\langle\bar{y},\zeta\rangle}\chi_G(\zeta)$, where $\chi_G(\zeta)$ is the characteristic function of G. Thus, $f(x) = K_{G^0}(x+i\bar{y})$, and we have proved in [1] that $f(x) \in L^p(\mathbf{R}^{n+1})$ for every $p > 1$. But $T_\Gamma(f)(x+i\bar{y}) = \int_{\Gamma^0} e^{-4\pi\langle\bar{y},\zeta\rangle+2\pi i\langle x,\zeta\rangle}\,d\zeta$ $= K_\Gamma(x+2i\bar{y})$, and by Theorem 1.2, $K_{\Gamma,2\bar{y}}(f) \in L^q(\mathbf{R}^{n+1})$ only if $q > \frac{2n}{n+1}$. Thus, if $(\frac{1}{p},\frac{1}{q}) \in \mathscr{D}_\Gamma$, necessarily, $q > \frac{2n}{n+1}$. It follows by duality that if $(\frac{1}{p},\frac{1}{q}) \in \mathscr{D}_\Gamma$, then $p < \frac{2n}{n-1}$. This concludes the proof of the proposition. □

3 Polygonal Cones and Fourier Multipliers

Let Γ be an open cone in $\mathbf{R}^{n+1}$ with vertex at the origin. After a rotation, we can assume $\Gamma = \{(\zeta_0,\zeta') : \zeta_0 \geq \phi(\zeta')\}$ where ϕ is a homogeneous function of degree 1. We let $m_y(\zeta) = e^{-2\pi\langle y,\zeta\rangle}\chi_{\Gamma^0}(\zeta)$. If there exists $C(y_0) > 0$ so that $||T_{\Gamma,y}(f)||_p \leq C(y)||f||_p$ for some $p \geq 1$, we say that m_y is a Fourier multiplier in $L^p(\mathbf{R}^{n+1},\mathbf{C})$, and we denote by $|||m_y|||_{p,p}$ the operator norm of $T_{\Gamma,y}$ in $L^p(\mathbf{R}^{n+1},\mathbf{C})$.

That is, $|||m_y|||_{p,p}$ is the smallest constant C for which

$$||L_{m_y}f||_p = \left\|\int_{\mathbf{R}^{n+1}} m_y(\zeta)\hat{f}(\zeta)e^{2\pi i\langle\zeta,x\rangle}\,d\zeta\right\|_p \leq C||f||_p$$

for every $f \in C_0^\infty(\mathbf{R}^{n+1},\mathbf{C})$.

By DeLeeuw's theorem, m_y cannot be a Fourier multiplier $L^p(\mathbf{R}^{n+1},\mathbf{C})$ if the intersection of Γ with planes which are parallel to $\zeta_0 = 0$ is not a Fourier multiplier in $L^p(\mathbf{R}^n,\mathbf{C})$. DeLeeuw's theorem can be stated as follows.

Proposition 3.1 *Suppose that $m(\zeta)$ is a Fourier multiplier for $L^p(\mathbf{R}^{n+1},\mathbf{C})$. Then, for every $\overline{\zeta}_0 \in \mathbf{R}$, (except perhaps a set of measure zero), $\tilde{m}(\zeta') = m(\overline{\zeta}_0,\zeta')$ is Fourier multiplier for $L^p(\mathbf{R}^n,\mathbf{C})$, and*

$$|||m|||_{p,p} \geq |||\tilde{m}|||_{p,p},$$

see [3] and also [6].

So, if Γ is the cone $\{(x_0,x') : x_0 \geq |x'|\}$, but also when the boundary of Γ is smooth and has $n-1$ non vanishing principal curvatures, m_y cannot be a Fourier multiplier for $L^p(\mathbf{R}^{n+1},\mathbf{C})$ unless $p = 2$; by Lemma 2.3, also the boundary of Γ^0 is smooth and has $n-1$ non vanishing principal curvatures, and a celebrate counterexample by C. Fefferman (see [F]) and its subsequent generalizations, show that m_y cannot be a Fourier multiplier for $L^p(\mathbf{R}^{n+1},\mathbf{C})$ unless $p = 2$. So, if Γ is as in Theorem 1.1, $\mathscr{D}_\Gamma$ cannot contain any point of the form $(\frac{1}{p},\frac{1}{p})$ with $p \neq 2$.

We say that Γ is a *polygonal cone* with N faces if Γ is the convex envelope of N linearly independent half lines passing through the origin of $\mathbf{R}^{n+1}$. Γ has empty interior if $N < n+1$.

The main result of this section is the following.

Theorem 3.2. *Let Γ be a polygonal cone with $N \leq n+1$ faces. Then, for every $y \in \Gamma$, and every $p > 1$*

$$m_y(\zeta) = \chi_{\Gamma^0} e^{-2\pi\langle y,\zeta\rangle}$$

is a Fourier multiplier for $L^p(\mathbf{R}^{n+1}, \mathbf{C})$, and

$$|||m_y|||_{p,p} = c_p^N \tag{22}$$

where

$$c_p = \frac{1}{\sin(\pi/p)} \tag{23}$$

is the (p,p) norm of the "half line multiplier" $h(t) = \chi_{(0,\infty)}(t)$.

The (p,p) norm of h has been evaluated by Hollembeck and Verbitskyi [5]. The $L^p(\mathbf{R}) \to L^p(\mathbf{R}, \mathbf{C})$ norm of L_h has been evaluated in [4, 10].

Proof. First of all we prove the result for cones Γ which are Cartesian products of N copies of $(0,\infty)$. We let $\Gamma_N = \{(x_0, x_1, \ldots, x_n) \ : \ x_j > 0 \text{ for every } 0 \leq j \leq N\}$. Clearly, $\Gamma_N^0 = (0,\infty)^N \times \mathbf{R}^{n+1-N}$.

We argue by induction on N. Let $\Gamma_1 = (0,\infty)$, and $\Gamma_1^0 = (0,\infty) \times \mathbf{R}^n$. For every $y_0 > 0$,

$$\begin{aligned} T_{\Gamma_1} f(x_0 + iy_0, x') &= \int_0^\infty \int_{\mathbf{R}^{n-1}} \hat{f}(\zeta_0, \zeta') e^{-2\pi y_0 x_0} e^{2\pi i x'\zeta'} d\zeta \\ &= \int_0^\infty \tilde{f}(\zeta_0, x') e^{-2\pi y_0 \zeta_0} e^{2\pi i x_0 \zeta_0} d\zeta_0, \end{aligned}$$

where $\tilde{f}(\zeta_0, x')$ denotes the Fourier Transform of $f(t, x')$ with respect to t. If we let $q_{y_0}(\zeta_0) = e^{-2\pi y_0 x_0} h(\zeta_0)$, where $h(t) = \chi_{(0,\infty)}(t)$, we can see at once that $T_{\Gamma_1} f(.+iy_0, x')$ is the multiplier operator associated to $q_{y_0}(\zeta_0)$ and applied to $x_0 \to f(x_0, x')$, with x' fixed.

We show that

$$|||q_{y_0}|||_{p,p} = |||h|||_{p,p}, \tag{24}$$

where the norm is in $L^p(\mathbf{R}, \mathbf{C})$. We let $m_{y_0}(x_0) = e^{-2\pi y_0 |x_0|}$. With this notation, $q_{y_0}(\zeta_0) = m_{y_0}(\zeta_0) h(\zeta_0)$. It is easy to show that the Fourier transform of $m_{y_0}(\zeta_0)$ is $\dfrac{y_0}{\pi(y_0^2 + x_0^2)}$, and $|||\widehat{m_{y_0}}|||_1 = 1$. By Young inequality,

$$||L_{m_{y_0}} f||_p = ||\widehat{m_{y_0}} * f||_p \leq ||f||_p$$

and so $|||m_{y_0}|||_{p,p} \leq 1$. Thus,

$$|||q_{y_0}|||_{p,p} = |||m_{y_0} h|||_{p,p} \leq |||m_{y_0}|||_{p,p}|||h|||_{p,p} \leq |||h|||_{p,p}. \tag{25}$$

Let us prove the opposite inequality.

It is easy to prove directly (or we can recall Proposition 2.1) that $|||q_{y_0}|||_{p,p}$ does not depend on y_0. By Fatou's lemma, for every $f \in L^p(\mathbf{R}, \mathbf{C})$,

$$\begin{aligned}
\lim_{y_0 \to 0} ||L_{q_{y_0}} f||_p^p &= \lim_{y_0 \to 0} \int_{\mathbf{R}} \left| \int_0^\infty e^{-y_0 \zeta_0} \hat{f}(\zeta_0) e^{i\zeta_0 x_0} d\zeta_0 \right|^p dx_0 \\
&\geq \int_{\mathbf{R}} \left| \int_0^\infty \hat{f}(\zeta_0) e^{i\zeta_0 x_0} d\zeta_0 \right|^p dx_0 = ||L_h f||_p^p.
\end{aligned}$$

Thus,

$$\lim_{y_0 \to 0} |||q_{y_0}|||_{p,p} \geq |||h|||_{p,p} = c_p$$

but since $|||q_{y_0}|||_{p,p}$ does not depend on y_0, we have proved that $|||q_{y_0}|||_{p,p} = |||h|||_{p,p}$. Consequently,

$$\begin{aligned}
||T_{\Gamma_1} f||_p &= \left\| \int_0^\infty \int_{\mathbf{R}^n} \hat{f}(\zeta_0, \zeta') e^{-2\pi y_0 x_0} e^{2\pi i x' \zeta'} d\zeta \right\|_p \\
&= \left(\int_{\mathbf{R}^n} \left(\int_{\mathbf{R}} \left| \int_0^\infty \tilde{f}(\zeta_0, x') e^{-2\pi y_0 \zeta_0} e^{2\pi i x_0 \zeta_0} d\zeta_0 \right|^p dx_0 \right) dx' \right)^{\frac{1}{p}} \\
&= \left(\int_{\mathbf{R}^n} ||L_{q_{y_0}} f(., x')||_p^p dx' \right)^{\frac{1}{p}},
\end{aligned}$$

where the norm inside the integral is in $L^p(\mathbf{R}, \mathbf{C})$. Consequently,

$$||T_{\Gamma_1} f||_p \leq |||q_{y_0}|||_{p,p} ||f||_p$$

and it is easy to see that this inequality is sharp.

We now consider $\Gamma = (0,\infty)^N$. The characteristic function of $\Gamma^0 = (0,\infty)^N \times \mathbf{R}^{n+1-N}$ is the product of the characteristic function of $\Gamma_j^0 = (0,\infty)_j \times \mathbf{R}^n$, where with this notation, we mean that $(0,\infty)$ is in the variable ζ_j. Thus, for every $y \in \Gamma$, the multiplier associated to $q_y^N(\zeta) = \chi_{\Gamma_0}(\zeta) e^{-2\pi\zeta . y}$, is the tensor product of N multipliers of the form of $q_{y_j}^j(\zeta_j) = \chi_{(0,\infty)}(\zeta_j) e^{-2\pi y_j \zeta_j}$ times the identity in the other variables, and so

$$|||q_y^N|||_{p,p} = \Pi_{j=1}^N |||q_{y_j}^j|||_{p,p} = c_p^N.$$

Finally, suppose that Γ is the convex envelope of $N \leq n+1$ linearly independent half lines passing through the origin. It is easy to see that the same is true also of Γ^0. There exists a linear and invertible transformation L on $\mathbf{R}^n$ which is such that

$L^{-1}(\Gamma^0) = (0,\infty)^N \times \mathbf{R}^{n+1-n}$. We let $G^0 = (0,\infty)^N \times \mathbf{R}^{n+1-n}$, and we assume without loss of generality that $\det(L) = 1$. Thus,

$$T_\Gamma f(x+iy) = \int_{\Gamma^0} e^{2\pi i\zeta(x+iy)} \hat{f}(\zeta)d\zeta = \int_{G^0} e^{2\pi iL(\zeta).(x+iy)} \hat{f}(L\zeta)d\zeta$$
$$= \int_{G^0} e^{2\pi i\zeta.L^*(x+iy)} \widehat{f\circ(L^{-1})^*)}(\zeta)d\zeta$$

and

$$||T_{\Gamma,y}f||_p = ||T_{G,L^*(y)}(f\circ(L^{-1})^*)||_p$$

from which it is easy to prove that $|||T_{\Gamma,y}|||_{p,p} = |||T_{G,L^*(y)}|||_{p,p} = c_p^N$.

It would be tempting to conclude that if Γ is any polygonal cone with N faces, (e.g., if $\Gamma \subset \mathbf{R}^3$ is a convex cone whose sections with the planes $\{\zeta_0 = t\}$ are polygons with N sides), then $|||T_{\Gamma,y}|||_{p,p} \le c_p^N$, but unfortunately this is not true.

For example, consider a cone in $\mathbf{R}^3$ whose sections with the planes $\{\zeta_0 = t\}$ are rectangles. It is proved in [2] that the (p,p) multiplier norm of the characteristic function of a rectangle is n_p^2, where

$$n_p = \begin{cases} \tan\left(\frac{\pi}{2p}\right) \text{ if } 1 < p \le 2 \\ \cot\left(\frac{\pi}{2p}\right) \text{ if } 2 \le p < \infty \end{cases}$$

is the (p,p) norm of the Hilbert transform. By Proposition 3.1, $n_p^2 \le |||T_{\Gamma,y}|||_{p,p}$, and so, if it was true that $|||T_{\Gamma,y}|||_{p,p} \le c_p^4$, it would be true also that $n_p^2 \le c_p^4$. But, when $1 < p < 2$,

$$n_p^2 = \tan\left(\frac{\pi}{2p}\right) \le \left(\sin\left(\frac{\pi}{p}\right)\right)^{-4} = \left(2\sin\left(\frac{\pi}{2p}\right)\cos\left(\frac{\pi}{2p}\right)\right)^{-4}$$

implies that

$$16\sin^6\left(\frac{\pi}{2p}\right)\cos^2\left(\frac{\pi}{2p}\right) \le 1,$$

and this inequality is not verified for all $p < 2$. □

However, we can provide an interesting upper bound for the (p,p) multiplier norm of the characteristic function of a triangle in $\mathbf{R}^2$. Let $P \subset \mathbf{R}^2$ be a triangle with vertices p_1, p_2, and p_3. Let Q_1, Q_2, and Q_3 be vectors in $\mathbf{R}^3$ which are such that $Q_j = (1, p_j)$. The Q_j's are linearly independent, and if we let Γ^0 be the convex envelope of the half lines tQ_j, we can see at once that P is congruent to $\Gamma^0 \cap \{\zeta_0 = 1\}$. By Proposition 3.1, $|||\chi_T|||_{p,p} \le ||T_\Gamma||_{p,p}$, and by Theorem 3.2, we can prove the following.

Corollary 3.3 *Let P be a triangle in $\boldsymbol{R}^3$. Then, for every $p > 1$,*

$$|||\chi_P|||_{p,p} \leq c_p^3, \tag{26}$$

where c_p is defined as in Eq. (23).

References

1. De Carli, L.: L^p estimates for the Cauchy transform of distributions with respect to convex cones. Rend. Sem. Mat. Univ. Padova **88**, 35–53 (1992)
2. De Carli, L., Laeng, E.: Truncations of weak- L^p functions and sharp L^p bounds for the segment multiplier. Collect. Math. **51**(3), 309–326 (2000)
3. de Leeuw, K.: On L^p multipliers. Ann. Math. **81**(2), 364-379 (1965)
4. Essén, M.: A superharmonic proof of the M. Riesz conjugate function theorem. Ark. Mat. **22**(2), 241–249 (1984)
5. Hollenbeck, B., Verbitsky, I.E.: Best constants for the Riesz projection. J. Funct. Anal. **175**, 370–392 (2000)
6. Jodeit, M.: A note on Fourier multipliers. Proc. of Am. Math. Soc. **27**(2), 423–424 (1971)
7. Mockenhaupt, G.: A note on the cone multiplier. Proc. Amer. Math. Soc. **117**(1), 145–152 (1993)
8. Sogge, C.D.: Fourier integrals in classical analysis. Cambridge University Press, Cambridge (1993)
9. Tao, T., Vargas, A.: A bilinear approach to cone multipliers. I. Restriction estimates. Geom. Funct. Anal. **10**(1), 185–215 (2000)
10. Verbitsky, I.E.: An estimate of the norm of a function in a Hardy space in terms of the norms of its real and imaginary parts. A.M.S. Transl. (2), 11–15 (1984) (Translation of Mat. Issled. Vyp. **54** 16–20 (1980))
11. Vladimirov, V.S.: Generalized Functions in Mathematical Physics. MIR, Moscow (1979)
12. Wakabayashi, S.: Classical microlocal analysis in the space of hyperfunctions. In: Lecture Notes in mathematics, vol. 1737. Springer, Berlin (2000)

Multiparameter Projection Theorems with Applications to Sums-Products and Finite Point Configurations in the Euclidean Setting

Burk Erdoğan, Derrick Hart, and Alex Iosevich

Abstract In this paper we study multi-parameter projection theorems for fractal sets. With the help of these estimates, we recover results about the size of $A \cdot A + \cdots + A \cdot A$, where A is a subset of the real line of a given Hausdorff dimension, $A + A = \{a + a' : a, a' \in A\}$ and $A \cdot A = \{a \cdot a' : a, a' \in A\}$. We also use projection results and inductive arguments to show that if a Hausdorff dimension of a subset of $\mathbb{R}^d$ is sufficiently large, then the $\binom{k+1}{2}$-dimensional Lebesgue measure of the set of k-simplexes determined by this set is positive. The sharpness of these results and connection with number theoretic estimates is also discussed.

1 Introduction

We start out by briefly reviewing the underpinnings of the sum–product results in the discrete setting. A classical conjecture in geometric combinatorics is that either the sum-set or the product-set of a finite subset of the integers is maximally large. More precisely, let $A \subset \mathbb{Z}$ of size N and define

$$A + A = \{a + a' : a, a' \in A\}; \; A \cdot A = \{a \cdot a' : a, a' \in A\}.$$

B. Erdoğan
Department of Mathematics, University of Illinois at Urbana-Champaign, Urbana-Champaign, Illinois, United States,
e-mail: berdogan@math.uiuc.edu

D. Hart
Department of Mathematics, Kansas State University, Manhattan, Kansas
e-mail: dnh0a9@gmail.com

A. Iosevich (✉)
Department of Mathematics, University of Rochester, Rochester, NY
e-mail: iosevich@math.rochester.edu

D. Bilyk et al. (eds.), *Recent Advances in Harmonic Analysis and Applications*, Springer Proceedings in Mathematics & Statistics 25, DOI 10.1007/978-1-4614-4565-4_11,

The Erdos-Szemeredi conjecture says that

$$\max\{\#(A+A), \#(A\cdot A)\} \gtrapprox N^2,$$

where here, and throughout, $X \lesssim Y$ means that there exists $C > 0$ such that $X \leq CY$ and $X \lessapprox Y$, with the controlling parameter N if for every $\varepsilon > 0$ there exists $C_\varepsilon > 0$ such that $X \leq C_\varepsilon N^\varepsilon Y$. The best currently known result is due to Jozsef Solymosi [18] who proved that

$$\max\{\#(A+A), \#(A\cdot A)\} \gtrapprox N^{\frac{4}{3}}. \tag{1}$$

For the finite field analogs of these problems, see, for example, [2, 7–11, 19].

The sum-product problem has also received considerable amount of attention in the Euclidean setting. The following result was proved by Edgar and Miller [3] and, independently, by Bourgain [1].

Theorem 1. *A Borel subring of the reals either has Hausdorff dimension* 0 *or is all of the real line.*

Bourgain [1] proved the following quantitative bound that was conjectured in [13].

Theorem 2. *Suppose that $A \subset \mathbb{R}$ is a (δ, σ)-set in the sense that A is a union of δ-intervals and for $0 < \varepsilon \ll 1$,*

$$|A \cap I| < \left(\frac{r}{\delta}\right)^{1-\sigma} \delta^{1-\varepsilon}$$

whenever I is an arbitrary interval of size $\delta \leq r \leq 1$. Suppose that $0 < \sigma < 1$ and $|A| > \delta^{\sigma+\varepsilon}$. Then

$$|A+A| + |A\cdot A| > \delta^{\sigma-c}$$

for an absolute constant $c = c(\sigma) > 0$.

One of the key steps in the proof is the study of the size of

$$A\cdot A - A\cdot A + A\cdot A - A\cdot A,$$

and this brings us to the main results of this chapter. Our goal is to show that if the Hausdorff dimension of $A \subset \mathbb{R}$ is sufficiently large, then

$$\mathscr{L}^1(a_1A + a_2A + \cdots + a_dA) > 0$$

for a generic choice of $(a_1, \ldots, a_d) \in A \times A \times \cdots \times A$. It is of note that much work has been done in this direction in the setting of finite fields. See [8, 9], and the references contained therein. In particular, it is the work in [8] that inspired some of the following results.

Our results are proved using generalized projections theorems, similar in flavor to the ones previously obtained by Peres and Schlag [16] and Solomyak [17].

Theorem 3. *Let $E,F \subset \mathbb{R}^d$, and $d \geq 2$ be of Hausdorff dimension s_E, s_F, respectively. Suppose that there exist Frostman measures μ_E, μ_F, supported on E and F, respectively, such that for δ sufficiently small and $|\xi|$ sufficiently large, there exist non-negative numbers γ_F and l_F such that the following conditions hold:*

$$|\widehat{\mu}_F(\xi)| \lessapprox |\xi|^{-\gamma_F} \tag{2}$$

and

$$\mu_F(T_\delta) \lessapprox \delta^{s_F - l_F}, \tag{3}$$

for every tube T_δ of length ≈ 1 and radius $\approx \delta$ emanating from the origin.

Define, for each $y \in F$, the projection set

$$\pi_y(E) = \{x \cdot y : x \in E\}.$$

Suppose that for some $0 < \alpha \leq 1$,

$$\max\left\{\frac{\min\{\gamma_F, s_E\}}{\alpha}, \frac{s_E + s_F - l_F + 1 - \alpha}{d}\right\} > 1. \tag{4}$$

Then

$$\dim_{\mathscr{H}}(\pi_y(E)) \geq \alpha$$

for μ_F-every $y \in F$.

If Eq. (4) holds with $\alpha = 1$, then

$$\mathscr{L}^1(\pi_y(E)) > 0$$

for μ_F-every $y \in F$.

Remark 1. Observe that the conditions of Theorem 3 always hold with $\gamma_F = 0$ and $l_F = 1$ since every tube T_δ can be decomposed into $\approx \delta^{-1}$ balls of radius δ.

Corollary 1. *Let $A \subset \mathbb{R}$ and let μ_A be a Frostman measure on A. Then the following hold*

- *Suppose that the Hausdorff dimension of A, denoted by $\dim_{\mathscr{H}}(A)$, is greater than $\frac{1}{2} + \varepsilon$ for some $0 < \varepsilon \leq \frac{1}{2}$. Then for $\mu_A \times \mu_A \times \cdots \times \mu_A$-every $(a_1, a_2, \ldots, a_d) \in A \times A \times \cdots \times A$,*

$$\dim_{\mathscr{H}}(a_1A + a_2A + \cdots + a_dA) \geq \min\left\{1, \frac{1}{2} + \varepsilon(2d-1)\right\}. \tag{5}$$

- *Suppose that the Hausdorff dimension of A is greater than*

$$\frac{1}{2}+\frac{1}{2(2d-1)}$$

for some $d \geq 2$. Then for $\mu_A \times \mu_A \times \cdots \times \mu_A$-every $(a_1, a_2, \ldots, a_d) \in A \times A \times \cdots \times A$,

$$\mathscr{L}^1(a_1A + a_2A + \cdots + a_dA) > 0. \tag{6}$$

- *Suppose that $F \subset \mathbb{R}^d$, $d \geq 2$, is star-like in the sense that the intersection of F with every tube of width δ containing the origin is contained in a fixed number of balls of radius δ. Assume that*

$$\dim_{\mathscr{H}}(A) + \frac{dim_{\mathscr{H}}(F)}{d} > 1.$$

Then for μ_F-every $x \in F$,

$$\mathscr{L}^1(x_1A + x_2A + \cdots + x_dA) > 0. \tag{7}$$

In particular, if $dim_{\mathscr{H}}(A) = \frac{1}{2} + \varepsilon$, for some $\varepsilon > 0$, and $dim_{\mathscr{H}}(F) > \frac{d}{2} - \varepsilon$, then Eq. (7) holds for μ_F-every $x \in F$.

- *Suppose that $F \subset \mathbb{R}^d$, $d \geq 2$, possesses a Borel measure μ_F such that Eq. (2) holds with $\gamma_F > 1$. Suppose that $dim_{\mathscr{H}}(A) > \frac{1}{2}$. Then, for μ_F-every $x \in F$,*

$$\mathscr{L}^1(x_1A + x_2A + \cdots + x_dA) > 0.$$

1.1 Applications to the Finite Point Configurations

Recall that the celebrated Falconer distance conjecture [6] says that if the Hausdorff dimension of $E \subset \mathbb{R}^d$, $d \geq 2$, is $> \frac{d}{2}$, then the Lebesgue measure of the set of distances $\{|x-y| : x, y \in E\}$ is positive. The best known result to date, due to Wolff [20] in two dimension and Erdogan [4] in higher dimensions, says that the Lebesgue measure of the distance set is indeed positive if the Hausdorff dimension of E is greater than $\frac{d}{2} + \frac{1}{3}$.

Corollary 2. *Let*

$$E \subset S^{d-1} = \left\{ x \in \mathbb{R}^d : \sqrt{x_1^2 + x_2^2 + \cdots + x_d^2} = 1 \right\}$$

of Hausdorff dimension $> \frac{d}{2}$. Let μ_E be a Frostman measure on E. Then

$$\mathscr{L}^1(\{|x-y| : x \in E\}) > 0$$

for μ-every $y \in E$.

Before stating our next result, we need the following definition. Let $T_k(E)$, $1 \leq k \leq d$, denote the $(k+1)$-fold Cartesian product of E with the equivalence relation where $(x^1,\dots,x^{k+1}) \sim (y^1,\dots,y^{k+1})$ if there exists a translation τ and an orthogonal transformation O such that $y^j = O(x^j)+\tau$.

In analogy with the Falconer distance conjecture, it is reasonable to ask how large the Hausdorff dimension of $E \subset \mathbb{R}^d$, $d \geq 2$, needs to be to ensure that the $\binom{k+1}{2}$-dimensional Lebesgue measure of $T_k(E)$ is positive.

Theorem 4. *Let $E \subset S_t = \{x \in \mathbb{R}^d : |x| = t\}$ for some $t > 0$. Suppose that $dim_{\mathscr{H}}(E) > \frac{d+k-1}{2}$. Then*

$$\mathscr{L}^{\binom{k+1}{2}}(T_k(E)) > 0.$$

Using a pigeon-holing argument, we obtain the following result for finite point configurations in $\mathbb{R}^d$.

Corollary 3. *Let $E \subset \mathbb{R}^d$ of Hausdorff dimension $> \frac{d+k+1}{2}$. Then*

$$\mathscr{L}^{\binom{k+1}{2}}(T_k(E)) > 0.$$

Remark 2. We do not know to what extent our results are sharp beyond the following observations. If the Hausdorff dimension of E is less than $\frac{d}{2}$, the classical example due to Falconer [6] shows that the set of distances may have measure 0, so, in particular, $\mathscr{L}^{\binom{k+1}{2}}(T_k(E))$ may be 0 for $k > 1$. See also [5, 15] for the description of the background material and [12, 14] for related counter-example construction. In two dimensions, one can generalize Falconer's example to show that if the Hausdorff dimension of E is less than $\frac{3}{2}$, then the three dimensional Lebesgue measure of $T_2(E)$ may be zero. In higher dimension, construction of examples of this type is fraught with serious number theoretic difficulties. We hope to address this issue in a systematic way in the sequel.

2 Proofs of Main Results

2.1 *Proof of the Projection Results (Theorem 3)*

Define the measure ν_y on $\pi_y(E)$ by the relation

$$\int g(s)d\nu_y(s) = \int g(x\cdot y)d\mu_E(x).$$

It follows that

$$\int\int |\widehat{\nu}_y(t)|^2 dt d\mu_F(y) = \int\int |\widehat{\mu}_E(ty)|^2 d\mu_F(y)dt.$$

We have

$$\int |\widehat{\mu}_E(ty)|^2 d\mu_F(y) = \int\int \widehat{\mu}_F(t(u-v))d\mu_E(u)d\mu_E(v)$$
$$= \int\int_{|u-v|\leq t^{-1}} \widehat{\mu}_F(t(u-v))d\mu_E(u)d\mu_E(v)$$
$$+ \int\int_{|u-v|>t^{-1}} \widehat{\mu}_F(t(u-v))d\mu_E(u)d\mu_E(v) = I + II.$$

Since μ_E is a Frostman measure,

$$|I| \lessapprox t^{-s_E}.$$

By Eq. (2),

$$|II| \lessapprox t^{-\gamma_F} \int\int |u-v|^{-\gamma_F} d\mu(u)d\mu(v) \lesssim t^{-\gamma_F}$$

as long as $\gamma_F \leq s_E$. If $\gamma_F > s_E$, then for any $\varepsilon > 0$,

$$|II| \lesssim t^{-s_E+\varepsilon} \int\int |u-v|^{-s_E+\varepsilon} d\mu(u)d\mu(v) \lessapprox t^{-s_E},$$

and we conclude that

$$\int |\widehat{\mu}_E(ty)|^2 d\mu_F(y) \lessapprox t^{-\min\{s_E,\gamma_F\}}. \tag{8}$$

It follows that

$$\int\int |\widehat{\nu}_y(t)|^2 t^{-1+\alpha} d\mu_F(y)dt < \infty$$

if $\min\{s_E, \gamma_F\} > \alpha$.

We now argue via the uncertainty principle. We may assume, by scaling and pigeon-holing, that $F \subset \{x \in \mathbb{R}^d : 1 \leq |x| \leq 2\}$. Let ϕ be a smooth cut-off function supported in the ball of radius 3 and identically equal to 1 in the ball of radius 2. It follows that

$$\int |\widehat{\nu}_y(t)|^2 t^{-1+\alpha} dt = \int\int |\widehat{\mu}_E(ty)|^2 \, t^{-1+\alpha} \, dt \, d\mu_F(y)$$
$$= \int\int \left|\widehat{\mu}_E * \widehat{\phi}(ty)\right|^2 d\mu_F(y) \, t^{-1+\alpha} \, dt$$
$$\lesssim \int\int\int |\widehat{\mu}_E(\xi)|^2 |\widehat{\phi}(ty-\xi)| d\mu_F(y) \, t^{-1+\alpha} \, dt d\xi$$
$$\lesssim \sum_m 2^{-mn} \int |\widehat{\mu}_E(\xi)|^2 \mu_F \times \mathscr{L}^1\{(y,t) : |ty-\xi| \leq 2^m\} |\xi|^{-1+\alpha} d\xi$$
$$\lesssim \sum_m 2^{-mn} \cdot 2^m \int |\widehat{\mu}_E(\xi)|^2 \mu_F \left(T_{2^m|\xi|^{-1}}(\xi)\right) |\xi|^{-1+\alpha} d\xi,$$

where $T_\delta(\xi)$ is the tube of width δ and length 10 emanating from the origin in the direction of $\frac{\xi}{|\xi|}$. By assumption, this expression is

$$\lessapprox \sum_m 2^{-mn} \cdot 2^m \cdot 2^{m(s_F - l_f)} \int |\widehat{\mu}_E(\xi)|^2 |\xi|^{-s_F + l_F - 1 + \alpha} d\xi \lesssim 1$$

if

$$s_F - l_F + 1 - \alpha > d - s_E$$

and n is chosen to be sufficiently large. Combining this with Eq. (8) we obtain the conclusion of Theorem 3.

2.2 *Proof of Applications to Sums and Products (Corollary 1)*

Let $A \subset \mathbb{R}$ have dimension greater than $s_A := \frac{1}{2} + \frac{1}{2(2d-1)}$. Note that we can find a probability measure, μ_A, supported on A satisfying

$$\mu_A(B(x,r)) \leq Cr^{s_A}, \quad x \in \mathbb{R}, r > 0 \tag{9}$$

Let $E = A \times A \times \cdots \times A$. Define $\mu_E = \mu_A \times \mu_A \times \cdots \times \mu_A$.

Lemma 1. *With the notation above,*

$$\mu_E(T_\delta) \lessapprox \delta^{(d-1)s_A}, \tag{10}$$

where $\dim_{\mathcal{H}}(A) = s_A$.

Proof. Let $l_\xi = \{s\xi : s \in \mathbb{R}\}$ and assume without loss of generality that ξ_1 is the largest coordinate of ξ in absolute value. In particular, $\xi_1 \neq 0$. Define the function

$$\Psi : A \to (A \times A \times \cdots \times A) \cap l_\xi$$

by the relation

$$\Psi(a) = \left(a, a\frac{\xi_2}{\xi_1}, \ldots, a\frac{\xi_d}{\xi_1}\right).$$

Note that

$$\begin{aligned}
\mu(T_\delta(\xi)) &= \mu_A \times \cdots \times \mu_A(T_\delta(\xi)) \\
&\leq \int_{-10}^{10} \mu_A \times \cdots \times \mu_A(B(\Psi(x_1), \delta)) d\mu_A(x_1) \\
&\lessapprox \int_{-10}^{10} \delta^{(d-1)s_A} d\mu_A(x_1) \lesssim \delta^{(d-1)s_A}.
\end{aligned}$$

It follows that if $E = F = A \times A \times \cdots \times A$, then the assumptions of Theorem 3 are satisfied with $s_E = s_F = ds_A$, $\gamma_F = 0$, and $l_F = \frac{s_E}{d}$. The conclusion of the first part of Corollary 1 follows in view of Theorem 3.

The second conclusion of Corollary 1 follows, in view of Theorem 3, if we observe that if F is star-like, then

$$\mu_F(T_\delta) \lessapprox \delta^{-s_F}.$$

The third conclusion of Corollary 1 follows from Theorem 3 since we may always take $l_F = 1$. This holds since every tube δ is contained in a union of δ^{-1} balls of radius δ. □

2.3 *Proof of the Spherical Configuration Result (Theorem 4)*

Let $y = (y^1, y^2, \dots, y^k)$, $y^j \in E$ and define

$$\pi_y(E) = \{x \cdot y^1, \dots, x \cdot y^k : x \in E\}.$$

Define a measure on $\pi_y(E)$ by the relation

$$\int g(s) d\nu_y(s) = \int g(x \cdot y^1, \dots, x \cdot y^k) d\mu(x),$$

where $s = (s_1, \dots, s_k)$ and μ is a Frostman measure on E. It follows that

$$\widehat{\nu}_y(t) = \widehat{\mu}(t \cdot y),$$

where $t = (t_1, \dots, t_k)$ and

$$t \cdot y = t_1 y^1 + t_2 y^2 + \cdots + t_k y^k.$$

It follows that

$$\int \int |\widehat{\nu}_y(t)|^2 dt d\mu^*(y) = \int \int |\widehat{\mu}(t \cdot y)|^2 dt d\mu^*(y),$$

where

$$d\mu^*(y) = d\mu(y^1) d\mu(y^2) \dots d\mu(y^k).$$

Arguing as above, this quantity is bounded by

$$\int \int \int |\widehat{\mu}(\xi)|^2 |\widehat{\phi}(t \cdot y - \xi)| dt d\mu^* d\xi.$$

It is not difficult to see that

$$\int \int |\widehat{\phi}(t \cdot y - \xi)| dt d\mu^* \lesssim |\xi|^{-s+k-1}$$

since once we fix a linearly independent collection $y^1, y^2, \ldots, y^{k-1}$, y^k is contained in the intersection of E with a k-dimensional plane. Since E is also a subset of a sphere, the claim follows. If $y^1, \ldots, y^{k-1}$ are not linearly independent, the estimate still holds, but the easiest way to proceed is to observe that since the Hausdorff dimension of E is greater than k by assumption, there exist $k-1$ disjoint subsets $E_1, E_2, \ldots, E_{k-1}$ of E and a constant $c > 0$ such that $\mu(E_j) \geq c$ and any collection $y^1, \ldots, y^{k-1}$, $y^j \in E_j$, is linearly independent. This establishes our claim with μ^* replaced by the product measure restricted to E_js, which results in the same conclusion.

Plugging this in we get

$$\int |\widehat{\mu}(\xi)|^2 |\xi|^{-s+k-1} d\xi < \infty$$

if

$$s > \frac{d}{2} + \frac{k-1}{2}.$$

This implies that for μ^k almost every k-tuple $y = (y^1, y^2, \ldots, y^k) \in E^k$, ν_y is absolutely continuous w.r.t $\mathscr{L}^k$, and hence its support, $\pi_y(E)$, is of positive $\mathscr{L}^k$ measure.

Let

$$\mathscr{E}_k = \{y = (y^1, \ldots, y^k) \in E^k : \mathscr{L}^k(\pi_y(E)) > 0\}.$$

We just proved that $\mu^k(\mathscr{E}_k) = \mu^k(E^k) > 0$.

Consider the set

$$P_{k-1} = \left\{(y^1, \ldots, y^{k-1}) \in E^{k-1} : \mu\left(\{x : (y^1, \ldots, y^{k-1}, x) \in \mathscr{E}_k\}\right) = \mu(E)\right\}.$$

By the discussion above, and Fubini, $\mu^{k-1}(P_{k-1}) = \mu(E)^{k-1} > 0$.

For each $y = (y^1, \ldots, y^{k-1}) \in P_{k-1}$, let $F_y = \{x \in E : (y^1, \ldots, y^{k-1}, x) \in \mathscr{E}_k\}$, and define

$$\pi_y(F_y) = \{(x \cdot y^1, \ldots, x \cdot y^{k-1}) : x \in F_y\}.$$

As above we construct the measure ν_y supported on $\pi_y(F_y)$, and we have

$$\widehat{\nu_y}(t) = \widehat{\mu \chi_{F_y}}(t \cdot y) = \widehat{\mu}(t \cdot y),$$

since $\mu(E \backslash F_y) = 0$. By the argument above, we conclude that $\pi_y(F_y)$ is of positive $\mathscr{L}^{k-1}$ measure. Therefore, by Fubini, for μ^{k-1} a.e. $(y^1, \ldots, y^{k-1}) \in E^{k-1}$,

$$\mathscr{L}^{k+k-1}\{y^1 \cdot y, \ldots, y^{k-1} \cdot y, y^1 \cdot x, \ldots, y^{k-1} \cdot x, y \cdot x : x, y \in E\} > 0.$$

The result now follows from this induction step.

3 Proof of the Euclidean Configuration Result (3)

We shall make use of the following intersection result. See Theorem 13.11 in [15].

Theorem 5. *Let $a,b > 0, a+b > d$, and $b > \frac{d+1}{2}$. If A,B are Borel subsets in $\mathbb{R}^d$ with $\mathscr{H}^a(A) > 0$ and $\mathscr{H}^b(B) > 0$, then for almost all $g \in O(d)$,*

$$\mathscr{L}^d\{z \in \mathbb{R}^d : dim_{\mathscr{H}}(A \cap (\tau_z \circ g)B) \geq a+b-d\} > 0,$$

where τ_z denotes the translation by z.

In the special case when B is the unit sphere in dimension 4 or higher, we get the following corollary.

Corollary 4. *Let $E \subset \mathbb{R}^d$, $d \geq 4$, of Hausdorff dimension $s > 1$. Then*

$$\mathscr{L}^d\{z \in \mathbb{R}^d : dim_{\mathscr{H}}(E \cap (S^{d-1}+z)) \geq s-1\} > 0.$$

It follows from the corollary that if $E \subset \mathbb{R}^d$, $d \geq 4$, is of Hausdorff dimension $> \frac{d+k+1}{2}$, there exists $z \in \mathbb{R}^d$, such that the Hausdorff dimension of $E \cap (z+S^{d-1})$ is $> \frac{d+k-1}{2}$. Now observe that if $x,y \in z+S^{d-1}$, means that $x = x'+z$, $y = y'+z$, where $x', y' \in S^{d-1}$. It follows that

$$|x-y|^2 = |x'-y'|^2 = 2-2x' \cdot y'.$$

In other words, the problem of simplexes determined by elements of $E \cap (z+S^{d-1})$ reduces to Theorem 4 and thus Theorem 3 is proved.

References

1. Bourgain, J.: On the Erdős-Volkmann and Katz-Tao ring conjectures. GAFA, **13**, 334–364 (2003)
2. Bourgain, J., Katz, N., Tao, T.: A sum–product estimate in finite fields, and applications. Geom. Func. Anal. **14**, 27–57 (2004)
3. Edgar, G., Miller, C.: Borel sub-rings of the reals. PAMS, **131**, 1121–1129 (2002)
4. Erdoğan, B.: A bilinear Fourier extension theorem and applications to the distance set problem, IMRN **23**, 1411–1425 (2006)
5. Falconer, K.: The Geometry of Fractal Sets. Cambridge University Press, Cambridge (1985)
6. Falconer, K.: On the Hausdorff dimensions of distance sets. Mathematika **32**, 206–212 (1986)
7. Hart, D., Iosevich, A.: Ubiquity of simplices in subsets of vectors spaces over finite fields. Anal. Mathematica **34**, 29–38 (2008)
8. Hart, D., Iosevich, A.: Sums and products in finite fields: An integral geometric viewpoint. Contemp. Math. Radon transforms, geometry, and wavelets **464**, 129–135 (2008)
9. Hart, D., Iosevich, A., Alex, D.K., Rudnev, M.: Averages over hyperplanes, sum-product theory in vector spaces over finite fields and the Erdős-Falconer distance conjecture. Transactions of the AMS, **363**, 3255–3275 (2011).

10. Hart, D., Iosevich, A., Solymosi, J.: Sum-product estimates in finite fields via Kloosterman sums. Int. Math. Res. Not. IMRN **5**, p. 4 (2007)
11. Iosevich, A., Rudnev, M.: Erdös distance problem in vector spaces over finite fields. Trans. Amer. Math. Soc. **359**(12), 6127–6142 (2007)
12. Iosevich, A., Senger, S.: Sharpness of Falconer's estimate in continuous and arithmetic settings, geometric incidence theorems and distribution of lattice points in convex domains (2010). (http://arxiv.org/pdf/1006.1397)
13. Katz, N., Tao, T.: Some connections between Falconer's distance conjecture and sets of Furstenberg type. New York J. Math **7**, 149–187 (2001)
14. Mattila, P.: On the Hausdorff dimension and capacities of intersections. Mathematika **32**, 213–217 (1985)
15. Mattila, P.: Geometry of Sets and Measures in Euclidean Space, vol. 44. Cambridge University Press, Cambridge (1995)
16. Peres, Y., Schlag, W.: Smoothness of projections, Bernoulli convolutions and the dimension of exceptions. Duke Math J. **102**, 193–251 (2000)
17. Solomyak, B.: Measure and dimension of some fractal families. Math. Proc. Cambridge Phil. Soc. **124**, 531–546 (1998)
18. Solymosi, J.: Bounding multiplicative energy by the sumset. Adv. Math. **222**(2), 402–408 (2009)
19. Tao, T., Vu, V.: Additive Combinatorics. Cambridge University Press, Cambridge (2006)
20. Wolff, T.: Decay of circular means of Fourier transforms of measures. Int. Math. Res. Not. **10**, 547–567 (1999)

Riesz Potentials, Bessel Potentials, and Fractional Derivatives on Besov-Lipschitz Spaces for the Gaussian Measure

A. Eduardo Gatto, Ebner Pineda, and Wilfredo O. Urbina

Abstract Gaussian Lipschitz spaces $Lip_{\alpha}(\gamma_d)$ and the boundedness properties of Riesz potentials, Bessel potentials, and fractional derivatives there were studied in detail in Gatto and Urbina (On Gaussian Lipschitz Spaces and the Boundedness of Fractional Integrals and Fractional Derivatives on them, 2009. Preprint. arXiv:0911.3962v2). In this chapter we will study the boundedness of those operators on Gaussian Besov-Lipschitz spaces $B^{\alpha}_{p,q}(\gamma_d)$. Also, these results can be extended to the case of Laguerre or Jacobi expansions and even further to the general framework of diffusions semigroups.

1 Introduction

On $\mathbb{R}^d$ let us consider the Gaussian measure

$$\gamma_d(x) = \frac{e^{-|x|^2}}{\pi^{d/2}}dx, x \in \mathbb{R}^d \tag{1}$$

A.E. Gatto
Department of Mathematical Sciences, DePaul University, Chicago, IL 60614, USA
e-mail: aegatto@depaul.edu

E. Pineda
Departamento de Matemática, Decanato de Ciencia y Tecnologia, UCLA Apartado 400
Barquisimeto 3001 Venezuela
e-mail: epineda@uicm.ucla.edu.ve

W.O. Urbina (✉)
Department of Mathematics and Actuarial Sciences, Roosevelt University,
Chicago, IL 60605, USA
e-mail: wurbinaromero@roosevelt.edu

D. Bilyk et al. (eds.), *Recent Advances in Harmonic Analysis and Applications*, Springer Proceedings in Mathematics & Statistics 25, DOI 10.1007/978-1-4614-4565-4_12,

and the Ornstein-Uhlenbeck differential operator

$$L = \frac{1}{2}\triangle_x - \langle x, \nabla_x \rangle. \tag{2}$$

Let $\nu = (\nu_1, \ldots, \nu_d)$ be a multi-index such that $\nu_i \geq 0, i = 1, \cdots, d$, let $\nu! = \prod_{i=1}^d \nu_i!$, $|\nu| = \sum_{i=1}^d \nu_i$, $\partial_i = \frac{\partial}{\partial x_i}$, for each $1 \leq i \leq d$ $\partial^\nu = \partial_1^{\nu_1} \ldots \partial_d^{\nu_d}$, and consider the normalized Hermite polynomials of order ν in d variables,

$$h_\nu(x) = \frac{1}{\left(2^{|\nu|}\nu!\right)^{1/2}} \prod_{i=1}^{d} (-1)^{\nu_i} e^{x_i^2} \frac{\partial^{\nu_i}}{\partial x_i^{\nu_i}} (e^{-x_i^2}). \tag{3}$$

It is well known that the Hermite polynomials are eigenfunctions of the operator L,

$$Lh_\nu(x) = -|\nu| h_\nu(x). \tag{4}$$

Given a function $f \in L^1(\gamma_d)$ its ν-Fourier–Hermite coefficient is defined by

$$\hat{f}(\nu) = < f, h_\nu >_{\gamma_d} = \int_{\mathbb{R}^d} f(x) h_\nu(x) \gamma_d(dx).$$

Let C_n be the closed subspace of $L^2(\gamma_d)$ generated by the linear combinations of $\{h_\nu : |\nu| = n\}$. By the orthogonality of the Hermite polynomials with respect to γ_d it is easy to see that $\{C_n\}$ is an orthogonal decomposition of $L^2(\gamma_d)$,

$$L^2(\gamma_d) = \bigoplus_{n=0}^{\infty} C_n,$$

this decomposition is called the Wiener chaos.

Let J_n be the orthogonal projection of $L^2(\gamma_d)$ onto C_n, then if $f \in L^2(\gamma_d)$,

$$J_n f = \sum_{|\nu|=n} \hat{f}(\nu) h_\nu.$$

Let us define the Ornstein–Uhlenbeck semigroup $\{T_t\}_{t\geq 0}$ as

$$\begin{aligned} T_t f(x) &= \frac{1}{(1-e^{-2t})^{d/2}} \int_{\mathbb{R}^d} e^{-\frac{e^{-2t}(|x|^2+|y|^2)-2e^{-t}\langle x,y\rangle}{1-e^{-2t}}} f(y)\gamma_d(dy) \\ &= \frac{1}{\pi^{d/2}(1-e^{-2t})^{d/2}} \int_{\mathbb{R}^d} e^{-\frac{|y-e^{-t}x|^2}{1-e^{-2t}}} f(y)dy. \end{aligned} \tag{5}$$

The family $\{T_t\}_{t\geq 0}$ is a strongly continuous Markov semigroup on $L^p(\gamma_d)$, $1 \leq p \leq \infty$, with infinitesimal generator L. Also, by a change of variable we can write,

$$T_t f(x) = \int_{\mathbb{R}^d} f(\sqrt{1-e^{-2t}}u + e^{-t}x)\gamma_d(du). \tag{6}$$

Now, by Bochner subordination formula, see Stein [8] p. 61, we define the Poisson–Hermite semigroup $\{P_t\}_{t\geq 0}$ as

$$P_t f(x) = \frac{1}{\sqrt{\pi}} \int_0^{\infty} \frac{e^{-u}}{\sqrt{u}} T_{t^2/4u} f(x) du. \tag{7}$$

From Eq. (5) we obtain, after the change of variable $r = e^{-t^2/4u}$,

$$\begin{aligned} P_t f(x) &= \frac{1}{2\pi^{(d+1)/2}} \int_{\mathbb{R}^d} \int_0^1 t \frac{\exp\left(t^2/4\log r\right)}{(-\log r)^{3/2}} \frac{\exp\left(\frac{-|y-rx|^2}{1-r^2}\right)}{(1-r^2)^{d/2}} \frac{dr}{r} f(y) dy \\ &= \int_{\mathbb{R}^d} p(t,x,y) f(y) dy, \end{aligned} \tag{8}$$

with

$$p(t,x,y) = \frac{1}{2\pi^{(d+1)/2}} \int_0^1 t \frac{\exp\left(t^2/4\log r\right)}{(-\log r)^{3/2}} \frac{\exp\left(\frac{-|y-rx|^2}{1-r^2}\right)}{(1-r^2)^{d/2}} \frac{dr}{r}. \tag{9}$$

Also, by the change of variable $s = t^2/4u$, we have

$$P_t f(x) = \frac{1}{\sqrt{\pi}} \int_0^{\infty} \frac{e^{-u}}{\sqrt{u}} T_{t^2/4u} f(x) du = \int_0^{\infty} T_s f(x) \mu_t^{(1/2)}(ds), \tag{10}$$

where the measure

$$\mu_t^{(1/2)}(ds) = \frac{t}{2\sqrt{\pi}} \frac{e^{-t^2/4s}}{s^{3/2}} ds, \tag{11}$$

is called the one-side stable measure on $(0,\infty)$ of order $1/2$.

The family $\{P_t\}_{t\geq 0}$ is also a strongly continuous semigroup on $L^p(\gamma_d)$, $1 \leq p \leq \infty$, with infinitesimal generator $-(-L)^{1/2}$. In what follows, often, we are going to use the notation

$$u(x,t) = P_t f(x),$$

and

$$u^{(k)}(x,t) = \frac{\partial^k}{\partial t^k} P_t f(x).$$

Observe that by Eq. (4), we have that

$$T_t h_\nu(x) = e^{-t|\nu|} h_\nu(x), \tag{12}$$

and

$$P_t h_\nu(x) = e^{-t\sqrt{|\nu|}} h_\nu(x), \tag{13}$$

i.e., the Hermite polynomials are eigenfunctions of T_t and P_t for any $t \geq 0$.

The operators that we are going to consider in this chapter are the following:

- For $\beta > 0$, the fractional integral or Riesz potential of order β, I_β, with respect to the Gaussian measure, is defined formally as

$$I_\beta = (-L)^{-\beta/2} \Pi_0, \tag{14}$$

where, $\Pi_0 f = f - \int_{\mathbb{R}^d} f(y)\gamma_d(dy)$, for $f \in L^2(\gamma_d)$. That means that for the Hermite polynomials $\{h_\nu\}$, for $|\nu| > 0$,

$$I_\beta h_\nu(x) = \frac{1}{|\nu|^{\beta/2}} h_\nu(x), \tag{15}$$

and for $\nu = 0$, $I_\beta(h_0) = 0$. Then by linearity can be extended to any polynomial.
Now, it is easy to see that if f is a polynomial,

$$I_\beta f(x) = \frac{1}{\Gamma(\beta)} \int_0^\infty t^{\beta-1}(P_t f(x) - P_\infty f(x))\, dt. \tag{16}$$

Moreover, by P. A. Meyer's multiplier theorem, see [4] or [11], I_β admits a continuous extension to $L^p(\gamma_d)$, $1 < p < \infty$, and then Eq. (16) can be extended for $f \in L^p(\gamma_d)$.

- The Bessel potential of order $\beta > 0$, $\mathcal{J}_\beta$, associated to the Gaussian measure is defined formally as

$$\mathcal{J}_\beta = (I + \sqrt{-L})^{-\beta}, \tag{17}$$

meaning that for the Hermite polynomials, we have

$$\mathcal{J}_\beta h_\nu(x) = \frac{1}{(1 + \sqrt{|\nu|})^\beta} h_\nu(x).$$

Again, by linearity can be extended to any polynomial, and Meyer's theorem allows us to extend Bessel potentials to a continuous operator on $L^p(\gamma_d)$, $1 < p < \infty$. It can be proved that the Bessel potentials can be represented as

$$\mathcal{J}_\beta f(x) = \frac{1}{\Gamma(\beta)} \int_0^{+\infty} t^\beta e^{-t} P_t f(x) \frac{dt}{t}. \tag{18}$$

- The Riesz fractional derivate of order $\beta > 0$ with respect to the Gaussian measure, D^β, is defined formally as

$$D^\beta = (-L)^{\beta/2}, \tag{19}$$

meaning that for the Hermite polynomials, we have

$$D^\beta h_\nu(x) = |\nu|^{\beta/2} h_\nu(x), \tag{20}$$

thus by linearity can be extended to any polynomial.

The Riesz fractional derivate D^β with respect to the Gaussian measure was first introduced in [3]. For more details we refer to that article. Also, see [5] for improved and simpler proofs of some results contained there. In the case of $0 < \beta < 1$, we have the following integral representation:

$$D^\beta f = \frac{1}{c_\beta} \int_0^\infty t^{-\beta-1}(P_t - I) f dt, \tag{21}$$

where $c_\beta = \int_0^\infty u^{-\beta-1}(e^{-u} - 1)du$. Moreover, for $f \in C_B^2(\mathbb{R}^d)$, i.e., the set of two times continuously differentiable functions with bounded derivatives, then it can be proved using integration by parts (for details, see [3]) that

$$D^\beta f = \frac{1}{\beta c_\beta} \int_0^\infty t^{-\beta} \frac{\partial}{\partial t} P_t f dt. \tag{22}$$

Moreover, if $\beta \geq 1$, let k be the smallest integer greater than β, i.e., $k-1 \leq \beta < k$, then the fractional derivative D^β can be represented as

$$D^\beta f = \frac{1}{c_\beta^k} \int_0^\infty t^{-\beta-1}(P_t - I)^k f dt, \tag{23}$$

where $c_\beta^k = \int_0^\infty u^{-\beta-1}(e^{-u} - 1)^k du$, see [7]. Now, if f is a polynomial, by the linearity of the operators I_β and D^β, Eqs. (15) and (20), we get

$$\Pi_0 f = I_\beta(D^\beta f) = D^\beta(I_\beta f). \tag{24}$$

- We can also consider a Bessel fractional derivative $\mathcal{D}^\beta$, defined formally as

$$\mathcal{D}^\beta = (I + \sqrt{-L})^\beta,$$

which means that for the Hermite polynomials, we have

$$\mathcal{D}^\beta h_\nu(x) = (1 + \sqrt{|\nu|}))^\beta h_\nu(x). \tag{25}$$

In the case of $0 < \beta < 1$ we have the following integral representation:

$$\mathcal{D}^\beta f = \frac{1}{c_\beta} \int_0^\infty t^{-\beta-1}(e^{-t} P_t - I) f dt, \tag{26}$$

where, as before, $c_\beta = \int_0^\infty u^{-\beta-1}(e^{-u} - 1)du$. Moreover, if $\beta > 1$, let k be the smallest integer greater than β, i.e., $k-1 \leq \beta < k$, then the fractional derivative $\mathcal{D}^\beta$ can be represented as

$$\mathcal{D}^\beta f = \frac{1}{c_\beta^k} \int_0^\infty t^{-\beta-1}(e^{-t} P_t - I)^k f dt, \tag{27}$$

where $c_\beta^k = \int_0^\infty u^{-\beta-1}(e^{-u} - 1)^k du$, see [7].

The Gaussian Besov-Lipschitz $B^{\alpha}_{p,q}(\gamma_d)$ spaces were introduced in [6], see also [5], as follows.

Definition 1. Let $\alpha \geq 0$, k be the smallest integer greater than α, and $1 \leq p,q \leq \infty$. For $1 \leq q < \infty$ the Gaussian Besov-Lipschitz space $B^{\alpha}_{p,q}(\gamma_d)$ are defined as the set of functions $f \in L^p(\gamma_d)$ for which

$$\left(\int_0^{\infty} (t^{k-\alpha} \left\| \frac{\partial^k P_t f}{\partial t^k} \right\|_{p,\gamma_d})^q \frac{dt}{t} \right)^{1/q} < \infty. \tag{28}$$

The norm of $f \in B^{\alpha}_{p,q}(\gamma_d)$ is defined as

$$\|f\|_{B^{\alpha}_{p,q}} := \|f\|_{p,\gamma_d} + \left(\int_0^{\infty} (t^{k-\alpha} \left\| \frac{\partial^k P_t f}{\partial t^k} \right\|_{p,\gamma_d})^q \frac{dt}{t} \right)^{1/q}. \tag{29}$$

For $q = \infty$ the Gaussian Besov-Lipschitz space $B^{\alpha}_{p,\infty}(\gamma_d)$ is defined as the set of functions $f \in L^p(\gamma_d)$ for which exists a constant A such that

$$\| \frac{\partial^k P_t f}{\partial t^k} \|_{p,\gamma_d} \leq A t^{-k+\alpha},$$

and then the norm of $f \in B^{\alpha}_{p,\infty}(\gamma_d)$ is defined as

$$\|f\|_{B^{\alpha}_{p,\infty}} := \|f\|_{p,\gamma_d} + A_k(f), \tag{30}$$

where $A_k(f)$ is the smallest constant A appearing in the above inequality. In particular, the space $B^{\alpha}_{\infty,\infty}(\gamma_d)$ is the Gaussian Lipschitz space $Lip_{\alpha}(\gamma_d)$.

The definition of $B^{\alpha}_{p,q}(\gamma_d)$ does not depend on which $k > \alpha$ is chosen and the resulting norms are equivalent; for the proof of this result and other properties of these spaces, see [6].

In what follows, we need the following technical result about $L^p(\gamma_d)$-norms of the derivatives of the Poisson–Hermite semigroup, see [6], Lemma 2.2.

Lemma 1. *Suppose $f \in L^p(\gamma_d)$, $1 \leq p < \infty$; then for any integer k, the function $\| \frac{\partial^k P_t f}{\partial t^k} \|_{p,\gamma_d}$ is a non-increasing function of t, for $0 < t < +\infty$. Moreover,*

$$\| \frac{\partial^k P_t f}{\partial t^k} \|_{p,\gamma_d} \leq C \|f\|_{p,\gamma_d} t^{-k}, t > 0. \tag{31}$$

Also, we will need some inclusion relations among the Gaussian Besov-Lipschitz spaces, see [6].

Proposition 1. *The inclusion $B^{\alpha_1}_{p,q_1}(\gamma_d) \subset B^{\alpha_2}_{p,q_2}(\gamma_d)$ holds if either:*

(i) $\alpha_1 > \alpha_2 > 0$ *where q_1 and q_2 need not to be related.*
(ii) If $\alpha_1 = \alpha_2$ *and* $q_1 \leq q_2$.

In [2] Gaussian Lipschitz spaces $Lip_\alpha(\gamma_d)$ were considered, and the boundedness of Riesz potentials, Bessel potentials, and fractional derivatives on them were studied. In the next section, we are going to extend those results for Gaussian Besov-Lipschitz spaces, but not including them. Thus, the main purpose of this chapter is to study the boundedness of Gaussian fractional integrals and derivatives associated to Hermite polynomial expansions on Gaussian Besov-Lipschitz spaces $B^{\alpha}_{p,q}(\gamma_d)$. To get these results, we introduce formulas for these operators in terms of the Hermite-Poisson semigroup as well as the Gaussian Besov-Lipschitz spaces. This approach was originally developed for the classical Poisson integral, see Stein [8], Chap. V Sect. 5. These proofs can also be extended to the case of Laguerre and Jacobi expansions. These results can be also obtained using abstract interpolation theory on the Poisson–Hermite semigroup, see [9].

As usual in what follows, C represents a constant that is not necessarily the same in each occurrence.

2 Main Results and Proofs

In the case of the Lipschitz spaces only a truncated version of the Riesz potentials is bounded from $Lip_\alpha(\gamma_d)$ to $Lip_{\alpha+\beta}(\gamma_d)$, see [2] Theorem 3.2. Now, we will study the boundedness properties of the Riesz potentials on Besov-Lipschitz spaces, and we will see that in this case the results are actually better.

In what follows we will need Hardy's inequalities, so for completeness we will write them here, see [8] p. 272,

$$\int_0^{+\infty} \left(\int_0^x f(y)dy \right)^p x^{-r-1} dx \leq \frac{p}{r} \int_0^{+\infty} (yf(y))^p y^{-r-1} dy, \tag{32}$$

and

$$\int_0^{+\infty} \left(\int_x^{\infty} f(y)dy \right)^p x^{r-1} dx \leq \frac{p}{r} \int_0^{+\infty} (yf(y))^p y^{r-1} dy, \tag{33}$$

where $f \geq 0, p \geq 1$ and $r > 0$.

Theorem 1. *Let* $\alpha \geq 0, \beta > 0,\ 1 < p < \infty$, *and* $1 \leq q \leq \infty$ *then* I_β *is bounded from* $B^{\alpha}_{p,q}(\gamma_d)$ *into* $B^{\alpha+\beta}_{p,q}(\gamma_d)$.

Proof. Let $k > \alpha+\beta$ a fixed integer, $f \in B^{\alpha}_{p,q}(\gamma_d)$, using the integral representation of Riesz potentials (16), the semigroup property of $\{P_t\}$, and the fact that $P_\infty f(x)$ is a constant and the semigroup is conservative, we get

$$P_t I_\beta f(x) = \frac{1}{\Gamma(\beta)} \int_0^{+\infty} s^{\beta-1} P_t(P_s f(x) - P_\infty f(x)) ds$$
$$= \frac{1}{\Gamma(\beta)} \int_0^{+\infty} s^{\beta-1} (P_{t+s} f(x) - P_\infty f(x)) ds. \tag{34}$$

Now, using again that $P_\infty f(x)$ is a constant and the chain rule,

$$\frac{\partial^k}{\partial t^k}(P_t I_\beta f)(x) = \frac{1}{\Gamma(\beta)} \int_0^{+\infty} s^{\beta-1} \frac{\partial^k}{\partial t^k}(P_{t+s} f(x) - P_\infty f(x))\, ds$$
$$= \frac{1}{\Gamma(\beta)} \int_0^{+\infty} s^{\beta-1} u^{(k)}(x, t+s)\, ds. \tag{35}$$

Now, by Minkowski's integral inequality,

$$\|\frac{\partial^k}{\partial t^k} P_t I_\beta f\|_{p,\gamma} \leq \frac{1}{\Gamma(\beta)} \int_0^{+\infty} s^{\beta-1} \|u^{(k)}(\cdot, t+s)\|_{p,\gamma} ds. \tag{36}$$

Then, if $1 \leq q < \infty$,

$$\left(\int_0^{+\infty} \left(t^{k-(\alpha+\beta)} \|\frac{\partial^k}{\partial t^k}(P_t I_\beta f)\|_{p,\gamma}\right)^q \frac{dt}{t}\right)^{\frac{1}{q}}$$
$$\leq \frac{1}{\Gamma(\beta)} \left(\int_0^{+\infty} t^{(k-(\alpha+\beta))q} \left(\int_0^{+\infty} s^{\beta-1} \|u^{(k)}(\cdot, t+s)\|_{p,\gamma} ds\right)^q \frac{dt}{t}\right)^{\frac{1}{q}}$$
$$\leq C_\beta \left(\int_0^{+\infty} t^{(k-(\alpha+\beta))q} \left(\int_0^{t} s^{\beta-1} \|u^{(k)}(\cdot, t+s)\|_{p,\gamma} ds\right)^q \frac{dt}{t}\right)^{\frac{1}{q}}$$
$$+ C_\beta \left(\int_0^{+\infty} t^{(k-(\alpha+\beta))q} \left(\int_t^{+\infty} s^{\beta-1} \|u^{(k)}(\cdot, t+s)\|_{p,\gamma} ds\right)^q \frac{dt}{t}\right)^{\frac{1}{q}}$$
$$= (I) + (II).$$

Now, as $\beta > 0$ using Lemma 1, as $t+s > t$,

$$(I) \leq C_\beta \left(\int_0^{+\infty} t^{(k-(\alpha+\beta))q} \left(\int_0^{t} s^{\beta-1} \|u^{(k)}(\cdot, t)\|_{p,\gamma} ds\right)^q \frac{dt}{t}\right)^{\frac{1}{q}}$$
$$= C_\beta \left(\int_0^{+\infty} t^{(k-(\alpha+\beta))q} \|\frac{\partial^k P_t f}{\partial t^k}\|^q_{p,\gamma} (\frac{t^\beta}{\beta})^q \frac{dt}{t}\right)^{\frac{1}{q}}$$
$$= C'_\beta \left(\int_0^{+\infty} \left(t^{k-\alpha} \|\frac{\partial^k P_t f}{\partial t^k}\|_{p,\gamma}\right)^q \frac{dt}{t}\right)^{\frac{1}{q}} < +\infty$$

since $f \in B_p^{\alpha,q}(\gamma_d)$.

On the other hand, as $k > \alpha + \beta$ using again Lemma 1, since $t + s > s$, and Hardy's inequality (33), we obtain

$$
\begin{aligned}
(II) &\leq C_\beta \left(\int_0^{+\infty} t^{(k-(\alpha+\beta))q} \left(\int_t^{+\infty} s^\beta \|u^{(k)}(\cdot,s)\|_{p,\gamma} \frac{ds}{s} \right)^q \frac{dt}{t} \right)^{\frac{1}{q}} \\
&\leq \frac{C_\beta}{k-(\alpha+\beta)} \left(\int_0^{+\infty} \left(s^{k-\alpha} \| \frac{\partial^k P_s f}{\partial s^k} \|_{p,\gamma} \right) \right)^q \frac{ds}{s}^{\frac{1}{q}} < +\infty
\end{aligned}
$$

since $f \in B^{\alpha}_{p,q}(\gamma_d)$. Therefore, $I_\beta f \in B^{\alpha+\beta}_{p,q}(\gamma_d)$ and moreover

$$
\begin{aligned}
\|I_\beta f\|_{B^{\alpha+\beta}_{p,q}} &= \|I_\beta f\|_{p,\gamma} + \left(\int_0^{+\infty} \left(t^{k-(\alpha+\beta)} \| \frac{\partial^k}{\partial t^k} (P_t I_\beta f) \|_{p,\gamma} \right)^q \frac{dt}{t} \right)^{\frac{1}{q}} \\
&\leq C\|f\|_{p,\gamma} + C_{\alpha,\beta} \left(\int_0^{+\infty} \left(t^{k-\alpha} \| \frac{\partial^k P_t f}{\partial t^k} \|_{p,\gamma} \right)^q \frac{dt}{t} \right)^{\frac{1}{q}} \\
&\leq C\|f\|_{B^{\alpha}_{p,q}}.
\end{aligned}
$$

Now, if $q = \infty$, Eq. (36) can be written as

$$
\begin{aligned}
\| \frac{\partial^k}{\partial t^k} P_t I_\beta f \|_{p,\gamma} &\leq \frac{1}{\Gamma(\beta)} \int_0^{+\infty} s^{\beta-1} \|u^{(k)}(\cdot,t+s)\|_{p,\gamma} ds \\
&= \frac{1}{\Gamma(\beta)} \int_0^t s^{\beta-1} \|u^{(k)}(\cdot,t+s)\|_{p,\gamma} ds \\
&\quad + \frac{1}{\Gamma(\beta)} \int_t^{\infty} s^{\beta-1} \|u^{(k)}(\cdot,t+s)\|_{p,\gamma} ds \\
&= (I) + (II).
\end{aligned}
$$

Now, using that $\beta > 0$, Lemma 1, as $t + s > t$ and since $f \in B^{\alpha}_{p,\infty}(\gamma_d)$,

$$
\begin{aligned}
(I) &\leq \frac{1}{\Gamma(\beta)} \| \frac{\partial^k P_t f}{\partial t^k} \|_{p,\gamma} \int_0^t s^{\beta-1} ds \leq \frac{1}{\Gamma(\beta)} \frac{t^\beta}{\beta} A_k(f) t^{-k+\alpha} \\
&= C_\beta A_k(f) t^{-k+\alpha+\beta}.
\end{aligned}
$$

On the other hand, since $k > \alpha + \beta$, using Lemma 1, as $t + s > s$ and since $f \in B^{\alpha}_{p,\infty}(\gamma_d)$, we get

$$(II) \leq \frac{1}{\Gamma(\beta)} \int_t^\infty s^{\beta-1} \| \frac{\partial^k P_s f}{\partial s^k} \|_{p,\gamma} ds \leq \frac{A_k(f)}{\Gamma(\beta)} \int_t^\infty s^{-k+\alpha+\beta-1} ds$$
$$= \frac{A_k(f)}{\Gamma(\beta)} \frac{t^{-k+\alpha+\beta}}{k-(\alpha+\beta)} = C_{k,\alpha,\beta} t^{-k+\alpha+\beta}.$$

Therefore

$$\| \frac{\partial^k}{\partial t^k} P_t I_\beta f \|_{p,\gamma} \leq C A_k(f) t^{-k+\alpha+\beta}, \quad t > 0,$$

and this implies that $I_\beta f \in B^{\alpha+\beta}_{p,\infty}(\gamma_d)$ and $A_k(I_\beta f) \leq C A_k(f)$.

Moreover, as I_β is bounded operator on $L^p(\gamma_d), 1 < p < \infty$,

$$\|I_\beta f\|_{B^{\alpha+\beta}_{p,\infty}} = \|I_\beta f\|_{p,\gamma} + A_k(I_\beta f)$$
$$\leq \|f\|_{p,\gamma} + C A_k(f) \leq C \|f\|_{B^\alpha_{p,\infty}}. \qquad \square$$

Now, we want to study the boundedness properties of the Bessel potentials on Besov-Lipschitz spaces. In [2], Theorem 3.1, the following result was proved.

Theorem 2. *Let $\alpha \geq 0, \beta > 0$, then $\mathscr{J}_\beta$ is bounded from $Lip_\alpha(\gamma_d)$ into $Lip_{\alpha+\beta}(\gamma_d)$.*

Also in [6], Theorem 2.4, it was proved that

Theorem 3. *Let $\alpha \geq 0, \beta > 0$, then for $1 \leq p, q < \infty$, $\mathscr{J}_\beta$ is bounded from $B^\alpha_{p,q}(\gamma_d)$ into $B^{\alpha+\beta}_{p,q}(\gamma_d)$.*

Therefore the following result is the only case that was missing.

Theorem 4. *Let $\alpha \geq 0, \beta > 0$ then for $1 \leq p < \infty$ $\mathscr{J}_\beta$ is bounded from $B^\alpha_{p,\infty}(\gamma_d)$ into $B^{\alpha+\beta}_{p,\infty}(\gamma_d)$.*

Proof. Let $k > \alpha + \beta$ a fixed integer, $f \in B^\alpha_{p,\infty}(\gamma_d)$, by using the representation of Bessel potential (18), we get

$$P_t(\mathscr{J}_\beta f)(x) = \frac{1}{\Gamma(\beta)} \int_0^{+\infty} s^\beta e^{-s} P_{t+s} f(x) \frac{ds}{s};$$

thus using the chain rule, we obtain

$$\frac{\partial^k}{\partial t^k} P_t(\mathscr{J}_\beta f)(x) = \frac{1}{\Gamma(\beta)} \int_0^{+\infty} s^\beta e^{-s} u^{(k)}(x, t+s) \frac{ds}{s};$$

this implies, using Minkowski's integral inequality,

$$\begin{aligned}\|\frac{\partial^k}{\partial t^k}P_t(\mathscr{I}_\beta f)\|_{p,\gamma} &\leq \frac{1}{\Gamma(\beta)}\int_0^{+\infty} s^\beta e^{-s}\|u^{(k)}(\cdot,t+s)\|_{p,\gamma}\frac{ds}{s}\\ &= \frac{1}{\Gamma(\beta)}\int_0^{t} s^\beta e^{-s}\|u^{(k)}(\cdot,t+s)\|_{p,\gamma}\frac{ds}{s}\\ &\quad+\frac{1}{\Gamma(\beta)}\int_t^{\infty} s^\beta e^{-s}\|u^{(k)}(\cdot,t+s)\|_{p,\gamma}\frac{ds}{s}\\ &= (I)+(II).\end{aligned}$$

Now, as $\beta > 0$, using Lemma 1 (as $t+s>t$) and since $f \in B^{\alpha}_{p,\infty}(\gamma_d)$,

$$\begin{aligned}(I) &\leq \frac{1}{\Gamma(\beta)}\|\frac{\partial^k P_t f}{\partial t^k}\|_{p,\gamma}\int_0^t s^\beta e^{-s}\frac{ds}{s} \leq \frac{1}{\Gamma(\beta)}\|\frac{\partial^k P_t f}{\partial t^k}\|_{p,\gamma}\int_0^t s^{\beta-1}ds\\ &\leq \frac{1}{\Gamma(\beta)}\frac{t^\beta}{\beta}A_k(f)t^{-k+\alpha} = C_\beta A_k(f)t^{-k+\alpha+\beta}.\end{aligned}$$

On the other hand, as $k > \alpha+\beta$ using Lemma 1 as $t+s>s$ and since $f \in B^{\alpha}_{p,\infty}(\gamma_d)$,

$$\begin{aligned}(II) &\leq \frac{1}{\Gamma(\beta)}\int_t^{\infty} s^\beta e^{-s}\|\frac{\partial^k P_s f}{\partial s^k}\|_{p,\gamma}\frac{ds}{s} \leq \frac{A_k(f)}{\Gamma(\beta)}\int_t^{\infty} s^\beta e^{-s}s^{-k+\alpha}\frac{ds}{s}\\ &\leq \frac{A_k(f)}{\Gamma(\beta)}\int_t^{\infty} s^{-k+\alpha+\beta-1}ds = \frac{A_k(f)}{\Gamma(\beta)}\frac{t^{-k+\alpha+\beta}}{k-(\alpha+\beta)} = C_{k,\alpha,\beta}A_k(f)t^{-k+\alpha+\beta}.\end{aligned}$$

Therefore

$$\|\frac{\partial^k}{\partial t^k}P_t(\mathscr{I}_\beta f)\|_{p,\gamma} \leq CA_k(f)t^{-k+\alpha+\beta},$$

then $\mathscr{I}_\beta f \in B^{\alpha+\beta}_{p,\infty}(\gamma_d)$ and $A_k(\mathscr{I}_\beta f) \leq CA_k(f)$. Thus,

$$\begin{aligned}\|\mathscr{I}_\beta f\|_{B^{\alpha+\beta}_{p,\infty}} &= \|\mathscr{I}_\beta f\|_{p,\gamma}+A_k(\mathscr{I}_\beta f)\\ &\leq \|f\|_{p,\gamma}+CA_k(f) \leq C\|f\|_{B^{\alpha}_{p,\infty}}.\end{aligned}$$

□

Now, we will study the boundedness of the (Riesz) fractional derivative D^β on Besov-Lipschitz spaces. We will use the representation (20) of the fractional derivative and Hardy's inequalities.

Theorem 5. *Let $0 < \beta < \alpha < 1$, $1 \le p < \infty$ and $1 \le q \le \infty$ then D^β is bounded from $B^\alpha_{p,q}(\gamma_d)$ into $B^{\alpha-\beta}_{p,q}(\gamma_d)$.*

Proof. Let $f \in B^\alpha_{p,q}(\gamma_d)$, using Hardy's inequality (32), with $p = 1$, and the Fundamental Theorem of Calculus,

$$\begin{aligned} |D^\beta f(x)| &\le \frac{1}{c_\beta}\int_0^{+\infty} s^{-\beta-1}|P_s f(x) - f(x)|ds \\ &\le \frac{1}{c_\beta}\int_0^{+\infty} s^{-\beta-1}\int_0^s |\frac{\partial}{\partial r}P_r f(x)|drds \\ &\le \frac{1}{c_\beta \beta}\int_0^{+\infty} r^{1-\beta}|\frac{\partial}{\partial r}P_r f(x)|\frac{dr}{r}. \end{aligned} \tag{37}$$

Thus, using Minkowski's integral inequality,

$$\|D^\beta f\|_{p,\gamma} \le C_\beta \int_0^{+\infty} r^{1-\beta}\|\frac{\partial}{\partial r}P_r f\|_{p,\gamma}\frac{dr}{r} < \infty, \tag{38}$$

since $B^\alpha_{p,q}(\gamma_d) \subset B^\beta_{p,1}(\gamma_d)$, $1 \le q \le \infty$ as $\alpha > \beta$, i.e., $D_\beta f \in L^p(\gamma_d)$.

Now, by analogous argument,

$$\begin{aligned} \frac{\partial}{\partial t}P_t(D^\beta f)(x) &= \frac{1}{c_\beta}\int_0^{+\infty} s^{-\beta-1}[\frac{\partial}{\partial t}P_{t+s}f(x) - \frac{\partial}{\partial t}P_t f(x)]ds \\ &= \frac{1}{c_\beta}\int_0^{+\infty} s^{-\beta-1}\int_t^{t+s} u^{(2)}(x,r)drds, \end{aligned}$$

and again, by Minkowski's integral inequality,

$$\|\frac{\partial}{\partial t}P_t(D^\beta f)\|_p \le \frac{1}{c_\beta}\int_0^{+\infty} s^{-\beta-1}\int_t^{t+s}\|u^{(2)}(\cdot,r)\|_p drds. \tag{39}$$

Then, if $1 \le q < \infty$, by Eq. (39)

$$\begin{aligned} &\int_0^\infty \left(t^{1-(\alpha-\beta)}\|\frac{\partial}{\partial t}P_t(D_\beta f)\|_{p,\gamma}\right)^q \frac{dt}{t} \\ &\le C_\beta \int_0^\infty \left(t^{1-(\alpha-\beta)}\int_0^{+\infty} s^{-\beta-1}\int_t^{t+s}\|u^{(2)}(\cdot,r)\|_{p,\gamma}drds\right)^q \frac{dt}{t} \\ &= C_\beta \int_0^\infty \left(t^{1-(\alpha-\beta)}\int_0^{t} s^{-\beta-1}\int_t^{t+s}\|u^{(2)}(\cdot,r)\|_{p}drds\right)^q \frac{dt}{t} \\ &+C_\beta \int_0^\infty \left(t^{1-(\alpha-\beta)}\int_t^{+\infty} s^{-\beta-1}\int_t^{t+s}\|u^{(2)}(\cdot,r)\|_{p}drds\right)^q \frac{dt}{t} \\ &= (I) + (II). \end{aligned}$$

Now, since $r > t$ using Lemma 1 and the fact that $0 < \beta < 1$,

$$(I) \leq C_\beta \int_0^\infty \left(t^{1-(\alpha-\beta)} \int_0^t s^{-\beta} ds \, \|u^{(2)}(\cdot,r)\|_{p,\gamma} \right)^q \frac{dt}{t}$$
$$= C_{\beta,q} \int_0^\infty \left(t^{2-\alpha} \| \frac{\partial^2}{\partial r^2} P_r f\|_{p,\gamma} \right)^q \frac{dt}{t}.$$

On the other hand, as $r > t$ using Hardy's inequality (33), since $(1-\alpha)q > 0$, we get

$$(II) \leq C_\beta \int_0^\infty t^{(1-(\alpha-\beta))q} \left(\int_t^{+\infty} s^{-\beta-1} ds \int_t^\infty \|u^{(2)}(\cdot,r)\|_{p,\gamma} dr \right)^q \frac{dt}{t}$$
$$= C'_\beta \int_0^\infty t^{(1-\alpha)q} \left(\int_t^\infty \|u^{(2)}(\cdot,r)\|_{p,\gamma} dr \right)^q \frac{dt}{t}$$
$$\leq \frac{C'_\beta}{(1-\alpha)} \int_0^\infty \left(r^{2-\alpha} \| \frac{\partial^2}{\partial r^2} P_r f\|_{p,\gamma} \right)^q \frac{dr}{r}.$$

Thus,

$$\left(\int_0^\infty \left(t^{1-\alpha+\beta} \| \frac{\partial}{\partial t} P_t D_\beta f\|_{p,\gamma} \right)^q \frac{dt}{t} \right)^{1/q} \leq C \left(\int_0^\infty \left(t^{2-\alpha} \| \frac{\partial^2}{\partial t^2} P_t f\|_{p,\gamma} \right)^q \frac{dt}{t} \right)^{1/q} < \infty,$$

as $B^\alpha_{p,q}(\gamma_d)$. Then, $D_\beta f \in B^{\alpha-\beta}_{p,q}(\gamma_d)$ and

$$\|D_\beta f\|_{B^{\alpha-\beta}_{p,q}} = \|D_\beta f\|_{p,\gamma} + \left(\int_0^\infty \left(t^{1-\alpha+\beta} \| \frac{\partial}{\partial t} P_t D_\beta f\|_{p,\gamma} \right)^q \frac{dt}{t} \right)^{1/q}$$
$$\leq C_1 \|f\|_{B^\alpha_{p,q}} + C_2 \left(\int_0^\infty \left(t^{2-\alpha} \| \frac{\partial^2}{\partial t^2} P_t f\|_{p,\gamma} \right)^q \frac{dt}{t} \right)^{1/q}$$
$$\leq C \|f\|_{B^\alpha_{p,q}}.$$

Therefore $D_\beta f : B^\alpha_{p,q} \to B^{\alpha-\beta}_{p,q}$ is bounded.
Now, if $q = \infty$, inequality (39) can be written as

$$\| \frac{\partial}{\partial t} P_t (D_\beta f)\|_{p,\gamma} \leq \frac{1}{c_\beta} \int_0^{+\infty} s^{-\beta-1} \int_t^{t+s} \| \frac{\partial^2}{\partial r^2} P_r f\|_p dr ds$$
$$= \frac{1}{c_\beta} \int_0^t s^{-\beta-1} \int_t^{t+s} \| \frac{\partial^2}{\partial r^2} P_r f\|_p dr ds$$
$$+ \frac{1}{c_\beta} \int_t^{+\infty} s^{-\beta-1} \int_t^{t+s} \| \frac{\partial^2}{\partial r^2} P_r f\|_p dr ds = (I) + (II).$$

Now, by Lemma 1, since $r > t$,

$$(I) \leq \frac{1}{c_\beta}\int_0^t s^{-\beta}\|\frac{\partial^2}{\partial t^2}P_t f\|_p ds = C_\beta\|\frac{\partial^2}{\partial t^2}P_t f\|_{p,\gamma}t^{1-\beta}$$
$$\leq C_\beta A(f)t^{-2+\alpha}t^{1-\beta} = C_\beta A(f)t^{-1+\alpha-\beta},$$

and by Lemma 1, since $r > t$, and the fact that $f \in B^\alpha_{p,\infty}$,

$$(II) \leq \frac{1}{c_\beta}\int_t^{+\infty} s^{-\beta-1}\int_t^\infty \|\frac{\partial^2}{\partial r^2}P_r f\|_p dr ds$$
$$\leq C_\beta t^{-\beta}\int_t^\infty \|\frac{\partial^2}{\partial r^2}P_r f\|_{p,\gamma}dr \leq C_\beta A(f)t^{-\beta}\int_t^\infty r^{-2+\alpha}dr$$
$$= C_{\alpha,\beta}A(f)t^{-1+\alpha-\beta}.$$

Thus,

$$\|\frac{\partial}{\partial t}P_t(D_\beta f)\|_{p,\gamma} \leq CA(f)t^{-1+\alpha-\beta}, \quad t > 0,$$

i.e., $D_\beta f \in B^{\alpha-\beta}_{p,\infty}(\gamma_d)$ then $A(D_\beta f) \leq CA(f)$, and

$$\|D_\beta f\|_{B^{\alpha-\beta}_{p,\infty}} = \|D_\beta f\|_{p,\gamma} + A(D_\beta f)$$
$$\leq C_1\|f\|_{B^\alpha_{p,\infty}} + C_2 A(f) \leq C\|f\|_{B^\alpha_{p,\infty}}.$$

Therefore $D_\beta : B^\alpha_{p,\infty} \to B^{\alpha-\beta}_{p,\infty}$ is bounded. □

Now, we will study the boundedness of the Bessel fractional derivative on Besov-Lipschitz spaces, for $0 < \beta < \alpha < 1$.

Theorem 6. *Let $0 < \beta < \alpha < 1$, $1 \leq p < \infty$, and $1 \leq q \leq \infty$ then $\mathscr{D}_\beta$ is bounded from $B^\alpha_{p,q}(\gamma_d)$ into $B^{\alpha-\beta}_{p,q}(\gamma_d)$.*

Proof. Let $f \in L^p(\gamma_d)$, and using the Fundamental Theorem of Calculus, we can write,

$$|\mathscr{D}_\beta f(x)| \leq \frac{1}{c_\beta}\int_0^{+\infty} s^{-\beta-1}|e^{-s}P_s f(x) - f(x)|ds$$
$$\leq \frac{1}{c_\beta}\int_0^{+\infty} s^{-\beta-1}e^{-s}|P_s f(x) - f(x)|ds + \frac{1}{c_\beta}\int_0^{+\infty} s^{-\beta-1}|e^{-s}-1||f(x)|ds$$
$$\leq \frac{1}{c_\beta}\int_0^{+\infty} s^{-\beta-1}|\int_0^s \frac{\partial}{\partial r}P_r f(x)dr|\,ds + \frac{1}{c_\beta}\int_0^{+\infty} s^{-\beta-1}|e^{-s}-1||f(x)|ds$$
$$\leq \frac{1}{c_\beta}\int_0^{+\infty} s^{-\beta-1}\int_0^s |\frac{\partial}{\partial r}P_r f(x)|dr ds + \frac{1}{c_\beta}\int_0^{+\infty} s^{-\beta-1}|e^{-s}-1||f(x)|ds$$

$$= \frac{1}{c_\beta}\int_0^{+\infty} s^{-\beta-1}\int_0^s |\frac{\partial}{\partial r}P_r f(x)|dr\,ds + |f(x)|\frac{1}{c_\beta}\int_0^{+\infty} s^{-\beta-1}|-\int_0^s e^{-r}dr|ds$$
$$= \frac{1}{c_\beta}\int_0^{+\infty} s^{-\beta-1}\int_0^s |\frac{\partial}{\partial r}P_r f(x)|dr\,ds + \frac{1}{c_\beta}|f(x)|\int_0^{+\infty} s^{-\beta-1}\int_0^s e^{-r}dr\,ds.$$

Now, using Hardy's inequality (32) with $p=1$, in both integrals, we have

$$|\mathscr{D}_\beta f(x)| \le \frac{1}{c_\beta}\int_0^{+\infty} s^{-\beta-1}\int_0^s |\frac{\partial}{\partial r}P_r f(x)|dr\,ds + \frac{1}{c_\beta}|f(x)|\int_0^{+\infty} s^{-\beta-1}\int_0^s e^{-r}dr\,ds$$
$$\le \frac{1}{\beta c_\beta}\int_0^{+\infty} r|\frac{\partial}{\partial r}P_r f(x)|r^{-\beta-1}dr + \frac{1}{\beta c_\beta}|f(x)|\int_0^{+\infty} re^{-r}r^{-\beta-1}dr$$
$$= \frac{1}{\beta c_\beta}\int_0^{+\infty} r^{1-\beta}|\frac{\partial}{\partial r}P_r f(x)|\frac{dr}{r} + \frac{1}{\beta c_\beta}|f(x)|\int_0^{+\infty} r^{(1-\beta)-1}e^{-r}dr$$
$$= \frac{1}{\beta c_\beta}\int_0^{+\infty} r^{1-\beta}|\frac{\partial}{\partial r}P_r f(x)|\frac{dr}{r} + \frac{1}{\beta c_\beta}\Gamma(1-\beta)|f(x)|.$$

Therefore using the Minkowski's integral inequality

$$\|\mathscr{D}_\beta f\|_p \le \frac{1}{\beta c_\beta}\int_0^{+\infty} r^{1-\beta}\|\frac{\partial}{\partial r}P_r f\|_p\frac{dr}{r} + \frac{1}{\beta c_\beta}\Gamma(1-\beta)\|f\|_p < C_1\|f\|_{B^\alpha_{p,q}} < \infty,$$

since $f \in B^\alpha_{p,q}(\gamma_d) \subset B^\beta_{p,1}(\gamma_d)$, $1 \le q \le \infty$ as $\alpha > \beta$, i.e., $D_\beta f \in L^p(\gamma_d)$.

On the other hand, using the Fundamental Theorem of Calculus and using Hardy's inequality (32) with $p=1$, in the second integral, we have

$$|\frac{\partial}{\partial t}P_t(\mathscr{D}_\beta f)(x)| \le \frac{1}{c_\beta}\int_0^\infty s^{-\beta-1}|e^{-s}\frac{\partial}{\partial t}P_{t+s}f(x) - \frac{\partial}{\partial t}P_t f(x)|ds$$
$$\le \frac{1}{c_\beta}\int_0^\infty s^{-\beta-1}e^{-s}|\frac{\partial}{\partial t}P_{t+s}f(x) - \frac{\partial}{\partial t}P_t f(x)|ds$$
$$+\frac{1}{c_\beta}\int_0^\infty s^{-\beta-1}|e^{-s}-1||\frac{\partial}{\partial t}P_t f(x)|ds$$
$$\le \frac{1}{c_\beta}\int_0^\infty s^{-\beta-1}\int_t^{t+s}|\frac{\partial^2}{\partial r^2}P_r f(x)|dr\,ds$$
$$+\frac{1}{c_\beta}|\frac{\partial}{\partial t}P_t f(x)|\int_0^\infty s^{-\beta-1}\int_0^s e^{-r}dr\,ds,$$

$$\leq \frac{1}{c_\beta}\int_0^\infty s^{-\beta-1}\int_t^{t+s}|\frac{\partial^2}{\partial r^2}P_rf(x)|drds$$
$$+\frac{1}{\beta c_\beta}|\frac{\partial}{\partial t}P_tf(x)|\int_0^\infty r^{(1-\beta)-1}e^{-r}dr$$
$$=\frac{1}{c_\beta}\int_0^\infty s^{-\beta-1}\int_t^{t+s}|\frac{\partial^2}{\partial r^2}P_rf(x)|drds+\frac{\Gamma(1-\beta)}{\beta c_\beta}|\frac{\partial}{\partial t}P_tf(x)|.$$

Therefore by Minkowski's integral inequality

$$\|\frac{\partial}{\partial t}P_t(\mathscr{D}_\beta f)\|_{p,\gamma}\leq \frac{1}{c_\beta}\int_0^\infty s^{-\beta-1}\int_t^{t+s}\|\frac{\partial^2}{\partial r^2}P_rf\|_{p,\gamma}drds+\frac{\Gamma(1-\beta)}{\beta c_\beta}\|\frac{\partial}{\partial t}P_tf\|_{p,\gamma}. \tag{40}$$

Then, if $1\leq q<\infty$, we get

$$\left(\int_0^\infty\left(t^{1-(\alpha-\beta)}\|\frac{\partial}{\partial t}P_t\mathscr{D}_\beta f\|_{p,\gamma}\right)^q\frac{dt}{t}\right)^{1/q}$$
$$\leq\frac{1}{c_\beta}\left(\int_0^\infty\left(t^{1-(\alpha-\beta)}\int_0^\infty s^{-\beta-1}\int_t^{t+s}\|\frac{\partial^2}{\partial r^2}P_rf\|_{p,\gamma}drds\right)^q\frac{dt}{t}\right)^{1/q}$$
$$+\frac{\Gamma(1-\beta)}{\beta c_\beta}\left(\int_0^\infty\left(t^{1-(\alpha-\beta)}\|\frac{\partial}{\partial t}P_tf\|_{p,\gamma}\right)^q\frac{dt}{t}\right)^{1/q}$$
$$=(I)+(II).$$

Now, the first term is the same as the one considered in the second part of the proof of Theorem 5 thus by the same argument

$$(I)\leq C_\beta\left(\int_0^\infty\left(t^{2-\alpha}\|\frac{\partial^2}{\partial t^2}P_tf\|_{p,\gamma}\right)^q\frac{dt}{t}\right)^{1/q}<\|f\|_{B^\alpha_{p,q}}<\infty,$$

since $f\in B^\alpha_{p,q}(\gamma_d)$, and for the second term, trivially

$$(II)\leq C\|f\|_{B^{\alpha-\beta}_{p,q}}\leq C\|f\|_{B^\alpha_{p,q}}$$

since $\alpha>\alpha-\beta$ and the inclusion relation, Proposition 1.

Thus, if $1\leq q<\infty$,

$$\left(\int_0^\infty\left(t^{1-(\alpha-\beta)}\|\frac{\partial}{\partial t}P_t(\mathscr{D}_\beta f)\|_{p,\gamma}\right)^q\frac{dt}{t}\right)^{1/q}\leq C_2\|f\|_{B^\alpha_{p,q}},$$

i.e., $\mathscr{D}_\beta f \in B^{\alpha-\beta}_{p,q}(\gamma_d)$ and moreover

$$\begin{aligned}\|\mathscr{D}_\beta f\|_{B^{\alpha-\beta}_{p,q}} &= \|\mathscr{D}_\beta f\|_{p,\gamma} + \left(\int_0^\infty \left(t^{1-\alpha+\beta}\|\frac{\partial}{\partial t}P_t\mathscr{D}_\beta f\|_{p,\gamma}\right)^q \frac{dt}{t}\right)^{1/q}\\ &\leq C_1\|f\|_{B^\alpha_{p,q}} + C_2\left(\int_0^\infty \left(t^{2-\alpha}\|\frac{\partial^2}{\partial t^2}P_t f\|_{p,\gamma}\right)^q \frac{dt}{t}\right)^{1/q}\\ &\leq C\|f\|_{B^\alpha_{p,q}}.\end{aligned}$$

If $q = \infty$, using the same argument as in Theorem 5, inequality (40) can be written as

$$\begin{aligned}\|\frac{\partial}{\partial t}P_t\mathscr{D}_\beta f\|_{p,\gamma} &\leq \frac{1}{c_\beta}\int_0^\infty s^{-\beta-1}\int_t^{t+s}\|\frac{\partial^2}{\partial r^2}P_r f\|_{p,\gamma}dr\,ds + \frac{\Gamma(1-\beta)}{\beta c_\beta}\|\frac{\partial}{\partial t}P_t f\|_{p,\gamma}\\ &\leq C_{\alpha,\beta}A(f)t^{-1+\alpha-\beta} + \frac{\Gamma(1-\beta)}{\beta c_\beta}A(f)t^{-1+\alpha-\beta}\\ &\leq C_{\alpha,\beta}A(f)t^{-1+\alpha-\beta}, \quad t>0,\end{aligned}$$

i.e., $\mathscr{D}_\beta f \in B^{\alpha-\beta}_{p,\infty}(\gamma_d)$ and $A(\mathscr{D}_\beta f) \leq C_{\alpha,\beta}A(f)$, thus

$$\begin{aligned}\|\mathscr{D}_\beta f\|_{B^{\alpha-\beta}_{p,\infty}} &= \|\mathscr{D}_\beta f\|_{p,\gamma} + A(\mathscr{D}_\beta f)\\ &\leq C_1\|f\|_{B^\alpha_{p,\infty}} + C_2A(f) \leq C\|f\|_{B^\alpha_{p,\infty}}.\end{aligned}$$

□

Now, we will consider the general case for fractional derivatives, removing the condition that the indexes must be less than 1. We need to consider forward differences. Remember for a given function f, the k-th order forward difference of f starting at t with increment s is defined as,

$$\Delta^k_s(f,t) = \sum_{j=0}^k \binom{k}{j}(-1)^j f(t+(k-j)s).$$

The forward differences have the following properties (see Appendix in [2]), and we will need the following technical result.

Lemma 2. *For any positive integer k:*

(i) $\Delta^k_s(f,t) = \Delta^{k-1}_s(\Delta_s(f,\cdot),t) = \Delta_s(\Delta^{k-1}_s(f,\cdot),t)$.

(ii) $\Delta^k_s(f,t) = \int_t^{t+s}\int_{v_1}^{v_1+s}\cdots\int_{v_{k-2}}^{v_{k-2}+s}\int_{v_{k-1}}^{v_{k-1}+s} f^{(k)}(v_k)dv_k dv_{k-1}\ldots dv_2 dv_1$. *For any positive integer k,*

$$\frac{\partial}{\partial s}(\Delta^k_s(f,t)) = k\Delta^{k-1}_s(f',t+s), \tag{41}$$

and for any integer $j > 0$,

$$\frac{\partial^j}{\partial t^j}(\Delta_s^k(f,t)) = \Delta_s^k(f^{(j)},t). \tag{42}$$

Observe that, using the Binomial Theorem and the semigroup property of $\{P_t\}$, we have

$$\begin{aligned}(P_t - I)^k f(x) &= \sum_{j=0}^{k} \binom{k}{j} P_t^{k-j}(-I)^j f(x) = \sum_{j=0}^{k} \binom{k}{j} (-1)^j P_t^{k-j} f(x)\\ &= \sum_{j=0}^{k} \binom{k}{j} (-1)^j P_{(k-j)t} f(x) = \sum_{j=0}^{k} \binom{k}{j} (-1)^j u(x,(k-j)t)\\ &= \Delta_t^k(u(x,\cdot),0),\end{aligned} \tag{43}$$

where as usual, $u(x,t) = P_t f(x)$.

Additionally we will need in what follows the following result.

Lemma 3. *Let $f \in L^p(\gamma_d)$, $1 \le p < \infty$, and $k,n \in \mathbb{N}$ then*

$$\|\Delta_s^k(u^{(n)},t)\|_{p,\gamma_d} \le s^k \|u^{(k+n)}(\cdot,t)\|_{p,\gamma_d}.$$

Proof. From (ii) of Lemma 2, we have

$$\Delta_s^k(u^{(n)}(x,\cdot),t) = \int_t^{t+s}\int_{v_1}^{v_1+s}\cdots\int_{v_{k-2}}^{v_{k-2}+s}\int_{v_{k-1}}^{v_{k-1}+s} u^{(k+n)}(x,v_k)\,dv_k dv_{k-1}\dots dv_2 dv_1,$$

then, using Minkowski's integral inequality k-times and Lemma 1,

$$\begin{aligned}\|\Delta_s^k(u^{(n)},t)\|_{p,\gamma_d} &\le \int_t^{t+s}\int_{v_1}^{v_1+s}\cdots\int_{v_{k-2}}^{v_{k-2}+s}\int_{v_{k-1}}^{v_{k-1}+s} \|u^{(k+n)}(\cdot,v_k)\|_{p,\gamma_d} dv_k dv_{k-1}\dots dv_2 dv_1\\ &\le s^k\|u^{(k+n)}(\cdot,t)\|_{p,\gamma_d} = s^k\|\frac{\partial^{k+n}}{\partial t^{k+n}}u(\cdot,t)\|_{p,\gamma_d}.\end{aligned}$$

□

Let us start with the case of the Riesz derivative.

Theorem 7. *Let $0 < \beta < \alpha$, $1 \le p < \infty$, and $1 \le q \le \infty$ then D^β is bounded from $B_{p,q}^{\alpha}(\gamma_d)$ into $B_{p,q}^{\alpha-\beta}(\gamma_d)$.*

Proof. Let $f \in B_{p,q}^{\alpha}(\gamma_d)$, using Eq. (43), Hardy's inequality (32) $p = 1$, the Fundamental Theorem of Calculus and (iii) of Lemma 2, we get

$$\begin{aligned}|D^\beta f(x)| &\le \frac{1}{c_\beta}\int_0^{+\infty} s^{-\beta-1}|\Delta_s^k(u(x,\cdot),0)|\,ds\\ &\le \frac{1}{c_\beta}\int_0^{+\infty} s^{-\beta-1}\int_0^s |\frac{\partial}{\partial r}\Delta_r^k(u(x,\cdot),0)|\,dr\,ds\end{aligned}$$

$$\leq \frac{1}{\beta c_\beta} \int_0^{+\infty} r^{-\beta} |\frac{\partial}{\partial r} \Delta_r^k(u(x,\cdot),0)| dr$$
$$= \frac{k}{\beta c_\beta} \int_0^{+\infty} r^{-\beta} |\Delta_r^{k-1}(u'(x,\cdot),r)| dr.$$

Now, using Minkowski's integral inequality and Lemma 3,

$$\|D_\beta f\|_{p,\gamma} \leq \frac{k}{\beta c_\beta} \int_0^{+\infty} r^{-\beta} \|\Delta_r^{k-1}(u',r)\|_{p,\gamma} dr$$
$$\leq \frac{k}{\beta c_\beta} \int_0^{+\infty} r^{k-\beta} \|\frac{\partial^k}{\partial r^k} P_r f\|_{p,\gamma} \frac{dr}{r} < \infty,$$

since $f \in B^\alpha_{p,q}(\gamma_d) \subset B^\beta_{p,1}(\gamma_d)$, as $\alpha > \beta$. Therefore $D_\beta f \in L^p(\gamma_d)$.

On the other hand,

$$P_t[(P_s - I)^k f(x)] = P_t(\Delta_s^k(u(x,\cdot),0)) = P_t(\sum_{j=0}^{k} \binom{k}{j} (-1)^j P_{(k-j)s} f(x))$$
$$= \sum_{j=0}^{k} \binom{k}{j} (-1)^j P_{t+(k-j)s} f(x) = \Delta_s^k(u(x,\cdot),t).$$

Thus, if n is the smaller integer greater than α, i.e., $n-1 \leq \alpha < n$, then by Lemma 2 (iv),

$$\frac{\partial^n}{\partial t^n} P_t(D_\beta f)(x) = \frac{1}{c_\beta} \int_0^{+\infty} s^{-\beta-1} \frac{\partial^n}{\partial t^n} (\Delta_s^k(u(x,\cdot),t)$$
$$= \frac{1}{c_\beta} \int_0^{+\infty} s^{-\beta-1} \Delta_s^k(u^{(n)}(x,\cdot),t) ds,$$

and therefore, by Minkowski's integral inequality,

$$\|\frac{\partial^n}{\partial t^n} P_t(D_\beta f)\|_{p,\gamma} \leq \frac{1}{c_\beta} \int_0^{+\infty} s^{-\beta-1} \|\Delta_s^k(u^{(n)},t)\|_{p,\gamma} ds. \tag{44}$$

Now, if $1 \leq q < \infty$, by Eq. (44),

$$\left(\int_0^\infty \left(t^{n-(\alpha-\beta)} \|\frac{\partial^n}{\partial t^n} P_t(D_\beta f)\|_{p,\gamma}\right)^q \frac{dt}{t}\right)^{1/q}$$
$$\leq \frac{1}{c_\beta} \left(\int_0^\infty \left(t^{n-(\alpha-\beta)} \int_0^{+\infty} s^{-\beta-1} \|\Delta_s^k(u^{(n)},t)\|_{p,\gamma} ds\right)^q \frac{dt}{t}\right)^{1/q}$$
$$\leq \frac{1}{c_\beta} \left(\int_0^\infty \left(t^{n-(\alpha-\beta)} \int_0^{t} s^{-\beta-1} \|\Delta_s^k(u^{(n)},t)\|_{p,\gamma} ds\right)^q \frac{dt}{t}\right)^{1/q}$$

$$+\frac{1}{c_\beta}\left(\int_0^\infty\left(t^{n-(\alpha-\beta)}\int_t^{+\infty}s^{-\beta-1}\|\Delta_s^k(u^{(n)},t)\|_{p,\gamma}ds\right)^q\frac{dt}{t}\right)^{1/q}$$
$$=(I)+(II).$$

Then, by Lemma 3,

$$(I)\leq\frac{1}{c_\beta}\left(\int_0^\infty\left(t^{n-(\alpha-\beta)}\|\frac{\partial^{n+k}}{\partial t^{n+k}}P_tf\|_{p,\gamma}\int_0^t s^{k-\beta-1}ds\right)^q\frac{dt}{t}\right)^{1/q}$$
$$=\frac{1}{c_\beta(k-\beta)}\left(\int_0^\infty\left(t^{n+k-\alpha}\|u^{(n+k)}(\cdot,t)\|_{p,\gamma}\right)^q\frac{dt}{t}\right)^{1/q}<\infty,$$

since $f\in B^\alpha_{p,q}(\gamma_d)$, and by Lemma 1,

$$(II)\leq\frac{1}{c_\beta}\left(\int_0^\infty\left(t^{n-(\alpha-\beta)}\int_t^{+\infty}s^{-\beta-1}\left(\sum_{j=0}^k\binom{k}{j}\|u^{(n)}(\cdot,t+(k-j)s)\|_{p,\gamma}\right)ds\right)^q\frac{dt}{t}\right)^{1/q}$$
$$\leq\frac{1}{c_\beta}\left(\int_0^\infty\left(t^{n-(\alpha-\beta)}\int_t^{+\infty}s^{-\beta-1}\left(\sum_{j=0}^k\binom{k}{j}\|u^{(n)}(\cdot,t)\|_{p,\gamma}\right)ds\right)^q\frac{dt}{t}\right)^{1/q}$$
$$=\frac{2^k}{c_\beta}\left(\int_0^\infty\left(t^{n-(\alpha-\beta)}\|\frac{\partial^n}{\partial t^n}P_tf\|_{p,\gamma}\int_t^{+\infty}s^{-\beta-1}ds\right)^q\frac{dt}{t}\right)^{1/q}$$
$$=\frac{2^k}{c_\beta\beta}\left(\int_0^\infty\left(t^{n-\alpha}\|\frac{\partial^n}{\partial t^n}P_tf\|_{p,\gamma}\right)^q\frac{dt}{t}\right)^{1/q}<\infty,$$

since $f\in B^\alpha_{p,q}(\gamma_d)$. Therefore, if $1\leq q<\infty$, $D_\beta f\in B^{\alpha-\beta}_{p,q}(\gamma_d)$, and moreover

$$\|D_\beta f\|_{B^{\alpha-\beta}_{p,q}}=\|D_\beta f\|_{p,\gamma}+\left(\int_0^\infty\left(t^{n-\alpha+\beta}\|\frac{\partial^n}{\partial t^n}P_t(D_\beta f)\|_{p,\gamma}\right)^q\frac{dt}{t}\right)^{1/q}$$
$$\leq C_1\|f\|_{B^\alpha_{p,q}}+C_2\|f\|_{B^\alpha_{p,q}}\leq C\|f\|_{B^\alpha_{p,q}}$$

Thus, $D_\beta f: B^\alpha_{p,q}\to B^{\alpha-\beta}_{p,q}$ is bounded.

If $q=\infty$, inequality (44) can be written as

$$\|\frac{\partial^n}{\partial t^n}P_t(D_\beta f)\|_{p,\gamma}\leq\frac{1}{c_\beta}\int_0^t s^{-\beta-1}\|\Delta_s^k(u^{(n)},t)\|_{p,\gamma}ds$$
$$+\frac{1}{c_\beta}\int_t^{+\infty}s^{-\beta-1}\|\Delta_s^k(u^{(n)},t)\|_{p,\gamma}ds$$
$$=(I)+(II),$$

and then as $f \in B^{\alpha}_{p,\infty}$, by Lemma 3,

$$(I) \leq \frac{1}{c_\beta}\int_0^t s^{-\beta-1}s^k\|u^{(n+k)}\|_{p,\gamma}ds = C_\beta\|\frac{\partial^{n+k}}{\partial t^{n+k}}P_tf\|_{p,\gamma}t^{k-\beta}$$
$$\leq C_\beta A(f)t^{-n-k+\alpha}t^{k-\beta} = C_\beta A(f)t^{-n+\alpha-\beta},$$

and as above, by Lemma 1,

$$(II) \leq \frac{1}{c_\beta}\int_t^{+\infty} s^{-\beta-1}\left(\sum_{j=0}^{k}\binom{k}{j}\|u^{(n)}(\cdot,t+(k-j)s)\|_{p,\gamma}\right)ds$$
$$\leq C_\beta\int_t^{+\infty} s^{-\beta-1}\left(\sum_{j=0}^{k}\binom{k}{j}\|u^{(n)}(\cdot,t)\|_{p,\gamma}\right)ds = C_\beta t^{-\beta}\|\frac{\partial^{n}}{\partial t^{n}}P_tf\|_{p,\gamma}$$
$$\leq C_\beta A(f)t^{-n+\alpha}t^{-\beta} = C_\beta A(f)t^{-n+\alpha-\beta}.$$

□

There is an alternative proof of the fact that $D_\beta f \in L^p(\gamma_d)$ without using Hardy's inequality following the same scheme as in the proof of (i) Theorem 3.5 in [2], using the inclusion $B^{\alpha}_{p,q} \subset B^{\beta+\varepsilon}_{p,\infty}$ with $\beta+\varepsilon < k$.

Theorem 8. *Let $0 < \beta < \alpha$, $1 \leq p < \infty$, and $1 \leq q \leq \infty$ then $\mathscr{D}_\beta$ is bounded from $B^{\alpha}_{p,q}(\gamma_d)$ into $B^{\alpha-\beta}_{p,q}(\gamma_d)$.*

Proof. Let $f \in B^{\alpha}_{p,q}(\gamma_d)$, and set $v(x,t) = e^{-t}u(x,t)$ then using the Hardy's inequality (32), the Fundamental Theorem of Calculus and (iii) of Lemma 2,

$$|\mathscr{D}_\beta f(x)| \leq \frac{1}{c_\beta}\int_0^{+\infty} s^{-\beta-1}|\Delta_s^k(v(x,\cdot),0)|ds$$
$$\leq \frac{1}{c_\beta}\int_0^{+\infty} s^{-\beta-1}\int_0^s |\frac{\partial}{\partial r}\Delta_r^k(v(x,\cdot),0)|drds$$
$$\leq \frac{k}{\beta c_\beta}\int_0^{+\infty} r^{-\beta}|\Delta_r^{k-1}(v'(x,\cdot),r)|dr,$$

and this implies by Minkowski's integral inequality

$$\|\mathscr{D}_\beta f\|_{p,\gamma_d} \leq \frac{k}{\beta c_\beta}\int_0^{+\infty} r^{-\beta}\|\Delta_r^{k-1}(v',r)\|_{p,\gamma}dr.$$

Now, using Lemma 2

$$\|\Delta_r^{k-1}(v',r)\|_{p,\gamma} \leq \int_r^{2r}\int_{v_1}^{v_1+r}\cdots\int_{v_{k-2}}^{v_{k-2}+r}\|v^{(k)}(\cdot,v_{k-1})\|_{p,\gamma}dv_{k-1}dv_{k-2}\dots dv_2dv_1$$

and by Leibnitz's differentiation rule for the product,

$$\begin{aligned}\|v^{(k)}(\cdot,v_{k-1})\|_{p,\gamma} &= \|\sum_{j=0}^{k}\binom{k}{j}(e^{-v_{k-1}})^{(j)}u^{(k-j)}(\cdot,v_{k-1})\|_{p,\gamma_d}\\ &\leq \sum_{j=0}^{k}\binom{k}{j}e^{-v_{k-1}}\|u^{(k-j)}(\cdot,v_{k-1})\|_{p,\gamma}.\end{aligned}$$

Then

$$\begin{aligned}&\|\Delta_r^{k-1}(v',r)\|_{p,\gamma}\\ &\leq \sum_{j=0}^{k}\binom{k}{j}\int_r^{2r}\int_{v_1}^{v_1+r}\cdots\\ &\int_{v_{k-2}}^{v_{k-2}+r}e^{-v_{k-1}}\|u^{(k-j)}(\cdot,v_{k-1})\|_{p,\gamma}dv_{k-1}dv_{k-2}\dots dv_2dv_1\\ &\leq \sum_{j=0}^{k}\binom{k}{j}r^{k-1}e^{-r}\|u^{(k-j)}(\cdot,r)\|_{p,\gamma}.\end{aligned}$$

Therefore

$$\begin{aligned}\|\mathscr{D}_\beta f\|_{p,\gamma} &\leq \frac{k}{\beta c_\beta}\sum_{j=0}^{k}\binom{k}{j}\int_0^{+\infty}r^{k-\beta-1}e^{-r}\|u^{(k-j)}(\cdot,r)\|_{p,\gamma}dr\\ &= \frac{k}{\beta c_\beta}\sum_{j=0}^{k-1}\binom{k}{j}\int_0^{+\infty}r^{(k-j)-(\beta-j)-1}e^{-r}\|\frac{\partial^{k-j}}{\partial r^{k-j}}P_rf\|_{p,\gamma}dr\\ &\quad+\frac{k}{\beta c_\beta}\int_0^{+\infty}r^{k-\beta-1}e^{-r}\|P_rf\|_{p,\gamma}dr\\ &\leq \frac{k}{\beta c_\beta}\sum_{j=0}^{k-1}\binom{k}{j}\int_0^{+\infty}r^{(k-j)-(\beta-j)-1}\|\frac{\partial^{k-j}}{\partial r^{k-j}}P_rf\|_{p,\gamma}dr\\ &\quad+\frac{k}{\beta c_\beta}\int_0^{+\infty}r^{k-\beta-1}e^{-r}\|f\|_{p,\gamma}dr.\end{aligned}$$

Thus,

$$\|\mathscr{D}_\beta f\|_{p,\gamma} \leq \frac{k}{\beta c_\beta}\sum_{j=0}^{k-1}\binom{k}{j}\int_0^{+\infty} r^{k-j-(\beta-j)}\|\frac{\partial^{k-j}}{\partial r^{k-j}}P_r f\|_{p,\gamma}\frac{dr}{r} + \frac{k\Gamma(k-\beta)}{\beta c_\beta}\|f\|_{p,\gamma} < \infty,$$

since $B^{\alpha}_{p,q}(\gamma_d) \subset B^{\beta-j}_{p,1}(\gamma_d)$ as $\alpha > \beta > \beta - j$, for $j \in \{0,\ldots,k-1\}$, then $\mathscr{D}_\beta f \in L^p(\gamma_d)$.

On the other hand,

$$P_t(e^{-s}P_s - I)^k f(x) = \sum_{j=0}^{k}\binom{k}{j}(-1)^j e^{-s(k-j)}u(x,t+(k-j)s).$$

Let n be the smaller integer greater than α, i.e., $n-1 \leq \alpha < n$, and we have

$$\begin{aligned}
\frac{\partial^n}{\partial t^n}P_t(\mathscr{D}_\beta f)(x) &= \frac{1}{c_\beta}\int_0^{+\infty} s^{-\beta-1}\sum_{j=0}^{k}\binom{k}{j}(-1)^j e^{-s(k-j)}u^{(n)}(x,t+(k-j)s)ds \\
&= \frac{e^t}{c_\beta}\int_0^{+\infty} s^{-\beta-1}\sum_{j=0}^{k}\binom{k}{j}(-1)^j e^{-(t+s(k-j))}u^{(n)}(x,t+(k-j)s)ds \\
&= \frac{e^t}{c_\beta}\int_0^{+\infty} s^{-\beta-1}\Delta_s^k(w(x,\cdot),t)ds,
\end{aligned}$$

where $w(x,t) = e^{-t}u^{(n)}(x,t)$. Now, using the Fundamental Theorem of Calculus,

$$\begin{aligned}
\frac{\partial^n}{\partial t^n}P_t(\mathscr{D}_\beta f)(x) &= \frac{e^t}{c_\beta}\int_0^{+\infty} s^{-\beta-1}\Delta_s^k(w(x,\cdot),t)ds \\
&= \frac{e^t}{c_\beta}\int_0^{+\infty} s^{-\beta-1}\int_0^s \frac{\partial}{\partial r}\Delta_r^k(w(x,\cdot),t)drds.
\end{aligned}$$

Then, using Hardy's inequality (32), and (iii) of Lemma 2,

$$\begin{aligned}
\left|\frac{\partial^n}{\partial t^n}P_t(D_\beta f)(x)\right| &\leq \frac{e^t}{c_\beta}\int_0^{+\infty} s^{-\beta-1}\int_0^s \left|\frac{\partial}{\partial r}\Delta_r^k(w(x,\cdot),t)\right|drds \\
&\leq \frac{e^t}{c_\beta\beta}\int_0^{+\infty} r\left|\frac{\partial}{\partial r}\Delta_r^k(w(x,\cdot),t)\right|r^{-\beta-1}dr \\
&= \frac{ke^t}{c_\beta\beta}\int_0^{+\infty} r^{-\beta}|\Delta_r^{k-1}(w'(x,\cdot),t+r)|dr,
\end{aligned}$$

and by Minkowski's integral inequality we get

$$\|\frac{\partial^n}{\partial t^n}P_t(\mathscr{D}_\beta f)\|_{p,\gamma} \leq \frac{ke^t}{\beta c_\beta}\int_0^{+\infty} r^{-\beta}\|\Delta_r^{k-1}(w',t+r)\|_{p,\gamma}dr.$$

Now, by analogous argument as above, Lemma 2 and Leibnitz's product rule give us

$$\|\Delta_r^{k-1}(w',t+r)\|_{p,\gamma} \leq \sum_{j=0}^{k}\binom{k}{j} r^{k-1}e^{-(t+r)}\|u^{(k+n-j)}(\cdot,t+r)\|_{p,\gamma},$$

and this implies that

$$\begin{aligned}\|\frac{\partial^n}{\partial t^n}P_t(\mathscr{D}_\beta f)\|_{p,\gamma} &\leq e^t\frac{k}{c_\beta\beta}\int_0^{+\infty} r^{-\beta}\left(\sum_{j=0}^{k}\binom{k}{j} r^{k-1}e^{-(t+r)}\|u^{(k+n-j)}(\cdot,t+r)\|_{p,\gamma}\right)dr\\ &= \frac{k}{c_\beta\beta}\sum_{j=0}^{k}\binom{k}{j}\int_0^{+\infty} r^{k-\beta-1}e^{-r}\|u^{(k+n-j)}(\cdot,t+r)\|_{p,\gamma}dr.\end{aligned}$$

Thus,

$$\|\frac{\partial^n}{\partial t^n}P_t(\mathscr{D}_\beta f)\|_{p,\gamma} \leq \frac{k}{c_\beta\beta}\sum_{j=0}^{k}\binom{k}{j}\int_0^{+\infty} r^{k-\beta-1}e^{-r}\|u^{(k+n-j)}(\cdot,t+r)\|_{p,\gamma}dr. \quad (45)$$

Now if $1 \leq q < \infty$, using Eq. (45) we have,

$$\begin{aligned}&\left(\int_0^{\infty}\left(t^{n-(\alpha-\beta)}\|\frac{\partial^n}{\partial t^n}P_t(\mathscr{D}_\beta f)\|_{p,\gamma}\right)^q\frac{dt}{t}\right)^{1/q}\\ &\leq \frac{k}{c_\beta\beta}\sum_{j=0}^{k}\binom{k}{j}\left(\int_0^{\infty}\left(t^{n-(\alpha-\beta)}\int_0^{+\infty} r^{k-\beta-1}e^{-r}\|u^{(k+n-j)}(\cdot,t+r)\|_{p,\gamma}dr\right)^q\frac{dt}{t}\right)^{1/q}.\end{aligned}$$

For each $1 \leq j \leq k$, $0 < \alpha-\beta+k-j \leq \alpha$ and by Lemma 1

$$\begin{aligned}&\left(\int_0^{\infty}\left(t^{n-(\alpha-\beta)}\int_0^{\infty} r^{k-\beta-1}e^{-r}\|u^{(k+n-j)}(\cdot,t+r)\|_{p,\gamma}dr\right)^q\frac{dt}{t}\right)^{1/q}\\ &\leq \left(\int_0^{\infty}\left(t^{n-(\alpha-\beta)}\|u^{(n+k-j)}(\cdot,t)\|_{p,\gamma}\int_0^{+\infty} r^{k-\beta-1}e^{-r}dr\right)^q\frac{dt}{t}\right)^{1/q}\\ &= \Gamma(k-\beta)\left(\int_0^{\infty}\left(t^{n+(k-j)-(\alpha-\beta+k-j)}\|u^{(n+k-j)}(\cdot,t)\|_{p,\gamma}\right)^q\frac{dt}{t}\right)^{1/q} < \infty,\end{aligned}$$

as $f \in B^{\alpha}_{p,q}(\gamma_d) \subset B^{\alpha-\beta+(k-j)}_{p,q}(\gamma_d)$ for any $1 \leq j \leq k$.

Now, for the case $j=0$,

$$\left(\int_0^\infty \left(t^{n-(\alpha-\beta)}\int_0^{+\infty} r^{k-\beta-1}e^{-r}\|u^{(n+k)}(\cdot,t+r)\|_{p,\gamma}dr\right)^q \frac{dt}{t}\right)^{1/q}$$
$$\leq \left(\int_0^\infty \left(t^{n-(\alpha-\beta)}\int_0^{t} r^{k-\beta-1}e^{-r}\|u^{(n+k)}(\cdot,t+r)\|_{p,\gamma}dr\right)^q \frac{dt}{t}\right)^{1/q}$$
$$+\left(\int_0^\infty \left(t^{n-(\alpha-\beta)}\int_t^{+\infty} r^{k-\beta-1}e^{-r}\|u^{(n+k)}(\cdot,t+r)\|_{p,\gamma}dr\right)^q \frac{dt}{t}\right)^{1/q}$$
$$=(I)+(II).$$

Using Lemma 1, and $k>\beta$,

$$(I)\leq \left(\int_0^\infty \left(t^{n-(\alpha-\beta)}\int_0^{t} r^{k-\beta-1}\|u^{(n+k)}(\cdot,t)\|_{p,\gamma}dr\right)^q \frac{dt}{t}\right)^{1/q}$$
$$=\left(\int_0^\infty \left(t^{n-(\alpha-\beta)}\|u^{(n+k)}(\cdot,t)\|_{p,\gamma}\int_0^{t} r^{k-\beta-1}dr\right)^q \frac{dt}{t}\right)^{1/q}$$
$$=\frac{1}{k-\beta}\left(\int_0^\infty \left(t^{n+k-\alpha}\|u^{(n+k)}(\cdot,t)\|_{p,\gamma}\right)^q \frac{dt}{t}\right)^{1/q}<\infty,$$

since $f\in B^{\alpha}_{p,q}(\gamma_d)$ and $n+k>\alpha$, and for the second term, using Lemma 1 and Hardy's inequality (33),

$$(II)\leq \left(\int_0^\infty \left(t^{n-(\alpha-\beta)}\int_t^{+\infty} r^{k-\beta-1}\|u^{(n+k)}(\cdot,r)\|_{p,\gamma}dr\right)^q \frac{dt}{t}\right)^{1/q}$$
$$\leq \frac{1}{n-(\alpha-\beta)}\left(\int_0^\infty \left(r^{n+k-\alpha}\|u^{(n+k)}(\cdot,r)\|_{p,\gamma}\right)^q \frac{dr}{r}\right)^{1/q}<\infty,$$

since $f\in B^{\alpha}_{p,q}(\gamma_d)$.

Therefore $\mathscr{D}_\beta f\in B^{\alpha-\beta}_{p,q}(\gamma_d)$. Moreover,

$$\|\mathscr{D}_\beta f\|_{B^{\alpha-\beta}_{p,q}} = \|\mathscr{D}_\beta f\|_{p,\gamma}+\left(\int_0^\infty \left(t^{n-(\alpha-\beta)}\|\frac{\partial^n}{\partial t^n}P_t\mathscr{D}_\beta f\|_{p,\gamma}\right)^q \frac{dt}{t}\right)^{1/q}$$
$$\leq C_1\|f\|_{p,\gamma}+\frac{k}{c_\beta\beta}\sum_{j=0}^{k}\binom{k}{j}C_2\left(\int_0^\infty \left(r^{n-\alpha}\|\frac{\partial^n}{\partial r^n}P_r f\|_{p,\gamma}\right)^q \frac{dr}{r}\right)^{1/q}$$
$$\leq C\|f\|_{B^{\alpha}_{p,q}}.$$

Finally, if $q=\infty$, from the inequality (45),

$$\|\frac{\partial^n}{\partial t^n}P_t(\mathscr{D}_\beta f)\|_{p,\gamma}\leq \frac{k}{c_\beta\beta}\sum_{j=0}^{k}\binom{k}{j}\int_0^{+\infty} r^{k-\beta-1}e^{-r}\|u^{(k+n-j)}(\cdot,t+r)\|_{p,\gamma}dr,$$

and then, the argument is essentially similar to the previous case, as we did in the last part of the proof of Theorem 7. □

Observation Let us observe that if instead of considering the *Ornstein-Uhlenbeck operator* (2) and the *Poisson–Hermite semigroup* (7), we consider the *Laguerre differential operator* in $\mathbb{R}^d_+$,

$$\mathscr{L}^\alpha = \sum_{i=1}^{d} \left[x_i \frac{\partial^2}{\partial x_i^2} + (\alpha_i + 1 - x_i)\frac{\partial}{\partial x_i} \right], \tag{46}$$

and the corresponding *Poisson–Laguerre semigroup*, or if we consider the *Jacobi differential operator* in $(-1,1)^d$,

$$\mathscr{L}^{\alpha,\beta} = -\sum_{i=1}^{d} \left[(1-x_i^2)\frac{\partial^2}{\partial x_i^2} + (\beta_i - \alpha_i - (\alpha_i + \beta_i + 2)x_i)\frac{\partial}{\partial x_i} \right], \tag{47}$$

and the corresponding *Poisson–Jacobi semigroup* (for details, we refer to [10]), the arguments are completely analogous. That is to say, we can defined in analogous manner *Laguerre–Besov–Lipschitz spaces* and *Jacobi–Besov–Lipschitz spaces* then prove that the corresponding notions of fractional integrals and fractional derivatives behave similarly. In order to see this it is more convenient to use the representation (7) of P_t in terms of the one-sided stable measure $\mu_t^{(1/2)}(ds)$, see [6].

References

1. Forzani, L., Scotto, R., Urbina, W.: Riesz and Bessel Potentials, the g_k functions and an Area function, for the Gaussian measure γ_d. Revista de la Unión Matemática Argentina (UMA) **42**(1), 17–37 (2000)
2. Gatto, A.E., Urbina, W.: On Gaussian Lipschitz Spaces and the Boundedness of Fractional Integrals and Fractional Derivatives on them (2012). Sent for publication. arXiv:0911.3962v2
3. López, I., Urbina, W.: Fractional Differentiation for the Gaussian Measure and Applications. Bull. Sci. Math **128**, 587–603 (2004)
4. Meyer, P.A.: Transformations de Riesz pour les lois Gaussiennes. Lectures Notes in Math, vol. 1059, pp. 179–193. Springer, Berlin (1984)
5. Pineda, E.: Tópicos en Análisis Armónico Gaussiano: Comportamiento en la frontera y Espacios de funciones para la medida Gaussiana. Doctoral Thesis, Facultad de Ciencias, UCV, Caracas (2009)
6. Pineda, E., Urbina, W.: Some results on Gaussian Besov-Lipschitz and Gaussian Triebel-Lizorkin spaces. J. Approx. Theor. **161**(2), 529–564 (2009)
7. Samko, S., Kilbas, A., Marichev, O.: Fractional Integrals and Derivatives: Theory and Applications. Gordon and Breach Science Publishers, Philadelphia (1992)
8. Stein, E.: Singular Integrals and Differentiability Properties of Functions. Princeton Univ. Press. Princeton, New Jersey (1970)
9. Triebel, H.: Theory of Function Spaces II. Birkhäuser, Basel (1992)
10. Urbina, W.: Operators Semigroups associated to Classical Orthogonal Polynomials and Functional Inequalities. Lecture Notes of the French Mathematical Society (SMF) (2008)
11. Watanabe, S.: Lecture on Stochastic Differential Equations and Malliavin Calculus. Tata Institute. Springer, Berlin (1984)

Maximal Operators Associated to Sets of Directions of Hausdorff and Minkowski Dimension Zero

Paul Hagelstein

Abstract For any set $\Omega \subset [0, \pi/2)$ we let R_Ω be the set of rectangles in $\mathbb{R}^2$ oriented in one of the directions of Ω. The geometric maximal operator associated to Ω is given by

$$M_\Omega f(x) = \sup_{x \in R \in R_\Omega} \frac{1}{|R|} \int_R |f(y)|\, dy\,.$$

In this chapter we show that if M_Ω is bounded on $L^p(\mathbb{R}^2)$ for $1 < p \leq \infty$, then Ω must be countable and of Hausdorff and Minkowski dimension zero. We shall see that the converse does not hold, however, by exhibiting an example of a countable set Ω of Hausdorff and Minkowski dimension zero for which the associated maximal operator M_Ω is unbounded on $L^p(\mathbb{R}^2)$ for $1 \leq p < \infty$. All of these results will be seen to be consequences of a recent theorem of Bateman (Duke Math. J. 147:55–77, 2009) regarding geometric maximal operators and N-lacunary sets.

1 Introduction

Let $\Omega \subset [0, \pi/2)$ correspond to a set of directions in the plane. Denote by R_Ω the set of rectangles in $\mathbb{R}^2$ oriented in one of the directions of Ω. The geometric maximal operator associated to Ω is given by

$$M_\Omega f(x) = \sup_{x \in R \in R_\Omega} \frac{1}{|R|} \int_R |f(y)|\, dy\,.$$

The L^p-boundedness properties of M_Ω are, as expected, dependent on Ω. If $\Omega = [0, \pi/2)$ itself, corresponding to the set of all directions, M_Ω is the so-called *Kakeya*

P. Hagelstein (✉)
Baylor University, Waco, TX 76798, USA
e-mail: paul_hagelstein@baylor.edu

D. Bilyk et al. (eds.), *Recent Advances in Harmonic Analysis and Applications*, Springer Proceedings in Mathematics & Statistics 25, DOI 10.1007/978-1-4614-4565-4_13,

or *Nikodym* maximal operator and is unbounded for $1 \leq p < \infty$, a result following readily from the papers of Nikodym [12] and Busemann and Feller [3]. The most fundamental positive result along these lines goes back to Jessen, Marcinkiewicz, and Zygmund, who proved in [8] that if $\Omega = \{0\}$, then M_Ω is bounded on $L^p(\mathbb{R}^2)$ for $1 < p \leq \infty$. (This corresponds to the L^p-boundedness of the so-called *strong maximal operator.*) The techniques of Jessen, Marcinkiewicz, and Zygmund were rather straightforward, their proof being based on bounding the strong maximal operator by a composition of L^p-bounded one-parameter maximal operators oriented in the horizontal and vertical directions. Together with simple sublinearity considerations, the Jessen–Marcinkiewicz–Zygmund result also implies that M_Ω is bounded for $1 < p \leq \infty$ provided Ω has a finite number of elements. The first positive result corresponding to a case where Ω has an infinite number of directions is due to Strömberg, who proved in [14] that if $\Omega = \{2^{-j} : j = 1, 2, \ldots\}$, then M_Ω is bounded on $L^p(\mathbb{R}^2)$ for $p > 2$. (The corresponding maximal operator in this case is often referred to as the *lacunary maximal operator* M_{lac}.) Strömberg's argument was refined by Córdoba and Fefferman, who used covering lemma type arguments in [4] to prove that M_{lac} is of weak type $(2,2)$. Using Fourier analytic techniques and complex interpolation, Nagel, Stein, and Wainger extended Córdoba and Fefferman's result by showing in [11] that M_{lac} is bounded on $L^p(\mathbb{R}^2)$ for $1 < p \leq \infty$.

In 1996 Nets Katz proved in [10] the unexpected result that if $\mathscr{C}$ is the ternary Cantor set, then $M_\mathscr{C}$ is *not* bounded on $L^2(\mathbb{R}^2)$. This result was extended over a decade later by Bateman and Katz, who proved in [2] that $M_\mathscr{C}$ is not bounded on $L^p(\mathbb{R}^2)$ for $1 \leq p < \infty$.

As is well known, although the ternary Cantor set is of Lebesgue measure zero, it does have positive Hausdorff dimension $(\log 2/\log 3)$. In fact, Hare proved in [6] that the maximal operator associated to any Cantor set of positive Hausdorff dimension is unbounded on $L^2(\mathbb{R}^2)$. These results naturally lead one to consider the possibility that the L^p-boundedness of M_Ω for $1 < p \leq \infty$ implies that the Hausdorff dimension of Ω is zero. Such a consideration is explicitly mentioned in, e.g., the work of Hare and Rönning on Max(p) sets and density bases [7]. The question of whether the L^p-boundedness of M_Ω for $1 < p \leq \infty$ implies that the Minkowski dimension of Ω is zero is also of interest. In this chapter we will show that Ω is indeed of Hausdorff and Minkowski dimension zero if M_Ω is L^p-bounded for $1 < p \leq \infty$. The Hausdorff dimension zero result will be seen to follow from the surprising fact that the L^p-boundedness of M_Ω implies the *countability* of Ω. We shall see that the converse does not hold, however, by exhibiting a countable set Ω of Hausdorff and Minkowski dimension zero such that the corresponding maximal operator M_Ω is not bounded on $L^p(\mathbb{R}^2)$ for $1 \leq p < \infty$. The proofs will rely on elementary considerations in geometric measure theory together with a remarkable recent result of Bateman relating L^p-boundedness properties of M_Ω to N-lacunary sets.

2 Maximal Operators Associated to Sets of Directions of Hausdorff Dimension Zero

In this section we will show that if M_Ω is bounded on $L^p(\mathbb{R}^2)$ for $1 < p \leq \infty$, then Ω must necessarily be of Hausdorff dimension zero. Before we prove this result, however, it would be useful for us to state Bateman's main theorem in [1] as well as recall the notion of lacunary sets of finite order. We mention that, before Bateman's paper [1], lacunary sets of finite order appeared in the harmonic analysis context in the work of Sjögren and Sjölin on Littlewood-Paley decompositions [13] as well as the work of Karagulyan and Lacey [9] on directional maximal operators.

Let E be a closed set of measure 0 in $\mathbb{R}$. For any point $t \in \mathbb{R}$, we let $d_E(t)$ denote the distance from t to E. A closed set E' of measure 0 is said to be a *successor* of E if there exists a constant $c > 0$, called the *successor constant*, such that $x, y \in E'$ and $x \neq y$ imply $|x-y| \geq c\, d_E(x)$. A lacunary set of order 0 is a one-point set. Proceeding inductively, we say a set is *lacunary of order* N if it is a successor of a lacunary set of order $N-1$.

A few examples may be instructive here. $\{0\} \cup \{2^{-j} : j \in \mathbb{Z}_+\}$ is the classic example of a lacunary set of order 1.

$$\{0\} \cup \{2^{-j} : j \in \mathbb{Z}_+\} \cup \left\{2^{-j} + 2^{-k} : j,k \in \mathbb{Z}_+\right\}$$

forms a lacunary set of order 2. The set

$$\{0\} \cup \{2^{-j} : j \in \mathbb{Z}_+\} \cup \left\{2^{-j} + 2^{-j}(1-2^{-j})^k : j,k \in \mathbb{Z}_+\right\}$$

does *not* form a lacunary set of order 2, even though it is the union of a lacunary set L together with a union of lacunary sets each converging to a different element of L.

Bateman proved the following:

Theorem 1 ([1]). *Let $1 < p < \infty$. Then M_Ω is bounded on $L^p(\mathbb{R}^2)$ if and only if there exist $N_1, N_2 < \infty$ such that Ω is covered by N_1 lacunary sets of order N_2.*

Bateman's theorem enables us to in short order show that the L^p-boundedness of M_Ω for $1 < p \leq \infty$ implies that Ω is not only of Hausdorff dimension zero but also countable.

Theorem 2. *If M_Ω is bounded on $L^p(\mathbb{R}^2)$ for $1 < p \leq \infty$, then Ω is countable and of Hausdorff dimension zero.*

Proof. Suppose M_Ω is bounded on $L^p(\mathbb{R}^2)$ for $1 < p \leq \infty$. Then there exist $N_1, N_2 < \infty$ such that Ω is covered by N_1 lacunary sets of order N_2. As each of these lacunary sets is countable (being a countable union of countable sets) we have that Ω is countable. As any countable set is of Hausdorff dimension zero, the result follows. □

The converse of the above theorem does *not* hold. For a straightforward example, let Ω be the set of all rational numbers in $(0,\pi/2)$. Since Ω is countable it is of Hausdorff dimension zero. However, M_Ω corresponds to the Kakeya or Nikodym maximal operator and is accordingly unbounded on $L^p(\mathbb{R}^2)$ for $1 \leq p < \infty$.

We remark that another known example of a set of zero Hausdorff dimension for which M_Ω is unbounded on $L^p(\mathbb{R}^2)$ for $1 \leq p < \infty$ is

$$\Omega = \{1/2, 1/3, 1/4, \ldots\} \cup \{0\}.$$

M_Ω was shown to be unbounded on $L^p(\mathbb{R}^2)$ for $1 \leq p < \infty$ by de Guzmán in [5] and, being countable, is of Hausdorff dimension zero. It is interesting to note that, although the set $\{1/2, 1/3, 1/4, \ldots\} \cup \{0\}$ is clearly not lacunary of order 1, it is not a priori obvious that it should not be lacunary of higher order—that this is the case that can be viewed as a consequence of the combined work of Bateman and de Guzmán.

3 Maximal Operators Associated to Sets of Directions of Minkowski Dimension Zero

In this section we show that if M_Ω is bounded on $L^p(\mathbb{R}^2)$ for $1 < p \leq \infty$, then the Minkowski dimension of Ω must be zero. Moreover we will exhibit a set Ω of Minkowski dimension zero but such that M_Ω is unbounded on $L^p(\mathbb{R}^2)$ for $1 < p < \infty$.

It is helpful at the outset to fix our notation regarding concepts involving the Minkowski dimension of sets within $\mathbb{R}$. Let A be a non-empty-bounded subset of $\mathbb{R}$. For $0 < \varepsilon < \infty$, let $N(A,\varepsilon)$ be the smallest number of intervals of length 2ε needed to cover A:

$$N(A,\varepsilon) = \min\left\{ k : A \subset \bigcup_{i=1}^{k} [x_i - \varepsilon, x_i + \varepsilon] \text{ for some } x_i \in \mathbb{R} \right\}.$$

The upper and lower Minkowski dimensions of A are given respectively by

$$\overline{\dim}_M A = \inf\left\{ s : \limsup_{\varepsilon \downarrow 0} N(A,\varepsilon)\varepsilon^s = 0 \right\}$$

and

$$\underline{\dim}_M A = \inf\left\{ s : \liminf_{\varepsilon \downarrow 0} N(A,\varepsilon)\varepsilon^s = 0 \right\}.$$

If the upper and lower Minkowski dimensions of A agree, the Minkowski dimension of A is given by

$$\dim_M A = \overline{\dim}_M A = \underline{\dim}_M A\ .$$

Note that both the upper and lower Minkowski dimensions of A are greater than or equal to the Hausdorff dimension of A.

Theorem 3. *If M_Ω is bounded on $L^p(\mathbb{R}^2)$ for $1 < p \leq \infty$, then Ω must be of Minkowski dimension zero.*

Proof. By Theorem 1, it suffices to show that any N-lacunary set in $[0,\pi/2)$ must be of Minkowski dimension zero. As a single element in $[0,\pi/2)$ has Minkowski dimension zero, by induction, it suffices to show that any closed set in $[0,\pi/2)$ that is a successor of a set of Minkowski dimension zero also has Minkowski dimension zero.

Let L be a closed set in $[0,\pi/2)$ such that $\dim_M L = 0$, and let S be a successor of L. We let λ denote the successor constant associated to S. In particular we have that if $x,y \in S$ with $x \neq y$, then $|x-y| \geq \lambda d_L(x)$. We assume $0 < \lambda < 1$ without loss of generality.

For each $x \in S \backslash L$, let p_x denote a point in L such that $|x - p_x| = d_L(x)$ (Note such a p_x exists as L is closed.). Let I_x denote the open interval connecting x and p_x. Observe that if $x,y \in S\backslash L$, then either $I_x = I_y$, $I_x \subset I_y$, or $I_y \subset I_x$.

Fix now an $x \in S\backslash L$ and the associated $p_x \in L$. Without loss of generality we assume $I_x = (p_x, x)$. Assuming without loss of generality that $I_x \cap S \neq \emptyset$, let $y_1 \in I_x \cap S$ be such that $y_1 - p_x$ is maximal. Note that since $x - y_1 \geq \lambda(x - p_x)$ we must have $y_1 - p_x \leq (1-\lambda)(x-p_x)$. If $(p_x, y_1) \cap S$ is nonempty, we let $y_2 \in (p_x, y_1)$ be such that $y_2 - p_x$ is maximal. Since $y_2 - y_1 \geq \lambda(y_2 - p_x)$ we then have $y_2 - p_x \leq (1-\lambda)(y_1 - p_x) \leq (1-\lambda)^2(x-p_x)$. Continuing in this way, we yield that $S \cap I_x$ is the (possibly finite) set $\{y_1, y_2, \ldots\}$ such that $|y_k - p_x| \leq (1-\lambda)^k |I_x|$ for each k. As $|I_x| \leq \frac{\pi}{2}$, we then have that, for $0 < \varepsilon < \frac{1}{10}$, less than $\frac{2\log\varepsilon}{\log(1-\lambda)}$ of the y_j in I_x are such that $|y_j - p_x| > \varepsilon$.

Let now $0 < \varepsilon < \frac{1}{10}$. There exist $z_1, \ldots, z_{N(L,\varepsilon)}$ such that

$$L \subset \bigcup_{i=1}^{N(L,\varepsilon)} [z_i - \varepsilon, z_i + \varepsilon].$$

Note the closure of every interval I_x (as it includes p_x) intersects $[z_i - \varepsilon, z_i + \varepsilon]$ for at least one i. We now estimate the number of intervals of length 4ε necessary to cover, for a fixed i, $\cup_{x \in S\backslash L}\{y \in S : y \in \bar{I}_x \text{ with } p_x \in [z_i - \varepsilon, z_i + \varepsilon]\}$. Well, the one interval $[z_i - 2\varepsilon, z_i + 2\varepsilon]$ can cover $\cup_{x \in S\backslash L}\{y \in S : y \in \bar{I}_x \text{ with } \bar{I}_x \subset [z_i - \varepsilon, z_i + \varepsilon]\}$. Note that at most two distinct maximal (by inclusion) intervals in the collection $\{\bar{I}_x : p_x \in [z_i - \varepsilon, z_i + \varepsilon]\}$ intersect $[z_i - \varepsilon, z_i + \varepsilon]$ but are only partially covered by it. Now, suppose $\bar{I}_x \cap [z_i - \varepsilon, z_i + \varepsilon] \neq \emptyset$, where $p_x \in [z_i - \varepsilon, z_i + \varepsilon]$. So $|p_x - z_i| < \varepsilon$. As $y \notin [z_i - 2\varepsilon, z_i + 2\varepsilon]$ implies $|y - p_x| > \varepsilon$ we then have

$$\#\{y \in \bar{I}_x : |y - z_i| > 2\varepsilon\} \leq \frac{2\log\varepsilon}{\log(1-\lambda)}.$$

So

$$\cup_{x \in S\backslash L}\{y \in S : y \in \bar{I}_x \text{ with } p_x \in [z_i - \varepsilon, z_i + \varepsilon]\}$$

can be covered by $1 + \frac{4\log\varepsilon}{\log(1-\lambda)}$ intervals of length 4ε. As there are $N(L,\varepsilon)$ z_i's, we then have

$$N(S,2\varepsilon) \leq N(L,\varepsilon)\left[1+\frac{4\log\varepsilon}{\log(1-\lambda)}\right].$$

As

$$\limsup_{\varepsilon\downarrow 0} N(L,\varepsilon)\cdot\varepsilon^s = 0$$

for any $s>0$ we have that

$$\limsup_{\varepsilon\downarrow 0} N(S,\varepsilon)\cdot\varepsilon^s = 0$$

as well, so $\dim_M(S)=0$. □

We now show there exists a set $\Omega \subset [0,\pi/2)$ of Minkowski dimension zero such that M_Ω is unbounded on $L^p(\mathbb{R}^2)$ for $1\leq p<\infty$.

Theorem 4. *There exists a countable set $\Omega \subset [0,\pi/2)$ of Minkowski dimension zero for which the associated maximal operator M_Ω is unbounded on $L^p(\mathbb{R}^2)$ for $1\leq p<\infty$.*

Proof. By Theorem 1, it suffices to construct a countable set Ω that is not a finite union of N-lacunary sets for any positive integer N even though it is of Minkowski dimension zero. This is done as follows.

We define the sets B_k for $k=1,2,\dots$ inductively by

$$B_1 = \{0\}\cup\left\{\left(\frac{1}{5}\right)^j : j=1,2,3,\dots\right\},$$

$$B_k = B_{k-1}\cup\bigcup_{j=1}^{\infty}\left(\frac{1}{5^{5^k+j}}+\frac{1}{5^{5^k+j}}B_{k-1}\right),\ k=2,3,\dots .$$

We set $B=\cup_{k=1}^{\infty}B_k$. Note B is countable. Observe that each B_k is k-lacunary but not n-lacunary for any $n<k$. Accordingly B is not lacunary of order N for any finite N.

Suppose now $0<\varepsilon<\frac{1}{1000}$ and $A\subset[0,1]$ is a nonempty set such that $N(A,\varepsilon)\leq(\log\varepsilon^{-1})^k$. Define $\tilde{A}\subset[0,1]$ by

$$\tilde{A} = A\cup\left(\frac{1}{125}+\frac{1}{125}A\right)\cup\left(\frac{1}{625}+\frac{1}{625}A\right)\cup\left(\frac{1}{3125}+\frac{1}{3125}A\right)\cup\cdots .$$

Let j be the smallest positive integer such that $5^{-j}<\varepsilon$. Note $j\leq\log(\varepsilon^{-1})$. We have

$$\begin{aligned}N(\tilde{A},\varepsilon) &\leq 1+N(A,\varepsilon)+N\left(\frac{1}{125}+\frac{1}{125}A,\varepsilon\right)+\cdots+N(5^{-j}+5^{-j}A,\varepsilon)\\ &\leq N(A,\varepsilon)(j)\end{aligned}$$

$$\leq \left(\log \varepsilon^{-1}\right)^k \left(\log \varepsilon^{-1}\right)$$
$$\leq \left(\log \varepsilon^{-1}\right)^{k+1} .$$

Now $N(B_1, \varepsilon) \leq 1 - \frac{\log \varepsilon}{\log 5} < \log \varepsilon^{-1}$. As $B_k \subset \tilde{B}_{k-1}$ for any $k \geq 2$ we have $N(B_k, \varepsilon) \leq \left(\log \varepsilon^{-1}\right)^k$. As $B_k = B_{k+j}$ on $[5^{-5^k}, 1]$ for any positive j we have $N(B, \varepsilon) \leq 1 + N(B_k, \varepsilon)$ for any positive integer k satisfying $5^{-5^k} < \varepsilon$, and in particular for any integer $k \geq \log\log \varepsilon^{-1}$. So

$$N(B, \varepsilon) \leq 1 + \left(\log \varepsilon^{-1}\right)^{1+\log\log \varepsilon^{-1}} .$$

Now if $s > 0$,

$$\lim_{\varepsilon \downarrow 0} \varepsilon^s \left(1 + \left(\log \varepsilon^{-1}\right)^{1+\log\log \varepsilon^{-1}}\right)$$
$$= \lim_{\varepsilon \downarrow 0} \varepsilon^s \left(\log \varepsilon^{-1}\right)^{1+\log\log \varepsilon^{-1}}$$
$$= \lim_{x \to \infty} x \frac{e^{(\log x)^2}}{e^{xs}}$$
$$= 0 .$$

Accordingly $\dim_M B = 0$, and the desired result follows. □

Acknowledgments This work was partially supported by a grant from the Simons Foundation (#208831 to Paul Hagelstein). The author also wishes to thank Alex Iosevich and the referee for helpful comments and suggestions regarding this chapter.

References

1. Bateman, M.: Kakeya sets and directional maximal operators in the plane. Duke Math. J. **147**, 55–77 (2009)
2. Bateman, M., Katz, N.H.: Kakeya sets in Cantor directions. Math. Res. Lett. **15**, 73–81 (2008)
3. Busemann, H., Feller, W.: Zur Differentiation des Lebesguesche Integrale. Fund. Math. **22**, 226–256 (1934)
4. Córdoba, A., Fefferman, R.: On differentiation of integrals. Proc. Nat. Acad. Sci. U.S.A. **74**, 2211–2213 (1977)
5. de Guzmán, M.: Real Variable Methods in Fourier Analysis. North Holland, Amsterdam (1981)
6. Hare, K.E.: Maximal operators and Cantor sets. Canad. Math. Bull. **43**, 330–342 (2000)
7. Hare, K.E., Rönning, J.-O.: The size of Max(p) sets and density bases. J. Fourier Anal. Appl. **8**, 259–268 (2002)
8. Jessen, B., Marcinkiewicz, J., Zygmund, A.: Note on the differentiability of multiple integrals. Fund. Math. **25**, 217–234 (1935)

9. Karagulyan, G.A., Lacey, M.T.: An estimate for maximal operators associated with generalized lacunary sets (Russian. English, Russian summary). Izv. Nats. Akad. Nauk Armenii Mat. **39**, 73–82 (2004); translation in J. Contemp. Math. Anal. **39** (2004), 50–59 (2005)
10. Katz, N.H.: A counterexample for maximal operators over a Cantor set of directions. Math. Res. Lett. **3**, 527–536 (1996)
11. Nagel, A., Stein, E.M., Wainger, S.: Differentiation in lacunary directions. Proc. Nat. Acad. Sci. U.S.A. **75**, 1060–1062 (1978)
12. Nikodym, O.: Sur les ensembles accessibles. Fund. Math. **10**, 116–168 (1927)
13. Sjögren, P., Sjölin, P.: Littlewood-Paley decompositions and Fourier multipliers with singularities on certain sets. Ann. Inst. Fourier (Grenoble) **31**, 157–175 (1981)
14. Strömberg, J.-O.: Weak estimates on maximal functions with rectangles in certain directions. Ark. Mat. **15**, 229–240 (1977)

Distance Graphs in Vector Spaces Over Finite Fields

Derrick Hart, Alex Iosevich, Doowon Koh, Steven Senger, and Ignacio Uriarte-Tuero

Abstract In this chapter we systematically study various properties of the distance graph in $\mathbb{F}_q^d$, the d-dimensional vector space over the finite field $\mathbb{F}_q$ with q elements. In the process we compute the diameter of distance graphs and show that sufficiently large subsets of d-dimensional vector spaces over finite fields contain every possible finite configuration.

1 Introduction

Let $\mathbb{F}_q$ be the finite field with q elements. We assume that the characteristic of $\mathbb{F}_q$ is greater than two. For each $r \in \mathbb{F}_q^*$, the multiplicative group of $\mathbb{F}_q$, the distance graph $G_q^\Delta(r)$ in $\mathbb{F}_q^d$ is obtained by taking $\mathbb{F}_q^d$ and connecting two vertices corresponding to $x, y \in \mathbb{F}_q^d$ by an edge if $||x-y|| = r$ where

$$||x|| = x_1^2 + x_2^2 + \cdots + x_d^2.$$

D. Hart
Rutgers University, 110 Frelinghuysen Rd., Piscataway, NJ 08854-8019, USA

A. Iosevich (✉)
University of Rochester, 915 Hylan Building RC, Rochester, NY 14627, USA
e-mail: iosevich@math.rochester.edu

D. Koh
Chungbuk National University, 410 Sungbong-Ro, Heungdeok-Gu, Cheongju, Chungbuk 361-763, Korea

S. Senger
University of Delaware, 501 Ewing Hall, Newark, DE 19716, USA

I. Uriarte-Tuero
Michigan State University, East Lansing, MI 48824, USA

D. Bilyk et al. (eds.), *Recent Advances in Harmonic Analysis and Applications*, Springer Proceedings in Mathematics & Statistics 25, DOI 10.1007/978-1-4614-4565-4_14,

More generally, consider the set of colors $L_q = \{c_q^r : r \in \mathbb{F}_q^*\}$ corresponding to elements of $\mathbb{F}_q^*$. We connect two vertices corresponding to points $x, y \in \mathbb{F}_q^d$ by a c_q^r-colored edge if $||x-y|| = r$. We denote by G_q^Δ the resulting almost complete graph with the implied edges and the coloring set L_q. When q runs over powers of odd primes, we obtain a family of the almost complete distance graphs $\{G_q^\Delta\}$. For each $r \in \mathbb{F}_q^*$, the single-colored distance graph $G_q^\Delta(r)$ can be considered as a sub-graph of the almost complete distance graph G_q^Δ with $q-1$ colors.

The main goal of this chapter is a systematic study of the distance graph, including its diameter and pseudorandomness properties. In the course of this investigation we prove sharp estimates for intersections of algebraic and non-algebraic varieties in $\mathbb{F}_q^d$ and the existence of arbitrary k-point configurations in sufficiently large subsets thereof.

1.1 Diameter of the Distance Graph and Related Objects

Consider the almost complete distance graph G_q^Δ with the coloring set $L_q = \{c_q^r : r \in \mathbb{F}_q^*\}$ defined as before. For each $r \in \mathbb{F}_q^*$, we also consider the c_q^r-colored distance graph $G_q^\Delta(r)$ defined as before. Given a fixed color c_q^r in L_q, we define the diameter of the c_q^r-colored distance graph $G_q^\Delta(r)$ as follows. Given vertices x, y in $G_q^\Delta(r)$, define a *path* of length k from x to y to be a sequence $\{x^1, \ldots, x^{k+1}\}$, where x^js are distinct, $x^1 = x$, $x^{k+1} = y$, each x^j is a vertex in $G_q^\Delta(r)$ and x^i is connected to x^{i+1} by a c_q^r-colored edge for every $1 \leq i \leq k$. We say that a path from x to y is optimal if it is a path and its length is as small as possible. Define the *diameter* of $G_q^\Delta(r)$, denoted by $\mathbb{D}(G_q^\Delta(r))$, to be the largest length of the optimal path between any two vertices in $G_q^\Delta(r)$. We also define the diameter of the almost complete distance graph G_q^Δ, denoted by $\mathbb{D}(G_q^\Delta)$, as follows:

$$\mathbb{D}(G_q^\Delta) = \max_{r \in \mathbb{F}_q^*} \left(\mathbb{D}(G_q^\Delta(r)) \right).$$

Our first result in this direction is actually about more general families of graphs. For a fixed $d \geq 2$, consider a family $\{U_q\}$ where $U_q \subset \mathbb{F}_q^d$ and q is over all odd prime powers. We say that the family $\{U_q\}$ is Salem if there exists a uniform constant $C > 0$ such that for every $\xi \in \mathbb{F}_q^d \setminus \{(0, \cdots, 0)\}$,

$$|\widehat{U_q}(\xi)| \leq Cq^{-d}|U_q|^{\frac{1}{2}}$$

with a uniform constant $C > 0$ independent of q, where the Fourier transform with respect to a non-trivial additive character χ is defined and briefly reviewed in Eq. (8) and the lines that follow. We shall also see below (Lemma 3) that the family $\{S_{t_q}\}$ of spheres S_{t_q} given by

$$S_{t_q} = \{x \in \mathbb{F}_q^d : x_1^2 + \cdots + x_d^2 = t_q \in \mathbb{F}_q^*\} \tag{1}$$

is Salem, and that the number of elements of the sphere $S_r, r \neq 0$, in $\mathbb{F}_q^d$ is $\approx q^{d-1}$.

Given a set $U_q \subset \mathbb{F}_q^d$, define $G_q^{U_q}$ to the graph with vertices in $\mathbb{F}_q^d$ and two vertices, corresponding to $x, y \in \mathbb{F}_q^d$ connected by an edge if $x - y \in U_q$. We do not attach a coloring scheme in this context.

Theorem 1. *Suppose that $\{U_q\}$ is Salem and $|U_q| \geq Cq^{\frac{2d}{3}}$ with a sufficiently large constant $C > 0$. Then the diameter of $G_q^{U_q}$ is ≤ 3 provided that q is sufficiently large.*

Corollary 1. *Suppose that q is sufficiently large. For each $r \in \mathbb{F}_q^*$, the single-colored graph $G_q^{\Delta}(r)$ has diameter ≤ 3 if $d \geq 4$. In other words, we have $\mathbb{D}(G_q^{\Delta}) \leq 3$ if $d \geq 4$.*

It follows that the diameter of $G_q^{\Delta}(r)$ is ≤ 3 provided that $|S_r| \geq Cq^{\frac{2d}{3}}$ with a sufficiently large constant $C > 0$. Since $|S_r| \approx q^{d-1}$ by Theorem 6, this holds if $d \geq 4$, which completes the proof of the corollary. We can do a bit better, however.

Theorem 2. (1) *If $d \geq 4$, then $\mathbb{D}(G_q^{\Delta}) = 2$ for all $q \geq 3$.*

(2) *Suppose that $d = 3$ and Ψ is the quadratic character of $\mathbb{F}_q$. For each $r \in \mathbb{F}_q^*$ $q \geq 3$, we have*

$$\mathbb{D}(G_q^{\Delta}(r)) = \begin{cases} 2 \text{ if } & \Psi(-r) = 1 \\ 3 \text{ if } & \Psi(-r) = -1. \end{cases}$$

Namely, $\mathbb{D}(G_q^{\Delta}) = 3$.

(3) *If $d = 2$, then we have*

$$\mathbb{D}(G_q^{\Delta}) = \begin{cases} 2 \text{ if } & q = 3 \\ 3 \text{ if } & q \neq 3, 5, 9. \\ 4 \text{ if } & q = 5, 9 \end{cases}$$

Moreover, $\mathbb{D}(G_q^{\Delta}(r)) \neq 2$ for any $r \in \mathbb{F}_q^*$ when $q \equiv 3 \pmod 4$ and $q \neq 3$.

Remark 1. Recently, the authors [10] also studied the diameter of distance graphs in two dimensions. They proved that if $d = 2, q \equiv 1 \pmod 4$, then $\mathbb{D}(G_q^{\Delta})$ is three or four. The third item in our Theorem 2 makes their work clear. Moreover, although aforementioned authors [10] claimed that if $d = 2, q \equiv 3 \pmod 4$, then $\mathbb{D}(G_q^{\Delta}) = 3$, their argument does not justify their claim if $q = 3$. Let us see this. In the proof of Theorem 2 in [10], they claim that if $g(x) = (4 - x)x$, then there exist $u, v \neq 0 \in \mathbb{F}_q$ such that $g(u)$ is a square in $\mathbb{F}_q$, while $g(v)$ is not. However, $g(x)$ can not be -1 if $x \in \mathbb{F}_3^*$. Since $-1 \in \mathbb{F}_q$ is the unique non-square number, the claim is not true. In fact, $\mathbb{D}(G_3^{\Delta}) = 2$ as our proof of Theorem 2 says. Similar questions were recently addressed in [1].

1.2 Kaleidoscopic Pseudorandomness

We say that the family of graphs $\{G_j\}_{j=1}^{\infty}$ with the set of colors

$$L_j = \{c_j^1, c_j^2, \ldots, c_j^{|L_j|}\}$$

and the edge set $\mathscr{E}_j = \cup_{i=1}^{|L_j|}\mathscr{E}_j^i$, with $\mathscr{E}_j^i$ corresponding to the color c_j^i, is *kaleidoscopically pseudo-random* if there exist constants $C, C' > 0$ such that the following conditions are satisfied:

$$|G_j| \to \infty \text{ as } j \to \infty. \tag{2}$$

$$\frac{1}{C'}|\mathscr{E}_j^{i'}| \leq |\mathscr{E}_j^i| \leq C'|\mathscr{E}_j^{i'}| \text{ for all } 1 \leq i, i' \leq |L_j|. \tag{3}$$

- $\{G_j\}_{j=1}^{\infty}$ is asymptotically complete in the sense that

$$\lim_{j\to\infty} \frac{\binom{|G_j|}{2} - \sum_{i=1}^{|L_j|}|\mathscr{E}_j^i|}{\binom{|G_j|}{2}} = 0. \tag{4}$$

- If $1 \leq k-1 \leq n$ and $L_j' \subset L_j$, with $|L_j'| \leq |L_j| - \binom{k}{2} + n$, then any sub-graph H of G_j of size

$$\geq C|G_j|^{\frac{k-1}{k}}|L_j|^{\frac{n}{k}}, \tag{5}$$

  contains every possible sub-graph with k vertices and n edges with an arbitrary edge color distribution from L_j'.

See, for example, a survey by Krivelevich and Sudakov [5] for related notions of pseudo-random graphs, examples and applications. The first result of this chapter is the following.

Theorem 3. *If the dimension d is odd, then the above defined family of the almost complete distance graphs, $\{G_q^{\Delta}\}$, is kaleidoscopically pseudo-random.*

This is our main graph-theoretic result. A somewhat stronger, though more technical, version of Theorem 3 is Theorem 4 below.

The proof shows that in this setting, we could replace the constant C', in the definition of kaleidoscopic pseudorandomness, with $(1 + o(1))$. The constant, C, that the proof yields will depend on the number of vertices exponentially. That is, C will be proportional to B^k, for some constant B.

We actually prove a little more as the arguments below indicate. We shall see that under the set of hypotheses corresponding to kaleidoscopic pseudorandomness, every finite geometric configuration in $\mathbb{F}_q^d$ is realized. See [9, 11] where related questions are studied using graph-theoretic methods.

The first item in the definition of kaleidoscopic pseudo-randomness above Eq. (2) is automatic as the $|G_j| = q^d$, by construction. The second and third items, Eqs. (3) and (4), respectively, are easy special cases of the following calculation. It is implicit in [8] or [3] and is a direct result from Theorem 6.

Lemma 1. *For any $r \in \mathbb{F}_q$,*

$$|\{(x,y) \in \mathbb{F}_q^d \times \mathbb{F}_q^d : ||x-y|| = r\}| = \begin{cases} (2+o(1))\,q^{2d-1} & \text{if } d=2, r=0 \\ (1+o(1))\,q^{2d-1} & \text{otherwise,} \end{cases}$$

where $o(1)$ means that the quantity goes to 0 as $q \to \infty$.

We are now ready to address the meat of our definition of kaleidoscopic pseudo-randomness, which is the fourth item (5).

Definition 1. Given $L' \subset \mathbb{F}_q^*$, $E \subset \mathbb{F}_q^d$, and a set J of ordered pairs of $\{1,2,\ldots,k\}$ of the form (i,j), where $i < j$, a k-point J-configuration in E is a set of k points $\{x^1, x^2, \ldots, x^k\} \subset E$ such that

$$\begin{cases} ||x^i - x^j|| = a_{ij} \in L' & \text{for all } (i,j) \in J \\ ||x^i - x^j|| \neq 0 & \text{for all } (i,j) \notin J. \end{cases}$$

Denote the set of all k-point J-configurations by $\mathrm{T}_k^J(E)$, which also depends on the choice of L', but our result below is independent of the choice L'.

The item (5) follows from the following geometric estimate:

Theorem 4. *Let $E \subset \mathbb{F}_q^d$ and $d \geq 3$ is odd. Suppose that $1 \leq k-1 \leq n \leq d$ and*

$$|E| \geq C q^{d\left(\frac{k-1}{k}\right)} q^{\frac{n}{k}} \tag{6}$$

with a sufficiently large constant $C > 0$. Then for any

$$J \subset \{1,2,\ldots,k\}^2 \setminus \{(i,i) : 1 \leq i \leq k\}$$

with $|J| = n$, we have

$$|\mathrm{T}_k^J(E)| = (1+o(1))|E|^k q^{-n}.$$

Our proof uses geometric and character sum machinery similar to the one used in [3], and the proof is similar to the proof of its earlier version in [2]. In the former paper, Theorem 4 is proved in the case $k = 2$ and $n = 1$, and in the latter article Theorem 4 is demonstrated in the case of general k and $n = \binom{k}{2}$. Thus Theorem 4 and, consequently, Theorem 3 may be viewed as filling the gap between these results.

2 Classical Exponential Sums and Fourier Decay Estimates

In this section, we collect the well-known facts related to exponential sums and apply them for Fourier decay estimates. Such facts shall be used in the next sections. Let χ be a non-trivial additive character of $\mathbb{F}_q$ and Ψ a quadratic character of $\mathbb{F}_q$, that is, $\Psi(ab) = \Psi(a)\Psi(b)$ and $\Psi^2(a) = 1$ for all $a, b \in \mathbb{F}_q^*$ but $\Psi \not\equiv 1$. For each $a \in \mathbb{F}_q$, the Gauss sum $G_a(\Psi, \chi)$ is defined by

$$G_a(\Psi,\chi) = \sum_{s \in \mathbb{F}_q^*} \Psi(s)\chi(as).$$

The magnitude of the Gauss sum is given by the relation

$$|G_a(\Psi,\chi)| = \begin{cases} q^{\frac{1}{2}} & \text{if} \quad a \neq 0 \\ 0 & \text{if} \quad a = 0. \end{cases}$$

Remark 2. Here, and throughout this chapter, we denote by χ and Ψ the canonical additive character and the quadratic character of $\mathbb{F}_q$, respectively. Recall that if Ψ is the quadratic character of $\mathbb{F}_q$, then $\Psi(0) = 0$, and $\Psi(s) = 1$ if s is a square number in $\mathbb{F}_q^*$ and $\Psi(s) = -1$ otherwise.

The following theorem provided us the explicit formula of the Gauss sum $G_1(\Psi, \chi)$. For the nice proof, see ([7], p. 199).

Theorem 5. *Let $\mathbb{F}_q$ be a finite field with $q = p^l$, where p is an odd prime and $l \in \mathbb{N}$. Then we have*

$$G_1(\Psi,\chi) = \begin{cases} (-1)^{l-1} q^{\frac{1}{2}} & \text{if} \quad p = 1 \ (mod\, 4) \\ (-1)^{l-1} i^l q^{\frac{1}{2}} & \text{if} \quad p = 3 \ (mod\, 4). \end{cases}$$

Also, we have

$$\sum_{s \in \mathbb{F}_q} \chi(as^2) = \Psi(a) G_1(\Psi,\chi) \quad \text{for any} \quad a \neq 0. \tag{7}$$

For a nice proof for this equality and the magnitude of Gauss sums, see [7] or [4]. As the direct application of the equality in Eq. (7), we have the following estimate.

Lemma 2. *For $\beta \in \mathbb{F}_q^k$ and $t \neq 0$, we have*

$$\sum_{\alpha \in \mathbb{F}_q^k} \chi(t\alpha \cdot \alpha + \beta \cdot \alpha) = \chi\left(\frac{\|\beta\|}{-4t}\right) \Psi^k(t) \left(G_1(\Psi,\chi)\right)^k.$$

Proof. It follows that

$$\sum_{\alpha\in\mathbb{F}_q^k}\chi(t\alpha\cdot\alpha+\beta\cdot\alpha)=\prod_{j=1}^{k}\sum_{\alpha_j\in\mathbb{F}_q}\chi(t\alpha_j^2+\beta_j\alpha_j).$$

Completing the square in α_j-variables, applying a change of variables, $\alpha_j+\frac{\beta_j}{2t}\to\alpha_j$, and using the inequality in Eq. (7), the proof is complete. □

Due to the explicit formula for the Gauss sum $G(\Psi,\chi)$, we can count the number of the elements in the sphere $S_r\subset\mathbb{F}_q^d$ defined as before. The following theorem enables us to see the exact number of the elements of the sphere S_r which depends on the radius r, dimensions, and the size of the underlining finite field $\mathbb{F}_q$. This result can be found in [8].

Theorem 6. *Let $S_r\subset\mathbb{F}_q^d$ be the sphere defined as in Eq. (1). For each $r\neq 0$, we have*

$$|S_r|=\begin{cases} q^{d-1}-q^{\frac{d-2}{2}}\Psi\left((-1)^{\frac{d}{2}}\right) & \text{if } d \text{ is even}\\ q^{d-1}+q^{\frac{d-1}{2}}\Psi\left((-1)^{\frac{d-1}{2}}r\right) & \text{if } d \text{ is odd.}\end{cases}$$

Moreover,

$$|S_0|=\begin{cases} q^{d-1}+(q-1)q^{\frac{d-2}{2}}\Psi\left((-1)^{\frac{d}{2}}\right) & \text{if } d \text{ is even}\\ q^{d-1} & \text{if } d \text{ is odd.}\end{cases}$$

Recall that given a function $f:\mathbb{F}_q^m\to\mathbb{C}$, the Fourier transform with respect to a non-trivial additive character χ on $\mathbb{F}_q$ is given by the relation

$$\widehat{f}(\xi)=q^{-m}\sum_{x\in\mathbb{F}_q^m}\chi(-x\cdot\xi)f(x). \tag{8}$$

Also recall that

$$f(x)=\sum_{\xi\in\mathbb{F}_q^m}\chi(x\cdot\xi)\widehat{f}(\xi) \tag{9}$$

and

$$\sum_{\xi\in\mathbb{F}_q^m}|\widehat{f}(\xi)|^2=q^{-m}\sum_{x\in\mathbb{F}_q^m}|f(x)|^2. \tag{10}$$

We shall also need the following estimates based on classical Gauss and Kloosterman sum bounds, where $S_t(x)$ is the indicator function of S_t evaluated at x.

Lemma 3. *With the notation above, for any $t\neq 0$, $\xi\neq(0,\ldots,0)$,*

$$|\widehat{S_t}(\xi)|\le 2q^{-\frac{d+1}{2}}. \tag{11}$$

Moreover, if d is odd, then

$$\left|\sum_{t\neq 0}\widehat{S}_t(\xi)\right| \leq q^{-\frac{d+1}{2}} \tag{12}$$

and

$$\widehat{S}_t(0,\ldots,0) = q^{-d}|S_t| = (1+o(1))q^{-1}, \tag{13}$$

where $o(1)$ means that the quantity goes to 0 as $q \to \infty$ and here, throughout the chapter, we identify a set with its characteristic function.

3 Proof of Theorem 1

We shall deduce Theorem 1 from the following estimate.

Theorem 7. *$U,E,F \subset \mathbb{F}_q^d$ such that $\{U\}$ is Salem and*

$$|E||F| \geq C\frac{q^{2d}}{|U|}$$

with a sufficiently large constant $C > 0$. Then

$$\nu_U = \{(x,y) \in E \times F : x - y \in U\} > 0.$$

Taking Theorem 7 for granted, for a moment, Theorem 1 follows instantly. Indeed, take x,y with $x \neq y$ in $\mathbb{F}_q^d$. Let $E = U + x$ and $F = U + y$. It follows that $|E| = |F| = |U|$, so $|E||F| = |U|^2$. We conclude from Theorem 7 that if $|U| \geq Cq^{\frac{2d}{3}}$ with a sufficiently large constant $C > 0$, then there exists $x' \in U + x$ and $y' \in U + y$ such that $x' - y' \in U$. This implies that the diameter of G_q^U is at most three as desired.

To prove Theorem 7, observe that

$$\begin{aligned}
\nu_U &= \sum_{x,y} E(x)F(y)U(x-y) \\
&= \sum_{x,y}\sum_{m} \widehat{U}(m)\chi((x-y)\cdot m)E(x)F(y) \\
&= q^{2d}\sum_{m} \overline{\widehat{E}(m)}\widehat{F}(m)\widehat{U}(m) \\
&= |E||F||U|q^{-d} + q^{2d}\sum_{m\neq(0,\ldots,0)} \overline{\widehat{E}(m)}\widehat{F}(m)\widehat{U}(m) = I + II.
\end{aligned}$$

By assumption,

$$|II| \le Cq^{2d} \times q^{-d}|U|^{\frac{1}{2}} \times \sum_{m\neq(0,\dots,0)} \overline{|\widehat{E}(m)|}|\widehat{F}(m)|$$

$$\le Cq^{d}|U|^{\frac{1}{2}}\left(\sum|\widehat{E}(m)|^{2}\right)^{\frac{1}{2}} \times \left(\sum|\widehat{F}(m)|^{2}\right)^{\frac{1}{2}}$$

$$= C|U|^{\frac{1}{2}}|E|^{\frac{1}{2}}|F|^{\frac{1}{2}}$$

by Eq. (10). Comparing I and II we complete the proof of Theorem 7.

4 Proof of Theorem 2

In this section, we provide the proof of Theorem 2. We first introduce the known explicit formula for the number of intersection points of two different spheres with the same non-zero radius. We just need to know $|S_r \cap (S_r + b)|$ for all $r \in \mathbb{F}_q^*$, $b \in \mathbb{F}_q^d \setminus \{(0,\dots,0)\}$. It is clear that $|S_r \cap (S_r + b)|$ is the same as the number of common solutions in $\mathbb{F}_q^d$ of the equations

$$\begin{cases} x_1^2 + \cdots + x_d^2 = r \\ b_1x_1 + \cdots + b_dx_d = 2^{-1}\|b\|, \end{cases}$$

where $x = (x_1,\dots,x_d) \in \mathbb{F}_q^d$ and $b = (b_1,\dots,b_d) \neq (0,\dots,0)$. The number of common solutions is well known. For example, see ([7], p. 341) where the statement of the following theorem is implicitly given. See also [6].

Theorem 8. *For each $r \in \mathbb{F}_q^*$, and $b = (b_1,\dots,b_d) \in \mathbb{F}_q^d \setminus \{(0,\dots,0)\}, d \ge 2$, we have the following:*
If $\|b\| \neq 0$ and $r = 4^{-1}\|b\|$, then

$$|S_r \cap (S_r + b)| = \begin{cases} q^{d-2} & \text{if } d \text{ is even} \\ q^{d-2} + q^{(d-3)/2}(q-1)\Psi\left((-1)^{(d-1)/2}\|b\|\right) & \text{if } d \text{ is odd.} \end{cases}$$

If $\|b\| \neq 0$ and $r \neq 4^{-1}\|b\|$, then

$$|S_r \cap (S_r + b)| = \begin{cases} q^{d-2} + q^{(d-2)/2}\Psi\left((-1)^{d/2}(4^{-1}\|b\|^2 - r\|b\|)\right) & \text{if } d \text{ is even} \\ q^{d-2} - q^{(d-3)/2}\Psi\left((-1)^{(d-1)/2}\|b\|\right) & \text{if } d \text{ is odd.} \end{cases}$$

Moreover, if $\|b\| = 0$, then we have

$$|S_r \cap (S_r + b)| = \begin{cases} q^{d-2} - q^{(d-2)/2}\Psi\left((-1)^{d/2}\right) & \text{if } d \text{ is even} \\ q^{d-2} + q^{(d-1)/2}\Psi\left((-1)^{(d-1)/2}r\right) & \text{if } d \text{ is odd.} \end{cases}$$

To complete the proof of Theorem 2, we also need the following reduction.

Lemma 4. *For each $a,b \in \mathbb{F}_q^d$ with $a \neq b$, and $r \neq 0$, we have*

$$|\{(x,y) \in (S_r+a) \times (S_r+b) : \|x-y\| = r\}| = q^{-d}|S_r|^3 + I + II,$$

where

$$I = \begin{cases} q^{-3}(G_1(\Psi,\chi))^d & \text{if } d \text{ is even} \\ -q^{-3}(G_1(\Psi,\chi))^{d+3}\Psi(r) & \text{if } d \text{ is odd,} \end{cases}$$

$$II = q^{-3}(G_1(\Psi,\chi))^{2d} \sum_{u,v,w\neq 0: u-v+w\neq 0} \Psi^d(uvw)\Psi^d(-(u-v+w))\chi\left(\frac{\|a-b\|}{u-v+w}\right)\chi\left(-r\left(\frac{1}{u}-\frac{1}{v}+\frac{1}{w}\right)\right). \tag{14}$$

Proof. For each $a,b \in \mathbb{F}_q^d$ with $a \neq b$, we have

$$\begin{aligned}
&|\{(x,y) \in (S_r+a) \times (S_r+b) : \|x-y\| = r\}| \\
&= \sum_{x,y} S_r(x)S_r(y)S_r(x-y+a-b) \\
&= \sum_{x,y}\sum_{m} S_r(x)S_r(y)\chi((x-y+a-b)\cdot m)\widehat{S_r}(m) \\
&= q^{-d}|S_r|^3 + q^{2d}\sum_{m\neq(0,\ldots,0)} \chi((a-b)\cdot m)\widehat{S_r}(m)|\widehat{S_r}(m)|^2.
\end{aligned}$$

Thus, it suffices to show that

$$q^{2d}\sum_{m\neq(0,\ldots,0)} \chi((a-b)\cdot m)\widehat{S_r}(m)|\widehat{S_r}(m)|^2 = I + II, \tag{15}$$

where I and II are defined as in statement of Lemma 4. By orthogonality of χ and Lemma 2, notice that if $m \neq (0,\cdots,0)$, then $\widehat{S_r}(m)$ is given by

$$\widehat{S_r}(m) = q^{-d-1}(G_1(\Psi,\chi))^d \sum_{s\neq 0}\chi\left(\frac{\|m\|}{-4s} - sr\right)\Psi^d(s).$$

Plugging this into Eq. (15), the left-hand side value in Eq. (15) is given by

$$\begin{aligned}
&q^{-3}(G_1(\Psi,\chi))^d \sum_{m\neq(0,\ldots,0)}\sum_{u,v,w\neq 0} \Psi^d(uvw)\chi((a-b)\cdot m)\chi\left(\frac{\|m\|}{-4}\left(\frac{1}{u}-\frac{1}{v}+\frac{1}{w}\right)\right)\chi(-r(u-v+w)) \\
&= q^{-3}(G_1(\Psi,\chi))^d \sum_{m}\sum_{u,v,w\neq 0} \Psi^d(uvw)\chi((a-b)\cdot m)\chi\left(\frac{\|m\|}{-4}\left(\frac{1}{u}-\frac{1}{v}+\frac{1}{w}\right)\right)\chi(-r(u-v+w)) \\
&\quad -q^{-3}(G_1(\Psi,\chi))^d \sum_{u,v,w\neq 0} \Psi^d(uvw)\chi(-r(u-v+w)) = A + B.
\end{aligned}$$

To complete the proof, it is enough to show that $B = I$ and $A = II$. The value B above is given by

$$\begin{cases} q^{-3}(G_1(\Psi,\chi))^d & \text{if } \; d \; \text{ is even} \\ -q^{-3}(G_1(\Psi,\chi))^{d+3}\Psi(r) & \text{if } \; d \; \text{ is odd} \end{cases} \tag{16}$$

That $B = I$ easily follows from properties of the definition of the Gauss sum. It remains to show that $A = II$. Applying a change of variables, $u^{-1} \to u, v^{-1} \to v, w^{-1} \to w$, the value A above is given by

$$q^{-3}(G_1(\Psi,\chi))^d \sum_m \sum_{u,v,w\neq 0} \Psi^d(uvw)\chi((a-b)\cdot m)\chi\left(\frac{\|m\|}{-4}(u-v+w)\right)$$
$$\chi\left(-r\left(\frac{1}{u}-\frac{1}{v}+\frac{1}{w}\right)\right).$$

Since $a \neq b$, the sum over $m \in \mathbb{F}_q^d$ vanishes if $u-v+w=0$. Thus, we may assume that $u-v+w \neq 0$. Therefore using Lemma 2, the value A above takes the form

$$q^{-3}(G_1(\Psi,\chi))^{2d} \sum_{u,v,w\neq 0: u-v+w\neq 0} \Psi^d(uvw)\Psi^d(-(u-v+w))\chi\left(\frac{\|a-b\|}{u-v+w}\right)$$
$$\chi\left(-r\left(\frac{1}{u}-\frac{1}{v}+\frac{1}{w}\right)\right),$$

where we used $\Psi(4^{-1}) = 1$. Thus, $A = II$. This completes the proof. □

4.1 *The Proof of the First and Second Part of Theorem 2*

We first prove the first part of Theorem 2. From Theorem 8, we see that if $d \geq 4$, then for each $r \in \mathbb{F}_q^*$, $|S_r \cap (S_r + b)| \geq 1$ for all $b \in \mathbb{F}_q^d$ and for all $q \geq 3$. This implies that if $d \geq 4$, then two different spheres in $\mathbb{F}_q^d$ with the same non-zero radius $t \neq 0$ always intersect. Thus, the first part of Theorem 2 immediately follows, because it is clear that the diameter of G_q^Δ is not one. We now prove the second part of Theorem 2. Suppose $d = 3$ and $b \in \mathbb{F}_q^d \setminus \{(0,\ldots,0)\}$. Then, Theorem 8 says that for each $r \in \mathbb{F}_q^*$,

$$|S_r \cap (S_r + b)| = \begin{cases} q + (q-1)\Psi((-1)\|b\|) & \text{if } \; \|b\| \neq 0, \; r = 4^{-1}\|b\| \\ q - \Psi((-1)\|b\|) & \text{if } \; \|b\| \neq 0, \; r \neq 4^{-1}\|b\| \\ q + q\Psi(-r) & \text{if } \; \|b\| = 0. \end{cases} \tag{17}$$

Thus, if $\Psi(-r) = 1$, then $|S_r \cap (S_r + b)| \geq 1$ for all $q \geq 3$ and for all $b \in \mathbb{F}_q^3$. This implies that if $\Psi(-r) = 1$, then $\mathbb{D}(G_q^{\Delta}(r)) = 2$ for all $q \geq 3$. On the other hand, if $\Psi(-r) = -1$, then Eq. (17) shows that $|S_r \cap (S_r + b)| = 0$ if $\|b\| = 0$. Thus, there exist two different disjoint circles with the same non-zero radius and so the diameter of $G_q^{\Delta}(r)$ can not be two. In order to complete the proof of the second part of Theorem 2, it therefore suffices to show that

$$\mathbb{D}(G_q^{\Delta}(r)) \leq 3 \quad \text{for all} \quad r \in \mathbb{F}_q^*,\ q \geq 3.$$

This follows from the following claim: for every $a, b \in \mathbb{F}_q^3$ with $a \neq b, r \in \mathbb{F}_q^*$ and $q \geq 3$, we have

$$|\{(x,y) \in (S_r + a) \times (S_r + b) : \|x - y\| = r\}| > 0.$$

Let us prove the claim. Since $d = 3$, Lemma 4 yields the following:

$$|\{(x,y) \in (S_r + a) \times (S_r + b) : \|x - y\| = r\}| \geq q^{-3}|S_r|^3 - |I| - |II|, \tag{18}$$

where I and II are given as in Lemma 4. From Theorem 6, we see that if $d = 3$ and $r \neq 0$, then

$$|S_r| \geq q^2 - q, \tag{19}$$

and using the fact that the absolute value of the Gauss sum $G_1(\Psi, \chi)$ is exactly $q^{\frac{1}{2}}$, we see

$$|I| = 1. \tag{20}$$

In order to estimate the value $|II|$, use the trivial estimation along with the bound of the Gauss sum, and then we obtain

$$|II| \leq \sum_{u,v,w \neq 0 : u - v + w \neq 0,} 1,$$

We observe the following:

$$\sum_{u,v,w \neq 0 : u - v + w \neq 0} 1 = (q-1)^2 + (q-1)(q-2)^2. \tag{21}$$

Thus, Eq. (21) holds. From Eqs. (19)–(21) along with Eq. (18), we obtain that if $d = 3$, then

$$\begin{aligned} |\{(x,y) \in (S_r + a) \times (S_r + b) : \|x - y\| = r\}| &\geq q^{-3}(q^2 - q)^3 - 1 - (q-1)^2 - (q-1)(q-2)^2 \\ &= (q-1)\left(q - 2 - (q-1)^{-1}\right), \end{aligned}$$

which is greater than zero if $q \geq 3$. This proves that the diameter of G_q^{Δ} in three dimension is less than equal to three. Thus, we complete the proof of the second part of Theorem 2.

4.2 *The Proof of the Third Part of Theorem 2*

We first observe from Theorem 8 that if $d = 2$, then for each $r \in \mathbb{F}_q^*$,

$$|S_r \cap (S_r + b)| = \begin{cases} 1 & \text{if} \quad \|b\| \neq 0,\ r = 4^{-1}\|b\| \\ 1 + \Psi((-1)(4^{-1}\|b\|^2 - r\|b\|)) & \text{if} \quad \|b\| \neq 0,\ r \neq 4^{-1}\|b\| \\ 1 - \Psi(-1) & \text{if} \quad \|b\| = 0. \end{cases} \tag{22}$$

Case A *(The diameter of G_3^Δ is exactly two in two dimensions).* Suppose $d = 2, q = 3$. We shall show that $\mathbb{D}(G_3^\Delta) = 2$. It suffices to show that for every $r \in \mathbb{F}_3^*$, we have

$$|S_r \cap (S_r + b)| \geq 1 \quad \text{for all } b \in \mathbb{F}_3^2 \setminus \{(0,0)\}.$$

However, since $-1 \in \mathbb{F}_3$ is not a square number, $\Psi(-1) = -1$. Using this fact along with Eq. (22), it is enough to see that for all $r \in \mathbb{F}_3^*$ and $b \in \mathbb{F}_3^2 \setminus \{(0,0)\}$ such that $\|b\| \neq 0$ and $r \neq 4^{-1}\|b\|$, we have

$$\Psi((-1)(4^{-1}\|b\|^2 - r\|b\|)) = 1. \tag{23}$$

Case B *(The diameter of G_q^Δ in two dimensions is greater than two if $q \neq 3$).* Here, we show that

$$\mathbb{D}(G_q^\Delta) \neq 2 \quad \text{if} \quad q \neq 3,\ d = 2. \tag{24}$$

If $q \equiv 1\ (mod\ 4)$, then $\Psi(-1) = 1$, because $-1 \in \mathbb{F}_q$ is a square number. In this case, Eq. (22) says that there exist two disjoint circles with the same nonzero radius. Thus, Eq. (24) holds if $q \equiv 1\ (mod\ 4)$. We now consider the case where $q \equiv 3\ (mod\ 4)$. From Eq. (22), we see that for each $r \in \mathbb{F}_q^*$,

$$|S_r \cap (S_r + b)| = 1 + \Psi\left((-1)(4^{-1}\|b\|^2 - r\|b\|)\right) \quad \text{for all} \quad b \in \mathbb{F}_q^2 \quad \text{with} \quad \|b\| \neq 0,\ 4r.$$

In the case when $q \equiv 3\ (mod\ 4)$, the claim (24) holds if we can show that for each $r \in \mathbb{F}_q^*$, there exists $b \in \mathbb{F}_q^2$ with $\|b\| \neq 0,\ 4r$ such that

$$\Psi\left((-1)(4^{-1}\|b\|^2 - r\|b\|)\right) = -1. \tag{25}$$

By contradiction, if we assume that Eq. (25) is not true, then we must have that for all $b \in \mathbb{F}_q^2$ with $\|b\| \neq 0, 4r$, which, by the definition of Ψ, would imply

$$\Psi\left((-1)(4^{-1}\|b\|^2 - r\|b\|)\right) = 1, \tag{26}$$

where we also used the fact that $\Psi\left((-1)(4^{-1}\|b\|^2 - r\|b\|)\right) \neq 0$ if $\|b\| \neq 0, 4r$. From Theorem 6, we see that for each $s \in \mathbb{F}_q$, there exists $b \in \mathbb{F}_q^2$ such that $\|b\| = s$. Using this along with Eq. (26), we must have

$$\sum_{s \in \mathbb{F}_q \setminus \{0,4r\}} \Psi\left((-1)(4^{-1}s^2 - rs)\right) = (q-2).$$

Since $\Psi(0)=0$, we see that $\Psi(((-1)(4^{-1}s^2-rs))=0$ for $s=0,4r$. Thus, above equality is same as the following:

$$\sum_{s\in\mathbb{F}_q}\Psi\left((-1)(4^{-1}s^2-rs)\right)=(q-2).$$

However, this is impossible if $q\geq 5$ due to the following theorem (see [7], p. 225).

Theorem 9. *Let Ψ be a multiplicative character of $\mathbb{F}_q$ of order $k>1$ and let $g\in\mathbb{F}_q[x]$ be a monic polynomial of positive degree that is not an kth power of a polynomial. Let e be the number of distinct roots of g in its splitting field over $\mathbb{F}_q$. Then for every $t\in\mathbb{F}_q$ we have*

$$\left|\sum_{s\in\mathbb{F}_q}\Psi(tg(s))\right|\leq(e-1)q^{1/2}.$$

To see this, note from Theorem 9 that we must have

$$\left|\sum_{s\in\mathbb{F}_q}\Psi\left((-1)(4^{-1}s^2-rs)\right)\right|\leq q^{1/2},$$

where we used $r\neq 0$. We finish proving that $\mathbb{D}(G_q^{\Delta})\neq 2$ if $q\neq 3$ $d=2$.

Case C *(In two dimensions, the diameter of G_q^{Δ} is three unless q is* 3,5,9, *or* 13*).* Suppose that $d=2$ and $q\neq 3,5,9,13$, as the case $q=13$ will be handled separately. We shall prove that the diameter of G_q^{Δ} is exactly three unless $q=3,5,9,13$. Since we have already seen that the diameter of $\mathbb{D}(G_q^{\Delta})$ is not two for $q\neq 3$, and the diameter of $\mathbb{D}(G_3^{\Delta})$ is two, it suffices to show that the diameter of $\mathbb{D}(G_q^{\Delta})$ is less than equal to three for $q\neq 5,9,13$. This follows from the following theorem:

Theorem 10. *For each $r\in\mathbb{F}_q^*$, if $a,b\in\mathbb{F}_q^2$ with $a\neq b$, and $q\neq 5,9,13$, then we have*

$$|\{(x,y)\in(S_r+a)\times(S_r+b):\|x-y\|=r\}|>0.$$

Proof. If $d=2$, then $\Psi^d(s)=1$ for $s\neq 0$. Therefore, Lemma 4 yields the following estimation:

$$|\{(x,y)\in(S_r+a)\times(S_r+b):\|x-y\|=r\}|=q^{-2}|S_r|^3+q^{-3}\left(G_1(\Psi,\chi)\right)^2+M, \quad (27)$$

where M is given by

$$M=q^{-3}\left(G_1(\Psi,\chi)\right)^4\sum_{\substack{u,v,w\neq 0\\ :u-v+w\neq 0}}\chi\left((u-v+w)^{-1}\|a-b\|\right)\chi\left(-r(u^{-1}-v^{-1}+w^{-1})\right).$$

Fix $u \neq 0$. Putting $u^{-1}v = s, u^{-1}w = t$, we see that

$$M = q^{-3}(G_1(\Psi,\chi))^4 \sum_{u\neq 0} \sum_{s\neq 0} \sum_{t\neq 0: s-t\neq 1} \chi\left(-r\left(\frac{1}{u} + \frac{s-t}{ust}\right)\right) \chi\left(\frac{\|a-b\|}{u(1-s+t)}\right).$$

Using a change of variables, $u^{-1} \to u$, we have

$$M = q^{-3}(G_1(\Psi,\chi))^4 \sum_{u\neq 0} \sum_{s\neq 0} \sum_{t\neq 0: s-t\neq 1} \chi\left(\left(-r + \frac{-rs+rt}{st} + \frac{\|a-b\|}{1-s+t}\right) u\right).$$

Note that the sum over $u \neq 0$ is -1 if $-r+(-rs+rt)/st+\|a-b\|/(1-s+t) \neq 0$, and $q-1$ otherwise.

Thus, the value M can be written by

$$\begin{aligned} M = & -q^{-3}(G_1(\Psi,\chi))^4 \sum_{s\neq 0} \sum_{\substack{t\neq 0: s-t\neq 1 \\ -r+(-rs+rt)/st+\|a-b\|/(1-s+t)\neq 0}} 1 \\ & + q^{-3}(G_1(\Psi,\chi))^4 (q-1) \sum_{s\neq 0} \sum_{\substack{t\neq 0: s-t\neq 1 \\ -r+(-rs+rt)/st+\|a-b\|/(1-s+t)=0}} 1 \\ = & \, q^{-2}(G_1(\Psi,\chi))^4 \sum_{s\neq 0} \sum_{\substack{t\neq 0: s-t\neq 1 \\ -r+(-rs+rt)/st+\|a-b\|/(1-s+t)=0}} 1 \\ & - q^{-3}(G_1(\Psi,\chi))^4 \sum_{s\neq 0} \sum_{t\neq 0: s-t\neq 1} 1. \end{aligned}$$

Notice

$$\sum_{s\neq 0} \sum_{t\neq 0: s-t\neq 1} 1 = (q-2)^2 + (q-1).$$

Thus the term M is given by

$$\begin{aligned} M = & \, q^{-2}(G_1(\Psi,\chi))^4 \sum_{s\neq 0} \sum_{\substack{t\neq 0: s-t\neq 1 \\ -r+(-rs+rt)/st+\|a-b\|/(1-s+t)=0}} 1 \\ & - q^{-3}\left((q-2)^2 + (q-1)\right)(G_1(\Psi,\chi))^4. \end{aligned}$$

By Theorem 6 and Theorem 5, we see that $|S_r| = q - \Psi(-1)$ if $d = 2$, and $(G_1(\Psi,\chi))^4 = q^2$, respectively. Thus, we aim to show that the following value is positive:

$$\begin{aligned} & |\{(x,y) \in (S_r+a) \times (S_r+b) : \|x-y\| = r\}| \\ & = q^{-2}(q-\Psi(-1))^3 + q^{-3}(G_1(\Psi,\chi))^2 - q^{-1}(q^2-3q+3) \\ & \quad + \sum_{s\neq 0} \sum_{\substack{t\neq 0: s-t\neq 1 \\ -r+(-rs+rt)/st+\|a-b\|/(1-s+t)=0}} 1. \end{aligned} \tag{28}$$

Case I. Suppose that $q = p^l$ for some odd prime $p \equiv 3\,(mod\,4)$ with l odd.

Then $q \equiv 3\,(mod\,4)$, which means that -1 is not a square number in $\mathbb{F}_q$ and so $\Psi(-1) = -1$. We also note from Theorem 5 that $(G_1(\Psi,\chi))^2 = -q$. Thus, the value in Eq. (28) can be estimated as follows:

$$|\{(x,y) \in (S_r+a) \times (S_r+b) : \|x-y\| = r\}|$$

$$\geq q^{-2}(q+1)^3 - q^{-2} - q^{-1}\,(q^2-3q+3) = 6$$

which is greater than zero. This completes the proof of Theorem 10 in the case when $q \equiv 3\ (mod\ 4)$.

Case II. Suppose that $q = p^l$ for some odd prime $p \equiv 3\,(mod\,4)$ with l even, or $q = p^l$ with $p \equiv 1\,(mod\,4)$.

Then, $q \equiv 1\,(mod\,4)$, which implies that -1 is a square number in $\mathbb{F}_q$ and so $\Psi(-1) = 1$. Moreover, $(G_1(\Psi,\chi))^2 = q$ by Theorem 5. From these observations, the value in Eq. (28) is given by

$$\begin{aligned}&|\{(x,y) \in (S_r+a) \times (S_r+b) : \|x-y\| = r\}|\\ &= q^{-2}(q-1)^3 + q^{-2} - q^{-1}(q^2-3q+3) + R(a,b,r) = R(a,b,r),\end{aligned}$$

where

$$R(a,b,r) = \sum_{s\neq 0} \sum_{\substack{t\neq 0: s-t\neq 1\\ -r+(-rs+rt)/st+\|a-b\|/(1-s+t)=0}} 1.$$

To complete the proof, it suffices to show that $R(a,b,r) > 0$.

Case II-1: Suppose that $\|a-b\| = 0$. Then, we have

$$R(a,b,r) = \sum_{s\neq 0} \sum_{\substack{t\neq 0: s-t\neq 1,\\ -st-s+t=0}} 1.$$

If $p \neq 3$ for $l \in \mathbb{N}$, then it is clear that $R(a,b,r) \geq 1$, because if we choose $s = -1, t = -2^{-1}$ then $s-t \neq 1$ and $-st-s+t = 0$. Thus, we may assume that $q = 3^l$ with l even. Since $\mathbb{F}_{3^2}$ can be considered as a subfield of any finite field $\mathbb{F}_{3^l}$ with l even up to isomorphism, it is enough to show that $R(a,b,r) \geq 1$ for a subfield, $\mathbb{F}_9$. Let i be such that $i^2 = -1$. Taking $s = i, t = 1-i$, we see that $s-t \neq 1$ and $-st-s+t = 0$. Thus, we conclude that $R(a,b,r) \geq 1$ as desired.

Case II-2: Assume that $\|a-b\| \neq 0$. Letting $c = \frac{\|a-b\|}{r} \neq 0$, we have

$$R(a,b,r) = \sum_{s\neq 0} \sum_{t\neq 0: s-t\neq 1, T^*(s,t,c)=0} 1,$$

where $T^*(s,t,c)$ is defined by

$$\begin{aligned} T^*(s,t,c) =& (c-3)st + s^2t - st^2 - s + t + s^2 + t^2 \\ =& (t+1)s^2 + \left(t(c-3) - t^2 - 1\right)s + (t+t^2). \end{aligned} \tag{29}$$

Notice that if $s-t=1$, then $t \neq 0,1$, so $T^*(t+1,t,c) = t(t+1)c \neq 0$. That is, we see that

$$R(a,b,r) = \sum_{s\neq 0} \sum_{t\neq 0: T^*(s,t,c)=0} 1,$$

as the extra condition $s-1 \neq 1$ contributes nothing to the sum.

Now, notice that $T^*(s,-1,c) = (1-c)s$, so if $c=1$, we get that $R(a,b,r) > 0$. So it is enough to consider the case that $c \neq 1$. But in this case, $T^*(s,-1,c) = 0$ if and only if $s=0$. Moreover, if $t \neq 0,-1$, then $T^*(0,t,c) = t(t+1) \neq 0$. So we can further reduce our consideration to

$$R(a,b,r) = \sum_{t\neq 0,-1} \sum_{s: T^*(s,t,c)=0} 1,$$

as, again, the other terms contribute nothing to our sum. Now, $R(a,b,r) = 0$ only if the discriminant of T^*, viewed as a polynomial in s, is never a square. That is, if

$$\sum_{t\neq 0,-1: g(t,c)\neq 0} \Psi(g(t,c)) = -(q-2), \tag{30}$$

where $g(t,c) = \left((c-3)t - t^2 - 1\right)^2 - 4t(t+1)^2$.

Since Ψ is the quadratic character of $\mathbb{F}_q$, the number $\Psi(g(t,c))$ takes $+1$ or -1, because $\Psi(s) = \pm 1$ for $s \neq 0$. Thus, if Eq. (30) holds, then it must be true that $\Psi(g(t,c)) = -1$ for all $t \neq 0,-1$. This implies that $g(t,c)$ is not a square number for all $t \neq 0,-1$, and the following estimate holds

$$\left| \sum_{t\in\mathbb{F}_q} \Psi(g(t,c)) \right| \geq |-(q-2)| - 2 = q-4. \tag{31}$$

However, this is an impossible result if $q \geq 17$, because Eq. (31) violates the conclusion of Theorem 9. To see this, note from Theorem 9 that it must be true that

$$\left| \sum_{t\in\mathbb{F}_q} \Psi(g(t,c)) \right| \leq 3q^{\frac{1}{2}},$$

because $g(t,c)$ is the polynomial of degree four in terms of t variables. Thus, if $q \geq 17$, then the inequality in Eq. (31) is not true and so we conclude that if $q \geq 17$ and $q \equiv 1\ (mod\ 4)$(assumption of *Case* II), then

$$|\{(x,y) \in (S_r + a) \times (S_r + b) : \|x-y\| = r \neq 0\}| = R(a,b,r) > 0.$$

Combining this and the result of *Case* I, the proof of Theorem 10 is complete.

□

The remaining values of q, namely 5,9, and 13, will make use of the following lemma, which follows from the above exposition:

Lemma 5. *Suppose that $q \equiv 1 \pmod 4$. For each $r \in \mathbb{F}_q^*$ and $a, b \in \mathbb{F}_q^2$ with $a \neq b$, let*

$$|\{(x,y) \in (S_r + a) \times (S_r + b) : \|x - y\| = r\}| = R(a,b,r).$$

Assuming that $\|a-b\| \neq 0$ and letting $c = \frac{\|a-b\|}{r}$, we have the following formula:

$$R(a,b,r) = \sum_{t \neq 0,-1: g(t,c) \neq 0} \Psi(g(t,c)) + q - 2 + (q-1)\delta_0(1-c),$$

where Ψ is the quadratic character on $\mathbb{F}_q$, $\delta_0(u) = 1$ if $u = 0$ and 0 otherwise, and $g(t,c)$ is given by

$$g(t,c) = \left((c-3)t - t^2 - 1\right)^2 - 4t(t+1)^2.$$

Moreover, if $\|a-b\| = 0$, then $R(a,b,c) \geq 1$.

The following theorem is from [10].

Theorem 11 ([10]). *Let G_q^{Δ} be a distance graph in two dimensions. Then, $\mathbb{D}(G_q^{\Delta})$ is never two if $q \neq 3$. Moreover, $\mathbb{D}(G_q^{\Delta})$ is three for $3 \neq q \equiv 3 \pmod 4$, and $\mathbb{D}(G_q^{\Delta})$ is three or four for $q \equiv 1 \pmod 4$.*

Claim. The diameter of G_5^{Δ} is exactly four in two dimensions.

Proof. Here, we show that $\mathbb{D}(G_5^{\Delta}) = 4$. Given $r \in \mathbb{F}_5^*$, we shall find two different points $a, b \in \mathbb{F}_5^2$ such that we cannot get to the point a from the point b by one, two, or three steps where the length of each step is r. Given $r \in \mathbb{F}_q^*$, choose $a = (0,0), b \in \mathbb{F}_5^2$ with $\|b\| = 3r$. Since $r \neq 3r = \|a-b\| = \|b\|$, it is impossible to get to the point b from the origin a by one step. We now show that two steps are not enough to reach from the origin into the point b. By Eq. (22), it suffices to show that for $\|b\| = 3r$,

$$|S_r \cap (S_r + b)| = 1 + \Psi((-1)(4^{-1}\|b\|^2 - r\|b\|)) = 0.$$

However, it is clear because $3 \in \mathbb{F}_5$ is not a square number and so $\Psi(3) = -1$. Finally, we show that we cannot reach from the origin to the point $b \in \mathbb{F}_5^2$ by three steps. Since $\|a-b\| = \|b\| = 3r$, using Lemma 5, it is enough to show that

$$R(a,b,r) = \sum_{t \neq 0,-1: g(t,3) \neq 0} \Psi(g(t,3)) + 3 = 0,$$

where $g(t,3) = 4t(t+1)^2 - \left(-t^2 - 1\right)^2$. This follows by the simple observation: for $t = 1,2,3, g(t,3) \equiv 2 \pmod 5$ and $\Psi(2) = -1$ because $2 \in \mathbb{F}_5$ is not a square number. By our observations and Theorem 11, we conclude that $\mathbb{D}(G_5^{\Delta}) = 4$, and the proof of Claim is complete. □

Claim. The diameter of G_9^{Δ} is exactly four in two dimensions.

Proof. Here, we prove that $\mathbb{D}(G_9^{\Delta}) = 4$. The idea is same as before. First, we note that

$$\mathbb{F}_{3^2} \cong \mathbb{Z}_3[i]/(i^2+1) \cong \{\alpha+\beta i : \alpha, \beta \in \mathbb{Z}_3\},$$

where $i^2 = -1$. In addition, note that $\{1+i, 2+i, 1+2i, 2+2i\}$ is a set of nonsquare numbers in $\mathbb{F}_9$. For each $r \in \mathbb{F}_9^*$, choose $a = (0,0)$ and $b \in \mathbb{F}_9^2$ such that $\|a-b\| = \|b\| = r(1+i)$. Then, it is clear that one step is not enough to get to the point b from the origin a. Using Eq. (22), we see that

$$|S_r \cap (S_r + b)| = 1 + \Psi(1+2i) = 0,$$

because the characteristic of $\mathbb{F}_9$ is three, and $(1+2i)$ is not a square number in $\mathbb{F}_9$ and so $\Psi(1+2i) = -1$. This implies that two steps are not enough to reach from the origin a to the point $b \in \mathbb{F}_9^2$. By Theorem 11, it therefore suffices to show that we can not reach from the origin a to the point $b \in \mathbb{F}_9^2$ by three steps. To show this, we just need to show from Lemma 5 that for every $t \in \mathbb{F}_9 \setminus \{0, -1\}, g(t, (1+i)) = 4t(t+1)^2 - \left((-2+i)t - t^2 - 1\right)^2$ is not a square number in $\mathbb{F}_9$. However, this follows from the following calculation:

$$g(1,(1+i)) = g(i,(1+i)) = g(2i,(1+i)) = g((2+2i),(1+i)) = 1+2i,$$
$$g((1+i),(1+i)) = 1+i, \;\; g((2+i),(1+i)) = 2+2i, \;\; g((1+2i),(1+i)) = 2+i.$$

This completes the proof of Claim. □

Claim. The diameter of G_{13}^{Δ} is exactly three in two dimensions.

Proof. Here, we prove $\mathbb{D}(G_{13}^{\Delta}) = 3$. From Theorem 11, the diameter of G_{13}^{Δ} is never two. Applying Lemma 5, it therefore suffices to show that for each $c \in \mathbb{F}_{13} \setminus \{0,1\}$, there exists $t \in \mathbb{F}_{13} \setminus \{0,-1\}$ such that $g(t,c) = 4t(t+1)^2 - \left((c-3)t - t^2 - 1\right)^2$ is zero or a square number in $\mathbb{F}_{13}$. Note that the set $\{1,3,4,9,10,12\}$ is the set of square numbers in $\mathbb{F}_{13}$. Thus, the proof of Claim is complete by the following observation:

$$g(2,2) = 10, \; g(1,3) = 12, \; g(3,4) = 0, \; g(1,5) = 3, \; g(3,6) = 9,$$

$$g(1,7) = 12, \; g(4,8) = 1, \; g(1,9) = 0, \; g(1,10) = 4, \; g(2,11) = 3, \; g(4,12) = 0.$$

□

This completes the proof of Theorem 2.

5 Proof of the "Kaleidoscopic" Result (Theorem 4)

Recall that given $S \subset \mathbb{F}_q^d$, $S(x)$ shall denote its characteristic function. Let T_k^J denote the set of k-point J-configurations in E. Assume, inductively, that for every $J' \subset J$,

$$|T_{k-1}^{J'}| = (1+o(1))|E|^{k-1}q^{-|J'|} \tag{32}$$

if

$$|E| \geq Cq^{d\left(\frac{k-2}{k-1}\right)}q^{\frac{|J'|}{k-1}}.$$

The initialization step is the following: Observe that

$$|T_1^J| = |T_1^{\emptyset}| = |E| = |E|q^{-0},$$

and this needs to hold if

$$|E| \geq Cq^{d\left(\frac{k-1}{k}\right)}q^{\frac{|J|}{k}} = C.$$

5.1 The Induction Step

We have, without loss of generality,

$$|T_k^J| = \sum_{x^1,\ldots,x^k} T_{k-1}^{J'}(x^1,\ldots,x^{k-1})E(x^k)\Pi_{j=1}^{l}S(x^j-x^k)\Pi_{i=l+1}^{k-1}\sum_{a_i\neq 0} S_{a_i}(x^i-x^k) \tag{33}$$

for some $1 \leq l \leq k-1$, depending on the degree of the vertex corresponding to x^k, where $T_{k-1}^{J'} \subset \mathbb{F}_q^{dk}$,

$$S_t = \{x \in \mathbb{F}_q^d : x_1^2 + x_2^2 + \cdots + x_d^2 = t\},$$

and $S \equiv S_1$. Technically, we should replace $\Pi_{j=1}^{l}S(x^j-x^k)$ by $\Pi_{j=1}^{l}S_{a_j}(x^j-x^k)$ for an arbitrary set of a_js, but this does not change the proof at all, and it would complicate the notation.

Using Eq. (9) and the definition of the Fourier transform, we see from Eq. (33) that

$$|T_k^J| = q^{kd}\sum_{\xi^1,\ldots,\xi^{k-1}:\xi^s\in\mathbb{F}_q^d} \widehat{T_{k-1}^{J'}}(\xi^1,\ldots,\xi^{k-1})\widehat{E}\left(\sum_{u=1}^{k-1}\xi^u\right)\Pi_{j=1}^{l}\widehat{S}(\xi^j)\Pi_{i=l+1}^{k-1}\sum_{a_i\neq 0}\widehat{S}_{a_i}(\xi^i)$$
$$= \mathit{Main} + \mathit{Remainder},$$

where Main is the term corresponding to taking $\xi^s = (0,\ldots,0)$ for every $1 \leq s \leq k-1$. It follows by Lemma 3 that

$$Main = (1+o(1))|T_{k-1}^{J'}||E|q^{-l}.$$

The Remainder is the sum of terms of the form $R_{U,V}$, where

$$U = \{j \in \{1,2,\ldots,l\} : \xi^j \neq (0,\ldots,0)\},$$

and

$$V = \{j \in \{l+1,\ldots,k-1\} : \xi^j \neq (0,\ldots,0)\}.$$

We first analyze the term where complements of U and V are empty sets. We get

$$R_{U,V} = \left| q^{kd} \sum_{\xi^1,\ldots,\xi^{k-1}:\xi^s \in \mathbb{F}_q^d;\xi^s \neq (0,\ldots,0)} \widehat{T_{k-1}^{J'}}(\xi^1,\ldots,\xi^{k-1})\widehat{E} \right.$$
$$\left. \times \left(\sum_{u=1}^{k-1} \xi^u\right) \Pi_{j=1}^{l} \widehat{S}(\xi^j) \Pi_{i=l+1}^{k-1} \sum_{a_i \neq 0} \widehat{S}_{a_i}(\xi^i) \right|. \tag{34}$$

Applying Lemma 3 to the Fourier transforms of spheres and applying Cauchy–Schwartz in the variables $\xi^1,\ldots,\xi^{k-1}$, followed by Eq. (10) to the first two terms in the sum, we see that

$$R_{U,V} = O\left(q^{kd} \times |T_{k-1}^{J'}|^{\frac{1}{2}} q^{-\frac{d(k-1)}{2}} \times |E|^{\frac{1}{2}} \times q^{-\frac{d}{2}} \times q^{\frac{d(k-2)}{2}} \times q^{-\frac{d+1}{2}l} \times q^{-\frac{d+1}{2}(k-1-l)}\right)$$
$$= O\left(|T_{k-1}^{J'}|^{\frac{1}{2}} \times |E|^{\frac{1}{2}} \times q^{\frac{d(k-1)}{2}} q^{-\frac{l}{2}} q^{-\frac{(k-1-l)}{2}}\right). \tag{35}$$

Applying the inductive hypothesis (32) and noting that l may be as large as $k-1$, we see that for some constant $C_0 < 1$,

$$R_{U,V} \leq C_0 \times Main$$

if

$$|E| \geq Cq^{d\left(\frac{k-1}{k}\right)} \times q^{\frac{|J|}{k}},$$

with C sufficiently large, as desired.

To estimate the general $R_{U,V}$, we need only consider Eq. (34) with some of the ξ^i or ξ^j fixed as $(0,0,\ldots,0)$. Running through the estimate as above will involve fewer variables over which to sum, and the estimates on the products of Fourier transforms of the indicator functions of the spheres will change, but it will yield a smaller magnitude than the bound from (35).

Technically speaking, we must still show that if $|T_k^{J_1}|$ satisfies the conjectured estimate for every J_1 with $|J_1| = n_1$, then so does $T_k^{J_2}$ with $|J_2| = n_2 > n_1$. However, this is apparent from the proof above.

References

1. Bannai, E., Shimabukuro, O., Tanaka, H.: Finite Euclidean graphs and Ramanujan graphs. Discrete Math. **309**, 6126–6134 (2009)
2. Hart, D., Iosevich, A.: Ubiquity of simplices in subsets of vector spaces over finite fields. Anal. Mathematika **34**, 29–38 (2007)
3. Iosevich, A., Rudnev, M.: Erdős distance problem in vector spaces over finite fields. Trans. Amer. Math. Soc. **12**, 6127–6142 (2007)
4. Iwaniec, H., Kowalski, E.: Analytic Number Theory, vol. 53. AMS Colloquium Publications, Providence (2004)
5. Krivilevich, M., Sudakov, B.: Pseudo-random graphs. In: More Sets, Graphs and Numbers, Bolyai Society Mathematical Studies vol. 15, pp. 199–262. Springer, New York (2006)
6. Kwok, W.M.: Character tables of association schemes of affine type. European J. Combin. **13**, 167–185 (1992)
7. Lidl, R., Niederreiter, H.: Finite Fields. Cambridge University Press, Cambridge (1997)
8. Medrano, A., Myers, P., Stark, H.M., Terras, A.: Finite analogs of Euclidean space. J. Comput. Appl. Math. **68**, 221–238 (1996)
9. Vinh, L.: Explicit Ramsey graphs and Erdös distance problem over finite Euclidean and non-Euclidean spaces, Math. Res. Lett. **15**(2), 375–388 (2008) arXiv:0711.3508
10. Vinh, L., Dung, D.: Explicit tough Ramsey graphs. In: Proceedings of the International Conference on Relations, Orders and Graphs: Interaction with Computer Science, Nouha Editions, pp. 139–146 (2008)
11. Vu, V.: Sum-Product estimates via directed expanders, Math. Res. Lett. 15(2), 375–388 (2008)

Remarks on Extremals in Minimal Support Inequalities

Steven M. Hudson

Abstract We consider solutions of the Schrödinger equation $-\Delta u(x) = V(x)u(x)$, where $u \equiv 0$ on the boundary of a bounded open region $D \subset \mathbf{R}^n$, and $V(x) \in L^r(D)$. Minimal support inequalities are sharp relationships between a norm of the potential $||V||_r$ and the measure of the support $|D|$. The well-known Faber–Krahn inequality is a special case. In most such results, including many variations of the Schrödinger equation above, one can show that extremals exist and are closely related to a Sobolev inequality. We provide a reason for this pattern and show that existence of extremals can be predicted under fairly general conditions.

Mathematics Subject Classification: Primary 35B05; Secondary 26D20.

1 Introduction

We consider various partial differential equations (pdes), most of them variations of Schrödinger's equation (1), and existence of nontrivial solutions u which vanish on the boundary of a bounded domain $D \subset R^n$, $n \geq 2$. Under the proper assumptions, usually on a potential V occurring in the pde, and existence of a nontrivial solution u, one can often establish a lower bound on the measure of the support $|D|$; see [2–4]. We refer to such results as *minimal support* inequalities.

However, if D is fixed, these results offer constraints on the potential or on the eigenvalues of a differential operator. For example, the well-known Faber–Krahn inequality states that for fixed $|D|$, if there is a non-trivial solution of $-\Delta u(x)=\lambda u(x)$, then $\lambda \geq \lambda_0$, where λ_0 is an explicit constant, the first eigenvalue

S.M. Hudson Ph.D. (✉)
Mathematics, Florida International University, Miami, FL, USA
e-mail: hudsons@fiu.edu

D. Bilyk et al. (eds.), *Recent Advances in Harmonic Analysis and Applications*, Springer Proceedings in Mathematics & Statistics 25, DOI 10.1007/978-1-4614-4565-4_15,

for the case that D is a ball. The Faber–Krahn inequality may be considered a special case of the minimal support inequalities proved in [2].

Perhaps surprisingly, these results usually share extremal functions with other inequalities, such as Sobolev's. The main purpose of this chapter is to explain this phenomenon and clarify when such shared extremals are predictable. This chapter might be considered a companion to [3], and we begin with the most basic minimal support inequality from that paper.

Let $u \in W_0^{1,2}(D)$ be a nontrivial solution of the time-independent Schrödinger equation

$$-\Delta u(x) + V(x)u(x) = 0 \tag{1}$$

where $V \in L^r(D)$ is the potential, with r specified later. Here, D is a bounded and measurable subset of $\mathbf{R}^n$, and $\Delta = \sum_{j=1}^n \frac{\partial^2}{\partial x_j^2}$. Let $1 \le q < \infty$, but if $n \ge 3$, let $q \le \frac{n}{n-2}$. Let $K_q(D)$ be the smallest possible constant in Sobolev's inequality

$$||u||_{L^{2q}(D)} \le K_q(D)||\nabla u||_{L^2(D)}. \tag{2}$$

By dilation and symmetrization, there is a constant K_q (independent of D) such that $K_q(D) \le K_q \cdot |D|^{\frac{1}{n}-\frac{1}{2r}}$, and equality is attained when D is a ball. At the critical index, $q = \frac{n}{n-2}$, there is an extremal for the case $D = R^n$, but a dilation argument shows that there is no extremal if D is bounded. We will assume throughout this chapter that D is bounded and that $q < \frac{n}{n-2}$; in this case there is at least one extremal, which we denote by $u_*(x)$; see [11].

Theorem 1.1. *Suppose that the Schrödinger equation (1) has a nontrivial solution in $W_0^{1,2}(D)$ for some $V \in L^r(D)$, with $r > \frac{n}{2}$. Define q by $\frac{1}{q} + \frac{1}{r} = 1$. Then,*

$$K_q^2(D)||V||_r \ge 1. \tag{3}$$

Corollary 1.2 *With the same assumptions,*

$$K_q^2|D|^{\frac{2}{n}-\frac{1}{r}}||V||_r \ge 1. \tag{4}$$

Clearly, the corollary follows from the theorem and the remarks above about $K_q(D)$. It explains the terminology *minimal support result*, but in this chapter, we are more interested in extremals than $|D|$ and will usually prefer the Eq. (3) notation, with the understanding that (in most cases) an analog of Eq. (4) follows from it.

A careful proof of this theorem appears in [3], and is outlined below. This reasoning applies in many other settings, and we will refer to it as the *minimal support sequence* for the problem. It uses the Sobolev inequality (2), Green's identity, and Hölder's inequality.

$$\begin{aligned} ||u||_{2q}^2 &\le K_q^2(D)\int_D |\nabla u(x)|^2 dx = -K_q^2(D)\int_D \Delta u(x)\, u(x) dx \\ &= -K_q^2(D)\int_D |u(x)|^2 V(x) dx \le K_q^2(D)||u^2||_q||V||_r = K_q^2(D)||u||_{2q}^2||V||_r \end{aligned} \tag{5}$$

which implies that $K_q^2(D)||V||_r \geq 1$. We will show that Theorem 1.1 has extremals in the sense that for each bounded domain D, there exists a potential V that makes the above an equality, and there exists a nontrivial u that satisfies Eq. (1) with this V. We will often refer to this u as the extremal for Theorem 1.1, with the understanding that V is determined from u and Eq. (1). If one prefers not to fix D (and therefore to focus on Corollary 1.2), then our comments on K_q show that equality is also attained in Eq. (4) by choosing D to be a ball in Theorem 1.1.

Actually, the extremal u_* mentioned above for the Sobolev inequality (2) is also extremal for Theorem 1.1. Clearly, the first inequality in the minimal support sequence is an equality when $u = u^*$. With some work [3], one can show that the potential V^* associated to u^* satisfies $|V^*|^r = c|u^*|^{2q}$. This apparent coincidence implies that the Holder step above is also an equality. Thus, u^* gives $K_q^2|D|^{\frac{2}{n}-\frac{1}{r}}||V||_r = 1$ and is an extremal for Theorem 1.1.

In Sect. 2, we offer an alternative approach to the extremals of Theorem 1.1, based on energy functionals. This does not rely on surprises or coincidences and in most cases can predict fairly quickly that extremals exist, though some checking of technical issues may still be required. In Sect. 3, we consider several other differential equations to show that this alternative method applies in approximately the same generality as the minimal support sequence. In Sect. 4, we state a general theorem that includes many of our pdes as special cases, and in Sect. 5 we explore some further results and open questions.

2 Energy Functionals

The alternative approach to extremals depends on the kinetic energy

$$T(u) = \int_D |\nabla u|^2 \, dx$$

and the potential energy

$$W(u) = \int_D V \cdot u^2 \, dx.$$

Let $E = \min T + W$ (where $||u||_2 = 1$), which is the ground energy state. The existence of a minimizer and the following related result are proved in [8] (both with $D = R^n$), see also [10].

Theorem 2.1. *Let $V \in L^\infty(R^n)$ be fixed. Suppose that u minimizes $T + W$ subject to the constraint $||u||_2 = 1$. Then u satisfies the Schrödinger equation*

$$-\Delta u(x) + V(x)u(x) = Eu. \tag{6}$$

We will not need to prove the existence of minimizers when $D \neq R^n$ because we can provide them directly, via Sobolev theory. But we will need the following slight generalization of Theorem 2.1, for general D and for $u \in L^{2q}(D)$, $1 \leq q < \infty$. Below, E has the same meaning as before, except that the minimum is over $u \in W_0^{1,2}$, with $||u||_{2q} = 1$.

Theorem 2.2. *Let $V \in L^r(D)$ be fixed, $1 < r \leq \infty$. Suppose that u minimizes $T + W$, subject to the constraint $||u||_{2q} = 1$, where $1/r + 1/q = 1$. Then u satisfies the nonlinear Schrödinger equation*

$$-\Delta u(x) + V(x)u(x) = Eu|u|^{2q-2}. \tag{7}$$

Of course, if $E = 0$ this reduces to Eq. (1). The variational proof is similar that of Theorem 2.1, but we provide it for completeness.

Proof. Let $F = F_u = (T + W)/||u||_{2q}^2$. Let $v = u + \varepsilon\phi$ for some $\phi \in C_c^\infty(D)$. Let δ be the operator $d/d\varepsilon$, followed by setting $\varepsilon = 0$. Then, if u minimizes $T + W$, it minimizes F, and $\delta F_u = 0$ for all such ϕ. From the quotient rule, applied to F, we get

$$\delta(T + W) + E\delta(||u||_{2q}^2) = 0$$

where $\delta(T) = 2\int \nabla u \nabla \phi \, dx = -2\int \triangle u\phi \, dx$, and $\delta(W) = 2\int Vu\phi \, dx$, and $\delta||u||_{2q}^2 = 2\int |u|^{2q}\phi/u \, dx$. We get the pde $-\triangle u + Vu = E|u|^{2q}/u = Eu|u|^{2q-2}$, as desired. □

This shows that the Schrödinger equation is a Euler–Lagrange equation. In passing, we note that Theorem 2.2 has a rather obvious converse.

Theorem 2.3. *Suppose u satisfies Eq. (7), where $E = \min T + W$ and $||u||_{2q} = 1$. Then u minimizes $T + W$, that is, $T(u) + W(u) = E$.*

Proof. $T(u) + W(u) = \int u(-\Delta u + Vu) \, dx = E\int |u|^{2q} \, dx = E$. □

While Eq. (1) arises from $T + W$, this proof illustrates how T and W might be reconstructed from a given pde; more on this in Sect. 4. To relate these results to extremals of Theorem 1.1, we consider the following *revised problem*, in which V is not fixed.

Theorem 2.4. *For each $C > 0$, there is a pair of functions (u_S, V_S) which minimizes $T + W$, subject to the constraints $||u_S||_{2q} = 1$ and $||V_S||_r = C$. Further, u_S is an extremal for the Sobolev inequality (2), and*

$$\int_D V_S \, u_S^2 \, dx = -||V_S||_r ||u_S^2||_q. \tag{8}$$

Proof. Choose u_S to minimize T; in other words, $u_S = u^*$ is a Sobolev extremal. Then, choose V_S to minimize W for this u_S. This implies that Eq. (8) holds. We claim that this pair minimizes $T + W$ and solves the revised problem. For, if (u, V) also satisfy the constraints, then

$$T(u) \geq T(u_S) \quad \text{and} \quad -W(u,V) \leq ||V||_r ||u^2||_q = ||V_S||_r ||u_S^2||_q = -W(u_S, V_S). \; \square$$

If we regard V_S as fixed, Theorem 2.2 implies the Schrödinger equation $-\Delta u + Vu = Eu|u|^{2q-2}$ holds for the pair (u_S, V_S), where $E = T(u_S) + W(u_S, V_S) = ||\nabla u_S||_2^2 - ||V||_r = K_q^{-2} - ||V||_r$. If we set $||V||_r = C = K_q^{-2}$, we get $E = 0$, and Eq. (1). So,

Corollary 2.5 *There is a pair (u,V) which satisfies Eq. (1) and makes the inequalities in Eqs. (5) and (3) equalities. Thus, this pair is extremal for Theorem 1.1.*

Proof. (summary) Solve the revised problem, with $C = K_q^{-2}$. The resulting pair (u,V) makes the minimal support sequence exact, a sequence of equalities. Theorem 2.2 shows they satisfy the given pde. □

3 Applications to Other Differential Equations

We can use energy functionals and an associated revised problem to establish extremals for a variation of Eq. (1), written in divergence form

$$-\operatorname{div}(a\nabla u)(x) + b(x)V(x)u(x) = 0 \tag{9}$$

where $0 < 1/M < a(x), b(x) < M < \infty$ are fixed bounded functions on D. Let $||V||_{r,b} = (\int_D |V(x|^r b(x)\ dx)^{1/r}$ be the weighted L^r norm; all such norms are taken over D. A compactness argument [3] shows that there is a sharp constant $K_q = K_{abqD}$ and a nontrivial extremal for the weighted Sobolev inequality

$$||u||_{2q,b} \leq K_q ||\nabla u||_{2,a}.$$

Theorem 3.1. *Suppose that Eq. (9) has a nontrivial solution u in $W_0^{1,2}(D)$ for some $V \in L^r(D)$, with $r > \frac{n}{2}$ and $n \geq 3$. Define q by $\frac{1}{q} + \frac{1}{r} = 1$. Then,*

$$K_q^2 ||V||_{rb} \geq 1, \tag{10}$$

and this is sharp.

This result and the one below are proved in [3], using a minimal support sequence similar to Eq. (5) above, and yield, exactness in the Holder step. Using energy functionals, we can provide an alternative proof of Theorem 3.2, which does not depend on any apparent coincidence.

Theorem 3.2. *Given D, a, b, q, and r as above, there is a pair (u,V) which satisfies Eq. (9) and makes Eq. (10) an equality. Thus, this pair is extremal for Theorem 3.1.*

Proof. This is similar to the proof of Corollary 2.5. Let $T = \int |\nabla u|^2\ a(x)\ dx$ and $W = \int V|u|^2\ b(x)\ dx$. Again, we minimize $T + W$ by first choosing u to minimize

T, subject to the constraint $\int |u|^{2q}\, b(x)\, dx = 1$; then by choosing V to minimize W, subject to the constraint $||V||_{r,b} = C = K^{-2}$. Reasoning as above, this pair makes the minimal support sequence for Theorem 3.1 exact and [if it satisfies Eq. (9)] is extremal for that theorem. The variational work is the following:

Set $v = u + \varepsilon\phi$ as above, and apply δ to $(T+W)/||v||^2_{2q,b}$; as before, we get $\delta(T+W) + (T+W)\delta||v||^2_{2q,b} = 0$. But $T+W = E = 0$, so $\delta(T+W) = 0$. And

$\delta(T) = 2\int \nabla u \nabla \phi a\, dx = -2\int \mathrm{div}[a\nabla u]\phi\, dx,$

$\delta(W) = 2\int V u \phi b\, dx.$

We get the pde $-\mathrm{div}[a\nabla u] + Vub = 0$, which is Eq. (9), so this completes the proof. □

Next, we study the p-Laplacian

$$\Delta_p u = \mathrm{div}(|\nabla u|^{p-2}\nabla u) \tag{11}$$

with $p > 1$. It is nonlinear, but energy functionals still work well. We study solutions of the following:

$$-\Delta_p u + V|u|^p/u = 0. \tag{12}$$

Let $V \in L^r(D)$ with $r > n/p$ and define q by $1/r + p/q = 1$ (so, $1/q > 1/p - 1/n$, the critical Sobolev value). Let $K = K_{qpD}$ be the sharp constant in the Sobolev inequality $||u||_q \leq K||\nabla u||_p$. Again, compactness arguments show that this has extremals.

Theorem 3.3. *If $V \in L^r$ and u in $W_0^{1,p}(D)$ satisfies Eq. (12), then*

$$K^p||V||_r \geq 1. \tag{13}$$

The proof in [4] is based on the minimal support sequence

$$||u||^p_{2q} \leq K^p||\nabla u||_p = -K^p \int u\Delta_p u\, dx \leq K^p||V||_r||u^p||_{q/p}.$$

Theorem 3.4. *For every $p > 1$ and every bounded $D \subset R^n$, Theorem 3.3 is sharp. Also, there is a pair (u, V) satisfying Eq. (12) with*

$$K^p||V||_r = 1. \tag{14}$$

Proof. To show extremals exist (and are the same as for a Sobolev inequality), we set $T = \int |\nabla u|^p\, dx$ and $W = \int V|u|^p\, dx$. The revised problem is to minimize $T+W$ subject to $\int_D |u|^q\, dx = 1$ and $||V||_r = K^{-p}$. As in the previous settings, this produces a pair (u, V) with equality in Eq. (14). The variational work

$$\delta(T) = p\int |\nabla u|^{p-2}\nabla u \nabla\phi\, dx = -p\int \operatorname{div}[|\nabla u|^{p-2}\nabla u]\phi\, dx,$$
$$\delta(W) = p\int V|u|^p/u\phi\, dx.$$

Again, $T+W=0$ reduces the Euler–Lagrange equation to $\delta(T+W)=0$, or $-\operatorname{div}(|\nabla u|^{p-2}\nabla u)+V|u|^p/u=0$, which is Eq. (12), as desired.

4 Towards a General Theory

We now ask how much of the previous work might be combined into a single general theorem about extremals of minimal support results. This might include different boundary conditions (e.g., Neumann instead of Dirichlet) or higher-order differential operators. For simplicity, we will assume the pde is linear, though our methods apply to the nonlinear pde (12), and to another nonlinear pde (16), to be discussed in the next section.

There are essentially two issues: (a) whether the pde can be considered a Euler–Lagrange equation for the minimization of some energy functional, $T+W$; and (b) whether there is an associated revised problem, which implies that this same u (combined with some associated potential V) is extremal for the minimal support sequence.

We first consider issue (a). Suppose that our pde can be written in the form $L(u)=R(u)$, where L and R are linear self-adjoint operators. For example, Eq. (1) can be rewritten as $\Delta u = Vu$. Typically, $R=R_V$ will depend on some potential V, which might be either scalar or vector-valued, and in some applications, R may involve several "potential" functions. Set $T=\int_D uL(u)\,dx$ and $W=-\int_D uR(u)\,dx$. Suppose that u is a nontrivial minimizer of $T+W$, making $(T+W)(u)=0$. Then, variational work gives $\delta(T)=\int uL(\phi)+\phi L(u)\,dx=\int 2\phi L(u)\,dx$ and $\delta(W)=-\int 2\phi R(u)\,dx$, so that $0=\delta(T+W)$. This shows that u satisfies the pde, $L(u)=R(u)$.

Most of the linear pdes considered in [3, 4] satisfy these assumptions. An example, due to De Carli [1], that does not fit is the pde

$$-\Delta u = \vec{V}\nabla u + u\operatorname{div}\vec{V}$$

where $\vec{V}$ is vector-valued (see also Eq. (17) below). The problem here is that R is not self-adjoint. The variational work above leads to a different pde

$$-\Delta u = u\operatorname{div}V. \tag{15}$$

This is typical; if R is not self-adjoint, it gets replaced in the variational work by $(R+R^*)/2$, which *is* self-adjoint. It is unclear whether the De Carli pde has a sharp minimal support result with extremals. The new pde (15) does have them (assuming div V can be prescribed), but it is unclear whether this observation is useful.

Next, to discuss issue (b), suppose there is some norm $||u||_N$ that satisfies an inequality (typically, a Sobolev inequality) of the form $||u||_N^2 \leq K^2 T(u)$. Assume there is nontrivial extremal u_S for this; it gives equality with the minimal value of K. Suppose that $R = R_V$ can be chosen (by choosing $V = V_S$) to give a minimal value of $W = -||u_S||_N ||V_S||_{N^*}$. Here, $||V||_{N^*}$ is a norm dual to $||\cdot||_N$. Usually, existence of such a V_S and $||\cdot||_{N^*}$ is simple Holder theory, but this idea may also be applicable to Orlicz spaces [7], Sobolev spaces, and pdes with first-order terms, such as Eq. (17). We fix $||V_S||_{N^*} = K^{-2}$, to make $T + W = 0$. Note that the other solutions of $L(u) = R(u)$ also satisfy $(T + W)(u) = 0$. The assumptions above imply a minimal support sequence similar to (5)

$$||u||_N^2 \leq K^2 T = -K^2 W \leq K^2 ||u||_N^2 ||V||_{N^*}$$

with equality for the pair (u_S, V_S). Under these assumptions, we have proved.

Theorem 4.1. *Suppose both sides of $L(u) = R(u)$ are linear and self-adjoint. Suppose some norm $||u||_N$ satisfies the assumptions above (mainly that $||u||_N^2 \leq K^2 \int uL(u)\, dx$ holds, with some extremal u_S). Then $K^2 ||V||_{N^*} \geq 1$, and there is a $V = V_S$ that, with $u = u_S$, makes this an equality.*

5 Further Results and Remarks

We now examine some other pdes and whether energy functionals give conclusions about minimal support extremals. In previous examples, we have always had $L(u) = -\Delta u$. We now consider briefly the higher-order pde

$$\Delta^2 u = Vu.$$

The analog of Eq. (5)

$$||u||_2^2 \leq K^2 \int (\Delta u)^2\, dx = K^2 \int u \Delta^2 u\, dx = K^2 \int u^2 V\, dx \leq K^2 ||u^2||_1 ||V||_\infty$$

holds, at least for smooth u with compact support in D. It gives $K^2 ||V||_\infty \geq 1$. The "Sobolev inequality" above holds on D because it holds on R^n [5, 9]. However, it is unclear to this author whether it has extremals. So, it is also unclear whether the minimal support inequality above does. The effectiveness of the energy functional approach depends rather heavily on Sobolev theory [6]. The nonlinear pde

$$-\Delta u = V|u|^\beta, \tag{16}$$

$\beta \geq 1$, can be handled with energy functionals. Even though $R(u) = V|u|^\beta$ is nonlinear, it is fairly easy to compute $\delta W = \delta \int uV|u|^\beta\, dx$ and reach the usual conclusions. However, we will leave this to the interested reader. At any rate, the methods in [3] are probably clearer for this problem.

The pde with first-order term

$$-\Delta u = Vu + \overrightarrow{U} \cdot \nabla u \tag{17}$$

is rather interesting. But in our context, it seems to reduce to Eq. (1). The analog of Eq. (5) gives

$$K^2 ||V - \frac{1}{2}\, \mathrm{div} \overrightarrow{U}\, ||_r \geq 1;$$

see [3] for more details. If none of u, V, and U are fixed, we can easily find an extremal for this by setting $\overrightarrow{U} = \overrightarrow{0}$. Then, our pde reduces to Eq. (1), for which we have already established extremals. If, for example, one regards $\overrightarrow{U} \neq \overrightarrow{0}$ as fixed (or dependent on u or V, as in the De Carli pde), this reasoning does not apply, and it is not clear in that case whether extremals exist. The author would like to thank Dmitriy Bilyk, Laura De Carli, and Alex Stokolos for the invitation and hospitality during the Statesboro conference.

References

1. De Carli, L.: personal communication (2011)
2. De Carli, L., Hudson, S.: A Faber-Krahn inequality for solutions of Schrödinger's equation. Adv. in Math. **230**, 2416–2427 (2012)
3. De Carli, L., Edward, J., Hudson, S., Leckband, M.: Minimal support results for Schrödinger equations, preprint (2011)
4. Edward, J., Hudson, S., Leckband, M.: Minimal support results for non-linear Schrödinger equations, preprint (2011)
5. Folland, G.: Introduction to Partial Differential Equations, 2nd edn. Princeton University Press, Princeton (1995)
6. Frank, R.L., Lieb, E.H., Seiringer, R.: Equivalence of Sobolev inequalities and Lieb-Thirring inequalities, in XVI th International Congress on Mathematical Physics, Proceedings of the ICMP held in Prague, World Scientific, Singapore, 2010. P. Exner (ed), 523–535, August 3–8, (2009)
7. Krasnosel'skiĭ, M., Rutickii, Y.: Convex Functions and Orlicz Spaces, Noordhoff Ltd, Groningen (1961)
8. Lieb, E. H., Loss, M: Analysis, 2nd edn. American Mathematical Society, Providence (2001)
9. Stein, E.: Singular Integrals and Differentiability Properties of Functions. Princeton University Press, Princeton (1972)
10. Struwe, M.: Variational Methods, 2nd edn. Springer, Berlin (1996)
11. Talenti, G.: Best constant in Sobolev inequality. Ann. Mat. Pura Appl. **110**(4), 353–372 (1976)

On Fubini Type Property in Lorentz Spaces

Viktor I. Kolyada

Dedicated to Professor Konstantin Oskolkov on the occasion of his 65th birthday.

Abstract We study Fubini-type property for Lorentz spaces $L^{p,r}(\mathbb{R}^2)$. This problem is twofold. First we assume that all linear sections of a function f in directions of coordinate axes belong to $L^{p,r}(\mathbb{R})$, and their one-dimensional $L^{p,r}$-norms belong to $L^{p,r}(\mathbb{R})$. We show that for $p \neq r$ it does not imply that $f \in L^{p,r}(\mathbb{R}^2)$ (this complements one result by Cwikel). Conversely, we assume that $f \in L^{p,r}(\mathbb{R}^2)$, and we show that then for $r < p$ almost all linear sections of f belong to $L^{p,r}(\mathbb{R})$, but for $p < r$ all linear sections may have infinite one-dimensional $L^{p,r}$-norms.

1 Introduction

Denote by $S_0(\mathbb{R}^n)$ the class of all measurable and almost everywhere finite functions f on $\mathbb{R}^n$ such that

$$\lambda_f(y) \equiv |\{x \in \mathbb{R}^n : |f(x)| > y\}| < \infty \quad \text{for each } y > 0.$$

A nonincreasing rearrangement of a function $f \in S_0(\mathbb{R}^n)$ is a nonnegative and nonincreasing function f^* on $\mathbb{R}_+ \equiv (0, +\infty)$ which is equimeasurable with $|f|$, that is, for any $y > 0$

$$|\{t \in \mathbb{R}_+ : f^*(t) > y\}| = \lambda_f(y).$$

V.I. Kolyada (✉)
Department of Mathematics, Karlstad University, Universitetsgatan 1 651 88 Karlstad, Sweden
e-mail: viktor.kolyada@kau.se

D. Bilyk et al. (eds.), *Recent Advances in Harmonic Analysis and Applications*, Springer Proceedings in Mathematics & Statistics 25, DOI 10.1007/978-1-4614-4565-4_16,

We shall assume in addition that the rearrangement f^* is left continuous on $(0,\infty)$. Under this condition it is defined uniquely by

$$f^*(t) = \inf\{y > 0 : \lambda_f(y) < t\}, \quad 0 < t < \infty.$$

In 1950 G. Lorentz introduced a scale of spaces, defined by two parameters, and containing the spaces L^p. Let $0 < p, r < \infty$. A function $f \in S_0(\mathbb{R}^n)$ belongs to the Lorentz space $L^{p,r}(\mathbb{R}^n)$ if

$$\|f\|_{L^{p,r}} \equiv \|f\|_{p,r} \equiv \left(\int_0^\infty \left(t^{1/p} f^*(t)\right)^r \frac{dt}{t}\right)^{1/r} < \infty.$$

For $0 < p < \infty$ the space $L^{p,\infty}(\mathbb{R}^n)$ is defined as the class of all $f \in S_0(\mathbb{R}^n)$ such that

$$\|f\|_{p,\infty} \equiv \sup_{t>0} t^{1/p} f^*(t) < \infty.$$

We have that $\|f\|_{p,p} = \|f\|_p$. For a fixed p, the Lorentz spaces $L^{p,r}$ strictly increase as the secondary index r increases, that is, the strict embedding $L^{p,r} \subset L^{p,s}$ $(r < s)$ holds (see [2, Chap. 4]).

The main objective of this note is to study the one-dimensional $L^{p,r}$-norms of the linear sections of a function $f \in L^{p,r}(\mathbb{R}^2)$. That is, we will study a kind of Fubini's property for spaces $L^{p,r}(\mathbb{R}^2)$. In a close connection with this problem, we consider also relations between mixed Lorentz norms and usual $L^{p,r}(\mathbb{R}^2)$-norm.

2 A Counterexample for Mixed Norm Spaces

Let $f \in S_0(\mathbb{R}^2)$. For any fixed $x \in \mathbb{R}$, the x-section of f (denoted by f_x) is the function of the variable y:

$$f_x(y) = f(x,y) \qquad (y \in \mathbb{R}).$$

Similarly, for a fixed $y \in \mathbb{R}$, the y-section of f is defined by

$$f_y(x) = f(x,y) \qquad (x \in \mathbb{R}).$$

Assume that almost all x-sections and almost all y-sections of a function $f(x,y)$ belong to $L^{p,r}(\mathbb{R})$. Furthermore, assume that both the functions

$$x \mapsto \|f_x\|_{L^{p,r}(\mathbb{R})} \quad \text{and} \quad y \mapsto \|f_y\|_{L^{p,r}(\mathbb{R})}$$

belong to $L^{q,s}(\mathbb{R})$. Then we write $f \in L^{q,s}[L^{p,r}]_{\text{sym}}$.

The space $L^{q,s}[L^{p,r}]_{\text{sym}}$ is invariant with respect to permutations of variables. It was first shown explicitly by Fournier [6] that mixed norm spaces of such type play an important role in Analysis. For the simplicity, we consider only two-dimensional case. A function $f \in S_0(\mathbb{R}^2)$ belongs to Fournier–Gagliardo mixed norm space $L^1(\mathbb{R})[L^\infty(\mathbb{R})]_{\mathrm{sym}}$ if

$$\int_{\mathbb{R}} ||f(x,\cdot)||_{L^\infty(\mathbb{R})}\,dx + \int_{\mathbb{R}} ||f(\cdot,y)||_{L^\infty(\mathbb{R})}\,dy < \infty.$$

It is easily seen that

$$W_1^1(\mathbb{R}^2) \subset L^1(\mathbb{R})[L^\infty(\mathbb{R})]_{\mathrm{sym}} \tag{1}$$

(where $W_1^1(\mathbb{R}^2)$ is the Sobolev space of all functions $f \in L^1(\mathbb{R}^2)$ for which both the first-order weak derivatives exist and belong to $L^1(\mathbb{R}^2)$).

Fournier [6] proved that

$$L^1(\mathbb{R})[L^\infty(\mathbb{R})]_{\mathrm{sym}} \subset L^{2,1}(\mathbb{R}^2). \tag{2}$$

Applying Eqs. (1) and (2), we obtain that

$$W_1^1(\mathbb{R}^2) \subset L^{2,1}(\mathbb{R}^2), \tag{3}$$

which yields a refinement of the classical Gagliardo–Nirenberg embedding $W_1^1(\mathbb{R}^2) \subset L^2(\mathbb{R}^2)$.

Different extensions of embedding Eq. (2) and their applications have been studied in the works [1, 3, 8–10]. In particular, it was proved in [1] that

$$L^1(\mathbb{R})[L^\infty(\mathbb{R})]_{\mathrm{sym}} \subset L^{p,1}(\mathbb{R})[L^{p',1}(\mathbb{R})]_{\mathrm{sym}} \tag{4}$$

for any $1 \le p \le \infty$. Let $p = 2$. Then Eqs. (1) and (4) imply that

$$W_1^1(\mathbb{R}^2) \subset L^{2,1}(\mathbb{R})[L^{2,1}(\mathbb{R})]_{\mathrm{sym}}. \tag{5}$$

We stress that Eq. (5) does not follow from Eq. (3). Indeed, it was shown by Cwikel [5] that if $p \neq r$, then neither of the spaces $L^{p,r}(\mathbb{R}^2)$ and $L_y^{p,r}(\mathbb{R})[L_x^{p,r}(\mathbb{R})]$ is contained in the other.

However, the *symmetric* mixed norm spaces may have much stronger properties than spaces with mixed norms taken in only one order. Therefore it is natural to ask whether the embedding

$$L^{p,r}[L^{p,r}]_{\mathrm{sym}} \subset L^{p,r}(\mathbb{R}^2) \tag{6}$$

is true.

If $p < r$, the solution is simple. Namely, the characteristic function of the set

$$E = \left\{(x,y) : 0 < y \le \frac{1}{x\ln(2/x)},\ \ 0 < x \le 1\right\}$$

belongs to $L^{p,r}[L^{p,r}]_{\text{sym}}$, but does not belong to $L^{p,r}(\mathbb{R}^2)$ (since $|E| = \infty$).

Let now $p > r$. Cwikel [5] constructed a function f such that

$$f \in L^{p,r}_y[L^{p,r}_x] \quad \text{but} \quad f \notin L^{p,r}(\mathbb{R}^2).$$

However, it can be easily verified that $f \notin L^{p,r}_x[L^{p,r}_y]$; hence, this counterexample cannot disprove the validity of Eq. (6).

Applying a different construction, we obtain the following result.

Theorem 2.1 *Let $0 < p, r < \infty$, and $p \neq r$. Then there exists a function $f \in L^{p,r}[L^{p,r}]_{\text{sym}}$ such that $f \notin L^{p,r}(\mathbb{R}^2)$.*

Proof. As we have seen, for $p < r$ this statement is immediate. We assume that $p > r$.

Set

$$n_k = [k^{-2p/r}2^k] \quad \text{and} \quad \alpha_k = \sqrt{n_k 2^{-k}}, \quad k \in \mathbb{N}.$$

Then $\alpha_k \le k^{-p/r}$. Since $p > r$, the series

$$\sum_{j=1}^{\infty} \alpha_j$$

converges. Set

$$\sigma_k = \sum_{j=k}^{\infty} \alpha_j \quad (k \in \mathbb{N}).$$

There exists $k_0 \in \mathbb{N}$ such that $2^{k/p}k^{-1/r}$ increases for $k \ge k_0$ and $n_k \ge 1$ for $k \ge k_0$. Denote

$$\Delta_k^{(j)} = \left[\sigma_{k+1} + j\frac{\alpha_k}{n_k}, \sigma_{k+1} + (j+1)\frac{\alpha_k}{n_k}\right)$$

and

$$Q_k^{(j)} = \Delta_k^{(j)} \times \Delta_k^{(j)},$$

where $k \ge k_0,\ j = 0, 1, \dots, n_k - 1$. Set also

$$Q_k = \bigcup_{j=0}^{n_k-1} Q_k^{(j)} \quad (k \ge k_0).$$

Since $\Delta_k^{(j)} \subset [\sigma_{k+1}, \sigma_k)$ for all $j = 0, 1, \dots, n_k - 1$, we have that

$$Q_k^{(j)} \subset [\sigma_{k+1}, \sigma_k) \times [\sigma_{k+1}, \sigma_k) \quad (j = 0, 1, \dots, n_k - 1)$$

and therefore

$$Q_k \subset [\sigma_{k+1}, \sigma_k) \times [\sigma_{k+1}, \sigma_k).$$

This implies that the sets Q_k are disjoint. Observe also that

$$\bigcup_{k=k_0}^{\infty} Q_k \subset (0,\sigma_{k_0}) \times (0,\sigma_{k_0}). \tag{7}$$

We have that

$$|Q_k| = n_k \left(\frac{\alpha_k}{n_k}\right)^2 = 2^{-k}.$$

Further, define

$$f(x,y) = \begin{cases} 2^{k/p}k^{-1/r} & \text{if} \quad (x,y) \in Q_k \quad (k \geq k_0) \\ 0 & \text{if} \quad (x,y) \notin \cup_{k=k_0}^{\infty} Q_k. \end{cases}$$

Since $2^{k/p}k^{-1/r}$ increases for $k \geq k_0$ and $|Q_k| = 2^{-k}$,

$$||f||_{L^{p,r}(\mathbb{R}^2)}^r \geq \sum_{k=k_0}^{\infty} \frac{2^{kr/p}}{k} \int_{2^{-k}}^{2^{-k+1}} t^{r/p-1}\, dt = \infty.$$

Further, for any $y \in [\sigma_{k+1}, \sigma_k)$,

$$\begin{aligned} \varphi(y)^r = ||f_y||_{L^{p,r}(\mathbb{R})}^r &= \frac{2^{kr/p}}{k} \int_0^{\alpha_k/n_k} t^{r/p-1}\, dt \\ &= \frac{p}{r}\frac{2^{kr/p}}{k}\left(\frac{\alpha_k}{n_k}\right)^{r/p} = \frac{p}{r}\left(\frac{2^k}{n_k}\right)^{r/(2p)} \frac{1}{k} \leq c_{p,r}. \end{aligned}$$

Thus,

$$||\varphi||_{L^{p,r}(\mathbb{R})}^r \leq c_{p,r} \sum_{k=k_0}^{\infty} \int_{\sigma_{k+1}}^{\sigma_k} t^{r/p-1}\, dt < \infty.$$

This implies that $f \in L_y^{p,r}[L_x^{p,r}]$. Since $f(x,y) = f(y,x)$, we have also that $f \in L_x^{p,r}[L_y^{p,r}]$. □

Remark 1. The function f constructed in Theorem 2.1 has a bounded support [see Eq. (7)], and $f \in L^\infty(\mathbb{R})[L^{p,r}(\mathbb{R})]_{\text{sym}}$. Thus, we have that for $0 < r < p$,

$$L^\infty(I)[L^{p,r}(I)]_{\text{sym}} \not\subset L^{p,r}(I^2) \quad (I = [0,1]). \tag{8}$$

This result is closely related to a problem studied in [4]. For $0 < r < p$ in [4, Theorem 2.6], it was constructed a sequence of functions $f_n(s,t)$ on I^2 such that

$$\sup_{t \in I} ||f_n(\cdot,t)||_{L^{p,r}} \leq 1 \quad \text{and} \quad \lim_{n\to\infty} ||f_n||_{L^{p,r}(I^2)} = \infty.$$

This implies that $L_t^{\infty}(I)[L_s^{p,r}(I)] \not\subset L^{p,r}(I^2)$. However, one can readily check that for the sequence $\{f_n\}$

$$\lim_{n\to\infty} ||f_n||_{L_s^{p,r}[L_t^{p,r}]} = \infty,$$

and thus neither Theorem 2.1 nor statement (8) can be proved with the use of this sequence.

3 Fubini-Type Property

Let $f \in L^p(\mathbb{R}^2)$. By Fubini's theorem, almost all cross sections of f belong to $L^p(\mathbb{R})$. We shall study to what extent a similar property is true for functions in the Lorentz space $L^{p,r}(\mathbb{R}^2)$. First we show that if $f \in L^{p,r}(\mathbb{R}^2)$, where $0 < r \le p$, then almost all cross sections of f belong to $L^{p,r}(\mathbb{R})$.

Theorem 3.1 *Let* $0 < r \le p < \infty$. *Then*

$$L^{p,r}(\mathbb{R}^2) \subset L^p(\mathbb{R})[L^{p,r}(\mathbb{R})].$$

Moreover, for any $f \in L^{p,r}(\mathbb{R}^2)$,

$$\left(\int_{\mathbb{R}} ||\varphi_y||^p_{L^p(\mathbb{R})}\, dy\right)^{1/p} \le ||f||_{L^{p,r}(\mathbb{R}^2)}. \tag{9}$$

Proof. For $\alpha \ge 0$, set

$$\lambda(\alpha) = \mathrm{mes}_2\left\{(x,y) \in \mathbb{R}^2 : |f(x,y)| > \alpha\right\}$$

and

$$\lambda_y(\alpha) = \mathrm{mes}_1\{x \in \mathbb{R} : |f_y(x)| > \alpha\}, \quad y \in \mathbb{R}.$$

Set also $\varphi(y) = ||f_y||_{L^{p,r}(\mathbb{R})}$.

By Fubini's theorem,

$$\lambda(\alpha) = \int_{\mathbb{R}} \lambda_y(\alpha)\, dy. \tag{10}$$

Further,

$$||f||^r_{L^{p,r}(\mathbb{R}^2)} = p\int_0^{\infty} \alpha^{r-1}\lambda(\alpha)^{r/p}\, d\alpha$$

and

$$\varphi(y)^r = ||f_y||^r_{L^{p,r}(\mathbb{R})} = p\int_0^{\infty} \alpha^{r-1}\lambda_y(\alpha)^{r/p}\, d\alpha$$

(see [7, Proposition 1.4.9]). We have $p/r \ge 1$. Thus, applying Minkowski inequality and using Eq. (10), we obtain

$$\left(\int_{\mathbb{R}} \varphi(y)^p \, dy\right)^{r/p} = p\left(\int_{\mathbb{R}}\left(\int_0^\infty \alpha^{r-1}\lambda_y(\alpha)^{r/p}\,d\alpha\right)^{p/r} dy\right)^{r/p}$$
$$\leq p\int_0^\infty \alpha^{r-1}\left(\int_{\mathbb{R}} \lambda_y(\alpha)\,dy\right)^{r/p} d\alpha$$
$$= p\int_0^\infty \alpha^{r-1}\lambda(\alpha)^{r/p}\,d\alpha = ||f||^r_{L^{p,r}(\mathbb{R}^2)}.$$

This implies Eq. (9). □

Remark 2. As it was already mentioned above, Cwikel [5] showed that if $p \neq r$, then there exists a function $f \in L^{p,r}(\mathbb{R}^2)$ such that $f \notin L_y^{p,r}(\mathbb{R})[L_x^{p,r}(\mathbb{R})]$. Thus, in Theorem 3.1 one cannot state that the function $\varphi(y) = ||f_y||_{L^{p,r}(\mathbb{R})}$ belongs to $L^{p,r}(\mathbb{R})$. Moreover, if $0 < s < p$, then this function may not belong to $L^{p,s}(\mathbb{R})$. Indeed, let $0 < r < p < \infty$ and $0 < s < p$. Set

$$g(y) = \frac{1}{y(\ln(2/y))^{p/s}}, \quad 0 < y \leq 1,$$

$$E = \{(x,y) : 0 < x \leq g(y),\ 0 < y \leq 1\},$$

and let f be a characteristic function of the set E. Since $|E| < \infty$, we have that $f \in L^{p,r}(\mathbb{R}^2)$. However, as it is easily seen, $f \notin L_y^{p,s}(\mathbb{R})[L_x^{p,r}(\mathbb{R})]$.

If $p < r$, then the Fubini-type property completely fails.

Theorem 3.2 *Let $0 < p < r \leq \infty$. Then there exists a function $f \in L^{p,r}(\mathbb{R}^2)$ such that $f_y \notin L^{p,\infty}(\mathbb{R})$ for any $y \in \mathbb{R}$.*

Proof. Set

$$A_n = \left[n, n + \frac{n}{2^{n+1}}\right], \quad n \in \mathbb{N}.$$

Further, for any $k = 0, 1, \ldots$, set

$$\beta_0^{(k)} = 0, \quad \beta_j^{(k)} = 2\sum_{i=0}^{j-1} \frac{1}{2^k + i} \quad (j = 1, \ldots, 2^k).$$

Then

$$\beta_{2^k}^{(k)} > 1 \quad (k = 0, 1, \ldots). \tag{11}$$

For $n \in \mathbb{N}$, write $n = 2^k + l$ $(l = 0, \ldots, 2^k - 1)$, and set

$$B_n = \left[\beta_l^{(k)}, \beta_{l+1}^{(k)}\right].$$

Observe that $\mathrm{mes}_1 B_n = 2/n$. Set also $Q_n = A_n \times B_n$. Then Q_n are disjoint, and $\mathrm{mes}_2 Q_n = 2^{-n}$.

Let $0 < \varepsilon < 1/p - 1/r$. Set $\mu_n = n^{-1/r-\varepsilon} 2^{n/p}$. Then

$$\mu_{n+1} \ge \mu_n, \ n \ge n_0.$$

Further, set

$$g(x,y) = \begin{cases} \mu_n & \text{if} \quad (x,y) \in Q_n, \quad n \ge n_0 \\ 0 & \text{if} \quad (x,y) \notin \cup_{n=n_0}^{\infty} Q_n. \end{cases}$$

Since $\{\mu_n\}$ increases and $\mathrm{mes}_2 Q_n = 2^{-n}$, we obtain that

$$g^*(t) = \sum_{n=n_0}^{\infty} \mu_n \chi_{[2^{-n}, 2^{-n+1})}(t) \quad (t > 0).$$

If $r = \infty$, then for $t \in [2^{-\nu}, 2^{-\nu+1})$ $(\nu \ge n_0)$,

$$t^{1/p} g^*(t) \le 2^{(1-\nu)/p} \mu_\nu = 2^{1/p} \nu^{-\varepsilon} \le c_p.$$

Thus, $g \in L^{p,\infty}(\mathbb{R}^2)$. Let $r < \infty$. Then

$$||g||_{L^{p,r}(\mathbb{R}^2)}^r = \sum_{n=n_0}^{\infty} \mu_n^r \int_{2^{-n}}^{2^{-n+1}} t^{r/p-1} dt \le 2^{r/p} \sum_{n=1}^{\infty} \frac{1}{n^{1+\varepsilon r}} < \infty,$$

and we have that $g \in L^{p,r}(\mathbb{R}^2)$.

Let now $0 \le y \le 1$. It follows from Eq. (11) that for any $k = 0, 1, \ldots$, there exists an integer l_k, $0 \le l_k < 2^k$, such that $y \in B_{n_k}$, where $n_k = 2^k + l_k$ $\quad (n_k = n_k(y))$. Thus,

$$g(x,y) \ge \mu_n \chi_{A_n}(x) \quad \text{for all} \quad n = n_k \ (k = 0, 1, \ldots).$$

From here,

$$||g_y||_{L^{p,\infty}(\mathbb{R})} \ge \mu_n \, \mathrm{mes}_1(A_n)^{1/p} = n^{1/p-1/r-\varepsilon} \quad (n = n_k).$$

This implies that $||g_y||_{L^{p,\infty}(\mathbb{R})} = \infty$ for any $y \in [0,1]$.

Finally, setting

$$f(x,y) = \sum_{k \in \mathbb{Z}} 2^{-|k|} g(x, y-k),$$

we have that $f \in L^{p,r}(\mathbb{R}^2)$ and $||f_y||_{L^{p,\infty}(\mathbb{R})} = \infty$ for any $y \in \mathbb{R}$. □

Acknowledgments The author is grateful to the referee for his/her useful remarks.

References

1. Algervik, R., Kolyada, V.I.: On Fournier-Gagliardo mixed norm spaces. Ann. Acad. Sci. Fenn. Math. **36**, 493–508 (2011)
2. Bennett, C., Sharpley, R.: Interpolation of Operators. Academic, Boston (1988)
3. Blei, R.C., Fournier, J.J.F.: Mixed-norm conditions and Lorentz norms, Commutative harmonic analysis (Canton, NY, 1987), 57–78, Contemp. Math., 91, Amer. Math. Soc., Providence, RI (1989)
4. Boccuto, A., Bukhvalov, A.V., Sambucini, A.R.: Some inequalities in classical spaces with mixed norms. Positivity **6**(4), 393–411 (2002)
5. Cwikel, M.: On $(L^{p_o}(A_o), L^{p_1}(A_1))_{\theta,q}$. Proc. Amer. Math. Soc. **44**, 286–292 (1974)
6. Fournier, J.: Mixed norms and rearrangements: Sobolev's inequality and Littlewood's inequality. Ann. Mat. Pura Appl. **148**(4), 51–76 (1987)
7. Grafakos, L.: Classical Fourier analysis. In: Graduate Texts in Mathematics, vol. 249, 2nd edn. Springer, New York (2008)
8. Kolyada, V.I.: Mixed norms and Sobolev type inequalities. In: Approximation and Probability, vol. 72, pp. 141–160. Banach Center Publ., Polish Acad. Sci. Warsaw (2006)
9. Kolyada, V.I.: Iterated rearrangement and Gagliardo-Sobolev type inequalities. J. Math. Anal. Appl. **387**, 335–348 (2012)
10. Milman, M.: Notes on interpolation of mixed norm spaces and applications. Quart. J. Math. Oxford Ser. (2) **42**(167), 325–334 (1991)

Some Applications of Equimeasurable Rearrangements

Anatolii A. Korenovskii and Alexander M. Stokolos

Abstract The goal of this chapter is to demonstrate the utility of equimeasurable rearrangements in studying certain function classes.

2000 MS Classification: Primary 26D15; Secondary 41B25, 46E30.

Introduction

The study of equimeasurable rearrangements of functions originated, to the best of our knowledge, in the works of Steiner [33] and Schwarz [31] (1881–1884). A systematic development of the properties and applications of rearrangements was continued in the famous monograph of Hardy et al. [13].

Since then, the equimeasurable rearrangements have been widely applied in many areas of analysis. The principal advantage of rearrangements is that in many cases they have a simpler form than the original function, yet retain its integrability properties. Presently, the equimeasurable rearrangements are not only a powerful tool for numerous applications but also an independent object of study. Depending on the specific task, various types of rearrangements could be used—increasing, decreasing, symmetric (radial), iterated, etc. Often, a rearrangement of the modulus of the original function is involved.

A.A. Korenovskii
Odessa Mechnikov National University, Odessa, Ukraine
e-mail: anakor@paco.net

A.M. Stokolos (✉)
Georgia Southern University, Statesboro, GA, USA
e-mail: astokolos@georgiasouthern.edu

D. Bilyk et al. (eds.), *Recent Advances in Harmonic Analysis and Applications*, Springer Proceedings in Mathematics & Statistics 25, DOI 10.1007/978-1-4614-4565-4_17,

In this chapter, we intend to demonstrate the effectiveness of rearrangements in studying functions from several standard classes. Specifically, our goal is to apply some known estimates for the rearrangements of functions to describe various properties of the functions themselves; consequently, we will not prove the rearrangement estimates (the proofs usually involve covering arguments). We do not seek to present the results in their utmost generality and so do not consider some known variants of common function classes. For simplicity, we will consider only Lebesgue measurable sets and functions. The measure will be denoted by $|\cdot|$.

We recall some definitions. For a function f measurable on a set $E \subset \mathbb{R}^d$, the nonincreasing equimeasurable rearrangement of its modulus can be defined by the equality

$$f^*(t) = \sup_{e \subset E,\, |e|=t} \inf_{x \in e} |f(x)|, \quad 0 < t < |E|$$

(cf. Kolyada [16]).

It can be shown that the function f^* (of one variable) is, as the name suggests, nonincreasing on $(0,|E|)$ and equimeasurable with $|f|$, i.e., the distribution functions of $|f|$ and f^* coincide

$$\lambda_f^*(y) \equiv |\{x \in E : |f(x)| > y\}| = |\{t \in (0,|E|) : f^*(t) > y\}|, \quad 0 < y < +\infty.$$

(In the case $|E| = +\infty$ one assumes, in addition, that $\lambda_f^*(y) < +\infty$ for all $y > 0$.) This equality means that in a certain sense the nonincreasing rearrangement f^* is the inverse of the distribution function λ_f^*. The nondecreasing equimeasurable rearrangement of $|f|$ is defined similarly:

$$f_*(t) = \inf_{e \subset E,\, |e|=t} \sup_{x \in e} |f(x)|, \quad 0 < t < |E|.$$

The function f_* is equimeasurable with $|f|$ in the sense that

$$\lambda_{*,f}(y) \equiv |\{x \in E : |f(x)| < y\}| = |\{t \in (0,|E|) : f_*(t) < y\}|, \quad 0 < y < +\infty.$$

If $|E| < +\infty$, then, for almost all t (more precisely, for all points of continuity), the following equality holds:

$$f^*(t) = f_*(|E| - t).$$

The equimeasurability of functions f^*, f_*, and $|f|$ implies that for any monotone function φ on $[0,+\infty)$, one has

$$\int_0^{|E|} \varphi(f^*(t))\, dt = \int_0^{|E|} \varphi(f_*(t))\, dt = \int_E \varphi(f(x))\, dx.$$

The following fundamental properties of equimeasurable rearrangements easy follow from their definitions:

$$\sup_{e\subset E,\,|e|=t} \frac{1}{|e|}\int_e |f(x)|\,dx = \frac{1}{t}\int_0^t f^*(u)\,du \equiv f^{**}(t), \quad 0<t\le |E|,$$

$$\inf_{e\subset E,\,|e|=t} \frac{1}{|e|}\int_e |f(x)|\,dx = \frac{1}{t}\int_0^t f_*(u)\,du \equiv f_{**}(t), \quad 0<t\le |E|,$$

and the supremum and infimum in the left-hand sides are attained (see [2], pp. 43–46).

The rearrangement operator f^{**}—in contrast with the operator f^*—has the subadditivity property,

$$(f+g)^{**}(t) \le f^{**}(t)+g^{**}(t), \quad 0<t\le |E|,$$

which is often important in applications.

1 Classes of Functions Defined by Conditions on Their Average Oscillations

1.1 Functions of Bounded Mean Oscillation

The first object of our study is the well-known class of functions of bounded mean oscillation (BMO), which first appeared in the paper of John and Nirenberg [14].

Let f be integrable on the cube $[0,1]^d \equiv Q_0 \subset \mathbb{R}^d$. The mean oscillation f on a cube $Q\subset Q_0$ is given by

$$\Omega(f,Q) = \frac{1}{|Q|}\int_Q |f(x)-f_Q|\,dx, \quad \text{where} \quad f_Q \equiv \frac{1}{|Q|}\int_Q f(x)\,dx.$$

The class BMO consists of all functions f such that

$$\|f\|_* \equiv \sup_{Q\subset Q_0} \Omega(f,Q) < +\infty,$$

where the supremum is taken over all cubes $Q\subset Q_0$.

In the paper [14], a fundamental result was established: if $f\in BMO$, then the distribution function of $|f-f_{Q_0}|$ is exponentially decreasing, i.e., there exist constants B and b, depending only on the dimension d, such that

$$\left|\left\{x\in Q_0 : \left|f(x)-f_{Q_0}\right| > \lambda\right\}\right| \le B\exp\left(-\frac{b\lambda}{\|f\|_*}\right), \quad 0<\lambda<+\infty. \tag{1}$$

This famous inequality—now known as the John–Nirenberg inequality—has numerous applications. In terms of rearrangements, it can be rewritten as

$$\left(f - f_{Q_0}\right)^* (t) \le \frac{\|f\|_*}{b} \log \frac{B}{t}, \quad 0 < t \le 1, \tag{2}$$

i.e., for $f \in BMO$, its nondecreasing rearrangement is growing no faster than the logarithm when $t \to +0$.

The standard direct proof of the John–Nirenberg inequality is based on the iterated application of a covering lemma and produces suboptimal constants. Slavin and Vasyunin [32] proved a sharp integral version of Eq. (1) for $d = 1$ by finding the explicit Bellman function for the inequality. Their proof was for the L^2-based BMO, whereby one uses the square average oscillation $\left(\frac{1}{|Q|}\int_Q |f - f_Q|^2\right)^{1/2}$ instead of $\Omega(f,Q)$. Later, Vasyunin [36] found the Bellman function for the sharp weak-form inequality and thus computed the best constants B and b in Eq. (1). His proof, again, was for $d = 1$ and the L^2-based BMO.

However, if one considers $d = 1$ and a function $\varphi \in BMO$ that is nonincreasing on $(0,1)$, then its logarithmic rate of growth as $t \to +0$ can be easily established using the following elementary inequality:

$$\frac{1}{t/a}\int_0^{t/a} \varphi(u)\,du - \frac{1}{t}\int_0^t \varphi(u)\,du$$
$$\le \frac{a}{2}\frac{1}{t}\int_0^t \left|\varphi(u) - \frac{1}{t}\int_0^t \varphi(v)\,dv\right| du, \quad 0 < t \le 1,\ a > 1. \tag{3}$$

Indeed, even under the weaker assumption

$$\|\varphi\|'_* \equiv \sup_{0<t\le 1} \Omega(\varphi,(0,t)) < +\infty,$$

the relation (3) immediately implies that

$$a^k \int_0^{a^{-k}} \varphi(t)\,dt \le k\frac{a}{2}\|\varphi\|'_* + \int_0^1 \varphi(t)\,dt, \quad k = 0,1,2,\ldots,$$

which implies (if one takes into account the monotonicity of φ) that

$$\varphi(t) \le \frac{1}{t}\int_0^t \varphi(u)\,du \le \left(\frac{1}{2}\frac{a}{\ln a}\ln\frac{1}{t} + \frac{a}{2} + 1\right)\|\varphi\|'_*, \quad 0 < t \le 1,\ a > 1.$$

It is easy to see that the smallest rate of increase on the right-hand side when $t \to +0$ is attained at $a = \mathrm{e}$, and so we have

$$\varphi(t) \le \frac{\mathrm{e}}{2}\|\varphi\|'_* \ln\frac{\exp(1 + 2/\mathrm{e})}{t}, \quad 0 < t \le 1.$$

Moreover, the function $\varphi(t) = \ln\frac{1}{t} - 1$, $0 < t \le 1$ shows that the factor $\frac{e}{2}$ in the right-hand side is sharp. Thus, we see that the John–Nirenberg inequality (1) [or (2)] becomes trivial for nonincreasing nonnegative functions. Since for any $f \in BMO$ this inequality gives an estimate of the rate of growth of $\left(f - f_{Q_0}\right)^*(t)$, it is natural to put $\varphi(t) = \left(f - f_{Q_0}\right)^*(t)$ and apply the above reasoning to the function φ. However, one would need an estimate of the BMO-norm of the rearrangement of the form

$$\|\varphi\|'_* \le c\|f\|_*, \tag{4}$$

in which case the constant $c \ge 1$ would determine the exponent b in the John–Nirenberg inequality (1).

Such an estimate, with $c > 1$, was obtained for $d = 1$ by Garsia and Rodemich in [9] and for $d > 1$ by Bennett et al. in [1]. Both proofs are based on the Calderon–Zygmund decomposition. In the one-dimensional case, the Calderon–Zygmund lemma can be replaced with a more precise analogue, the Riesz sunrise lemma. Its application allowed Klemes [15] to prove the estimate (4) with $c = 1$. However, in the left-hand side of Klemes's inequality, he used the equimeasurable rearrangement of the function itself rather than that of its modulus. Therefore, to use Eq. (4) in the proof of the John–Nirenberg inequality, we also need the following estimate:

$$\|\,|f|\,\|_* \le c'\|f\|_* \tag{5}$$

(With $c' = 2$ this inequality is trivial in any dimension, while for $d \ge 2$ the sharp constant c' is unknown.)

In the one-dimensional case, inequality (5) with $c' = 1$ was established in [18]. Together with Klemes's estimate, it leads to Eq. (4) with the constant $c = 1$. This means that for $d = 1$ the sharp constant in exponent in the John–Nirenberg inequality (1) is $b = 2/\mathrm{e}$.

As noted above, when $d \ge 2$, the best constants c and c' in Eqs. (4) and (5) are unknown. However, if in Eq. (5) instead of $\|\cdot\|_*$ one writes $\|\cdot\|_{*,R}$—the supremum of average oscillations not over all cubes $Q \subset Q_0$ but over all parallelepipeds $R \subset Q_0$ with sides parallel to the coordinate axes—then, as shown in [23], the corresponding analogue of Eq. (5) remains valid with $c' = 1$. Therefore, Eq. (4) is still valid with constant $c' = 1$ if $\|\cdot\|_{*,R}$ in the right-hand side is replaced with $\|\cdot\|_*$. Thus, an analogue of the John–Nirenberg inequality (1) holds with a constant $b = 2/\mathrm{e}$ and with the norm $\|\cdot\|_{*,R}$ in the exponent.

1.2 *The Gurov–Reshetnyak Class*

The class BMO is closely related to another class of functions, first defined by Gurov and Reshetnyak in [11]. In its simplest interpretation, the Gurov–Reshetnyak class $GR \equiv GR(\varepsilon)$ consists of all nonnegative functions f defined on the cube $Q_0 \equiv [0,1]^d$ such that

$$\Omega(f,Q) \leq \varepsilon \cdot f_Q \tag{6}$$

uniformly for all cubes $Q \subset Q_0$. Clearly, inequality (6) holds for any $f \in L^1(Q_0)$. if $\varepsilon \geq 2$. Therefore, the class $GR(\varepsilon)$ becomes nontrivial only for $0 < \varepsilon < 2$.

One of the fundamental properties of functions from $GR(\varepsilon)$, established in [11], is that, while they a priori are only assumed to be integrable on Q_0, they, in fact, possess a higher degree of summability. More precisely, for sufficiently small ε $(0 < \varepsilon < \varepsilon_0 \equiv \varepsilon_0(d))$ condition (6) implies that $f \in L^p(Q_0)$ for some $p = p(\varepsilon) > 1$. This important in applications result, along with its various refinements and generalizations, was later studied by Bojarski, Franciosi, Ivaniec, Milman, Moscariello, Wik, and others. It is now called the Gurov–Reshetnyak lemma.

Following the basic idea of this chapter and the template of the John–Nirenberg inequality of the previous section, let us consider the Gurov–Reshetnyak lemma for the case of nonnegative, nonincreasing function φ on $(0,1)$. Let us also replace condition (6) by a weaker condition:

$$\Omega(\varphi,(0,t)) \equiv \frac{1}{t}\int_0^t \left|\varphi(u) - \frac{1}{t}\int_0^t \varphi(v)\,dv\right| du \leq \varepsilon \frac{1}{t}\int_0^t \varphi(u)\,du, \quad 0 < t \leq 1. \tag{7}$$

Under these assumptions, the improvement of the exponent of integrability of φ follows in a completely elementary way. Indeed, for any $a > 1$ and $0 < t \leq 1$, inequality (3) and condition (7) imply that

$$\begin{aligned}\frac{1}{t/a}\int_0^{t/a} \varphi(u)\,du - \frac{1}{t}\int_0^t \varphi(u)\,du \\ \leq \frac{a}{2}\frac{1}{t}\int_0^t \left|\varphi(u) - \frac{1}{t}\int_0^t \varphi(v)\,dv\right| du \leq \frac{a}{2}\varepsilon\frac{1}{t}\int_0^t \varphi(u)\,du,\end{aligned}$$

i.e.,

$$\frac{1}{t/a}\int_0^{t/a} \varphi(u)\,du \leq \left(1 + \frac{a}{2}\varepsilon\right)\frac{1}{t}\int_0^t \varphi(u)\,du, \quad 0 < t \leq 1.$$

A repeated application of this inequality gives

$$a^k \int_0^{a^{-k}} \varphi(u)\,du \leq \left(1 + \frac{a}{2}\varepsilon\right)^k \int_0^1 \varphi(u)\,du, \quad k = 0, 1, \ldots,$$

and, using the monotonicity of φ, we obtain, for $a^{-k} < t \leq a^{-k+1}$,

$$\begin{aligned}\frac{1}{t}\int_0^t \varphi(u)\,du \leq a^k \int_0^{a^{-k}} \varphi(u)\,du \leq \left(1 + \frac{a}{2}\varepsilon\right)^k \int_0^1 \varphi(u)\,du \\ \leq \left(1 + \frac{a}{2}\varepsilon\right)^{1 + \frac{\ln\frac{1}{t}}{\ln a}} \int_0^1 \varphi(u)\,du = \left(1 + \frac{a}{2}\varepsilon\right) t^{-\frac{\ln\left(1 + \frac{a}{2}\varepsilon\right)}{\ln a}} \int_0^1 \varphi(u)\,du.\end{aligned}$$

Minimizing the factor in front of the integral, we set $a = \left(\frac{p_0}{p_0-1}\right)^{p_0}$, where $p_0 > 1$ is the root of the equation $\frac{p_0^{p_0}}{(p_0-1)^{p_0-1}} = \frac{2}{\varepsilon}$. This gives

$$\frac{1}{t}\int_0^t \varphi(u)\,du \leq \frac{p_0}{p_0-1} t^{-1/p_0} \int_0^1 \varphi(u)\,du, \quad 0 < t \leq 1, \tag{8}$$

and, using that $\frac{1}{t}\int_0^t \varphi(u)\,du \geq \varphi(t)$, we obtain, for any $0 < p < p_0$,

$$\int_0^1 \varphi^p(t)\,dt \leq \left(\frac{p_0}{p_0-1}\right)^p \frac{p_0}{p_0-p}\left(\int_0^1 \varphi(t)\,dt\right)^p.$$

This completes the proof of Gurov–Reshetnyak lemma for nonincreasing φ. The fact that the value of $p_0 \equiv p_0(\varepsilon)$ cannot be increased can be easily verified using the power function. Thus, we have shown the improved summability for any nonincreasing function φ that satisfies Eq. (7) for any fixed $0 < \varepsilon < 2$. In fact, we have shown even more: every function that satisfies the Gurov–Reshetnyak condition also satisfies a reverse Hölder inequality.

Another interesting property of Gurov–Reshetnyak functions is their summability to a certain negative power. Again, we consider this property first for monotone functions. Let ψ be a nonnegative, nondecreasing function on $(0,1)$. Then the appropriate analogue of inequality (3) is

$$\begin{aligned}&\frac{1}{t}\int_0^t \psi(u)\,du - \frac{1}{t/a}\int_0^{t/a} \psi(u)\,du \\ &\quad\leq \frac{a}{2}\frac{1}{t}\int_0^t \left|\psi(u) - \frac{1}{t}\int_0^t \psi(v)\,dv\right|\,du, \quad 0 < t \leq 1,\ a > 1.\end{aligned}$$

If we assume that ψ satisfies the Gurov–Reshetnyak condition (7), then, arguing as above, we get

$$\begin{aligned}&\frac{1}{t}\int_0^t \psi(u)\,du - \frac{1}{t/a}\int_0^{t/a} \psi(u)\,du \\ &\quad\leq \frac{a}{2}\frac{1}{t}\int_0^t \left|\psi(u) - \frac{1}{t}\int_0^t \psi(v)\,dv\right|\,du \leq \frac{a}{2}\varepsilon\frac{1}{t}\int_0^t \psi(u)\,du.\end{aligned}$$

From here, by analogy with the previous case, we come to the inequality

$$\frac{1}{t}\int_0^t \psi(u)\,du \geq \left(1 - \frac{a}{2}\varepsilon\right) t^{\frac{\ln \frac{1}{1-\frac{a}{2}\varepsilon}}{\ln a}} \int_0^1 \psi(t)\,dt.$$

Minimizing the factor in front of the integral, we set $a = \left(\frac{q_0+1}{q_0}\right)^{q_0}$, where $q_0 > 0$ is the root of the equation $\frac{q_0^{q_0}}{(q_0+1)^{q_0+1}} = \frac{\varepsilon}{2}$. Thus,

$$\frac{1}{t}\int_0^t \psi(u)\,du \geq \frac{q_0}{q_0+1} t^{1/q_0} \int_0^1 \psi(u)\,du,$$

and hence, for any $0 < q < q_0$,

$$\int_0^1 \psi^{-q}(t)\,dt \leq \left(\frac{q_0+1}{q_0}\right)^q \frac{q_0}{q_0-q}\left(\int_0^1 \psi(t)\,dt\right)^{-q}. \tag{9}$$

As above, the power function shows that $q_0 \equiv q_0(\varepsilon)$ cannot be increased, for any $\varepsilon \in (0,2)$.

As in the previous section, these arguments can be used to prove the Gurov–Reshetnyak lemma for arbitrary $f \in GR(\varepsilon)$, if one applies the equimeasurable rearrangement $\varphi(t) = f^*(t)$. Such an application will be possible if one can show that condition (6) implies the following estimate for the rearrangement:

$$\frac{1}{t}\int_0^t |f^*(u) - f^{**}(t)|\,dt \leq c\cdot\varepsilon\cdot f^{**}(t), \quad 0 < t \leq 1, \tag{10}$$

where one must clearly have $c \equiv c(d) \geq 1$.

Indeed, if it turns out that $\varepsilon_1 \equiv c\cdot\varepsilon < 2$, then for the number $p_1 > 1$, determined by the equation $\frac{p_1^{p_1}}{(p_1-1)^{p_1-1}} = \frac{2}{\varepsilon_1}$, the relation (8) implies that

$$f^{**}(t) \leq \frac{p_1}{p_1-1} t^{-1/p_1} \int_0^1 f^*(u)\,du, \quad 0 < t \leq 1, \tag{11}$$

and from here it follows immediately that $f \in L^p(Q_0)$ for all $p < p_1$.

Estimates similar to Eq. (10) for $f \in GR(\varepsilon)$ and $c > 1$ can be found in [7, 8, 20]. However, when $c \equiv c(d) > 1$, our approach produces an increase in the rate of summability of f only for sufficiently small ε $\left(\varepsilon < \frac{2}{c}\right)$. Moreover, it is not clear whether the critical value $p_1 \equiv p_1(\varepsilon, d)$ is, in fact, sharp. One would obviously prefer to have the estimate (10) with $c = 1$, which would immediately give the best possible critical exponent of integrability p_0 for any $0 < \varepsilon < 2$. In the one-dimensional case, the estimate (10) with $c = 1$ was obtained in [18].

We do not know the exact value of $c(d)$ in Eq. (10) for $d \geq 2$. An alternative approach suggested in [25] allows one to show that for any $d \geq 1$ and $0 < \varepsilon < 2$, the function $f \in GR(\varepsilon)$ admits an increased exponent of summability; however for $d \geq 2$, the exact critical exponent is not known.

If one considers the Gurov–Reshetnyak condition (6) over all parallelepipeds $R \subset Q_0$, rather than over all cubes $Q \subset Q_0$, then Eq. (10) remains true with $c = 1$

(see [21]) and, therefore, the sharp critical exponent of summability of the function from this class is, again, p_0. To find the negative critical exponent for all $f \in GR(\varepsilon)$, it is natural to use the nondecreasing equimeasurable rearrangement $\psi(t) = f_*(t)$. The appropriate analogue of inequality (10) then takes the form

$$\frac{1}{t}\int_0^t |f_*(u) - f_{**}(t)|\,du \le c\cdot\varepsilon\cdot f_{**}(t), \quad 0 < t \le 1.$$

For $c = 1$ and any $0 < \varepsilon < 2$, this inequality was proved in the one-dimensional case in [21], while for $d \ge 1$ and a function f satisfying condition (6) over all parallelepipeds $R \subset Q_0$, in [22]. Thus, in these two cases, the function f is summable to the power $-q$, $0 < q < q_0$, and the value q_0 cannot be increased.

2 The Self-Improvement Properties of Functions from the Gehring and Muckenhoupt Classes

We have already encountered reverse Hölder inequalities in Sect. 1.2. Indeed, take a cube Q_0 and let $f \in GR(\varepsilon)$ on Q_0 for a sufficiently small ε. Fix a cube $Q \subset Q_0$ and apply the reasoning from Sect. 1.2 to the restriction $f|_Q \equiv g$, writing Q instead of Q_0. Then Eq. (11) becomes

$$g^{**}(t) \le \frac{p_1}{p_1 - 1}\left(\frac{t}{|Q|}\right)^{-1/p_1}\frac{1}{|Q|}\int_0^{|Q|} g^*(u)\,du, \quad 0 < t \le |Q|.$$

Rising both sides to the power p, $1 < p < p_1$, and integrating with respect to t, we get

$$\begin{aligned}\int_Q f^p(x)\,dx &= \int_0^{|Q|}(g^*)^p(t)\,dt \le \int_0^{|Q|}(g^{**})^p(t)\,dt\\ &\le \left(\frac{p_1}{p_1-1}\right)^p\frac{p_1}{p_1-p}|Q|\left(\frac{1}{|Q|}\int_0^{|Q|} g^*(u)\,du\right)^p\\ &= \left(\frac{p_1}{p_1-1}\right)^p\frac{p_1}{p_1-p}|Q|\left(\frac{1}{|Q|}\int_Q f(x)\,dx\right)^p.\end{aligned}$$

Rising each part to the power $1/p$ produces the following reverse Hölder inequality:

$$\left(\frac{1}{|Q|}\int_Q f^p(x)\,dx\right)^{1/p} \le B\frac{1}{|Q|}\int_Q f(x)\,dx, \quad Q \subset Q_0, \tag{12}$$

where $B = \frac{p_1}{p_1-1}\left(\frac{p_1}{p_1-p}\right)^{1/p}$. This inequality is sometimes called Gehring's inequality. Similarly, letting in Eq. (9) $q = \frac{1}{p-1}$ and $\psi = (f|_Q)_*$, we obtain

$$\frac{1}{|Q|}\int_Q f(x)\,dx\left(\frac{1}{|Q|}\int_Q f^{-1/(p-1)}(x)\,dx\right)^{p-1} \leq B, \quad Q \subset Q_0, \tag{13}$$

where $B = \frac{q_0+1}{q_0}\left(\frac{q_0}{q_0-\frac{1}{p-1}}\right)^{p-1}$.

A nonnegative function f satisfying Eq. (12) is said to belong to the Gehring class $G_p \equiv G_p(B)$, while one satisfying Eq. (13) is said to belong to the Muckenhoupt class $A_p \equiv A_p(B)$. The A_p classes were introduced by Muckenhoupt [28], when studying the boundedness of the Hardy–Littlewood maximal operator on weighted Lebesgue spaces, and have subsequently been studied by many authors, mainly in conjunction with the behavior of various operators in weighted settings. The main property of these classes is that every function from A_p belongs to the class $A_{p-\varepsilon}$ for some $\varepsilon > 0$ (cf. [28]). The G_q classes were used by Gehring in [10], and then in the work of some other authors, in studying properties of quasi-conformal mappings. Like A_p, the Gehring classes exhibit self-improvement: any G_p function belongs to $G_{p+\varepsilon}$ for some $\varepsilon > 0$ (cf. [10]). In this section, we consider the question of finding the largest possible values of ε in the self-improvement properties of the Gehring and Muckenhoupt classes.

Coifman and Fefferman established in [3] that each Muckenhoupt class A_p is contained in some Gehring class G_q and vice versa. The arguments of the previous section do not allow one to obtain the embeddings $A_p \subset G_q$ and $G_q \subset A_p$ directly. Instead, these embeddings can be obtained as a special case of the self-improvement summability properties of functions satisfying reverse Hölder inequalities with arbitrary parameters. The lower bound of the negative exponent of summability of functions satisfying condition (14) and the upper bound of the positive exponent of summability of functions satisfying condition (19) were found by Malaksiano [26, 27]. In the one-dimensional case this allows to find the sharp dependence of the embedding exponents. Vasyunin [34, 35] used Bellman functions to obtain an exhaustive description of these embeddings for arbitrary parameters, including those for the A_∞ class; his results give both sharp values of class parameters and sharp embedding constants. An elementary proof can be found in [5, 6].

To start with, let φ be a nonnegative, nonincreasing function on $(0,1)$ that satisfies the Gehring inequality

$$\left(\frac{1}{t}\int_0^t \varphi^p(u)\,du\right)^{1/p} \leq B\cdot\frac{1}{t}\int_0^t \varphi(u)\,du, \quad 0 < t \leq 1. \tag{14}$$

Following D'Apuzzo and Sbordone [4], let us show how to compute the sharp bound on the exponent of summability of φ. This computation uses the following two well-known inequalities due to Hardy, the second of which is valid under the additional assumption that the function φ is nonincreasing:

$$\int_0^1 t^{p/q-1}\left(\frac{1}{t}\int_0^t \varphi(u)\,du\right)^p dt \le \left(\frac{q}{q-1}\right)^p \int_0^1 t^{p/q-1}\varphi^p(t)\,dt, \quad q\ge p\ge 1,\ q>1, \tag{15}$$

(cf. [13]) and

$$\left(\int_0^1 \varphi^r(t)\,dt\right)^{1/r} \le \frac{1}{r}\int_0^1 t^{1/r-1}\varphi(t)\,dt, \quad r>1, \tag{16}$$

(cf. [12]).

Applying, in order, inequality (15), condition (14), and Fubini's theorem, we obtain, for any $\varepsilon > 0$,

$$\begin{aligned}
&\left(\frac{p+\varepsilon}{p+\varepsilon-1}\right)^p \int_0^1 t^{p/(p+\varepsilon)-1}\varphi^p(t)\,dt \\
&\quad\ge \left(\frac{p+\varepsilon}{p+\varepsilon-1}\right)^p \int_0^1 t^{p/(p+\varepsilon)-1}\left(\frac{1}{t}\int_0^t \varphi(u)\,du\right)^p dt \\
&\quad\ge B^{-p}\int_0^1 t^{p/(p+\varepsilon)-2}\int_0^t \varphi^p(u)\,du\,dt \\
&\quad= B^{-p}\int_0^1 \varphi^p(u)\int_u^1 t^{p/(p+\varepsilon)-2}\,dt\,du \\
&\quad= B^{-p}\frac{p+\varepsilon}{\varepsilon}\left[\int_0^1 u^{p/(p+\varepsilon)-1}\varphi^p(u)\,du - \int_0^1 \varphi^p(u)\,du\right].
\end{aligned}$$

Under the assumption $0<\varepsilon<\varepsilon_0$, where ε_0 is the root of the equation

$$\left(\frac{p+\varepsilon_0}{\varepsilon_0}\right)^{1/p}\frac{p+\varepsilon_0-1}{p+\varepsilon_0} = B, \tag{17}$$

this inequality yields

$$\left[B^{-p}\frac{p+\varepsilon}{\varepsilon} - \left(\frac{p+\varepsilon}{p+\varepsilon-1}\right)^p\right]\int_0^1 t^{p/(p+\varepsilon)-1}\varphi^p(t)\,dt \le B^{-p}\frac{p+\varepsilon}{\varepsilon}\int_0^1 \varphi^p(t)\,dt.$$

Applying now Hardy's inequality (16) to the nonincreasing function φ^p with $r = \frac{p+\varepsilon}{p} > 1$, we get

$$\left(\int_0^1 \varphi^{p+\varepsilon}(t)\,dt\right)^{p/(p+\varepsilon)} \leq \frac{p}{p+\varepsilon}\int_0^1 t^{p/(p+\varepsilon)-1}\varphi^p(t)\,dt$$
$$\leq B_2\int_0^1 \varphi^p(t)\,dt \leq B_1^p\left(\int_0^1 \varphi(t)\,dt\right)^p,$$

where $B_1^p = B^p \cdot B_2$, $B_2 = \frac{p}{\varepsilon}\cdot\left(\frac{p+\varepsilon}{\varepsilon} - \left(B\frac{p+\varepsilon}{p+\varepsilon-1}\right)^p\right)^{-1}$, and the last inequality is valid because of Eq. (14). Rising everything to the power $1/p$ leads to the desired inequality:

$$\left(\int_0^1 \varphi^{p+\varepsilon}(t)\,dt\right)^{1/(p+\varepsilon)} \leq B_1\int_0^1 \varphi(t)\,dt. \tag{18}$$

The example of the power function shows that $\varphi^{p+\varepsilon}$ is not necessarily summable for $\varepsilon \geq \varepsilon_0$, and, therefore, the value of ε_0 determined by Eq. (17) is the best possible.

Now, consider a nonnegative, nondecreasing function ψ on $(0,1)$ satisfying the Muckenhoupt condition:

$$\frac{1}{t}\int_0^t \psi(u)\,du\left(\frac{1}{t}\int_0^t \psi^{-1/(p-1)}(u)\,du\right)^{p-1} \leq B, \quad 0 < t \leq 1, \tag{19}$$

for some $p > 1$. An argument similar to the one just employed but using a different inequality due to Hardy,

$$\int_0^1 t^{q/p-1}\left(\frac{1}{t}\int_0^t \psi(u)\,du\right)^{-q} dt \leq \left(\frac{p+1}{p}\right)^q\int_0^1 t^{q/p-1}\psi^{-q}(t)\,dt, \quad p \geq q > 0,$$

in place of Eq. (15), demonstrates that if ε_1 $(0 < \varepsilon_1 < p-1)$ is defined to be the root of the equation

$$\left(\frac{p-1}{\varepsilon_1}\right)^{p-1}\frac{1}{p-\varepsilon_1} = B, \tag{20}$$

then for any $0 < \varepsilon < \varepsilon_1$, the following inequality is valid:

$$\int_0^1 \psi(t)\,dt\left(\int_0^1 \psi^{-1/(p-\varepsilon-1)}(t)\,dt\right)^{p-\varepsilon-1} \leq B_1, \tag{21}$$

where $B_1 \equiv B_1(p,B,\varepsilon)$ (cf. [17]). In this case, inequality (16) should be applied to the nonincreasing function $\psi^{-1/(p-1)}$. As before, the example of the power function shows that the value ε_1 from Eq. (20) cannot be increased; more precisely, ψ is not necessarily in $A_{p-\varepsilon}$, if $\varepsilon \geq \varepsilon_1$.

Thus, we have seen how one can improve the summability of monotone functions from the Gehring and Muckenhoupt classes (more specifically, of nonincreasing $f \in G_p$ and nondecreasing $f \in A_p$). Therefore, it seems natural to use equimeasurable rearrangements to study the self-improvement properties of arbitrary G_p and A_p functions. Namely, one can argue as follows.

Let f be a function defined on a cube $Q_0 \subset \mathbb{R}^d$ and satisfying Gehring's condition (12) on Q_0 for some $p > 1$ and $B > 1$. Fix a sub-cube $Q \subset Q_0$, and let $\varphi = (f|_Q)^*$. If we show that φ—a nonincreasing function on $(0, |Q|]$—satisfies the following Eq. (15)-type condition:

$$\left(\frac{1}{t}\int_0^t \varphi^p(u)\,du\right)^{1/p} \leq c \cdot B \cdot \frac{1}{t}\int_0^t \varphi(u)\,du, \quad 0 < t \leq |Q|, \tag{22}$$

for some $c \equiv c(d)$ (it is clear that $c \geq 1$), then inequality (18) becomes

$$\left(\frac{1}{|Q|}\int_Q f^{p+\varepsilon}(x)\,dx\right)^{1/(p+\varepsilon)} \leq B' \frac{1}{|Q|}\int_Q f(x)\,dx, \quad Q \subset Q_0.$$

It means that $f \in G_{p+\varepsilon}(B')$, where $0 < \varepsilon < \varepsilon_0'$, and the value ε_0' is determined by Eq. (17), where instead of B we use $B' \equiv c \cdot B$.

Analogously, if f satisfies the Muckenhoupt condition (13) on Q_0 and Q is a sub-cube of Q_0, we let $\psi = (f|_Q)_*$. If we can show that ψ satisfies the Eq. (19)-type condition

$$\frac{1}{t}\int_0^t \psi(u)\,du \left(\frac{1}{t}\int_0^t \psi^{-1/(p-1)}(u)\,du\right)^{p-1} \leq c \cdot B, \quad 0 < t \leq |Q|, \tag{23}$$

with $c = c(d) \geq 1$, then inequality (21) turns into

$$\frac{1}{|Q|}\int_Q f(x)\,dx \left(\frac{1}{|Q|}\int_Q f^{-1/(p-\varepsilon-1)}(x)\,dx\right)^{p-\varepsilon-1} \leq B_1', \quad Q \subset Q_0.$$

This means that $f \in A_{p-\varepsilon}(B_1')$, where $0 < \varepsilon < \varepsilon_1'$, and ε_1' is defined by Eq. (20) with $B' \equiv c \cdot B$ used in place of B.

Thus, to show self-improvement of the classes $G_p(B)$ and $A_p(B)$, it suffices to obtain from Eqs. (12) and (13) the rearrangement estimates (22) and (23), respectively. Such estimates were established in the papers of Franciosi and Moscariello [8], Sbordone [29, 30], and Wik [37, 38].

However in these papers the constant $c(d) > 1$, and so it is difficult to judge the sharpness of values ε_0' and ε_1' obtained there. We do not know if the self-improvement bounds in these papers are sharp for $d \geq 2$; however, for $d = 1$, one can show that they are not.

We now describe the aforementioned sharp rearrangement estimate for functions satisfying a reverse Jensen inequality. First, recall that for a nonnegative, concave function Φ on $[0, +\infty)$, Jensen's inequality gives

$$\Phi\left(\frac{1}{|Q|}\int_Q f(x)\,dx\right) \le \frac{1}{|Q|}\int_Q \Phi(f(x))\,dx,$$

for any integrable, nonnegative function f on Q.

We say that f satisfies a reverse Jensen inequality on a cube Q_0 with a constant $C > 1$ if

$$\frac{1}{|Q|}\int_Q \Phi(f(x))\,dx \le C\cdot\Phi\left(\frac{1}{|Q|}\int_Q f(x)\,dx\right), \quad Q \subset Q_0. \tag{24}$$

For $\Phi(\tau) = \tau^p$ $(p > 1)$, this inequality gives Eq. (12) with $B = C^{1/p}$; while for $\Phi(\tau) = \tau^{-1/(p-1)}$ $(p > 1)$ we have Eq. (13) with $B = C^{p-1}$. It was shown in [17,19] that in the one-dimensional case condition (24) implies

$$\frac{1}{|I|}\int_I \Phi(f^*(t))\,dt \le C\cdot\Phi\left(\frac{1}{|I|}\int_I f^*(t)\,dt\right), \quad I \subset [0,|Q_0|], \tag{25}$$

where I is an arbitrary interval. Clearly, this inequality is still valid if we use f_* in place of f^*. Here it is important that the constant C on the right-hand side of Eq. (25) is the same as in condition (24). That means that for $d = 1$ one can take $c = 1$ in the right-hand sides of Eqs. (22) and (23) and thus get the sharp estimates of the rearrangements of functions from the Gehring and Muckenhoupt classes.

Estimates of the type (25), although with some additional restrictions on the function Φ and non-sharp constants, are contained in [30].

It might be interesting to mention that if one assumes that condition (24) is satisfied for all parallelepipeds $R \subset Q_0$ with sides parallel to the coordinate axes, then Eq. (25) is still valid with the same constant C. This immediately implies that if one assumes that Eqs. (12) and (13) hold not only over all cubes, but over all parallelepipeds, then the sharp ranges of self-improvement of such classes are the same as in the one-dimensional case. For details, see [24].

Acknowledgments The authors are grateful to Leonid Slavin for in-depth discussions on the subject and for his help in preparation of the manuscript.

References

1. Bennett, C., De Vore, R.A., Sharpley, R.: Weak-L^∞ and *BMO*. Ann. Math. **113**(2), 601–611 (1981)
2. Bennett, C., Sharpley, R.: Interpolation of Operators. Academic, New York (1988)
3. Coifman, R.R., Fefferman, C.: Weighted norm inequalities for maximal functions and singular integrals. Studia Math. **51**(3), 241–250 (1974)
4. D'Apuzzo, L., Sbordone, C.: Reverse Hölder inequalities. A sharp result. Rendiconti di Mat. **10**, ser. VII, 357–366 (1990)

5. Didenko, V.D., Korenovskyi, A.A.: Power means and the reverse Hölder inequality. Studia Mathematica **207**(1), 85–95 (2011)
6. Didenko, V.D., Korenovskyi, A.A.: Reverse Hölder inequality for power means. Ukrainian Math. Bull. **9**(1), 18–31 (2012)
7. Franciosi, M.: Weighted rearrangement and higher integrability results. Studia Math. **92**, 131–138 (1989)
8. Franciosi, M., Moscariello, G.: Higher integrability results. Manuscripta Math. **52**(1–3), 151–170 (1985)
9. Garsia, A.M., Rodemich, E.: Monotonicity of certain functionals under rearrangement. Ann. Inst. Fourier, Grenoble. **24**(2), 67–116 (1974)
10. Gehring, F.W.: The L^p-integrability of the partial derivatives of a quasiconformal mapping. Acta Math. **130**, 265–273 (1973)
11. Gurov, L., Reshetnyak, Yu.G.: An analogue of the concept of functions with bounded mean oscillation. Siberian Math. J. **17**(3), 417–422 (1976)
12. Hardy, G.H., Littlewoog, J.E., Pólya, G.: Some simple inequalities satisfied by convex functions. Messenger of Math. **58**, 145–152 (1929)
13. Hardy, G.H., Littlewood, J.E., Pólya, G.: Inequalities. Cambridge University Press, Cambridge (1934)
14. John, F., Nirenberg, L.: On functions of bounded mean oscillation. Comm. Pure Appl. Math. **14**(4), 415–426 (1961)
15. Klemes, I.: A mean oscillation inequality. Proc. Amer. Math. Soc. **93**(3), 497–500 (1985)
16. Kolyada, V.I.: On the embedding of certain classes of functions of several variables. Sibirsk. Math. Zh. **14**, 776–790 (1973)
17. Korenovskii, A.A.: The exact continuation of a reverse Hölder inequality and Muckenhoupt's condition. Math. Notes. **52**(6), 1192–1201 (1992)
18. Korenovskii, A.A.: On the connection between mean oscillation and exact integrability classes of functions. Math. USSR Sbornik. **71**(2), 561–567 (1992)
19. Korenovskii, A.A.: The reverse Hölder inequality, the Muckenhoupt condition, and equimeasurable rearrangement of functions. Russian Acad. Sci. Dokl. **45**(2), 301–304 (1992)
20. Korenovskii, A.: One refinement of the Gurov – Reshetnyak inequality. Ricerche di Mat. **45**(1), 197–204 (1996)
21. Korenovskii, A.A.: Relation between the Gurov – Reshetnyak and the Muckenhoupt function classes. Sbornik: Math. **194**(6), 919–926 (2003)
22. Korenovskii, A.A.: On classes of Gurov and Reshetnyak. The problems of theory of functions and related topics. In: Proceedings of the Institute of Mathematics of the National Academy of Sciences of Ukraine, vol. 1, no 1, pp. 189–206 (2004)
23. Korenovskii, A.A.: Riesz rising sun lemma for several variables and the John – Nirenberg inequality. Math. Notes. **77**(1), 48–60 (2005)
24. Korenovskii, A.A.: Estimate for a rearrangement of a function satisfying the "Reverse Jensen Inequality". Ukrainian Math. J. **57**(2), 186–199 (2005)
25. Korenovskyy, A.A., Lerner, A.K., Stokolos, A.M.: A note on the Gurov – Reshetnyak condition. Math. Research Letters. **9**(5–6), 579–583 (2002)
26. Malaksiano, N.A.: Exact Inclusions of Gehring Classes in Muckenhoupt Classes. Math. Notes. **70**(5–6), 673–681 (2001)
27. Malaksiano, N.A.: The precise embeddings of the one-dimensional Muckenhoupt classes in Gehring classes. Acta Sci. Math. (Szeged). **68**, 237–248 (2002)
28. Muckenhoupt, B.: Weighted inequalities for the Hardy maximal function. Trans. Amer. Math. Soc. **165**, 533–565 (1972)
29. Sbordone, C.: Rearrangement of function and reverse Hölder inequalities. Ennio De Giorgi Colloquium, Paris. Res. Notes Math. **125**, 139–148 (1983)
30. Sbordone, C.: Rearrangement of function and reverse Jensen inequalities. Proc. Symp. Pure Math. **45**(2), 325–329 (1986)
31. Schwarz, H.A.: Beweis des Satzes dass die Kugel kleinere Oberfläche besitzt, als jeder andere Körper gleichen Volumens, Göttinger Nachrichten, 1–13 [*Werke*, II, 327–340] (1884)

32. Slavin, L., Vasyunin, V.: Sharp results in the integral-form John-Nirenberg inequality, Trans. Amer. Math. Soc. **363**(8), 4135–4169 (2011)
33. Steiner, J.: Gesammelte Werke, 2 volumes, Prussian Academy of Sciences, Berlin, 1881–1882.
34. Vasyunin, V.: The sharp constant in the reverse Hölder inequality for the Muckenhoupt weights. St. Petersburg Math. J. **15**(1), 49–79 (2004)
35. Vasyunin, V.I.: Mutual estimates for L^p-norms and the Bellman function. (Russian) Zap. Nauchn. Sem. S.-Peterburg. Otdel. Mat. Inst. Steklov. (POMI) 355 (2008), Issledovaniya po Lineinym Operatoram i Teorii Funktsii. 36, 81–138, 237–238; translation in J. Math. Sci. (N. Y.) **156**(5), 766–798 (2009)
36. Vasyunin, V, Volberg A.: Sharp constants in the classical weak form of the John–Nirenberg inequality. 2012, arXiv:1204.1782v1 [math.AP]
37. Wik, I.: On Muckenhoupt's classes of weight functions. Dep. Math. Univ. Umea (Publ.) **3**, 1–13 (1987)
38. Wik, I.: On Muckenhoupt's classes of weight functions. Studia Math. **94**(3), 245–255 (1989)

Maximal Functions Measuring Smoothness

Veniamin G. Krotov

Dedicated to Professor Konstantin Oskolkov on the occasion of his 65th birthday.

Abstract The family of maximal operators measuring the local smoothness of $L^p(\mathbb{R}^n)$ functions has been introduced in the work of A. Calderón, K.I. Oskolkov, and V.I. Kolyada. These maximal operators turned out to be useful in the solutions of a series of important problems of function theory (embedding theorems of Sobolev type, characterization of Sobolev spaces in terms of the metric and measure, quantitative estimates of Luzin property). Our survey is dedicated to the history of such operators and the main results related to them. We also consider properties and applications of such operators in the general context of metric spaces with a doubling measure.

1 Calderón-Kolyada Maximal Functions

1.1 Notation

Let (X,d,μ) be a metric space with metric d and a regular Borel measure μ such that the measure of any ball

$$B(x,r) = \{y \in X : d(x,y) < r\}$$

is positive and finite. The balls $B \subset X$ will often be denoted just by one letter B; in this case r_B will stand for the radius of B.

V.G. Krotov (✉)
Belarusian State University, Nesavisimosti av., 4, Minsk, 220030, Belarus
e-mail: krotov@bsu.by

D. Bilyk et al. (eds.), *Recent Advances in Harmonic Analysis and Applications*, Springer Proceedings in Mathematics & Statistics 25, DOI 10.1007/978-1-4614-4565-4_18,

For $x \in X$ we set

$$\mathscr{B}(x) = \{B \subset X : x \in B,\ 0 < r_B < 2\}.$$

For a ball $B \subset X$, we denote by

$$f_B = \fint_B f\,d\mu = \frac{1}{\mu(B)} \int_B f\,d\mu$$

the average value of $f \in L^1_{\mathrm{loc}}(X)$ over the ball $B \subset X$; f_B is also often called the Steklov average of f.

For $0 \le \alpha < \beta < \infty$ we define $\Omega[\alpha,\beta]$ to be the set of positive increasing functions $\eta : (0,2] \to (0,+\infty)$, $\eta(+0) = 0$ with the property that $\eta(t)t^{-\alpha}$ is increasing and $\eta(t)t^{-\beta}$ is decreasing. Set also

$$\Omega[\alpha,\beta) = \bigcup_{\alpha<\beta'<\beta} \Omega[\alpha,\beta'], \quad \Omega[\alpha,\infty) = \bigcup_{\beta>\alpha} \Omega[\alpha,\beta).$$

Let us remark for the future that for any function $\eta \in \Omega[\alpha,\beta)$, $\alpha > 0$, the following inequalities hold[1]

$$\int_0^t \eta(s)\frac{ds}{s} \lesssim \eta(t), \quad \int_t^\infty \eta(s)s^{-\beta-1}\,ds \lesssim \eta(t)t^{-\beta}. \quad t > 0 \tag{1}$$

1.2 The Definition of Maximal Functions

For $q > 0$, $\eta \in \Omega[0,\infty)$, and $f \in L^q_{\mathrm{loc}}(X)$, let us introduce the maximal functions

$$\mathscr{N}^q_\eta f(x) = \sup_{B\in\mathscr{B}(x)} \frac{1}{\eta(r_B)} \left(\fint_B |f(x) - f(y)|^q\,d\mu(y)\right)^{1/q}. \tag{2}$$

In the case $q = 1$ we shall write $\mathscr{N}_\eta$ instead of $\mathscr{N}^1_\eta$ and similarly $\mathscr{N}^q_\alpha$—in the case $\eta(t) = t^\alpha$, $\alpha > 0$.

The operators $\mathscr{N}^q_\eta$ naturally give rise to function classes

$$C^{p,q}_\eta(X) = \left\{f \in L^p(X) : \|f\|_{C^{p,q}_\eta(X)} = \|f\|_{L^p(X)} + \left\|\mathscr{N}^q_\eta f\right\|_{L^p(X)} < \infty\right\}. \tag{3}$$

When $q = 1$ we write $C^p_\eta(X) = C^{p,1}_\eta(X)$ and $C^{p,q}_\alpha(X)$ in the case $\eta(t) = t^\alpha$, $\alpha > 0$.

---

[1] The notation $A \lesssim B$ will always mean that $A \le cB$, where the constant c may depend on some parameters, but these dependences are not essential.

The operators (2) and also the closely related operators

$$\mathscr{S}_{\eta}^{q} f(x) = \sup_{B \in \mathscr{B}(x)} \frac{1}{\eta(r_B)} \left(\fint_B |f(y) - f_B|^q \, d\mu(y) \right)^{1/q} \tag{4}$$

are the main objects of this chapter. Of course, when $q < 1$ in definition (4), we have to additionally assume that $f \in L^1_{\mathrm{loc}}(X)$. In the case of operators $\mathscr{S}_{\eta}^{q}$ we keep the same conventions that we have adopted for $\mathscr{N}_{\eta}^{q}$ when $q = 1$ and $\eta(t) = t^{\alpha}$. We shall primarily consider the case $q = 1$, but in some situations we shall devote our attention to arbitrary values of $q > 0$.

Maximal operators $\mathscr{N}_{\alpha}^{q}$, $0 < \alpha \leq 1$, have first appeared in the work of Calderón [1] for $X = \mathbb{R}^n$. The study of these (and some other[2]) maximal functions was continued in [2]. In particular, the paper [2] first introduced operators (4). Although here, it would be worthwhile to mention a particular case $f^{\sharp} = \mathscr{S}_0$ of the operators (4). This is the famous sharp maximal function of Fefferman and Stein [4] that appeared somewhat earlier.

One of the main results of the pioneering work of Calderón [1] is the following statement which characterizes the classical Sobolev spaces $W_1^p(\mathbb{R}^n)$.

Theorem 1. *Let $p > 1$ and $f \in L^p(\mathbb{R}^n)$. Then the conditions $f \in W_1^p(\mathbb{R}^n)$ and $\mathscr{N}_1 f \in L^p(\mathbb{R}^n)$ are equivalent.*

Theorem 1 demonstrates in particular that the scale of spaces $C_{\alpha}^{p}(\mathbb{R}^n)$ may be viewed as an alternative way of defining the Sobolev spaces $W_{\alpha}^{p}(\mathbb{R}^n)$ with the smoothness parameter $\alpha \in (0,1]$. We shall return to this discussion in Sect. 2.3.

Another interesting result shows that the function classes $C_{\alpha}^{p}(\mathbb{R}^n)$ obey a Sobolev-type inclusion. It was proved in [2] using the Hardy–Littlewood–Sobolev theorem on fractional integration.

Theorem 2. *If $q > p \geq 1$ and $0 < \beta < \alpha \leq 1$, then*

$$C_{\alpha}^{p}(\mathbb{R}^n) \subset C_{\beta}^{q}(\mathbb{R}^n), \quad \beta = \alpha - n\left(\frac{1}{p} - \frac{1}{q}\right).$$

Further developments of this theorem will be studied in Sect. 1.5 for $\mathbb{R}^n$ and in Sect. 2.5 for general metric spaces.

[2] More general operators were also investigated in [1, 2] where $\alpha > 0$ and in the definition of $\mathscr{N}_{\alpha}^{q}$ instead of $f(x)$ one subtracts a certain polynomial whose degree depends on α. The monograph [3] is devoted to the systematic study of the maximal functions of this type; hence, we shall not touch upon them here.

1.3 Measuring Local Smoothness

The next substantial period in the development of the maximal operators $\mathcal{N}_\eta$ is associated with the work of Oskolkov [5] and Kolyada [6,7], who demonstrated the usefulness of considering general functions $\eta \in \Omega[0,\infty)$ in definition (2).

The work of Oskolkov [5] investigated the rate of almost everywhere approximation of a function on $[0,1]$ by step functions (Haar polynomials) and piecewise monotonous functions, depending on the behavior of the L^p modulus of continuity of f. Besides, in [5] the author studied the problem of P.L. Ul'yanov about the characterization of Luzin's C-property in terms of the L^p-modulus of continuity.[3]

Let us recall that the L^p-modulus of continuity of a function $f \in L^p([0,1]^n)$ is defined as

$$\omega(\delta,f)_{L^p} = \sup_{0<h<\delta}\left(\int_{[0,1]^n}|f(x+h)-f(x)|^p\,dx\right)^{1/p}$$

(for the sake of simplicity, we consider all functions on $[0,1]^n$ to be continued 1-periodically in each variable). For $p \geq 1$ and $\omega \in \Omega[0,\infty)$ let us introduce additional function classes

$$H_p^\omega([0,1]^n) = \{f \in L^p([0,1]^n) : \omega(\delta,f)_{L^p} = O(\omega(\delta))\}.$$

The essence of the results in [5] lies in the estimates of the rate of almost everywhere approximation of the function $f \in L^p[0,1]$, $p \geq 1$, by its Steklov averages $|f(x) - f_{(a,b)}|$. When $x \in (a,b)$ and $b-a=\delta$ this quantity is majorized by the expression

$$\sup_{|I|=\delta,x\in I} \fint_I |f(x)-f(y)|\,dy \leq g(x)\eta(\delta), \tag{5}$$

where $\eta \in \Omega[0,\infty)$ is a special function which is constructed according to the L^p-modulus of continuity and the function $g \in L^p[0,1]$ for $p>1$ and $g \in L^1_{\text{weak}}[0,1]$ when $p=1$.

Let us describe the construction of the function η from inequality (5). It is very illustrative and has been subsequently applied to various problems by numerous authors.

Let

$$\Omega^*[0,1] = \left\{\eta \in \Omega[0,1] : \lim_{t\to+0}\frac{\eta(t)}{t} = +\infty\right\}.$$

For $\omega \in \Omega^*[0,1]$ we also denote $\widetilde{\omega}(t) = t/\omega(t)$ and introduce the so-called Oskolkov's breaking sequence

$$\delta_{\omega,0} = 1, \quad \delta_{\omega,k+1} = \min\left\{\delta : \max\left(\frac{\omega(\delta)}{\omega(\delta_{\omega,k})}, \frac{\widetilde{\omega}(\delta)}{\widetilde{\omega}(\delta_{\omega,k})}\right) = \frac{1}{2}\right\}.$$

[3]References to the earlier work on these questions may be found in [5].

Next we define the function

$$\theta(t) = 2^{1-k}, \quad t \in (\delta_{\omega,k+1}, \delta_{\omega,k}], \quad k = 0, 1, \ldots.$$

The following theorem provides a construction of the function η from inequality (5) in terms if the sequence $\delta_{\omega,k}$and the function θ.

Theorem 3 (Oskolkov [5]). *Assume $p > 1$, $\omega \in \Omega^*[0,1]$ and the function $f \in H_p^{\omega}[0,1]$. Let also a positive decreasing function ψ on $(0,1]$ satisfy the condition*

$$\int_0^1 \frac{dt}{t\psi(t)} < \infty.$$

Then inequality (5) holds if one sets

$$\eta(t) = \omega(t)\,(\psi(\theta(t)))^{1/p}. \tag{6}$$

The structure of the function (6) in the estimate (5) demonstrates that the rate of approximation (in this case, by the Steklov averages) is worse than in the standard direct and inverse theorems of the norm approximation theory. This worsening is characterized by the increasing factor $(\psi(\theta(t)))^{1/p}$ and cannot be improved—K.I. Oskolkov has shown (cf. [5, Theorem 3.11]) that Theorem 3 is sharp in its terms.

In the transparent partial case $\omega(t) = t^{\alpha}$, $0 < \alpha < 1$, Theorem 3 implies the following statement (cf. [5, Corollary 3.8]): *if $\omega(\delta, f)_{L^p} = O(\delta^{\alpha})$ $(p \geq 1, 0 < \alpha < 1)$, then for every $\varepsilon > 0$, the following relation holds for almost all $x \in [0,1]$:*

$$f(x) - \frac{1}{2h}\int_{x-h}^{x+h} f(u)\,du = o\left(h^{\alpha}\left(\ln\frac{1}{h}\right)^{\frac{1}{p}+\varepsilon}\right), \quad h \to +0.$$

On the other hand (cf. [5, Corollary 3.15]), *for any $p \geq 1$ and $0 < \alpha < 1$, there exists a function f such that $\omega(\delta, f)_{L^p} = O(\delta^{\alpha})$ and the following relation holds almost everywhere*:

$$\overline{\lim_{h\to+0}} \frac{1}{h^{\alpha}\left(\ln\frac{1}{h}\right)^{1/p}}\left| f(x) - \frac{1}{2h}\int_{x-h}^{x+h} f(u)\,du \right| = +\infty.$$

The operators $\mathcal{N}_{\eta}$ have not been explicitly introduced in [5]; however, the function g from inequality (5) is precisely the maximal operator $\mathcal{N}_{\eta}$. Besides, Eq. (5) immediately implies the bound

$$\mathcal{N}_{\eta} f \in L^p[0,1], \quad p > 1, \tag{7}$$

as well as the corresponding weak-type bound for $\mathcal{N}_{\eta} f$ when $p = 1$.

Let us stress a very important property of the maximal operators $\mathcal{N}_\eta$—they are extremely convenient in characterizing local properties of functions. For instance, $\mathcal{N}_\eta f$ measures the local smoothness of f which is evident from the simple inequality

$$|f(x)-f(y)| \le \eta(d(x,y))[\mathcal{N}_\eta f(x)+\mathcal{N}_\eta f(y)], \quad x,y \in X, \quad d(x,y)<2. \tag{8}$$

If $\lambda > 0$ and

$$E_\lambda = \{x \in X : g) \le \lambda\},$$

then Eq. (8) implies the inequality

$$|f(x)-f(y)| \le 2\lambda\eta(d(x,y)), \quad x,y \in E_\lambda, \quad d(x,y)<2$$

which amounts to the uniform continuity of f on the set E_λ and majorizes the uniform modulus of continuity of f on this set by $2\lambda\eta(\delta)$. Hence, estimates of the type (7) immediately lead to the quantitative bounds for the Luzin's C-property for the function f.

1.4 Bounds of $\mathcal{N}_\eta$ on $[0,1]^n$

In [6, 7] Kolyada considered the problem of transferring K.I. Oskolkov's estimate (7) to the multidimensional case. It is precisely in these papers that the maximal operators $\mathcal{N}_\eta$ for an arbitrary function $\eta \in \Omega[0,1]$ have first appeared (in the case of $X=[0,1]^n$). In our opinion, the names "Calderón–Kolyada maximal operators" for Eq. (2) and "Calderón–Kolyada classes" for Eq. (3) with arbitrary function $\eta \in \Omega[0,\infty)$ would be rather just and appropriate. And we keep the names "Calderón-Scott maximal operators" and "Calderón-Scott classes" in the case $\eta(t)=t^\alpha$.

The motivation for the consideration of such a wide spectrum of characteristics η in these papers was in increasing the possibilities for a finer classification of functions in terms of their local smoothness (see more on this subject in Sect. 2.1).

The use of the class of functions $\eta \in \Omega[0,1]$ (which is smaller than $\Omega[0,\infty)$) has been dictated by the specifics of the case $X=[0,1]^n$ under consideration. Namely, from the conditions $\eta(t)=o(t)$ for $t\to+0$ and $\mathcal{N}_\eta f \in L^p([0,1])$, $p>1$, one can deduce that the function f is equivalent to a constant.

We shall now quote a series of results from these very interesting papers (only the formulations were given in [6], while the proofs are contained in [7]).

The question of validity of estimate (7) for all functions $f \in H_p^\omega([0,1]^n)$ may be interpreted as a question of the relation between the function classes $H_p^\omega([0,1]^n)$ and $C_\eta^p([0,1]^n)$. It can be easily seen that inequality (8) immediately implies the embedding $C_\eta^p([0,1]^n) \subset H_p^\eta([0,1]^n)$, $\eta \in \Omega[0,1]$. The problem of the inverse embedding is substantially more difficult.

The following statement is a generalization of Theorem 3 to the multidimensional case, although it is formulated in a slightly different way.

Theorem 4. *Let $p > 1$, $\omega, \eta \in \Omega^*[0,1]$. Then, whenever*

$$\sum_{k=0}^{\infty} \left(\frac{\omega(\delta_{\omega,k})}{\eta(\delta_{\omega,k})} \right)^p < \infty,$$

one has the following embedding:

$$H_p^{\omega}([0,1]^n) \subset C_{\eta}^p([0,1]^n). \tag{9}$$

Analogous embeddings are proved in [7] also for the classes $C_{\eta}^q([0,1])$, where $q > p$. Denote by $E_p(\omega)$ the class of all nonnegative sequences ε_k, which satisfy the inequalities

$$2^{-sp} \sum_{k=1}^{s} 2^{kp} \varepsilon_k^p + \sum_{k=s}^{\infty} \varepsilon_k^p \leq \omega^p(2^{-s}), \quad s \in \mathbb{N}.$$

Theorem 5. *Let $q \geq p > 1$, $\omega, \eta \in \Omega[0,\infty)$. Then for the embedding*

$$H_p^{\omega}([0,1]^n) \subset C_{\eta}^q([0,1]^n) \tag{10}$$

to hold, the following condition is necessary and sufficient:

$$\sup_{\varepsilon \in E_p(\omega)} \sum_{k=1}^{\infty} 2^{kn(q/p-1)} \left(\frac{\varepsilon_k}{\eta(2^{-k})} \right)^q < \infty.$$

A more transparent condition

$$\int_0^1 \left(\frac{\omega(t)}{\eta(t)} \right)^q t^{n(1-q/p)} \frac{dt}{t} < \infty$$

is also sufficient for the embedding (10) (cf. [7, Proposition 4.6]), but it is less sharp.

1.5 Sobolev-Type Embedding Theorems

In [7], Kolyada also considered embedding theorems for two different classes $C_{\eta}^p([0,1]^n)$.

Theorem 6 (Kolyada [7]). *Let $\eta, \sigma \in \Omega[0,1]$, and $q > p \geq 1$. Then for the embedding*

$$C_{\eta}^p([0,1]^n) \subset C_{\sigma}^q([0,1]^n) \tag{11}$$

to hold, the following condition is necessary and sufficient:

$$\sup_{t\in(0,1]} \frac{\eta(t)}{\sigma(t)} t^{n(1/q-1/p)} < \infty. \tag{12}$$

The proof is based on the estimates of equimeasurable decreasing rearrangement of the maximal function $\mathcal{N}_\eta f$.

Setting $\eta(t) = t^\alpha$, $\sigma(t) = t^\beta$ $(0 < \beta < \alpha \leq 1)$ in Theorem 6, one obtains the embedding

$$C^p_\alpha([0,1]^n) \subset C^q_\beta([0,1]^n), \quad \beta = \alpha - n\left(\frac{1}{p} - \frac{1}{q}\right),$$

which in the case of $\mathbb{R}^n$ was already mentioned above in Theorem 2.

If $\eta(t) = \sigma(t)t^{n(1/p-1/q)}$, then one can deduce even more than embedding (11). Namely, in this case, the following theorem holds (cf. [7, Theorem 3.1]).

Theorem 7 (Kolyada [7]). *Let $q > p \geq 1$ and let the functions $\eta,\sigma \in \Omega[0,1]$ satisfy condition (12). Then*

$$\left\|\mathcal{N}_\eta f\right\|_{q,p} \lesssim \left\|\mathcal{N}_\eta f\right\|_p.$$

The norm on the left-hand side of this inequality is taken in the Lorentz space $L^{q,p}([0,1]^n)$.

2 Operators $\mathcal{N}_\eta$ on Metric Measure Spaces

2.1 *The Classes $C^p_\eta(X)$ on Metric Spaces*

In what follows we shall assume that the metric measure space (X,d,μ) satisfies the doubling condition

$$\mu(B(x,2r)) \lesssim \mu(B(x,r)), \quad x \in X, \quad r > 0. \tag{13}$$

In this case the space (X,d,μ) is called homogeneous [8].

We shall also assume that X is bounded, and for simplicity we set $\operatorname{diam} X = 1$. The boundedness condition is not always required, but we would like to avoid a discussion of minor technical questions which would take us away from the essential questions.

As we have already remarked in Sect. 1.4, in the definition of the classes $C^p_\eta([0,1]^n)$, it is natural to assume that $\eta \in \Omega[0,1]$. In general, it is possible that the classes $C^p_\eta(X)$ are nontrivial for some functions $\eta \in \Omega[0,\infty)$ satisfying the condition $\eta(t) = o(t)$ for $t \to +0$. For example, on the Koch curve K the classes $C^p_\eta(K)$ are nontrivial when $\eta \in \Omega[0,\ln 4/\ln 3]$ [9], such an effect is also described in [19]. This explains our assumptions on the function $\eta \in \Omega[0,\infty)$.

The role of the maximal functions $\mathcal{N}_\eta$ in the case of an arbitrary metric space is extended. They provide us with a possibility to measure smoothness of functions in terms of L^p spaces, while the more common objects usually used to this end (e.g., the L^p-modulus of continuity) are not meaningful.

And they indeed yield a complete classification of L^p-functions according to their smoothness—*for any function $f \in L^p(X)$, $p > 0$, there exists a function $\eta \in \Omega[0,\infty)$, such that $\mathcal{N}_\eta f \in L^p(X)$* (cf. [10, Theorem 2], and also [11, Theorem 2], where the proof is significantly simpler).

Moreover, numerous problems of function theory may be solved in terms of the maximal functions $\mathcal{N}_\eta$ and the classes $C_\eta^p(X)$. This will be the main theme of the rest of this survey.

2.2 The Relation Between Operators $\mathcal{N}_\eta$ and $\mathcal{S}_\eta$

The operators $\mathcal{N}_\eta$ and $\mathcal{S}_\eta$ are intimately related to each other—they satisfy the following simple inequality:

$$\mathcal{S}_\eta f(x) \le 2\mathcal{N}_\eta f(x), \quad f \in L^1_{\mathrm{loc}}(X), \tag{14}$$

and, in principle, their close definitions suggest that their roles in the investigation of function properties are similar.

However, they have a number of substantial differences. For example, the values of $\mathcal{S}_\eta f(x)$ do not depend on changing the values of the function on the set of measure 0, while the values of $\mathcal{N}_\eta f(x)$ do. The function $\mathcal{S}_\eta f$ is lower semicontinuous, but $\mathcal{N}_\eta f(x)$ is not. Generally speaking, $\mathcal{N}_\eta f(x)$ is more closely connected to the values of the function f, but $\mathcal{S}_\eta f$ is more convenient to work with due to the properties just mentioned.

With some additional assumptions on the function $\eta \in \Omega[0,\infty)$, one can deduce an inverse inequality to Eq. (14). Namely, *if the Bari condition holds, i.e.,*

$$\int_0^t \eta(s)\,\frac{ds}{s} \lesssim \eta(t), \tag{15}$$

then

$$\mathcal{N}_\eta f(x) \lesssim \mathcal{S}_\eta f(x), \quad f \in L^1_{\mathrm{loc}}(X).$$

(see, e.g., [22]).

We do not know whether the inequality

$$\|\mathcal{N}_\eta f\|_{L^p(X)} \lesssim \|\mathcal{S}_\eta f\|_{L^p(X)}$$

holds for all functions $\eta \in \Omega[0,\infty)$ and $p > 1$.

2.3 Sobolev Classes on Metric Spaces

The statement of Theorem 1 has an interesting property that it characterizes Sobolev spaces $W_1^p(\mathbb{R}^n)$ without using any features of the Euclidean spaces $\mathbb{R}^n$, other than the metric and the measure. This could give rise to the definition of Sobolev spaces $W_1^p(X)$ on arbitrary metric spaces X, equipped with a Borel measure. However, this fact remained unnoticed until the work of Hajlasz [12], who gave the following definition.

For a measurable function f on X, we denote by $D(f)$ the set of all nonnegative measurable functions g on X, such that there exists a set $E \subset X$ (depending on g), $\mu(E) = 0$ for which [compare this to Eq. (8)]

$$|f(x) - f(y)| \leq d(x,y)[g(x) + g(y)], \quad x,y \in X \setminus E. \tag{16}$$

The Hajlasz–Sobolev class $M_1^p(X)$ is defined as the set of (equivalence classes of) functions $f \in L^p(X)$, for which $D(f) \cap L^p(X) \neq \varnothing$. The norm in $M_1^p(X)$ is given by

$$\|f\|_{M_1^p(X)} = \|f\|_{L^p(X)} + \inf\left\{ \|g\|_{L^p(X)} : g \in D(f) \cap L^p(X) \right\}.$$

It was proved in [12] that $W_1^p(\mathbb{R}^n) = M_1^p(\mathbb{R}^n)$, $p > 1$. Moreover, the method of proof of the main part [that each function from $M_1^p(\mathbb{R}^n)$ belongs to $W_1^p(\mathbb{R}^n)$] is identical to the proof of Theorem 1 in the work of Calderón [1]. Therefore,

$$W_1^p(\mathbb{R}^n) = C_1^p(\mathbb{R}^n) = M_1^p(\mathbb{R}^n).$$

The Hajlasz–Sobolev classes $M_1^p(X)$ were studied in numerous papers; see [13–16] for more details.

2.4 Various Characterizations of $C_\eta^p(X)$

In the case of general metric spaces, the classes $C_\alpha^p(X)$ first appeared in [17, 18] and later in [19,20], while the classes $C_\eta^p(X)$ with an arbitrary function $\eta \in \Omega[0,\infty)$ were first considered in [21,22].

In [18] it was shown that Riesz potentials of L^p-functions on the space of homogeneous type belong to the class $C_\alpha^p(X)$.

The paper [20] is devoted to various characterizations of the classes $C_\alpha^p(X)$. The results of this work have been subsequently generalized in [22] to $C_\eta^p(X)$, $\eta \in \Omega[0,\infty)$.

Theorem 8 (Ivanishko [22]). *If $p > 1$, $\eta \in \Omega[0,\infty)$ and $f \in L^p(X)$, then the following statements are equivalent*

(1) $\mathcal{N}_\eta f \in L^p(X)$.
(2) There exists a function $g \in L^p(X)$ *and a set* $E \subset X$, $\mu(E) = 0$ *such that*

$$|f(x) - f(y)| \le \eta(d(x,y))[g(x) + g(y)], \quad x,y \in X \setminus E. \tag{17}$$

(3) The maximal function

$$\sup_{B \in \mathscr{B}(x)} \frac{|f(x) - f_B|}{\eta(r_B)}$$

belongs to $L^p(X)$.

Theorem 9 (Ivanishko [22]). *If* $p > 1$, $\eta \in \Omega[0,\infty)$, *and* $f \in L^p(X)$, *then the following statements are equivalent:*

(4) $\mathcal{S}_\eta f \in L^p(X)$.
(5) There exists a nonnegative function $g \in L^p(X)$ *and a number* $\lambda \ge 1$, *such that for any ball* $B \subset X$,

$$⨍_B |f(y) - f_B|\, d\mu(y) \le \eta(r_B) ⨍_{\lambda B} g\, d\mu.$$

(6) The maximal function

$$\sup_{B \in \mathscr{B}(x)} \frac{1}{\eta(r_B)} ⨍_B ⨍_B |f(y) - f(z)|\, d\mu(y)\, d\mu(z)$$

belongs to $L^p(X)$.

Condition (5) is related to inequalities of Poincaré type, which we shall discuss later in Sect. 5.

In view of the conditions given in Theorem 8, the following statement holds: *if a function* $\eta \in \Omega[0,\infty)$ *satisfies the Bari condition (15), then all the conditions (1)–(6) from Theorems 8 and 9 are equivalent.* In particular, this is true for $\eta(t) = t^\alpha$, $\alpha > 0$ [20].

We don't know whether conditions (1) and (4) are equivalent without the Bari condition (15) (see more on this issue in Sect. 2.2).

2.5 Embedding Theorems

Below we shall need a quantitative version of the doubling condition (13). Whenever Eq. (13) holds, there exists $\gamma > 0$ for which we have the inequality

$$\mu(B(x,s)) \lesssim \left(\frac{s}{r}\right)^\gamma \mu(B(x,r)), \quad x \in X,\ 0 < r \le s, \tag{18}$$

which is obtained by iterating the condition (13). The parameter γ is extremely important—it plays the role of the dimension of the space X. It should be minimized whenever possible.

The proof of Theorem 7 may be easily adapted to a much more general situation.

Theorem 10 (Ivanishko [21]). *Assume that condition (18) holds, $q > p \geq 1$, $\sigma \in \Omega[0,\infty)$, and*

$$\eta(t) = \sigma(t)t^{\gamma(1/p-1/q)}. \tag{19}$$

Then

$$\|\mathcal{N}_\eta f\|_{q,p} \lesssim \|\mathcal{N}_\eta f\|_p.$$

This implies, in particular, that for $q > p \geq 1$, the condition

$$\sup_{t\in(0,1]} \frac{\eta(t)}{\sigma(t)} t^{\gamma(1/q-1/p)} < \infty \tag{20}$$

is sufficient for the embedding

$$C^p_\eta(X) \subset C^q_\sigma(X), \tag{21}$$

Without additional restrictions the converse theorem is not true, at the very least because the doubling condition (18) may hold with a smaller value of γ. We shall say that (X,d,μ) is a γ-space if $\mu(B(x,r)) \asymp t^\gamma$ with implicit constants independent of $x \in X$ and $0 < r \leq 1$.

If (X,d,μ) is a γ-space and $\eta, \sigma \in \Omega[0,1]$, then Eq. (20) is also sufficient for embedding (21). As noticed in [23], this can be proved similarly to the case $X = [0,1]^n$ in the work of Kolyada [7, Proposition 3.2].

The paper [10] studied the question of compactness of the embeddings of the type (21). The main result in [10] has the following form.

Theorem 11. *Let $1 < p < q < \infty$, $\sigma \in \Omega[0,\infty)$, and η is defined by the identity (19). If the function $\sigma_0 \in \Omega[0,\infty)$ satisfies*

$$\lim_{r\to+0} \frac{\sigma(r)}{\sigma_0(r)} = 0, \tag{22}$$

then the embedding $C^p_\eta(X) \subset C^q_{\sigma_0}(X)$ is compact.

If (X,d,μ) is a γ-space and $\eta \in \Omega[0,1]$, then Eq. (22) is also sufficient for the compactness of the embedding $C^p_\eta(X) \subset C^q_{\sigma_0}(X)$ (cf. [10, Theorem 4]). Moreover, the additional restrictions on (X,d,μ) and η are substantial and cannot be excluded.

From the statement which immediately follows Theorem 10, it is easy to deduce that for $\eta(t) = t^{\gamma(1/p-1/q)}$, one has the embedding

$$C^p_\eta(X) \subset L^q(X). \tag{23}$$

If, however, $\overline{\lim\limits_{t\to+0}}\, \eta(t)t^{\gamma(1/q-1/p)} = \infty$, then embedding (23) does not hold [23].

In [10] it is shown that the condition $\lim_{t\to+0}\eta(t)t^{\gamma(1/q-1/p)}=0$ is sufficient, and with the additional assumptions that (X,d,μ) is a γ-space and $\eta\in\Omega[0,1]$, it is also necessary for the compactness of embedding (23) (cf. [10, Theorems 6, 7]).

2.6 Weighted Inequalities for $\mathscr{S}_\eta$

This circle of problems was investigated in [24], where weighted L^p estimates for the maximal operators $\mathscr{S}_\eta$ were proved. Let us quote some results from this work.

A nonnegative function ν, defined on a σ-algebra of Borel subsets of X, is called an outer measure if it is monotone and subadditive, i.e.,

$$G_1\subset G_2\Rightarrow \nu(G_1)\le\nu(G_2),\quad \nu\left(\bigcup_j G_j\right)\le\sum_j\nu(G_j).$$

For a Borel function f and an outer measure ν on X, we set

$$\|f\|_{L^p_\nu(X)}=\left(p\int_0^\infty\lambda^{p-1}\nu\{|f|>\lambda\}\,d\lambda\right)^{1/p},\qquad p>0.$$

Theorem 12. *Let ν be an outer measure and ω be a measure on X, satisfying the doubling condition (13)*

$$\varphi(t)=\sup_{x\in X}\frac{\omega(B(x,t))}{\nu(B(x,t))},\qquad t\in(0,2].$$

Let also $\sigma\in\Omega[0,\infty)$ and

$$\eta(t)=\sigma(t)\,[\varphi(t)]^{1/p}.$$

Then the following inequality holds:

$$\|\mathscr{S}_\sigma f\|_{L^p_\nu(X)}\lesssim\|\mathscr{S}_\eta f\|_{L^p_\omega(X)},\quad f\in L^1_{\mathrm{loc}}(X).\tag{24}$$

In particular, if $p>0$, $\delta>0$, $\sigma\in\Omega[0,\infty)$,

$$\eta(t)=t^{\frac{\delta}{p}}\sigma(t),$$

the measure ω and an outer measure ν satisfy the condition

$$\nu(B(x,t))\lesssim t^{-\delta}\omega(B(x,t)),\quad x\in X,\,t\in(0,2],\tag{25}$$

then inequality (24) holds.

In [24] there are also similar estimates for pairs of spaces $L^p_\omega(X)$ and $L^q_\nu(X)$.

Theorem 13. *Let $q \geq p > 0$ and $\gamma \geq \delta > 0$ and let the measure ω satisfy the doubling condition (18); the outer measure ν is related to ω by the condition*

$$\nu(B(x,t)) \lesssim t^{\delta-\gamma}\omega(B(x,t)), \quad x \in X, t > 0. \tag{26}$$

Assume also that $\sigma \in \Omega[0,\infty)$ and

$$\eta(t) = t^{\frac{\gamma}{p}-\frac{\delta}{q}}\sigma(t).$$

Then the following inequality holds:

$$\|\mathscr{S}_\sigma f\|_{L^q_\nu(X)} \lesssim \|\mathscr{S}_\eta f\|_{L^p_\omega(X)}, \quad f \in L^1_{\mathrm{loc}}(X). \tag{27}$$

3 Fine Properties of Functions

The properties of functions that we shall discuss below depend on changing the functions' values on sets of μ-measure zero; hence, we shall from now on assume that the values of locally integrable functions at each point $x \in X$ are defined by the identity

$$f(x) = \overline{\lim_{r\to+0}} \fint_{B(x,r)} f\,d\mu.$$

3.1 Lebesgue Points

Let the function $f \in L^1_{\mathrm{loc}}(X)$ and let $\Lambda(f)$ be the complement of the set of those points $x \in X$ where the limit

$$\lim_{r\to+0} \frac{1}{\mu(B(x,r))} \int_{B(x,r)} f\,d\mu \tag{28}$$

exists. The points where the limit (28) exists are usually called the Lebesgue points of the function f. At such points, the function has natural values, which are independent of redefining the functions on sets of measure zero. The set $\Lambda(f)$ may be called the exceptional Lebesgue set of the function f.

The classical theorem of Lebesgue says that $\mu(\Lambda(f)) = 0$ for any function $f \in L^1_{\mathrm{loc}}(X)$. For the spaces (X,d,μ) with the doubling property, this statement may be found in [15, Theorem 1.8].

When the function f is more regular, one can assert even more. We shall pay special attention to the results on Lebesgue points for the classes $C^p_\alpha(X)$ (for the estimates of the set $\Lambda(f)$ in the Euclidean setting, see [25] and the references cited therein). To this end, we shall need to introduce the capacities arising from these classes:

$$\mathrm{Cap}_{\alpha,p}(E) = \inf\left\{ \|f\|^p_{C^p_\alpha(X)} : f \in C^p_\alpha(X),\ \ f \geq 1 \text{ in neighborhood of } E \right\}.$$

They have been introduced and studied in [26] for $\alpha = 1$ and in [27, 32] for $0 < \alpha \leq 1$.

We also recall the definition of Hausdorff s-measure $\mathbb{H}^s$ and the Hausdorff dimension $\dim_{\mathbb{H}}$ of the set $E \subset X$ (see, e.g., [25, Sect. 5.1]):

$$\mathbb{H}^s_\delta(E) = \inf\left\{ \sum_{i=1}^{\infty} r_i^s : E \subset \bigcup_{i=1}^{\infty} B(x_i, r_i), r_i < \delta \right\}, \quad \mathbb{H}^s(E) = \lim_{\delta \to +0} \mathbb{H}^s_\delta(E),$$

$$\dim_{\mathbb{H}}(E) = \inf\left(s : \mathbb{H}^s(E) = 0\right\} = \sup\left\{s : \mathbb{H}^s(E) = \infty\right).$$

Theorem 14. *Let $\alpha > 0$, $1 < p < \gamma/\alpha$, and $f \in C^p_\alpha(X)$. Then there exists a set $E \subset X$ such that for any $x \in X \setminus E$, the limit*

$$\lim_{r \to +0} \fint_{B(x,r)} f\, d\mu = f^*(x)$$

exists. Moreover,

$$\lim_{r \to +0} \fint_{B(x,r)} |f - f^*(x)|^q\, d\mu = 0, \quad \frac{1}{q} = \frac{1}{p} - \frac{\alpha}{\gamma}, \tag{29}$$

and the following estimates hold:

(1) $\dim_{\mathbb{H}}(E) \leq \gamma - \alpha p$*, in particular,* $\dim_{\mathbb{H}}(\Lambda(f)) \leq \gamma - \alpha p$*, if* $\alpha > 0$.
(2) $\mathrm{Cap}_{\alpha,p}(E) = 0$*, in particular,* $\mathrm{Cap}_{\alpha,p}(\Lambda(f)) = 0$*, if* $0 < \alpha \leq 1$.

In the case $X = \mathbb{R}^n$, $\alpha = 1$ this result is due to Federer and Ziemer [28] (in [28] they consider the Sobolev class $W^p_1(\mathbb{R}^n)$, which coincides with $C^p_1(\mathbb{R}^n)$ in view of Theorem 1). This was the first statement of this sort.

The case of arbitrary X and $\alpha = 1$, but without property (29), has been considered in [29] and with property (29) (but for $\frac{1}{q} > \frac{1}{p} - \frac{\alpha}{\gamma}$)—in [30].

The general form of Theorem 14 is due to Prokhorovich [31, 32] for $0 < \alpha \leq 1$ and [33] for $\alpha > 0$.

During the work on the present survey, the author and Prokhorovich [34] proved the sharpness of Theorem 14 for $0 < \alpha \leq 1$.

Theorem 15. *Let $0 < \alpha \leq 1$ and $1 < p < n/\alpha$. Then there exists a function $f_0 \in C^p_\alpha(\mathbb{R}^n)$ for which:*

(1) $\dim_{\mathbb{H}}(\Lambda(f_0)) = n - \alpha p$.
(2) $\mathrm{Cap}_{\alpha+\varepsilon,p}(\Lambda(f_0)) > 0$, $\mathrm{Cap}_{\alpha,p+\varepsilon}(\Lambda(f_0)) > 0$ *for any* $\varepsilon > 0$.

Besides, in [34], the authors obtain the following statement about the exceptional Lebesgue set for the functions from the Sobolev spaces $W_l^p(\mathbb{R}^n)$ of higher order.

Theorem 16. *Let $1 < p < N/l$. Then there exists a function $f_0 \in W_l^p(\mathbb{R}^n)$ such that $\dim_{\mathbb{H}}(\Lambda(f_0)) = n - lp$ and $\mathrm{Cap}_{l,q}(\Lambda(f_0)) > 0$ for any $q > p$.*

For more information on the upper bounds on the Hausdorff dimension and the capacities of the set $\Lambda(f)$ for functions from $W_l^p(\mathbb{R}^n)$, and also for the spaces of Bessel potentials, the reader is referred, for example, to [25, Sect. 6.2] and the references cited therein.

The following result from [35] estimates, in particular, the complement of the set on which the Steklov averages converge to the function values at the rate

$$\lim_{r\to+0} r^{-\beta}\left(\fint_{B(x,r)} f\,d\mu - f(x)\right) = 0.$$

Theorem 17. *Let $\alpha > 0$, $1 < p < \gamma/\alpha$, and $0 < \beta < \alpha$. Then for any function $f \in C_\alpha^p(X)$, there exists a set $E \subset X$ such that:*

(1) $\mathbb{H}_\infty^{\gamma-(\alpha-\beta)p}(E) = 0$, in particular, $\dim_{\mathbb{H}}(E) \le \gamma - (\alpha-\beta)p$.
(2) For all $x \in X \setminus E$,

$$\lim_{r\to+0} r^{-\beta}\left(\fint_{B(x,r)} |f - f(x)|^q\,d\mu\right)^{1/q} = 0 \quad \textit{when} \quad \frac{1}{q} = \frac{1}{p} - \frac{\alpha}{\gamma}.$$

In this regard, we should remark that we do not know whether Theorem 14 can assert that $\mathbb{H}_\infty^{\gamma-\alpha p}(E) = 0$. In other words, whether the first statement of Theorem 17 holds for $\beta = 0$.

3.2 Luzin Approximation

As we have discussed earlier in Sects. 1.3 and 1.4, due to inequality (8), the maximal functions $\mathcal{N}_\eta$ are closely related to Luzin's theorem and measuring the local smoothness of functions defined on Euclidean spaces.

Here, we shall focus our attention on the properties of this sort for functions defined on metric spaces. First of all, let us remark that the local smoothness inequalities akin to Eq. (17) hold for any measurable function [11, Theorem 1].

Theorem 18. *For any measurable function f on X and there exist $\eta \in \Omega[0,\infty)$, a measurable nonnegative function g on X, and a set $E \subset X$, $\mu E = 0$, such that the following inequality holds:*

$$|f(x) - f(y)| \le [g(x) + g(y)]\eta(d(x,y)), \quad x,y \in X \setminus E. \tag{30}$$

Of course, the role of the function g in inequality (30) lies in the fact that it (as in the end of Sect. 1.4) controls Luzin's C-property of the function f. We shall return to this theorem in Sect. 4.1, where we shall give its generalizations related to compactness criteria in the space of measurable functions.

The construction of the function g in Theorem 18 is similar to the definition of the maximal function $\mathscr{N}_\eta^q$ [see Eq. (2)] but with the power function t^q being replaced by a positive monotone function φ with slow growth at infinity. Naturally, if $f \in L^p(X)$, $p > 0$, then in Theorem 18 one can set $g = \mathscr{N}_\eta^q f$ with $q < p$.

We shall further concentrate on the estimates of exceptional sets in Luzin's theorem for functions from the classes $C_\alpha^p(X)$.

Theorem 19. *Let* $0 < \beta \le \alpha \le 1$, $1 < p < \gamma/\alpha$, *and* $f \in C_\alpha^p(X)$.

Then for any $\varepsilon > 0$, *there exists a function* f_ε *and an open set* $O \subset X$ *such that:*

(1) $\mathrm{Cap}_{\alpha-\beta,p}(O) < \varepsilon$, $\mathbb{H}_\infty^{\gamma-(\alpha-\beta)p}(O) < \varepsilon$.
(2) $f = f_\varepsilon$ *on* $X \setminus O$.
(3) $f_\varepsilon \in C_\alpha^p(X) \cap H^\beta(X)$.
(4) $\|f - f_\varepsilon\|_{C_\alpha^p(X)} < \varepsilon$.

The statement uses the Hölder classes $H^\beta(X)$ defined in a traditional way:

$$H^\beta(X) = \left\{ \phi : \sup_{x \neq y, x, y \in X} [d(x,y)]^{-\beta} |\phi(x) - \phi(y)| < +\infty \right\}.$$

In addition, by $\mathrm{Cap}_{0,p}$, we mean the measure μ.

For $\beta = \alpha = 1$a similar result was obtained in [12]. The case $\beta \le \alpha = 1$ is substantially more difficult; it has been studied in [36].

The general form of Theorem 19 was proved in [37]. The proof relied on the general argument scheme from [36], as well as the weighted inequalities for the maximal functions $\mathscr{S}_\alpha$ from [24].

For more on the development of statements of this sort, see, for example, the bibliography in [37].

4 Compactness Criteria

In numerous situations compactness criteria in function spaces have the following form: the complete boundedness of a set[4] is equivalent to its boundedness plus a certain equicontinuity property with respect to the metric of the space X. A classical example is the historically first Arzelà-Ascoli compactness criterion for the space of continuous functions.

---

[4]A subset of a metric space is called completely bounded if for any $\varepsilon > 0$ it has s finite ε-net. In a complete space this property together with closeness is equivalent to compactness (Hausdorff criterion).

Since Luzin's property, as well as its quantitative form Eq. (30), is a certain manifestation of continuity, it would be natural to attempt to use such inequalities for the characterization of completely bounded sets in function spaces. Such a problem has been considered in [38], and the starting point for the author was Theorem 18. In the following two sections, we shall discuss the compactness criteria from [38].

4.1 Compactness Criteria in L^0

Let $L^0(X)$ be the set of all (equivalence classes of) measurable functions on X. It is a complete metric space with respect to the metric

$$d_{L^0}(f,g) = \int_X \varphi_0(f-g)\,d\mu, \quad \text{where} \quad \varphi_0(t) = \frac{|t|}{1+|t|},$$

and the convergence in this metric coincides with convergence in measure.

For $\eta \in \Omega[0,\infty)$ and the function $f \in L^0(X)$ we denote by $D_\eta(f)$ the class of functions $g \in L^0(X)$ for which there exists a set $E \subset X$, $\mu E = 0$ such that inequality (30) holds. Theorem 18 tells us that for any function $f \in L^0(X)$, the set $D_\eta(f)$ is not empty for some choice of the function $\eta \in \Omega[0,\infty)$.

In terms of the classes $D_\eta(f)$, the following compactness criterion holds in $L^0(X)$ [38, Theorem 4].

Theorem 20. *A set $S \subset L^0(X)$ is completely bounded if and only if*

$$\lim_{\lambda\to+\infty} \sup_{f\in S} \mu\{|f| > \lambda\} = 0 \tag{31}$$

and there exists a function $\eta \in \Omega[0,\infty)$ such that

$$\lim_{\lambda\to+\infty} \sup_{f\in S} \inf_{g\in D_\eta(f)} \mu\{g > \lambda\} = 0. \tag{32}$$

Condition (32) alone does not guarantee complete boundedness. Relation (31) is an adequate replacement of boundedness for the space $L^0(X)$ (which itself is bounded). Explicitly, Eq. (31) can be found in [39], where an analog of M. Riesz compactness criterion for $L^0([0,1]^n)$ is proved.

In [38, Theorem 4] one can find a number of other compactness criteria for $L^0(X)$. In particular, such a criterion may be formulated in terms of maximal functions of the form

$$\mathcal{N}_\eta^\varphi f(x) = \sup_{B\in\mathcal{B}(x)} \frac{1}{\eta(r_B)} \fint_B \varphi(f(x)-f(y))\,d\mu(y), \quad 0<t<2. \tag{33}$$

(see [38, Theorem 9]). A survey of earlier criteria for this space is contained in [40, Chap. 4, Sect. 16].

4.2 Compactness Criteria in L^p, $p > 0$

We now turn to the discussion of spaces $L^p(X)$, $p > 0$. The following statement is an analog of Theorem 20 (see [38, Theorem 3]).

Theorem 21. *Let $p > 0$ and let $S \subset L^p(X)$ be bounded. Then the following conditions are equivalent:*

(1) S is completely bounded.
(2) There exists a function $\eta \in \Omega[0,\infty)$ such that

$$\sup_{f \in S} \inf_{g \in D_\eta(f)} \|g\|_{L^p(X)} < +\infty. \tag{34}$$

The following result (cf. [38, Theorem 5]) translates Theorem 21 to the language of maximal functions $\mathcal{N}_\eta^q$.

Theorem 22. *Assume that $0 < q < p$ and the set $S \subset L^p(X)$is bounded. Then the following conditions are equivalent:*

(1) S is completely bounded.
(2) There exists a function $\eta \in \Omega[0,\infty)$ such that

$$\sup_{f \in S} \|\mathcal{N}_\eta^q f\|_{L^p(X)} < +\infty.$$

(3) The following condition holds:

$$\lim_{r \to +0} \sup_{f \in S} \int_X \left[\fint_{B(x,r)} |f(x) - f(y)|^q \, d\mu(y) \right]^{p/q} d\mu(x) = 0. \tag{35}$$

It is not hard to see that Theorem 22 fails if one sets $p = q$.

Theorem 22 implies that if $0 < q < p$, then completely bounded sets in $L^p(X)$ are precisely the bounded sets in $C_\eta^{p,q}(X)$, $\eta \in \Omega[0,\infty)$ and conversely [the classes $C_\eta^{p,q}(X)$ are defined in Eq. (3)].

5 Sobolev–Poincaré Inequalities

5.1 Approximation by Steklov Averages Revisited

At the end of our survey, we would like to return to the problem of estimating the rate

$$|f(x_0) - f_{B(x_0,r)}|$$

of approximations of function values by their Steklov averages. However, now these estimates will take a somewhat different form as they will be written in terms of maximal functions $\mathscr{S}_\eta$ [see Eq. (4)]. They lead to inequalities of Sobolev–Poincaré type.

The following theorem was obtained in [41]. Since this paper is hard to come by, we shall present this theorem with proof.

Theorem 23. *Let $p > 0$, $0 < \alpha < \gamma/p$, $\eta \in \Omega[\alpha, \gamma/p)$, and $f \in L^1_{\mathrm{loc}}(X)$. Then*

(1) If x_0 is a Lebesgue point of the function f, then

$$|f(x_0) - f_{B(x_0,r)}| \lesssim \eta(r)\,(\mathscr{S}_\eta f(x_0))^{1-\frac{\alpha p}{\gamma}} \left(⨍_{B(x_0,r)} (\mathscr{S}_\eta f)^p \, d\mu \right)^{\alpha/\gamma}. \tag{36}$$

(2) For any ball $B \subset X$

$$\left(⨍_B |f - f_B|^q \, d\mu \right)^{1/q} \lesssim \eta(r_B) \left(⨍_{2B} (\mathscr{S}_\eta f)^p \, d\mu \right)^{1/p}, \tag{37}$$

and for any Lebesgue point $x_0 \in B$ of the function f,

$$\left(⨍_B |f - f(x_0)|^q \, d\mu \right)^{1/q} \lesssim \eta(r) \left[\mathscr{S}_\eta f(x_0) + \left(⨍_{2B} (\mathscr{S}_\eta f)^p \, d\mu \right)^{1/p} \right], \tag{38}$$

where $\frac{1}{q} = \frac{1}{p} - \frac{\alpha}{\gamma}$ and the implicit constants on the right-hand sides of Eqs. (36)–(38) do not depend on f, x_0, and B.

Proof. Let us first of all notice two simple inequalities for the average values of functions, which we shall use below. Let $f \in L^1_{\mathrm{loc}}(X)$, $0 < r_1 \le r_2$. Then

(1) If $x_0 \in X$, then

$$|f_{B(x_0,r_1)} - f_{B(x_0,r_2)}| \lesssim \left(\frac{r_2}{r_1} \right)^\gamma ⨍_{B(x_0,r_2)} |f - f_{B(x_0,r_2)}| \, d\mu. \tag{39}$$

(2) If $B(x_1, r_1) \cap B(x_2, r_2) \ne \varnothing$, then for any point $x_0 \in B(x_1, r_1) \cap B(x_2, r_2)$, the following inequality holds:

$$|f_{B(x_1,r_1)} - f_{B(x_2,r_2)}| \lesssim \left(\frac{r_2}{r_1} \right)^\gamma ⨍_{B(x_0,2r_2)} |f - f_{B(x_0,2r_2)}| \, d\mu. \tag{40}$$

□

Let

$$u(x,s) = \frac{1}{\eta(s)} ⨍_{B(x,s)} |f - f_{B(x,s)}| \, d\mu.$$

Then Eq. (39) easily implies that

$$u(x,t) \lesssim u(x,s) \quad \text{when } t \le s \le 2t. \tag{41}$$

Denote $B_k = B(x_0, 2^{-k}r)$, $k \ge 0$, then

$$|f(x_0) - f_{B(x_0,r)}| = \left|\sum_{k=0}^{\infty}(f_{B_{k+1}} - f_{B_k})\right| \lesssim \sum_{k=0}^{\infty} \eta(2^{-k}r)u(x_0, 2^{-k}r). \tag{42}$$

Then, from the condition $\eta \in \Omega[\alpha, \gamma/p)$ and from Eq. (41), we can deduce

$$\begin{aligned}
|f(x_0) - f_{B(x_0,r)}| &\lesssim \eta(r)u(x_0,r) + \sum_{k=1}^{\infty} \eta(2^{-k}r)u(x_0, 2^{-k}r) \\
&\lesssim \eta(r)u(x_0,r) + \sum_{k=1}^{\infty} \int_{2^{-k}r}^{2^{-(k-1)}r} \eta(s)u(x_0,s)\frac{ds}{s} \\
&\lesssim \eta(r)u(x_0,r) + \int_0^r \eta(s)u(x_0,s)\frac{ds}{s}.
\end{aligned} \tag{43}$$

To estimate the last integral, we denote for simplicity

$$I = \left(\fint_{B(x_0,r)} (\mathscr{S}_\eta f)^p \, d\mu \right)^{1/p}$$

and consider two different cases:

1. Let $\mathscr{S}_\eta f(x_0) \le I$; then due to the condition $\eta \in \Omega[\alpha, \gamma/p)$ and Eq. (1),

$$\begin{aligned}
\int_0^r \eta(s)u(x_0,s)\frac{ds}{s} &\le \mathscr{S}_\eta f(x_0) \int_0^r \eta(s)\frac{ds}{s} \lesssim \eta(r)\mathscr{S}_\eta f(x_0) \\
&= c\eta(r)(\mathscr{S}_\eta f(x_0))^{1-\frac{\alpha p}{\gamma}}(\mathscr{S}_\eta f(x_0))^{\frac{\alpha p}{\gamma}} \lesssim \eta(r)(\mathscr{S}_\eta f(x_0))^{1-\frac{\alpha p}{\gamma}} I^{\frac{\alpha p}{\gamma}}.
\end{aligned} \tag{44}$$

2. Let $\mathscr{S}_\eta f(x_0) > I$. Set

$$t = r\left(\frac{I}{\mathscr{S}_\eta f(x_0)}\right)^{\frac{p}{\gamma}} < r$$

and break the last integral in Eq. (43) into two parts:

$$\int_0^r \eta(s)u(x_0,s)\frac{ds}{s} = \left(\int_0^t + \int_t^r\right) \eta(s)u(x_0,s)\frac{ds}{s} \equiv I_1 + I_2.$$

I_1 is estimated analogously to Eq. (44):

$$I_1 \lesssim \eta(t)\mathscr{S}_\eta f(x_0) \lesssim \eta\left(r\left(\frac{I}{\mathscr{S}_\eta f(x_0)}\right)^{p/\gamma}\right)\mathscr{S}_\eta f(x_0). \tag{45}$$

To estimate I_2, we notice that

$$u(x_0,s) \leq \mathscr{S}_\eta f(x), \quad x \in B(x_0,s).$$

Averaging this inequality over the ball $B(x_0,s)$ and using the doubling condition (18), we obtain

$$u(x_0,s) \leq \left(\fint_{B(x_0,s)} (\mathscr{S}_\eta f)^p \, d\mu\right)^{1/p} \leq \left(\frac{\mu B(x_0,r)}{\mu B(x_0,s)}\right)^{1/p} I \lesssim \left(\frac{r}{s}\right)^{\gamma/p} I. \tag{46}$$

Therefore, in view of Eq. (1),

$$\begin{aligned}
I_2 &\lesssim I r^{\gamma/p} \int_t^r \eta(s) s^{-\gamma/p-1} \, ds \lesssim I \left(\frac{r}{t}\right)^{\gamma/p} \eta(t) \\
&\lesssim I r^{\gamma/p} \eta\left(r\left(\frac{I}{\mathscr{S}_\eta f(x_0)}\right)^{p/\gamma}\right)\left(r\left(\frac{I}{\mathscr{S}_\eta f(x_0)}\right)^{p/\gamma}\right)^{-\gamma/p} \\
&\lesssim \eta\left(r\left(\frac{I}{\mathscr{S}_\eta f(x_0)}\right)^{p/\gamma}\right)\mathscr{S}_\eta f(x_0).
\end{aligned}$$

From this together with Eq. (45), it follows that in Case 2 under consideration, we have the following inequality:

$$\int_0^r \eta(s) u(x_0,s) \frac{ds}{s} \lesssim \eta\left(r\left(\frac{I}{\mathscr{S}_\eta f(x_0)}\right)^{p/\gamma}\right)\mathscr{S}_\eta f(x_0).$$

Therefore, in view of the fact that $\eta(t)t^{-\alpha}$ increases, we obtain

$$\int_0^r \eta(s) u(x_0,s) \frac{ds}{s} \lesssim \eta(r)\left(\frac{I}{\mathscr{S}_\eta f(x_0)}\right)^{\alpha p/\gamma}\mathscr{S}_\eta f(x_0) = \eta(r)(\mathscr{S}_\eta f(x_0))^{1-\frac{\alpha p}{\gamma}} I^{\frac{\alpha p}{\gamma}}.$$

The first term on the right-hand side of Eq. (43) may be estimated using inequality (46), where one should set $s = r$:

$$u(x_0,r) = [u(x_0,r)]^{1-\frac{\alpha p}{\gamma}}[u(x_0,r)]^{\frac{\alpha p}{\gamma}} \leq [\mathscr{S}_\eta f(x_0)]^{1-\frac{\alpha p}{\gamma}} I^{\frac{\alpha p}{\gamma}},$$

and inequality (36) is proved.

To prove Eq. (37) we write

$$\fint_B |f - f_B|^q \, d\mu \lesssim \fint_B |f(y) - f_{B(y,r)}|^q \, d\mu(y) + \fint_B |f_{B(y,r)} - f_B|^q \, d\mu(y) \equiv J_1 + J_2.$$

To estimate the integral J_1 we notice that $B(y,r) \subset B(x,2r)$ when $y \in B(x,r)$. Therefore, using the already proved inequality (36), we obtain

$$\begin{aligned} J_1 &\lesssim \eta^q(r) \fint_B (\mathscr{S}_\eta f(y))^{q(1-\alpha p/\gamma)} \left(\fint_{B(y,r)} (\mathscr{S}_\eta f)^p \, d\mu \right)^{q\alpha/\gamma} d\mu(y) \\ &\lesssim \eta^q(r) \left(\fint_{B(x,2r)} (\mathscr{S}_\eta f)^p \, d\mu \right)^{q\alpha/\gamma} \fint_B (\mathscr{S}_\eta f(y))^p \, d\mu(y) \\ &\lesssim \eta^q(r) \left(\fint_{B(x,2r)} (\mathscr{S}_\eta f)^p \, d\mu \right)^{q/p}. \end{aligned}$$

The integrand in J_2 may be bounded using Eq. (40)

$$|f_{B(y,r)} - f_B| \lesssim \fint_{B(x,2r)} |f - f_{B(x,2r)}| \, d\mu \lesssim \eta(r) \mathscr{S}_\eta f(z)$$

for all $y \in B$ and $z \in B(x,2r)$. Hence, averaging over $z \in B(x,2r)$ leads to inequalities

$$|f_{B(y,r)} - f_B| \lesssim \eta(r) \left(\fint_{B(x,2r)} (\mathscr{S}_\eta f)^p \, d\mu \right)^{1/p}, \quad y \in B, \tag{47}$$

and

$$J_2 \lesssim \eta^q(r) \left(\fint_{B(x,2r)} (\mathscr{S}_\eta f)^p \, d\mu \right)^{q/p}.$$

To prove Eq. (38) we write

$$\fint_B |f - f(x_0)|^q \, d\mu \lesssim \fint_B |f - f_B|^q \, d\mu + |f_B - f_{B(x_0,r)}|^q + |f(x_0) - f_{B(x_0,r)}|^q.$$

The first term has already been estimated in Eq. (37), and the second one may be bounded using Eqs. (40) and (47):

$$|f_B - f_{B(x_0,r)}| \lesssim \eta(r) \left(\fint_{B(x,2r)} (\mathscr{S}_\eta f)^p \, d\mu \right)^{1/p}.$$

The necessary estimate for the third term follows from inequalities (43) and (1):

$$|f(x_0) - f_{B(x_0,r)}| \lesssim \left(\eta(r) + \int_0^r \eta(s) \frac{ds}{s} \right) \mathscr{S}_\eta f(x_0) \lesssim \eta(r) \mathscr{S}_\eta f(x_0).$$

The theorem is proved.

In [42] the statement of Theorem 23 is proved with the help of the same method for the case when the Steklov averages f_B are replaced by rather general families of linear operators $A_B : L^1_{\text{loc}} \to L^1_{\text{loc}}$ (see also [17] about this subject).

In particular, from Theorem 23, one may deduce certain relations between the classes $C^{p,q}_\eta(X)$ for different values of q. We shall not discuss this here in full detail. We only remark, for instance, that for $p \geq 1$,

$$C^{p,q}_\alpha(X) = C^p_\alpha(X), \quad \frac{1}{q} > \frac{1}{p} - \frac{\alpha}{\gamma}.$$

5.2 The Case $\alpha p \geq \gamma$

In the case $\alpha p > \gamma$ the condition $\eta \in \Omega[\alpha, \gamma/p)$ in Theorem 23 may be naturally replaced by $\eta \in \Omega[\alpha, \infty)$. Then instead of Eq. (36) at every Lebesgue point $x_0 \in X$, the following inequality holds:

$$|f(x_0) - f_{B(x_0,r)}| \lesssim \eta(r) \left(\fint_{B(x_0,r)} (\mathscr{S}_\eta f)^p \, d\mu \right)^{1/p}.$$

This follows from inequalities (42) and (46).

From the last inequality it is easy to deduce the following statement: *if $p > 0$, $\alpha p > \gamma$, $\eta \in \Omega[\alpha, \infty)$, and $\mathscr{S}_\eta f \in L^p(X)$, then for any two Lebesgue points $x_1, x_2 \in X$ of the function f,*

$$|f(x_1) - f(x_2)| \lesssim \eta(r) \left(\fint_{B(x_1,3r)} (\mathscr{S}_\eta f)^p \, d\mu \right)^{1/p},$$

where $r = d(x_1, x_2)$.

The case $\alpha p = \gamma$, $\eta \in \Omega[\alpha, \infty)$ can also be investigated using Eq. (43), which leads to the corresponding exponential bound

$$|f(x_0) - f_{B(x_0,r)}| \lesssim \eta(r) \left(\fint_{B(x_0,r)} (\mathscr{S}_\eta f)^p \, d\mu \right)^{1/p}$$

$$\times \max \left(1, \ln \left[\mathscr{S}_\eta f(x_0) \left(\fint_{B(x_0,r)} (\mathscr{S}_\eta f)^p \, d\mu \right)^{-1/p} \right] \right).$$

This inequality easily implies that for any ball $B = B(x_0, r)$

$$\fint_B \exp\left(c\frac{|f - f_B|}{\eta(r_B)}\left(\fint_{2B} (\mathscr{S}_\eta f)^p\, d\mu\right)^{-1/p}\right) d\mu \lesssim 1.$$

(for the case $\eta(t) = t$, see [14, Sect. 6]). This generalizes Trudinger's inequality.

5.3 The Self-Improvement of Poincaré Inequality

The method of proof of Theorem 23 is also appropriate for the study of the "self-improvement" property of Poincaré inequality [14].

We shall say that a pair of functions $f \in L^1_{\mathrm{loc}}(X)$ and $g \in L^p(X)$ satisfies the (σ, η, p)-Poincaré inequality if for every ball $B \subset X$, we have

$$\fint_B |f - f_B|\, d\mu \le \eta(r_B)\left(\fint_{\sigma B} g^p\, d\mu\right)^{1/p}.$$

Theorem 24. *Let $p > 0$, $0 < \alpha < \gamma/p$, $\eta \in \Omega[\alpha, \gamma/p)$, and $\sigma \ge 1$ and assume that $f \in L^1_{\mathrm{loc}}(X)$ and $g \in L^p(X)$ satisfy the (σ, η, p)-Poincaré inequality.*

Then for each ball $B \subset X$:

(1) If $\frac{1}{q} = \frac{1}{p} - \frac{\alpha}{\gamma}$, then

$$\frac{\mu\{x \in B : |f(x) - f_B| > \lambda\}}{\mu B} \lesssim \left[\frac{\eta(r_B)}{\lambda}\left(\fint_{2\sigma B} g^p\, d\mu\right)^{1/p}\right]^q, \quad \lambda > 0. \tag{48}$$

(2) If $\frac{1}{q} > \frac{1}{p} - \frac{\alpha}{\gamma}$, then

$$\left(\fint_B |f - f_B|^q\, d\mu\right)^{1/q} \lesssim \eta(r_B)\left(\fint_{2\sigma B} g^p\, d\mu\right)^{1/p}. \tag{49}$$

To justify this let us set

$$v(x,s) = \left(\fint_{B(x,\sigma s)} g^p\, d\mu\right)^{1/p}$$

and copy the proof of inequality (36) in Theorem 23 with the function v in place of u. The role of $\mathscr{S}_\eta f$ will then be played by the modified Hardy–Littlewood maximal function

$$M_{p,r} g(x) = \sup_B \left(\fint_B g^p\, d\mu\right)^{1/p},$$

where the supremum is taken over all balls $B \in \mathscr{B}(x)$ with $r_B < \sigma r$.

For any Lebesgue point $x_0 \in X$ instead of Eq. (36) we obtain the inequality

$$|f(x_0) - f_{B(x_0,r)}| \lesssim \eta(r) \left(M_{p,r}g(x_0)\right)^{1-\frac{\alpha p}{\gamma}} \left(\fint_{B(x_0,\sigma r)} g^p \, d\mu \right)^{\alpha/\gamma}.$$

This estimate and the usual properties of the Hardy–Littlewood maximal function immediately imply a weak-type inequality (48), the passage to Eq. (49) from Eq. (48) is standard.

Let us remark that in the partial case $\alpha = 1$, $\eta(t) = t$ the statement of Theorem 24 is proved in [14].

References

1. Calderón, A.P.: Estimates for singular integral operators in terms of maximal functions. Studia Math. **44**, 561–582 (1972)
2. Calderón, A.P., Scott, R.: Sobolev type inequalities for $p > 0$. Studia Math. **62**, 75–92 (1978)
3. DeVore, R., Sharpley, R.: Maximal functions measuring local smoothness. Mem. Amer. Math. Soc. **47**, 1–115 (1984)
4. Fefferman, C., Stein, E.M.: H^p spaces of several variables. Acta Math. **129**(3–4), 137–193 (1972)
5. Oskolkov, K.I.: Approximation properties of summable functions on sets of full measure. Math USSR Sb **32**(4), 489–514 (1977)
6. Kolyada, V.I.: Estimates for maximal functions connected with local smoothness. Sovjet Math. Dokl. **35**, 345–348 (1987)
7. Kolyada, V.I.: Estimates of maximal functions measuring local smoothness. Analysis Math. **25**(1), 277–300 (1999)
8. Coifman, R.R., Weiss, G.: Analyse harmonique non-commutative sur certains espaces homogenés. In: Lecture Notes in Math, vol. 242, pp. 1–176. Springer, Berlin (1971)
9. Jonsson, A.: Haar wavelets of higher order on fractals and regularity of functions. J. Math. Anal. Appl. **290**(1), 86–104 (2004)
10. Ivanishko, I.A., Krotov, V.G.: Compactness of embeddings of Sobolev type on metric measure spaces. Math. Notes **86**(6), 775–788 (2009)
11. Krotov, V.G.: Quantitative form of the Luzin C-property. Ukrainian Math. J. **62**(3), 441–451 (2010)
12. Hajłasz, P.: Sobolev spaces on an arbitrary metric spaces. Potential Anal. **5**(4), 403–415 (1996)
13. Franchi, B., Hajłasz, P., Koskela, P.: Definitions of Sobolev classes on metric spaces. Ann. Inst. Fourier (Grenoble). **49**(6), 1903–1924 (1999)
14. Hajłasz, P., Koskela, P.: Sobolev met Poincaré. Mem. Amer. Math. Soc. **145**, 1–101 (2000)
15. Heinonen, J.: Lectures on Analysis on Metric Spaces. Springer, Berlin (2001)
16. Hajłasz, P.: Sobolev spaces on metric-measure spaces. Contemporary Math. **338**, 173–218 (2003)
17. Cifuentes, P., Dorronsoro, J.R., Sueiro, J.: Boundary tangential convergence on spaces of homogeneous type. Trans. Amer. Math. Soc. **332**(1), 331–350 (1992)
18. Gatto, E., Vagi, S.: On functions arising as potentials on spaces of homogeneous type. Proc. Amer. Math. Soc. **125**(4), 1149–1152 (1997)
19. Hu, J.: A note on Hajłasz-Sobolev spaces on fractals. J. Math. Anal. Appl. **280**(1), 91–101 (2003)

20. Yang, D.: New characterization of HajłaszSobolev spaces on metric spaces. Sci. China, Ser. 1. **46**(5), 675–689 (2003)
21. Ivanishko, I.A.: Estimates of Calderon-Kolyada maximal functions on spaces of homogeneous type. In: Proceedings of Institute of Mathematics of BeloRussian NAC, vol. 12, pp. 64–67 (2004) (In Russian)
22. Ivanishko, I.A.: Generalized Sobolev classes on metric measure spaces. Math. Notes. **77**(6), 865–869 (2005)
23. Ivanishko, I.A.: Classes of functions related to local smoothness on metric spaces with measure. Embedding theorems, PhD Thesis, Minsk (2010) (In Russian)
24. Krotov, V.G.: Weighted L^p-inequalities for sharp-maximal functions on metric spaces with measure. J. Contemporary Math. Anal. **41**(2), 22–38 (2006)
25. Adams, D.R., Hedberg, L.I.: Function Spaces and Potential Theory, 366 p. Springer, Berlin (1996)
26. Kinnunen, J., Martio, O.: The Sobolev capacity on metric spaces. Annales Academiæ Scientiarum Fennicæ Mathematica. **21**, 367–382 (1996)
27. Prokhorovich, M.A.: Capacity and Lebesque points for Sobolev classes. Bull. Natl. Acad. Sci. Belarus 1, 19–23 (2006) (In Russian)
28. Federer, H., Ziemer, C.: The Lebesgue sets of a function whose distribution derivatives are p-th power summable. Indiana Univ. Math. J. **22**(2), 139–158 (1972)
29. Hajłasz, P., Kinnunen, J.: Hölder quasicontinuity of Sobolev functions on metric spaces. Revista Matemática Iberoamericana. **14**(3), 601–622 (1998)
30. Kinnunen, J., Latvala, V.: Lebesgue points for Sobolev functions on metric spaces. Revista Matemática Iberoamericana. **18**(3), 685–700 (2002)
31. Prokhorovich, M.A.: Sobolev capacities on metric spaces with measure. Bull. Belarus Univ. 3, 106–111 (2007) (In Russian)
32. Prokhorovich, M.A.: Generalized Sobolev spaces on metric spaces with measures: delicate properties of functions. PhD Thesis. Minsk (2009) (In Russian)
33. Prokhorovich, M.A.: Hausdorff measures and Lebesgue points for the Sobolev classes W^p_α, $\alpha > 0$ on spaces of homogeneous type. Math. Notes. **85**(4), 584–589 (2011)
34. Krotov, V.G., Prokhorovich, M.A.: Estimates from below for the exceptional Lebesgue sets of functions from Sobolev classes in this volume
35. Krotov, V.G. Prokhorovich, M.A.: The rate of convergence of Steklov means on metric measure spaces and Hausdorff dimension. Math. Notes. **89**(1), 156–159 (2011)
36. Hajłasz, P., Kinnunen, J.: Hölder qasicontinuity of Sobolev functions on metric spaces. Revista Mat. Iberoamericana. **14**(3), 601–622 (1998)
37. Krotov, V.G. Prokhorovich, M.A.: The Luzin approximation of functions from classes W^p_α on metric spaces with measure. Russian Math. (IzvVUZ) **52**(5), 47–57 (2008)
38. Krotov, V.G.: Compactness criteria in L^p-spaces, $p \geq 0$, Mat. Sb., **203**(7), 129–148 (In Russian) (2012)
39. Katkovskaya, I.N.: Riesz compactness criterion for the space of measurable functions. Math. Notes. **89**(1), 145–149 (2011)
40. Dunford, N., Schwartz, J.: Linear Operators. Part 1 (General Theory), 874 p. Interscience publ., New York (1958)
41. Ivanishko, I.A., Krotov, V.G.: Generalized Poincaré-Sobolev inequality on metric spaces. In: Proceedings of Institute of Mathematics of BeloRussian NAC, vol. 14, no. 1, pp. 51–61 (2006) (In Russian)
42. Ignatieva, E.V.: SobolevPoincaré-type inequality on metric spaces. Math. Notes. **81**(1), 121–125 (2007)

Estimates for the Exceptional Lebesgue Sets of Functions from Sobolev Classes

Veniamin G. Krotov and Mikhail A. Prokhorovich

Dedicated to Professor Konstantin Oskolkov on the occasion of his 65th birthday.

Abstract We prove that known estimates for capacities and the Hausdorff dimension of exceptional Lebesgue sets of functions from Calderón classes are sharp. Our proof also gives a similar result for classical Sobolev spaces.

1 Introduction

Let (X,d,μ) be a metric space with metric d and a regular Borel measure μ such that the measure of any ball

$$B(x,r) = \{y \in X : d(x,y) < r\}$$

is positive and finite. In what follows, we shall assume that the metric measure space (X,d,μ) satisfies the doubling condition

$$\mu(B(x,2r)) \leq c\mu(B(x,r)), \quad x \in X, \quad r > 0, \tag{1}$$

where c is some positive constant independent of $x \in X$ and $r > 0$.[1]

Let $f \in L^1_{\mathrm{loc}}(X)$. If the limit

$$\lim_{r \to +0} \frac{1}{\mu(B(x,r))} \int_{B(x,r)} f \, \mathrm{d}\mu \tag{2}$$

[1]Here and thereafter we denote by c different positive constants, whose values play no role.

V.G. Krotov (✉) • M.A. Prokhorovich
Belarussian State University, Nesavisimosti av., 4, Minsk, 220030, Belarus
e-mail: krotov@bsu.by; prokhorovich@bsu.by

D. Bilyk et al. (eds.), *Recent Advances in Harmonic Analysis and Applications*, Springer Proceedings in Mathematics & Statistics 25, DOI 10.1007/978-1-4614-4565-4_19,

exists, then x is called a Lebesgue point[2] of the function f. Let $\Lambda(f)$ be the complement of the set of all Lebesgue points. The set $\Lambda(f)$ is the exceptional Lebesgue set of f.

The classical Lebesgue theorem says that $\mu(\Lambda(f)) = 0$ for any function $f \in L^1_{\mathrm{loc}}(X)$ (see [1, Theorem 1.8] for the spaces (X,d,μ) with the doubling property).

If the function f is more regular, then the exceptional set $\mathscr{L}(f)$ is smaller. The capacity Cap_p and the Hausdorff dimension $\dim_{\mathbb{H}}$ of the set $\mathscr{L}(f)$ for functions from Sobolev classes $W^p_1(\mathbb{R}^N)$ have been originally estimated by Federer and Ziemer [2]. For the further developments of results of such kind on $\mathbb{R}^N$ see [3, Sect. 6.2, 6.6], [4, Sect. 4.8], and [5, Sect. 3.3].

In our paper we consider the scale of Calderón classes C^p_α from this point of view. This scale contains the Sobolev space W^p_1 as a partial case.

2 Main Result

2.1 Calderón Classes

For $\alpha > 0$ consider the maximal functions

$$\mathscr{S}_\alpha f(x) = \sup_{B \ni x} r_B^{-\alpha} \frac{1}{\mu(B)} \int_B |f(y) - f_B| \, \mathrm{d}\mu(y), \quad f_B = \frac{1}{\mu(B)} \int_B f \, \mathrm{d}\mu,$$

where the supremum is taken over all balls B of radius $r_B \in (0,1)$ that contain the point $x \in X$. Define the classes

$$C^p_\alpha(X) = \left\{ f \in L^p : \|f\|^p_{C^p_\alpha} = \|f\|^p_{L^p} + \|\mathscr{S}_\alpha f\|^p_{L^p} < \infty \right\}, \quad \alpha > 0, \quad 1 < p < \infty. \tag{3}$$

For $X = \mathbb{R}^N$ these classes were introduced in [6] and in the general case in [7]. In [6] Calderón proved that $C^p_1(\mathbb{R}^n) = W^p_1(\mathbb{R}^n)$ (see also [8]).

Below we use the following equivalent description of the classes C^p_α that goes back to [9]. For a function $f \in L^p(X)$, denote by $D_\alpha(f)$ the class of all nonnegative μ-measurable functions g on X, for which there exists a set $E \subset X$, $\mu(E) = 0$, such that for $x, y \in X \setminus E$, the following inequality is true:

$$|f(x) - f(y)| \le [d(x,y)]^\alpha [g(x) + g(y)].$$

The elements of the set $D_\alpha(f)$ will be called the α-gradients of the function f.

---

[2] Often x is called a Lebesgue point, if (6) is true for $q = 1$.

The class $M^p_\alpha(X)$, $\alpha > 0$, is defined as the set of functions $f \in L^p(X)$, such that $D_\alpha(f) \cap L^p(X) \neq \varnothing$. The norm in $M^p_\alpha(X)$ is

$$\left(\|f\|^p_{L^p(X)} + \inf_g \|g\|^p_{L^p(X)} \right)^{1/p},$$

where the supremum is taken over all functions $g \in L^p(X) \cap D_\alpha(f)$.

The classes C^p_α and $M^p_\alpha(X)$ coincide, and their norms are equivalent [7, 10].

Note that functions from the classes $C^p_\alpha(X)$ are defined only μ-almost everywhere. However, we deal with the properties of such function that may depend on changing the values of functions on sets of μ-measure zero. Therefore we assume that the values of all locally summable functions at any point are defined by the equality

$$f(x) = \limsup_{r \to +0} \frac{1}{\mu(B(x,r))} \int_{B(x,r)} f \, d\mu.$$

2.2 *Capacity and Hausdorff Measure*

The capacities corresponding to the classes $C^p_\alpha(X)$ are defined in the following way:

$$\mathrm{Cap}_{\alpha,p}(E) = \inf \left\{ \|f\|^p_{C^p_\alpha(X)} : f \in C^p_\alpha(X),\ f \geq 1 \text{ in a neighborhood of } E \right\}.$$

For $\alpha = 1$ they were introduced and studied in [11] and in the case $0 < \alpha \leq 1$ in [12].

Recall also the definition of the Hausdorff s-measure $\mathbb{H}^s$ and the Hausdorff dimension $\dim_{\mathbb{H}}$ of $E \subset X$ (see, e.g., [1, Sect. 8.3])

$$\mathbb{H}^s_\delta(E) = \inf \left\{ \sum_{i=1}^\infty r_i^s : E \subset \bigcup_{i=1}^\infty B(x_i, r_i),\ r_i < \delta \right\}, \quad \mathbb{H}^s(E) = \lim_{\delta \to +0} \mathbb{H}^s_\delta(E),$$

$$\dim_{\mathbb{H}}(E) = \inf\{s : \mathbb{H}^s(E) = 0\} = \sup\{s : \mathbb{H}^s(E) = \infty\}.$$

The implication

$$\mathrm{Cap}_{\alpha,p}(E) = 0 \quad \Longrightarrow \quad \dim_{\mathbb{H}}(E) \leq N - \alpha p \tag{4}$$

shows the connection between the capacity and the Hausdorff dimension of the set $E \subset X$ (for general metric spaces with the doubling condition (5) see the proof in [11, Theorem 4.15] for $\alpha = 1$ and in [13, Theorem 2] for $0 < \alpha \leq 1$).

2.3 *Estimates of the Sets* $\Lambda(f)$

The doubling condition (1) can be written in the following quantitative form:

$$\mu(B(x,R)) \le c\left(\frac{R}{r}\right)^{\gamma}\mu(B(x,r)), \quad x\in X. \tag{5}$$

The number γ is called the doubling exponent of the measure and plays the role of dimension of the whole space X.

Theorem 1. *Let* $\alpha>0$, $1<p<\gamma/\alpha$, $f\in C^p_\alpha(X)$. *Then there exists a set* $E\subset X$, *such that for any* $x\in X\setminus E$, *there exists the limit*

$$\lim_{r\to+0}\frac{1}{\mu(B(x,r))}\int_{B(x,r)} f\,\mathrm{d}\mu = f^*(x);$$

moreover,

$$\lim_{r\to+0}\frac{1}{\mu(B(x,r))}\int_{B(x,r)} |f-f^*(x)|^q\,\mathrm{d}\mu = 0, \quad \frac{1}{q}=\frac{1}{p}-\frac{\alpha}{\gamma}, \tag{6}$$

and the following estimates hold

(1) $\dim_{\mathbb{H}}(E)\le\gamma-\alpha p$ *(for any* $\alpha>0$*).*
(2) $\mathrm{Cap}_{\alpha,p}(E)=0$ *(for* $0<\alpha\le 1$*).*

In the case $X=\mathbb{R}^n$, $\alpha=1$, this result is due to Federer and Ziemer [2] (recall that $W^p_1(\mathbb{R}^n)=C^p_1(\mathbb{R}^n)$). The case of arbitrary X and $\alpha=1$, but without property (6), has been considered by Hajłasz and Kinnunen [14], and with property (6)—in [15] (but only for $\frac{1}{q}>\frac{1}{p}-\frac{\alpha}{\gamma}$). The general form of Theorem 1 is due to Prokhorovich [12, 13] for $0<\alpha\le 1$ and [16] for $\alpha>0$.

We prove in this chapter that Theorem 1 is sharp.

Theorem 2. *Let* $0<\alpha\le 1$ *and* $1<p<N/\alpha$. *Then there exists a function* $f_0\in C^p_\alpha(\mathbb{R}^N)$ *such that*

(1) $\dim_{\mathbb{H}}(\Lambda(f_0))=N-\alpha p$.
(2) $\mathrm{Cap}_{\beta,q}(\Lambda(f_0))>0$ *for any pair* (β,q) *with* $\beta q>\alpha p$.

3 Auxiliary Tools

3.1 *The Properties of Gradients*

The following two properties of α-gradients were proved in [11, 15] for $\alpha=1$. For any $0<\alpha\le 1$ the proof is the same:

(1) Let $x_0 \in \mathbb{R}^N$, $0 < \alpha \le 1$, $0 < r_0 \le 1$, and

$$f(x) = \begin{cases} \dfrac{2r_0 - d(x,x_0)}{r_0}, & x \in B(x_0,2r_0) \setminus B(x_0,r_0), \\ 1, & x \in B(x_0,r_0), \\ 0, & x \in X \setminus B(x_0,2r_0). \end{cases} \tag{7}$$

Then for some constant c the function

$$g(x) = \begin{cases} cr_0^{-\alpha}, & x \in B(x_0,2r_0), \\ 0, & x \in X \setminus B(x_0,2r_0) \end{cases} \tag{8}$$

belongs to $D_\alpha(f)$.

(2) Let $\{f_i\}_{i\in\mathbb{N}}$ be a sequence of measurable functions and $g_i \in D_\alpha(f_i)$. If $\sup_i f_i < +\infty$ almost everywhere, then

$$\sup_{i\in\mathbb{N}} g_i \in D_\alpha\left(\sup_{i\in\mathbb{N}} f_i\right). \tag{9}$$

3.2 The Construction of the Set $\mathcal{Q}(\beta)$

The proof of Theorem 2 uses the following one-parametric construction of the Sierpiński carpet-type set.

Let

$$Q(x^0,r) = \left\{x \in \mathbb{R}^N : |x_k - x_k^0| \le \frac{r}{2},\ k = 1,\dots,N\right\}$$

be the cube centered at the point $x^0 \in \mathbb{R}^N$ with side length $r > 0$, $Q_0 = Q(0,1)$.

Fix a number $\beta \in \left(0,\frac{1}{2}\right)$ and build the sequence of sets $\{Q_i(\beta)\}_{i=0}^{\infty}$ by induction, beginning with $Q_0(\beta) = Q_0$. At the first step we delete from $Q_0(\beta)$ the cross

$$K_0(\beta) = \bigcup_{i=1}^{N} S_i(\beta), \quad \text{where} \quad S_i(\beta) = \left\{x \in Q_0 : |x_k| < \frac{1-2\beta}{2},\ i \ne k = 1,\dots,N\right\}.$$

We then obtain the set

$$Q_1(\beta) = Q_0(\beta) \setminus K_0(\beta) = \bigcup_{i=1}^{2^N} Q_{1,i}(\beta),$$

which is the sum of cubes $Q_{1,i}(\beta) \equiv Q\left(x^{1,i},\beta\right)$ with centers at some points $x^{1,i} \in Q_0$ and with side length β.

Denote by

$$K(x^0,\beta) = \beta K_0(\beta) + x^0$$

the cross $K_0(\beta)$ compressed $1/\beta$ times and shifted so that its center coincides with x^0. From any cube $Q_{1,i}(\beta)$ delete the cross $K(x^{1,i},\beta)$, we then obtain the set

$$Q_2(\beta) = \bigcup_{i=1}^{2^N} (Q_{1,i}(\beta) \setminus K(x^{1,i},\beta)) = \bigcup_{i=1}^{2^{2N}} Q_{2,i}(\beta),$$

which is the sum of cubes $Q_{2,i}(\beta) \equiv Q\left(x^{2,i},\beta^2\right)$ with centers at some points $x^{2,i} \in Q_0$ and with side length β^2.

On the picture we see the sets $Q_i(\beta)$ $(i = 0,1,2)$ for $\mathbb{R}^2$.

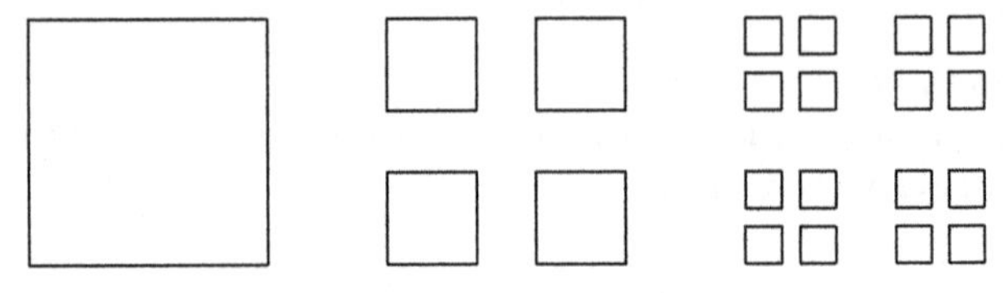

Repeating by induction we obtain the sequence of the sets $\{Q_i(\beta)\}_{i=0}^{\infty}$

$$Q_k(\beta) = \bigcup_{i=1}^{2^{(k-1)N}} \left(Q_{k-1,i}(\beta) \setminus K(x^{k-1,i},\beta^{k-1})\right) = \bigcup_{i=1}^{2^{kN}} Q_{k,i}(\beta), \tag{10}$$

$$\mu(Q_{k,i}(\beta)) = \beta^{kN}, \quad \operatorname{diam}(Q_{k,i}(\beta)) = \sqrt{N}\beta^k.$$

At the kth step we obtain the set $Q_k(\beta)$ of 2^{kN} N-dimensional cubes with centers at some points $x^{k,i}$ $(i = 1,\dots,2^{kN})$, with side length β^k, and of measure $\mu(Q_{k,i}(\beta)) = \beta^{kN}$.

Now we define the resulting set

$$\mathcal{Q}(\beta) = \bigcap_{k=1}^{\infty} Q_k(\beta) = \bigcap_{k=1}^{\infty} \bigcup_{i=1}^{2^{kN}} Q_{k,i}(\beta). \tag{11}$$

We shall use the notation $\mathcal{Q}(\beta)$, $Q_k(\beta)$, $Q_{k,i}(\beta)$, and $x^{k,i}$ throughout the chapter.

The Hausdorff dimension of the set $\mathcal{Q}(\beta)$ can be calculated by the following formula (see [17, theorem 9.3]):

$$\dim_{\mathbb{H}}(\mathcal{Q}(\beta)) = -N\ln 2/\ln\beta. \tag{12}$$

4 Proof of Theorem 2

Fix $\varepsilon > 0$ and build the function $f_\varepsilon \in C^p_\alpha([0,1]^N)$ such that

$$\dim_{\mathbb{H}}(\Lambda(f_\varepsilon)) > N - \alpha p - \varepsilon. \tag{13}$$

For $\beta_0 = \mathrm{e}^{-N\ln 2/(N-\alpha p-\varepsilon/2)}$, we can construct the set $\mathcal{Q}(\beta_0)$ (see (11)). By (12) Hausdorff dimension of the set $\mathcal{Q}(\beta_0)$ is equal

$$\dim_{\mathbb{H}}(\mathcal{Q}(\beta_0)) = -N\ln 2/\ln\beta_0 = N - \alpha p - \frac{\varepsilon}{2}. \tag{14}$$

Every set $Q_k(\beta_0)$ (see (10)) is contained in the union of 2^{kN} balls $\{B_{k,i}\}_{i=1}^{2^{kN}}$

$$Q_k(\beta_0) = \bigcup_{i=1}^{2^{kN}} Q_{k,i}(\beta_0) \subset \bigcup_{i=1}^{2^{kN}} B_{k,i},$$

$$Q_{k,i}(\beta_0) \subset B_{k,i} = B\left(x^{k,i}, \operatorname{diam}(Q_{k,i}(\beta_0))\right) = B\left(x^{k,i}, \sqrt{N}\beta_0^k\right).$$

Denote by $\mathcal{B}_{k,i}$ the ball whose center coincides with the center of the ball $B_{k,i}$ but of twice the radius.

For $k \in \mathbb{N}$ and $i = 1, \ldots, 2^{kN}$ build the pair of functions $(f_{k,i}, g_{k,i})$, $g_{k,i} \in D_\alpha(f_{k,i})$ according to (8), where

$$x_0 = x^{k,i}, \quad r_0 = \operatorname{diam}(Q_{k,i}(\beta_0)) = \sqrt{N}\beta_0^k.$$

Finally set

$$f_k = \max_{i=1,\ldots,2^{kN}} f_{k,i}, \quad g_k = \max_{i=1,\ldots,2^{kN}} g_{k,i}, \quad f_\varepsilon = \sum_{k=1}^{\infty} f_k, \quad g = \sum_{k=1}^{\infty} g_k.$$

To prove that $f_\varepsilon \in C_\alpha^p$, we show that $g \in D_\alpha(f_\varepsilon) \cap L^p(\mathbb{R}^N)$. By (9) $g_k \in D_\alpha(f_k)$. We estimate the norm $\|f_k\|_{C_\alpha^p}$

$$\|f_k\|_{C_\alpha^p} \le \left(\|f_k\|_{L^p}^p + c\|g_k\|_{L^p}^p\right)^{1/p} = \left[\int\limits_{\bigcup_{i=1}^{2^{kN}}\mathcal{B}_{k,i}} f_k^p \,\mathrm{d}\mu + \int\limits_{\bigcup_{i=1}^{2^{kN}}\mathcal{B}_{k,i}} g_k^p \,\mathrm{d}\mu\right]^{1/p}$$

$$\le c\left[\left(1 + \left(\sqrt{N}\beta_0^k\right)^{-\alpha p}\right)\mu\left(\bigcup_{i=1}^{2^{kN}}\mathcal{B}_{k,i}\right)\right]^{1/p} \le c\left[\left(\sqrt{N}\beta_0^k\right)^{-\alpha p}(2\beta_0)^{Nk}\right]^{1/p}.$$

The last inequality is true for sufficiently large $k > k_0$ (k_0 is defined as the smallest solution of the inequality $\sqrt{N}\beta_0^{k_0} < 1$).

Further

$$\|f_\varepsilon\|_{C_\alpha^p} \le \sum_{k=1}^{\infty}\|f_k\|_{C_\alpha^p} \le c\sum_{k=1}^{\infty}\left[2^{Nk}\beta_0^{(N-\alpha p)k}\right]^{1/p}.$$

This series converges because $2^N\beta_0^{(N-\alpha p)} = 2^N(\mathrm{e}^{-N\ln 2/(N-\alpha p-\varepsilon/2)})^{(N-\alpha p)} < 1$. Therefore the function f_ε belongs to C_α^p.

The function f_ε satisfies the condition

$$\lim_{r\to+0}\frac{1}{\mu(B(x,r))}\int_{B(x,r)} f_\varepsilon\,\mathrm{d}\mu = \infty$$

on the set $\mathcal{Q}(\beta_0)$.

Indeed if $x\in\mathcal{Q}(\beta_0)$, then $x\in Q_k(\beta_0)$ for any k. Therefore if $n\in\mathbb{N}$, then there exists $r_n>0$, such that for any $r<r_n$ and all $k=1,\dots,n$, the inclusion $B(x,r)\subset B_{k,i_k}$ holds for some i_k. Thus $f_k=1$ for $k=1,\dots,n$ and $f_\varepsilon\geq n$ on $B(x,r)$ if $r<r_n$. Hence

$$\lim_{r\to+0}\frac{1}{\mu(B(x,r))}\int_{B(x,r)} f_\varepsilon\,\mathrm{d}\mu = \infty. \tag{15}$$

By (15)

$$\mathcal{Q}(\beta_0)\subset\left\{x:\ \lim_{r\to+0}\frac{1}{\mu(B(x,r))}\int_{B(x,r)} f_\varepsilon\,\mathrm{d}\mu = \infty\right\}.$$

Therefore

$$\dim_{\mathbb{H}}(\Lambda(f_\varepsilon))\geq\dim_{\mathbb{H}}(\mathcal{Q}(\beta_0)) = N-\alpha p-\frac{\varepsilon}{2} > N-\alpha p-\varepsilon.$$

Now using the construction of function f_ε, we build the sequence of functions $\{f_{1/n}\}_{n=1}^\infty$, $\dim_{\mathbb{H}}(\Lambda(f_{1/n})) > N-\alpha p-1/n$. Then the function

$$f_0=\sum_{n=1}^\infty\frac{1}{2^n}\frac{f_{1/n}}{\|f_{1/n}\|_{C_\alpha^p}}$$

belongs to $C_\alpha^p(\mathbb{R}^N)$ and $\dim_{\mathbb{H}}(\Lambda(f_0)) = N-\alpha p$. Part (1) of Theorem 2 is proved. For the proof of part (2), assume that $\mathrm{Cap}_{\beta,q}(\Lambda(f_0))=0$. Then by (4), we obtain

$$\dim_{\mathbb{H}}(\Lambda(f_0))\leq N-\beta q < N-\alpha p.$$

5 Sobolev Classes $W_l^p(\mathbb{R}^N)$

In particular, for $\alpha=1$, Theorem 2 implies the corresponding result for the class $W_1^p(\mathbb{R}^N)$. Similarly to the proof of Theorem 2 we can prove the analogous statement for Sobolev spaces $W_l^p(\mathbb{R}^N)$ of any order. But these results are not new; see, for example, [5, exercise 3.6] and [18].[3]

[3]The authors would like to thank the referee for these references.

Theorem 3. *If $1 < p < N/l$, then there exists a function $f_0 \in W_l^p(\mathbb{R}^N)$ such that*

(1) $\dim_{\mathbb{H}}(\Lambda(f_0)) = N - lp$.
(2) $\mathrm{Cap}_{l,q}(\Lambda(f_0)) > 0$ *for any q with $q > p$.*

The distinction from the proof of Theorem 2 is only in the construction of the function f_ε. Here in the place of the function (7) we take a function $\varphi \in C_0^\infty(\mathbb{R}^n)$ such that

$$\varphi(x) = \begin{cases} 0, & |x| \geq 1 + \frac{1-2\beta}{3\beta}, \\ 0 \leq \varphi(x) \leq 1, & 1 \leq |x| \leq 1 + \frac{1-2\beta}{3\beta}, \\ 1, & |x| \leq 1. \end{cases} \tag{16}$$

Denote

$$\varphi_\delta^{k,i}(x) = \varphi(\delta^k x - x^{k,i}), \quad \delta = \frac{1}{\beta_0}, \quad f_k = \sum_{i=1}^{2^{kN}} \varphi_\delta^{k,i}, \quad f_\varepsilon = \sum_{k=1}^{\infty} f_k.$$

To prove that $f_\varepsilon \in W_l^p$ we estimate the norm $\|f_k\|_{W_l^p}$, keeping in mind that the supports of functions $\varphi^{k,i}$ and $\varphi^{k,j}$ are disjoint for $i \neq j$,

$$\|f_k\|_{W_l^p} = \sum_{|\nu|\leq l} \|D^\nu f_k\|_{L^p} = \sum_{|\nu|\leq l} \left(\int_{\mathbb{R}^n} \left| \sum_{i=1}^{2^{kN}} D^\nu \varphi_\delta^{k,i}(x) \right|^p \mathrm{d}x \right)^{1/p}$$

$$= \sum_{|\nu|\leq l} \left(\int_{\mathbb{R}^n} 2^{Nk} \left| D^\nu \varphi_\delta^{k,1}(x) \right|^p \mathrm{d}x \right)^{1/p} = \sum_{|\nu|\leq l} \left[2^{Nk} \beta_0^{(N-|\nu|p)k} \right]^{1/p} \|D^\nu \varphi\|_{L^p}.$$

Since $2^N \beta_0^{(N-lp)} = 2^N (\mathrm{e}^{-N\ln 2/(N-lp-\varepsilon/2)})^{(N-lp)} < 1$, we obtain $f_\varepsilon \in W_l^p$. The rest of the proof is the same.

References

1. Heinonen, J.: Lectures on Analysis on Metric Spaces. Springer, Berlin (2001)
2. Federer, H., Ziemer, W.: The Lebesgue sets of a function whose distribution derivatives are pth power summable. Indiana Univ. Math. J. **22**(2), 139–158 (1972)
3. Adams, D.R., Hedberg, L.I.: Function Spaces and Potential Theory, p. 366. Springer, Berlin, Heidelberg, New York (1996)
4. Evans, L.C., Gariepy, R.F.: Measure Theory and Fine Properties of Functions. CRC Press, Boca Raton, London, New York, Washington (1992)
5. Ziemer, W.P.: Weakly differentiable functions. Sobolev Spaces and Functions of Bounded Variation. Springer, New York (1989)
6. Calderón, A.P.: Estimates for singular integral operators in terms of maximal functions. Studia Math. **44**, 561–582 (1972)

7. Yang, D.: New characterization of Hajłasz-Sobolev spaces on metric spaces. Science in China, Ser. 1. **46**(5), 675–689 (2003)
8. DeVore, R., Sharpley, R.: Maximal functions measuring local smoothness. Memoirs Amer. Math. Soc. **47**, 1–115 (1984)
9. Hajłasz, P.: Sobolev spaces on an arbitrary metric spaces. Potential Anal. **5**(4), 403–415 (1996)
10. Ivanishko, I.A.: Generalized Sobolev classes on metric measure spaces. Math. Notes. **77**(6), 865–869 (2005)
11. Kinnunen, J., Martio, O.: The Sobolev capacity on metric spaces. Annales Academiæ Scientiarum Fennicæ Mathematica **21**, 367–382 (1996)
12. Prokhorovich, M.A.: Capacity and Lebesque points for Sobolev classes. Bulletin of the National Academy of Sciences of Belarus **1**, 19–23 (2006) (In Russian)
13. Prokhorovich, M.A.: Hausdorff dimension of Lebesgue sets for W_α^p classes on metric spaces. Math. Notes. **82**(1), 88-95 (2007)
14. Hajłasz, P., Kinnunen, J.: Hölder qasicontinuity of Sobolev functions on metric spaces. Revista Matemática Iberoamericana. **14**(3), 601–622 (1998)
15. Kinnunen, J., Latvala, V.: Lebesgue points for Sobolev functions on metric spaces. Revista Matemática Iberoamericana. **18**(3), 685–700 (2002)
16. Prokhorovich, M.A.: Hausdorff measures and Lebesgue points for the Sobolev classes W_α^p, $\alpha > 0$ on spaces of homogeneous type. Math. Notes. **85**(4), 584–589 (2009)
17. Falconer, K.: Fractal Geometry: Mathematical Foundations and Applications. Wiley, Chichester (1990)
18. Fonseca, I., Maly, J., Mingione, G.: Scalar minimizers with fractal singular sets. Arch. Rat. Mech. Anal. **172**(3), 295–307 (2004)

On the A_2 Inequality for Calderón–Zygmund Operators

Michael T. Lacey

Abstract We prove that for an $L^2(\mathbb{R}^d)$-bounded Calderón–Zygmund operator and weight $w \in A_2$, we have the inequality below due to Hytönen:

$$\|T\|_{L^2(w) \to L^2(w)} \leq C_T [w]_{A_2} .$$

Our proof will appeal to a distributional inequality used by several authors, adapted Haar functions, and standard stopping times.

1 Introduction: Main Theorem

We are interested in estimates for the norms of Calderón–Zygmund operators on weighted L^p-spaces, a question that has attracted significant interest recently; definitive estimates of this type were first obtained in [2], with a range of prior and subsequent contributions. In this chapter, we will concentrate on $p = 2$ and give a new proof, more elementary than some of the preceding proofs.

Let w be a weight on $\mathbb{R}^d$ with density also written as w. Assume $w > 0$ a. e. We define $\sigma = w^{-1}$, which is defined a. e., and set

$$[w]_{A_2} := \sup_Q \frac{w(Q)}{|Q|} \frac{\sigma(Q)}{|Q|} .$$

We give a new proof of

Theorem 1 ([2]). *Let T be an L^2-bounded Calderón–Zygmund operator, and $w \in A_2$. It then holds that*

$$\|Tf\|_{L^2(w)} \leq C_T [w]_{A_2} \|f\|_{L^2(w)} .$$

M.T. Lacey (✉)
School of Mathematics, Georgia Institute of Technology, Atlanta, GA 30332, USA
e-mail: lacey@math.gatech.edu

D. Bilyk et al. (eds.), *Recent Advances in Harmonic Analysis and Applications*, Springer Proceedings in Mathematics & Statistics 25, DOI 10.1007/978-1-4614-4565-4_20,

All proofs in this level of generality have used Hytönen's random Haar shift representation from [2]. So does this proof. After this point, two strategies of prior proofs are (a) fundamental appeal to two-weight inequalities, an approach initiated in [6] and further refined in [2–5], or (b) constructions of appropriate Bellman functions [9], extending the works of [1, 7]. In our approach, we borrow the distributional inequalities central to the two-weight approach but then combine them with adapted Haar functions from the Bellman approach. Then, the familiar stopping time considerations of Sect. 4 are sufficient to conclude the proof. More detailed histories of this question can be found in the introductions to [4, 5, 9].

As we will concentrate on the case of L^2 estimates, we will frequently use the notation $\|f\|_w := [\int f^2\, w(\mathrm{d}x)]^{1/2}$. At one or two points, an L^1 norm is needed, and this will be clearly indicated.

2 Haar Shift Operators

In this section, we introduce fundamental dyadic approximations of Calderón–Zygmund operators, the Haar shifts, and state reduction of the Main Theorem 1 to a similar statement, Theorem 2, in this dyadic model. In so doing, we are following the lead of [2].

Definition 1. A *dyadic grid* is a collection $\mathscr{D}$ of cubes so that for each Q we have

1. The set of cubes $\{Q' \in \mathscr{D} : |Q'| = |Q|\}$ partition $\mathbb{R}^d$, ignoring overlapping boundaries of cubes.
2. Q is a union of cubes in a collection $\mathrm{Child}(Q) \subset \mathscr{D}$, called the *children of* Q. There are 2^d children of Q, each of volume $|Q'| = 2^{-d}|Q|$.

We refer to any subset of a dyadic grid as simply a *grid*.

The standard choice for $\mathscr{D}$ consists of the cubes $2^k \prod_{s=1}^{d} [n_s, n_s + 1)$ for $k, n_1, \dots, n_d \in \mathbb{Z}$. But, the reduction we are stating here depends upon a random family of dyadic grids. This next definition is at slight variance with that of [2, 3, 5].

Definition 2. For integers $(m,n) \in \mathbb{Z}_+^2$, we say that a linear operator $\mathbb{S}$ is a *(generalized) Haar shift operator of complexity type* (m,n) if

$$\mathbb{S}f(x) = \sum_{Q \in \mathscr{D}} \mathbb{S}_Q f(x) = \sum_{Q \in \mathscr{D}} \int_Q s_Q(x,y)\, f(y)\, \mathrm{d}y,$$

where here and throughout $\ell(Q) = |Q|^{1/d}$, and these properties hold.

1. s_Q, the kernel of the component $\mathbb{S}_Q$, is supported on $Q \times Q$ and $\|s_Q\|_\infty \le \frac{1}{|Q|}$. It is easy to check that
$$\left| \sum_{Q \in \mathscr{D}} s_Q(x,y) \right| \lesssim \frac{1}{|x-y|^d}.$$

2. The kernel s_Q is constant on dyadic rectangles $R \times S \subset Q \times Q$ with $\ell(R) \leq 2^{-m}\ell(Q)$ and $\ell(S) \leq 2^{-n}\ell(Q)$.
3. For any subset $\mathscr{D}' \subset \mathscr{D}$, it holds that we have

$$\Big\| \sum_{Q \in \mathscr{D}'} \mathbb{S}_Q f \Big\|_2 \leq \|f\|_2 .$$

We say that the *complexity* of $\mathbb{S}$ is $\kappa := 1 + \max(m,n)$.

Note that the last property above is an statement about unconditionality of the sum in the operator norm. This is in fact a standard part of Calderón–Zygmund theory—and one that is automatic, depending upon exactly how the definition is formulated. This property is fundamental to the proofs of this chapter, and other results that we merely cite, justifying our inclusion of this property into the definition.

The main results of [5] (see [5, Theorem 4.1]; also [2, Theorem 4.2]) allow us to reduce the proof of the Main Theorem 1 to the verification of the following dyadic variant.

Theorem 2. *Let $\mathbb{S}$ be a Haar shift operator with complexity κ. For $w \in A_2$, we then have the estimates*

$$\|\mathbb{S}f\|_w \lesssim \kappa [w]_{A_2} \|f\|_w . \tag{1}$$

Indeed, any polynomial dependence on the complexity parameter κ would suffice for Theorem 1. (The linear bound in κ was shown in [3, Theorem 2.10] in even greater generality in L^p and maximal truncations. Later, and by different methods, it was shown by [9] as stated above.)

In the remainder of this chapter, $\mathbb{S}$ will denote a Haar shift operator of complexity κ, with *scales separated by* κ. Namely, we have for a subset $\mathscr{D}_\kappa \subset \mathscr{D}$,

$$\mathbb{S}f(x) = \sum_{Q \in \mathscr{D}_\kappa} \int_Q s_Q(x,y) f(y) \, dy, \tag{2}$$

and $\mathscr{D}_\kappa$ consists of all dyadic intervals with $\log_2 \ell(Q) = \ell \mod \kappa$, for some fixed integer $0 \leq \ell < \kappa$. In particular, if $Q', Q \in \mathscr{D}_\kappa$, and $Q' \subsetneq Q$, then $\int_Q s_Q(x,y) f(y)$ is constant on Q'. The dual statement is also true.

3 The Basic Inequalities

We make a remark here about the formulation of the inequalities that we will consider below. Recalling the dual weight $\sigma = w^{-1}$ to an A_2 weight, we will show that

$$\|\mathbb{S}(f\sigma)\|_w \lesssim \kappa [w]_{A_2} \|f\|_\sigma .$$

This is formally equivalent to the statement we are proving, namely (1); moreover, the inequality above is the natural way to phrase the inequality as it dualizes in the natural way: Interchange the roles of w and σ. Accordingly, we will, especially in the next section, use the notation $\langle f,g\rangle_w$ for the natural inner product on $L^2(w)$.

The arguments initiated in [6], further refined in [2, 3, 5], yield the following estimates for Haar shifts on intervals.

Lemma 1. *Let $w \in A_2$, $\mathbb{S}$ a Haar shift operator of complexity κ as in (2). For a cube Q, let $\mathscr{Q} \subset \mathscr{D}_\kappa$ be a collection of cubes contained in Q. We have*

$$\int_Q |\mathbb{S}_{\mathscr{Q}}(\sigma \mathbf{1}_Q)|\, w(\mathrm{d}x) \lesssim [w]_{A_2} |Q|\,, \tag{3}$$

$$\int_Q \mathbb{S}_{\mathscr{Q}}(\sigma \mathbf{1}_Q)^2\, w(\mathrm{d}x) \lesssim [w]_{A_2}^2 \sigma(Q)\,. \tag{4}$$

There are two estimates, one of an $L^1(w)$ norm, with the right-hand side being Lebesgue measure. The second is an $L^2(w)$ norm, with the right-hand side being in terms of σ. Indeed, the papers [2,3,5,6] are argued such that the second estimate (4), combined with a general two-weight theorem, implies the linear bound in A_2. Thus, the point of this chapter is that the general two-weight theorems are not needed.

We suppress the proof of the L^2 estimate, which is based upon a corona decomposition, and distributional estimate. The L^1 estimate follows from the same line of attack. The reader can consult for instance [5, Lemma 5.7] or [3, Section 11]. Of course, the L^2 estimate implies an estimate for the L^1 norm, and it is interesting to note that it is worse than what one gets by using the proof of the L^2 estimate.

4 Proof of the Weighted Estimate for the Haar Shift Operators

We will need the martingale difference operators associated with $\mathscr{D}_\kappa$ and weight σ. For cube $Q \in \mathscr{D}_\kappa$ consider the martingale difference operator

$$D_Q^\sigma f := \sum_{\substack{Q' \subset Q \\ \ell(Q') = 2^{-\kappa}\ell(Q)}} \mathbb{E}_{Q'}^\sigma f \mathbf{1}_{Q'} - \mathbb{E}_Q^\sigma f\,.$$

Here, $\mathbb{E}_Q^\sigma f = \sigma(Q)^{-1} \int_Q f\, \sigma \mathrm{d}x$. The operators $D_Q^\sigma f$ are self-adjoint contractions on $L^2(\sigma)$, and satisfy the fundamental orthogonality relationship is

$$\sum_{Q \in \mathscr{D}_\kappa} \|D_Q^\sigma f\|_\sigma^2 \leq \|f\|_\sigma^2\,, \tag{5}$$

which holds under minimal assumptions on σ, satisfied for a weight with density nonnegative almost everywhere.

Now, complexity shows that for any fixed Q, the components of the Haar shift operator are

$$\int_Q s_Q(x,y)f(y)\,\sigma(\mathrm{d}y) = \mathbb{E}_Q^\sigma f \int_Q s_Q(x,y)\,\sigma(\mathrm{d}y) + \int_Q s_Q(x,y) D_Q^\sigma f(y)\,\sigma(\mathrm{d}y).$$

Note that the bilinear form $\langle \mathbb{S}_\sigma f, g\rangle_w$ is the linear combination of the three terms below, and their duals:

$$\begin{aligned} \langle \mathbb{U}_\sigma f, g\rangle_w &:= \sum_{Q\in\mathscr{D}} \mathbb{E}_Q^\sigma f \int_Q\int_Q s_Q(x,y) g(x)\,\sigma(\mathrm{d}y) w(\mathrm{d}y) \\ \mathbb{V}_\sigma(f,g) &:= \sum_{Q\in\mathscr{D}} \mathbb{E}_Q^\sigma f \int_Q\int_Q s_Q(x,y) D_Q^w g(x)\,\sigma(\mathrm{d}y) w(\mathrm{d}y) \\ \mathbb{W}(f,g) &:= \int_Q s_Q(x,y) D_Q^\sigma f(y) D_Q^w g(x)\,\sigma(\mathrm{d}y) w(\mathrm{d}x). \end{aligned} \tag{6}$$

By dual, we mean that the roles of w and σ are interchanged, which is relevant to $\mathbb{U}_\sigma$ and $\mathbb{V}_\sigma$ above. We will show that each of these three bilinear forms is bounded by $[w]_{A_2}\|f\|_\sigma\|g\|_w$, which estimate also applies to their duals. Recall that we have assumed the separation of scales condition (2); hence, under this condition, we have proved Theorem 2 with absolute constant. This proves the theorem as stated, with κ bound in terms of complexity.

We prove the difficult estimate first, the estimate for $\mathbb{U}_\sigma$.

The Bound for $\mathbb{U}_\sigma$

The essential tool is this corona decomposition.

Definition 3. We say that $\mathscr{F} \subset \mathscr{D}_\kappa$ is a set of f-stopping cubes if these conditions are met:

1. If $F, F' \in \mathscr{F}$, $F' \subsetneq F$, then $\rho(F) := \mathbb{E}_F^\sigma |f| > 4\mathbb{E}_{Q_0}^\sigma |f|$.
2. Every cube $Q \in \mathscr{D}_\kappa$ is contained in some $F \in \mathscr{F}$.
3. Let $\mathscr{D}_F$ be those cubes for which F is the minimal element of $\mathscr{F}$ containing Q. For every $Q \in \mathscr{D}_F$, we have $\mathbb{E}_Q^\sigma |f| \le 4\mathbb{E}_F^\sigma |f|$.

It is easy to recursively construct such a collection $\mathscr{F}$, for $\sigma \in A_2$, which is the case we are considering.

A basic fact, a consequence of the maximal function estimate for general weights, that we have

$$\Big\| \sum_{F\in\mathscr{F}} \rho(F)\mathbf{1}_F \Big\|_\sigma^2 \lesssim \sum_{F\in\mathscr{F}} \rho(F)^2\sigma(F) \lesssim \|f\|_\sigma^2, \qquad \rho(F) = \mathbb{E}_F^\sigma |f|^2. \tag{7}$$

The collections $\mathscr{D}_F$ give a decomposition of $\mathbb{U}_\sigma$ via

$$\begin{aligned}\mathbb{U}_{\sigma,F} f &:= \sum_{Q \in \mathscr{D}_F} \mathbb{E}_Q^\sigma f \int_Q s_Q(x,y)\, \sigma(\mathrm{d}y)\\ &= \rho(F) \sum_{Q \in \mathscr{D}_F} [\mathbb{E}_Q^\sigma f \cdot \rho(F)^{-1}] \int_Q s_Q(x,y)\, \sigma(\mathrm{d}y)\,.\end{aligned}$$

Note that the products $|\mathbb{E}_Q^\sigma f \cdot \rho(F)^{-1}|$ are never more than 4, so by unconditionality of Haar shifts, the integral estimates of the previous section apply to the expressions above.

We abandon duality, expanding

$$\begin{aligned}\|U_\sigma f\|_w^2 &\le \Big\| \sum_{F \in \mathscr{F}} |\mathbb{U}_{\sigma,F} f| \Big\|_w^2 \le I + 2II,\\ I &:= \sum_{F \in \mathscr{F}} \big\|\mathbb{U}_{\sigma,F} f\big\|_w^2,\\ II &:= \sum_{F \in \mathscr{F}} \sum_{\substack{F' \in \mathscr{F}\\ F' \subsetneq F}} \int_{F'} \big|\mathbb{U}_{\sigma,F} f \mathbb{U}_{\sigma,F'} f\big|\, w(\mathrm{d}x)\,.\end{aligned}$$

These are the diagonal and off-diagonal terms. The diagonal is immediate from (4) and (7):

$$I \lesssim [w]_{A_2}^2 \sum_{F \in \mathscr{F}} \rho(F)^2 \sigma(F) \lesssim [w]_{A_2}^2 \|f\|_w^2\,.$$

The off-diagonal is as follows. By the separation of scales hypothesis, note that in the definition of II, that $\mathbb{U}_{\sigma,F} f$ is constant on F' in the display below. Hence, by (3), we have

$$\begin{aligned}II &\lesssim [w]_{A_2} \sum_{F \in \mathscr{F}} \sum_{\substack{F' \in \mathscr{F}\\ F' \subsetneq F}} \mathbb{E}_{F'}^\sigma \big|\mathbb{U}_{\sigma,F} f\big| \cdot \rho(F')|F'|\\ &\lesssim [w]_{A_2} \int \sum_{F \in \mathscr{F}} \big|\mathbb{U}_{\sigma,F} f\big| \cdot \phi\, \mathrm{d}x,\end{aligned}$$

where $\phi := \sum_{F \in \mathscr{F}} \rho(F) \mathbf{1}_F$, and using the identity $w \cdot \sigma \equiv 1$,

$$= [w]_{A_2} \int \sum_{F \in \mathscr{F}} \big|\mathbb{U}_{\sigma,F} f\big| \cdot \phi\, \sqrt{w(x)\sigma(x)} \mathrm{d}x \lesssim \Bigg\| \sum_{F \in \mathscr{F}} \big|\mathbb{U}_{\sigma,F} f\big| \Bigg\|_w \|\phi\|_\sigma\,.$$

We have, however, $\|\phi\|_\sigma \lesssim \|M^\sigma f\|_\sigma \lesssim \|f\|_\sigma$. Combining estimates, we see that we have proved

$$\Big\| \sum_{F \in \mathscr{F}} \big|\mathbb{U}_{\sigma,F} f\big| \Big\|_w^2 \lesssim [w]_{A_2}^2 \|f\|_\sigma^2 + [w]_{A_2} \Big\| \sum_{F \in \mathscr{F}} \big|\mathbb{U}_{\sigma,F} f\big| \Big\|_w \|f\|_\sigma,$$

which immediately implies our linear bound in A_2 for the term $\mathbb{U}_\sigma$.

The Remaining Estimates

Proof (the bound for $\mathbb{V}_{\sigma,k}(f,g)$). We consider $\mathbb{V}_{\sigma,k}(f,g)$, defined in (6). Using the orthogonality property of martingale differences (6), we see that

$$\left|\mathbb{V}_{\sigma,k}(f,g)\right| \le \sum_{Q\in\mathscr{D}} |\mathbb{E}_Q^\sigma f| \cdot \left| \left\langle \int_Q s_Q(x,y)\,\sigma(\mathrm{d}y), D_Q^w g \right\rangle_w \right|$$

$$\le \|g\|_w \left\| \left[\sum_{Q\in\mathscr{D}} \left[\mathbb{E}_Q^\sigma f \cdot \int_Q s_Q(x,y)\,\sigma(\mathrm{d}y) \right]^2 \right]^{1/2} \right\|_w \lesssim [w]_{A_2} \|f\|_\sigma \|g\|_w .$$

The last line follows from the bound already proved for the operator $\mathbb{U}_\sigma$ and the unconditionality. By a standard averaging over random choices of signs, we can deduce the linear in A_2 bound for the square function above.

Proof (the bound for $\mathbb{W}$). We insert $\sqrt{w\sigma}$ into the integrals below, and use the bound $|s_Q(x,y)| \le |Q|^{-1}$, to see that

$$\left| \int_Q \int_Q s_Q(x,y) D_Q^\sigma f(y) D_Q^w g(x)\,\mathrm{d}x\mathrm{d}y \right| \le \frac{1}{|Q|} \int_Q |D_Q^\sigma f(y)|\,\mathrm{d}y \int_Q |D_Q^w g(x)|\,\mathrm{d}x$$

$$\le \|D_Q^\sigma f\|_\sigma \|D_Q^w g\|_w \left[\frac{\sigma(Q)}{|Q|} \frac{w(Q)}{|Q|} \right]^{1/2}$$

$$\le [w]_{A_2} \|D_Q^\sigma f\|_\sigma \|D_Q^w g\|_w ,$$

since we always have $[w]_{A_2} \ge 1$. The martingale differences are pairwise orthogonal in $L^2(\sigma)$ and $L^2(w)$, so that a second application, in the variable Q, of the Cauchy–Schwartz inequality finishes this case.

Remark 1. Rather than consider the operators $\mathbb{V}_\sigma$ in (6), and the dual expression, we could have considered

$$\widetilde{\mathbb{V}}(f,g) := \sum_{Q\in\mathscr{D}_k} \mathbb{E}_Q^\sigma f \cdot \int_Q \int_Q s_Q(x,y)\sigma(\mathrm{d}y) w(\mathrm{d}x) \cdot \mathbb{E}_Q^w g .$$

It follows from unconditionality of Haar shift operators and the estimate (3) that we have the uniform estimate

$$\sum_{\substack{Q\in\mathscr{D}_\kappa \\ Q\subset Q_0}} \left| \int_Q \int_Q s_Q(x,y)\sigma(\mathrm{d}y) w(\mathrm{d}x) \right| \lesssim [w]_{A_2} |Q_0| , \qquad \mathscr{Q}_0 \in \mathscr{D}.$$

From this, it is easy to see that

$$
\begin{aligned}
|\widetilde{\mathbb{V}}(f,g)| &\lesssim [w]_{A_2} \int M^{\sigma} f \cdot M^{w} g \,\mathrm{d}x \\
&= [w]_{A_2} \int M^{\sigma} f \cdot M^{w} g \,(w(x)\sigma(s))^{1/2} \mathrm{d}x \\
&\leq [w]_{A_2} \|M^{\sigma} f\|_{\sigma} \|M^{w} g\|_{w} \lesssim [w]_{A_2} \|f\|_{\sigma} \|g\|_{w}.
\end{aligned}
$$

Compare to Sect. 4 of [8]. But, we do not prefer this proof as it obscures the central role of the operator $\mathbb{U}_{\sigma}$.

Acknowledgment Research supported in part by grant NSF-DMS 0968499.

References

1. Beznosova, O.V.: Linear bound for the dyadic paraproduct on weighted Lebesgue space $L_2(w)$. J. Funct. Anal. **255**(4), 994–1007 (2008). MR2433959
2. Hytönen, T.: The sharp weighted bound for general Calderon-Zygmund operators. Ann. Math. (2010), to appear. Available at http://arxiv.org/abs/1007.4330
3. Hytönen, T., Lacey, M.T., Martikainen, H., Orponen, T., Reguera, M.C., Sawyer, E.T., Uriarte-Tuero, I.: Weak And strong type estimates for maximal truncations of Calderón–Zygmund operators on A_p weighted spaces (2011). Available at http://www.arxiv.org/abs/1103.5229.
4. Hytönen, T., Pérez, C.: Sharp weighted bounds involving A_∞ (2011). Available at http://www.arxiv.org/abs/1103.5562
5. Hytönen, T., Pérez, C., Treil, S., Volberg, A.: Sharp weighted estimates of the dyadic shifts and A_2 conjecture. ArXiv e-prints (2010). Available at http://arxiv.org/abs/1010.0755
6. Lacey, M.T., Petermichl, S., Reguera, M.C.: Sharp A_2 inequality for Haar shift operators. Math. Ann. **348**(1), 127–141 (2010)
7. Petermichl, S.: The sharp bound for the Hilbert transform on weighted Lebesgue spaces in terms of the classical A_p characteristic. Amer. J. Math. **129**(5), 1355–1375 (2007)
8. Reznikov, A., Treil, S., Volberg, A.: A sharp weighted estimate of dyadic shifts of complexity 0 and 1. ArXiv e-prints (2011). Available at http://arxiv.org/abs/1103.5347
9. Treil, S.: Sharp A_2 estimates of Haar shifts via Bellman function (2011). Available at http://arxiv.org/abs/1105.2252

Quest for Negative Dependency Graphs

Linyuan Lu, Austin Mohr, and László Székely

Dedicated to our friend, Professor Konstantin Oskolkov, to mark this cairn of his journey searching beauty and boleti.

Abstract The Lovász local lemma is a well-known probabilistic technique commonly used to prove the existence of rare combinatorial objects. We explore the lopsided (or negative dependency graph) version of the lemma, which, while more general, appears infrequently in literature due to the lack of settings in which the additional generality has thus far been needed. We present a general framework (matchings in hypergraphs) from which many such settings arise naturally. We also prove a seemingly new generalization of Cayley's formula, which helps defining negative dependency graphs for extensions of forests into spanning trees. We formulate open problems regarding partitions and doubly stochastic matrices that are likely amenable to the use of the lopsided local lemma.

1 Introduction

The Lovász Local Lemma (hereinafter, "L3") says there is a positive probability that none of the events in some collection occurs, provided the dependencies among them are not too strong. First we state L3 [3,7,22,23]. Given events $A_1,\ldots,A_n$ in a probability space, define a *dependency graph* to be a simple graph G on $V(G)=[n]$ (i.e., the set of the names of the events, $\{1,2,\ldots,n\}$) satisfying that every event A_i is independent of the elements of the event algebra generated by $\{A_j : ij \notin E(G)\}$.

L. Lu • A. Mohr • L. Székely (✉)
University of South Carolina at Columbia, Department of Mathematics,
Columbia, SC 29208, USA
e-mail: lu@math.sc.edu; mohrat@email.sc.edu; szekely@math.sc.edu

D. Bilyk et al. (eds.), *Recent Advances in Harmonic Analysis and Applications*, Springer Proceedings in Mathematics & Statistics 25, DOI 10.1007/978-1-4614-4565-4_21,

Lemma 1 (Lovász Local Lemma). [3] *Let $A_1,\ldots,A_n$ be events with dependency graph G. If there are numbers $x_1,\ldots,x_n \in [0,1)$ such that for all i*

$$\Pr(A_i) \le x_i \prod_{ij \in E(G)} (1-x_j).$$

Then

$$\Pr\left(\bigwedge_{i=1}^{n} \overline{A_i}\right) \ge \prod_{i=1}^{n}(1-x_i) > 0.$$

The original lemma of Lovász in [7], which is a special case of Lemma 1, was used to prove the 2-colorability of a certain hypergraph, and in general, together with the extension L3 [22], to prove the *existence* of combinatorial objects, for which construction seemed hopeless. Surprisingly, there are recent *constructive* versions of the lemma (e.g., in [19]) that will produce the desired object explicitly for many coloration problems. Define $C_G(i) = \{j \in [n] \mid ij \notin E(G)\}$. Inspection of the—not too difficult—proof of L3 concludes that we use independence in the form of

$$\forall i \in [n]\ \forall S \subseteq C_G(i)\ \ \Pr\left(\bigwedge_{j\in S} \overline{A_j}\right) \ne 0 \to \Pr\left(A_i \,\middle|\, \bigwedge_{j\in S} \overline{A_j}\right) = \Pr(A_i). \tag{1}$$

Further analyzing the proof, it turns out that the inequality below is sufficient, instead of equality in (1):

$$\forall i \in [n]\ \forall S \subseteq C_G(i)\ \ \Pr\left(\bigwedge_{j\in S} \overline{A_j}\right) \ne 0 \to \Pr\left(A_i \,\middle|\, \bigwedge_{j\in S} \overline{A_j}\right) \le \Pr(A_i). \tag{2}$$

Formally, a *negative dependency graph* for given events $A_1,\ldots,A_n$ in a probability space is a simple graph G on $V(G) = [n]$ (i.e., the set of the names of the events, $\{1,2,\ldots,n\}$) satisfying (2). Observe that every negative dependency graph is a dependency graph, but not vice versa.

Lemma 2 (Lopsided Lovász Local Lemma, L4). *Lemma 1 holds with "dependency graph" relaxed to "negative dependency graph."*

The inequality in (2) explains the term "lopsided" in the literature for L4. Negative dependency graphs and the lopsided version of the lemma were first introduced in [8] by Erdős and Spencer to prove the existence of a certain latin transversal, and independently by Albert, Frieze, and Reed [2], following the original Lovász version [7]. L4 as stated above is due to Ku [14]. These papers did not investigate classes of problems where L4 could be applied. Few results are present in the literature making use of a negative dependency graph that is not also a dependency graph.

It is worth mentioning here the following useful equivalents of (2):

$$\forall i \in [n]\ \forall S \subseteq C_G(i)\ \ \Pr(A_i) \ne 0 \to \Pr\left(\bigwedge_{j\in S} \overline{A_j} \,\middle|\, A_i\right) \le \Pr\left(\bigwedge_{j\in S} \overline{A_j}\right), \tag{3}$$

or

$$\forall i \in [n]\ \forall S \subseteq C_G(i)\ \ \Pr\left(\bigwedge_{j\in S}\overline{A_j}\right) \neq 0 \rightarrow \Pr\left(\overline{A_i}\right) \leq \Pr\left(\bigwedge_{j\in S\cup\{i\}}\overline{A_j}\,\middle|\,\bigwedge_{j\in S}\overline{A_j}\right), \quad (4)$$

and also the following, which takes the form of a correlation inequality:

$$\forall i \in [n]\ \forall S \subseteq C_G(i)\ \ \Pr(A_i)\Pr\left(\bigvee_{j\in S} A_j\right) \leq \Pr\left(A_i \wedge \left(\bigvee_{j\in S} A_j\right)\right). \quad (5)$$

We note that even more general versions of the lemma hold practically with the same proof. Given events $A_1,\ldots,A_n$ in a probability space, define a *dependency digraph* $\overrightarrow{G}$ on $V(G) = [n]$ by requiring that every event A_i is independent of the elements of the event algebra generated by $\{A_j : j \in C_{\overrightarrow{G}}(i)\}$, with $C_{\overrightarrow{G}}(i) = \{j : ij \notin E(\overrightarrow{G})\}$, and a *negative dependency digraph graph* $\overrightarrow{G}$ by (2), using $\overrightarrow{G}$ and $C_{\overrightarrow{G}}(i)$ instead of G and $C_G(i)$ in (2).

Lemma 3 (Lopsided Lovász Local Lemma, digraph version $\overrightarrow{\mathbf{L4}}$ [3]). *Lemma 1 holds with "dependency graph" relaxed to "negative dependency digraph."*

2 Examples of Negative Dependency Graphs

2.1 Random Matchings in Complete Uniform Hypergraphs

A *matching* M in a hypergraph is a collection of pairwise vertex-disjoint hyperedges. The set of vertices covered by hyperedges of the matching M is denoted by $V(M)$. A matching is *perfect* if every vertex of the underlying hypergraph appears in some hyperedge of the matching. In what follows, we will restrict our attention to matchings of the complete k-uniform hypergraph on N vertices, commonly denoted K_N^k.

A matching of K_N^k is an *r-matching* if it consists of r vertex-disjoint hyperedges. For some fixed integer k, r, N satisfying $k \geq 2$, $r \geq 1$, and $N \geq rk$, let $\Omega_N^{k,r}$ be the uniform probability space over all r-matchings of K_N^k. An r-matching of K_N^k is *maximal* if $r = \lfloor \frac{N}{k} \rfloor$; it is *perfect* if $N = kr$. The space of maximal matchings of K_N^k is simply denoted by Ω_N^k.

Given a matching M in K_N^k, we will be interested in the canonical event A^M containing all r-matchings that extend M. More precisely,

$$A^M = A^M_{N,k,r} = \left\{M' \in \Omega_N^{k,r} \mid M \subseteq M'\right\}.$$

We will say that two matchings *conflict*, if they have edges that are neither identical nor disjoint, two *canonical events conflict*, whenever the matchings used to define them conflict.

Given a collection $\mathcal{M}$ of matchings (or other combinatorial objects, as we will see later), the *conflict graph* for the collection $\{A^M \mid M \in \mathcal{M}\}$ is the simple graph with vertex set $\mathcal{M}$ and edge set $\{M_1M_2 \mid M_1 \in \mathcal{M}$ and $M_2 \in \mathcal{M}$ are in conflict$\}$. The following theorem was proved in [17] in the special case $k = 2$ and N is even:

Theorem 1. *Let $\mathcal{M}$ be a collection of matchings in K_N^k. The conflict graph for the collection of canonical events $\{A^M \mid M \in \mathcal{M}\}$ is a negative dependency graph for the probability space Ω_N^k.*

Theorem 1 can be deduced from the following lemma, which deals with a more general probability space $\Omega_N^{k,r}$. The proof of the lemma is postponed.

Lemma 4. *In $\Omega_N^{k,r}$, fix a matching $M \in \mathcal{M}$ with $|M| < r$. Let $\mathcal{J}$ be any collection of matchings from $\mathcal{M}$ whose members do not conflict with M. If $|\cup_{M' \in \mathcal{J}} V(M' \setminus M)| \leq rk - |V(M)| + k - 1$ and $\Pr(\wedge_{M' \in \mathcal{J}} \overline{A^{M'}}) > 0$, then*

$$\Pr\left(A^M \,\middle|\, \bigwedge_{M' \in \mathcal{J}} \overline{A^{M'}}\right) \leq \Pr(A_M). \tag{6}$$

Proof of Theorem 1: Write $N = rk + t$. For the space Ω_N^k of maximum matchings, we have $0 \leq t \leq k - 1$. In this case, we have

$$|\cup_{M' \in \mathcal{J}} V(M' - M)| \leq N - |V(M)| = rk + t - |V(M)| \leq rk - |V(M)| + k - 1.$$

By Lemma 4, the conflict graph for the collection of canonical events $\{A^M \mid M \in \mathcal{M}\}$ is a negative dependency graph for Ω_N^k. □

The following example shows the conflict graph may not be a negative dependency graph of $\Omega_N^{k,r}$ for $r < \lfloor \frac{N}{k} \rfloor$. Take $M_1, M_2 \in \Omega_N^{k,r}$ so that $M_1 \cup M_2$ is a matching of size $r + 1$. This is possible since $r + 1 \leq \lfloor \frac{N}{k} \rfloor$. By our choice, M_1 and M_2 are not adjacent in the conflict graph. However, we have

$$Pr(\overline{A_{M_1}} | A_{M_2}) = 1 > Pr(\overline{A_{M_1}}),$$

contradicting (3). Nevertheless, we can add orientation and some edges to the conflict graph to turn it into a negative dependency *digraph*.

Theorem 2. *For integers $r \geq 1$, $k \geq 2$, and $N \geq (r+1)k$, let $\mathcal{M}$ be a collection of r-matchings in K_N^k. For any $M \in \mathcal{M}$, let S_M be an arbitrary set of $N - (r+1)k + 1$ vertices not in $V(M)$. Let G be a graph on the vertex set $\mathcal{M}$ where MM' is a directed edge of G if M' conflicts with M or M' contains some vertex in S_M. Then G is a negative dependency digraph for the collection of canonical events $\{A^M \mid M \in \mathcal{M}\}$ in the probability space $\Omega_N^{k,r}$.*

Throughout, we assume that the vertex set of K_N^k is $[N]$. We will view the identity map as an injection from $[N]$ into $[N+s]$ for $s \geq 0$ and also from $V(K_N^k)$ to $V(K_{N+s}^k)$ and from $E(K_N^k)$ to $E(K_{N+s}^k)$. To emphasize the difference in the probability space, we use $A_{N,k,r}^M$ to denote the canonical event induced by the matching M in $\Omega_N^{k,r}$ and use $\Pr_N^{k,r}(\bullet)$ to denote the probability in $\Omega_N^{k,r}$. To simplify our notation, we will write $\Pr(\bullet)$ for $\Pr_N^{k,r}(\bullet)$, if the probability space is $\Omega_N^{k,r}$, and spell out the full notation otherwise.

Lemma 5. *For any integers $r \geq 1$, $k \geq 2$, $N \geq rk$, $N_0 = \min\{N-k, rk-1\}$ and any collection $\mathscr{M}$ of matchings in $K_{N_0}^k$, we have*

$$\Pr_{N-k}^{k,r-1}\left(\bigwedge_{M\in\mathscr{M}} \overline{A_{N-k,k,r-1}^M}\right) \leq \Pr_N^{k,r}\left(\bigwedge_{M\in\mathscr{M}} \overline{A_{N,k,r}^M}\right). \tag{7}$$

Proof. Let $t = N - rk$. We have $0 \leq t \leq N-k$. Given an r-matching $M'' \in \Omega_N^{k,r}$, M'' can be viewed as a hypergraph with r pairwise disjoint hyperedges and t isolated vertices. The *right* end of a hyperedge H is $\max_{i\in H} i$. A hyperedge of M'' is called *rightmost* if it has the largest right end among all hyperedges in M''. A k-edge R can be the rightmost hyperedge of some matching $M'' \in \Omega_N^{k,r}$ if and only if the right end vertex of R is at least rk. For $i = 0,1,2,\ldots,t$, let $\mathscr{R}_i$ be the family of k-edges whose right end is $N-i$. Let $\mathscr{R} = \cup_{i=0}^t \mathscr{R}_i$. Consider the mapping $\psi\colon \Omega_N^{k,r} \to \mathscr{R}$, which maps M'' to its rightmost hyperedge. Clearly, $\cup_{R\in\mathscr{R}} \psi^{-1}(R)$ forms a partition of $\Omega_N^{k,r}$.

Fix an i $(0 \leq i \leq t)$ and a k-edge $R \in \mathscr{R}_i$. Easy calculation shows that

$$a_i := \Pr\left(\psi^{-1}(R)\right) = \frac{k!r(t)_i}{(N)_i(N-i)_k}.$$

Direct comparison of terms gives

$$a_0 \geq a_1 \geq \cdots \geq a_t. \tag{8}$$

Since $N_0 \leq kr-1$, the hyperedge R above is not in any matching $M' \in \mathscr{M}$. Define $\mathscr{M}'(R) = \{M' \in \mathscr{M} : V(M') \cap R = \emptyset\}$ and observe that

$$\bigwedge_{M'\in\mathscr{M}} \overline{A_{N,k,r}^{M'}} \wedge \psi^{-1}(R) = \bigwedge_{M'\in\mathscr{M}'(R)} \overline{A_{N,k,r}^{M'}} \wedge \psi^{-1}(R).$$

Let $F = \{N-k+1, N-k+2, \ldots, N\}$ and σ be any permutation of $[N]$ that maps $R \setminus F$ to $F \setminus R$, maps $F \setminus R$ to $R \setminus F$, and leaves other vertices as fixed points.

The permutation σ maps $\bigwedge_{M' \in \mathcal{M}'(R)} \overline{A_{N,k,r}^{M'}} \wedge \psi^{-1}(R)$ to $\bigwedge_{M' \in \mathcal{M}'(R)} \overline{A_{N,k,r}^{M'}} \wedge A_{N,k,r}^{F}$. We have

$$\begin{aligned}
\Pr\left(\bigwedge_{M' \in \mathcal{M}} \overline{A_{N,k,r}^{M'}}\right) &= \sum_{i=0}^{t} \sum_{R \in \mathcal{R}_i} \Pr\left(\bigwedge_{M' \in \mathcal{M}} \overline{A_{N,k,r}^{M'}} \wedge \psi^{-1}(R)\right) \\
&= \sum_{i=0}^{t} \sum_{R \in \mathcal{R}_i} \Pr\left(\bigwedge_{M' \in \mathcal{M}'(R)} \overline{A_{N,k,r}^{M'}} \wedge \psi^{-1}(R)\right) \\
&= \sum_{i=0}^{t} \sum_{R \in \mathcal{R}_i} \Pr\left(\bigwedge_{M' \in \mathcal{M}'(R)} \overline{A_{N,k,r}^{M'}} \wedge A_{N,k,r}^{F}\right) \\
&\geq \sum_{i=0}^{t} \sum_{R \in \mathcal{R}_i} \Pr\left(\bigwedge_{M' \in \mathcal{M}} \overline{A_{N,k,r}^{M'}} \wedge A_{N,k,r}^{F}\right) \\
&= \sum_{i=0}^{t} \sum_{R \in \mathcal{R}_i} \Pr\left(\bigwedge_{M' \in \mathcal{M}} \overline{A_{N,k,r}^{M'}} \,\middle|\, A_{N,k,r}^{F}\right) \Pr\left(A_{N,k,r}^{F}\right) \\
&= \Pr_{N-k}^{k,r-1}\left(\bigwedge_{M' \in \mathcal{M}} \overline{A_{N-k,k,r-1}^{M'}}\right) \sum_{i=0}^{t} \sum_{R \in \mathcal{R}_i} \Pr\left(A_{N,k,r}^{F}\right).
\end{aligned}$$

In the last step, we use the fact that

$$\Pr_{N}^{k,r}\left(\bigwedge_{M' \in \mathcal{M}} \overline{A_{N,k,r}^{M'}} \,\middle|\, A_{N,k,r}^{F}\right) = \Pr_{N-k}^{k,r-1}\left(\bigwedge_{M' \in \mathcal{M}} \overline{A_{N-k,k,r-1}^{M'}}\right).$$

Note that $\Pr(A_{N,k,r}^{F}) = a_0 \geq a_i = \Pr\left(\psi^{-1}(R)\right)$. We have

$$\sum_{i=0}^{t} \sum_{R \in \mathcal{R}_i} \Pr\left(A_{N,k,r}^{F}\right) \geq \sum_{i=0}^{t} \sum_{R \in \mathcal{R}_i} \Pr\left(\psi^{-1}(R)\right) = \Pr\left(\Omega_{N}^{k,r}\right) = 1.$$

Thus,

$$\Pr\left(\bigwedge_{M' \in \mathcal{M}} \overline{A_{N,k,r}^{M'}}\right) \geq \Pr_{N-k}^{k,r-1}\left(\bigwedge_{M' \in \mathcal{M}} \overline{A_{N-k,k,r-1}^{M'}}\right).$$

□

Proof of Lemma 4: Returning to the proof of Lemma 4, Fix a matching $M \in \mathcal{M}$ and let $\mathcal{J}$ be any collection of matchings from $\mathcal{M}$ that do not conflict with M. Our aim is to show

$$\Pr\left(A^{M} \,\middle|\, \bigwedge_{M' \in \mathcal{J}} \overline{A^{M'}}\right) \leq \Pr\left(A^{M}\right).$$

Observe that the inequality holds trivially when $\Pr\left(A^M\right) = 0$. Otherwise, the above formula is equivalent with the following (that is essentially (3)):

$$\Pr\left(\bigwedge_{M' \in \mathscr{J}} \overline{A^{M'}} \,\middle|\, A^M\right) \leq \Pr\left(\bigwedge_{M' \in \mathscr{J}} \overline{A^{M'}}\right).$$

Let $\mathscr{J}^M = \{M' \setminus M \mid M' \in \mathscr{J}\}$. If $M \in \mathscr{J}$, then the left-hand side of the estimate above is zero, and so we have nothing to do. Assume instead that $M \notin \mathscr{J}$. Since every matching M' in $\mathscr{J}$ is not in conflict with M, the vertex set of $M' \setminus M$ is nonempty and is disjoint from the vertex set of M. Let T be the set of vertices covered by the matching M and U be the set of vertices covered by at least one matching $F \in \mathscr{J}^M$. We have $T \cap U = \emptyset$. Let π be a permutation of $[N]$ mapping T to $\{N - |T| + 1, N - |T| + 2, \ldots, N\}$. We have $\pi(T) \cap \pi(U) = \emptyset$. Thus, $\pi(U) \subseteq [N - |T|]$. Define $\pi(\mathscr{J}^M)$ to be the collection $\{\pi(F) \mid F \in \mathscr{J}^M\}$. We obtain

$$\begin{aligned}
\Pr\left(\bigwedge_{M' \in \mathscr{J}} \overline{A^{M'}} \,\middle|\, A^M\right) &= \frac{\Pr\left(\bigwedge_{M' \in \mathscr{J}} \overline{A^{M'}} \wedge A^M\right)}{\Pr\left(A^M\right)} \\
&= \frac{\Pr\left(\bigwedge_{M' \in \mathscr{J}} \overline{A^{M' \setminus M}} \wedge A^M\right)}{\Pr\left(A^M\right)} \\
&= \frac{\Pr\left(\bigwedge_{F \in \mathscr{J}^M} \overline{A^F} \wedge A^M\right)}{\Pr\left(A^M\right)} \\
&= \Pr\left(\bigwedge_{F \in \mathscr{J}^M} \overline{A^F} \,\middle|\, A^M\right) \\
&= \Pr\left(\bigwedge_{\pi(F) \in \pi(\mathscr{J}^M)} \overline{A_{N,k,r}^{\pi(F)}} \,\middle|\, A^{\pi(M)}\right) \\
&= \Pr{}_{N-jk}^{k,r-j}\left(\bigwedge_{\pi(F) \in \pi(\mathscr{J}^M)} \overline{A_{N-jk,k,r-j}^{\pi(F)}}\right) && \text{(with } j = |M| < r\text{)} \\
&\leq \Pr\left(\bigwedge_{\pi(F) \in \pi(\mathscr{J}^M)} \overline{A_{N,k,r}^{\pi(F)}}\right) && \text{(by Lemma 5)} \\
&= \Pr\left(\bigwedge_{F \in \mathscr{J}^M} \overline{A_{N,k,r}^{F}}\right)
\end{aligned}$$

$$= \Pr\left(\bigwedge_{M' \in \mathscr{I}} \overline{A_{N,k,r}^{M' \setminus M}} \right)$$

$$\leq \Pr\left(\bigwedge_{M' \in \mathscr{I}} \overline{A_{N,k,r}^{M'}} \right).$$

□

2.2 Random Matchings in Complete Multipartite Graphs

Theorem 1 shows how a general class of negative dependency graphs can arise from the space of random perfect matchings of K_N^k. A similar result was shown in [16] for the uniform probability space of maximum matchings of a complete bipartite graph $K_{s,t}$, with the same definition of "conflict" and "canonical event" as above. This can be viewed as the uniform probability space of random injections from an s-element set into a t-element set (for $s \leq t$), providing a plethora of applications.

This generalizes to multipartite matchings as follows. For details see [18]. Let us be given disjoint sets $U_1, \ldots, U_m$ with $|U_1| \leq |U_i|$ for $1 < i$. Call *edges* the sets H, if $H \subseteq \cup_{i=1}^m U_i$ and for all i, $|H \cap U_i| = 1$. A *matching* is a set of disjoint edges. A matching is of maximum size if it covers all elements of U_1. Consider the uniform probability measure on set of all maximum size matchings. Given a matching M, let A^M denote the event of all maximum size matchings that contain all edges of M. We say that two matchings, M_1 and M_2, are in *conflict* if they contain edges that are neither identical nor disjoint.

Theorem 3. *[18] Let $\mathscr{M}$ be a collection of multipartite matchings on $U_1, \ldots, U_m$. The conflict graph for the collection of canonical events $\{A^M \mid M \in \mathscr{M}\}$ is a negative dependency graph.*

2.3 Spanning Trees in Complete Graphs

The various matching spaces we have mentioned have in common that a partial matching does not conflict with any element of its corresponding canonical event. Indeed, the proof of Theorem 1 relies heavily on this fact by reducing the problem of extending a given partial matching to the problem of finding matchings on the unmatched vertices only.

Consider now the uniform probability space of all spanning trees of K_N. Given a forest F (i.e., a cycle-free subset of the edges of K_N), the canonical event A^F is the collection of all spanning trees of K_N containing F. We say that two forests *conflict* whenever there are a pair of edges, one in the first forest and one in the second, that intersect in exactly one vertex. In other words, two forests F and F' do *not* conflict if for every connected component $C \subseteq F$ and $C' \subseteq F'$, C and C' are either identical or disjoint.

Theorem 4. *Let $\mathscr{F}$ be a collection of forests in K_N. The conflict graph for the collection of canonical events $\{A^F \mid F \in \mathscr{F}\}$ is a negative dependency graph.*

Notice that the spanning tree setting stands in stark contrast to the matching and partition settings; a forest conflicts with *every* spanning tree in its corresponding canonical event! The proof of Theorem 4 hinges on two lemmata. The first is a direct generalization of Cayley's theorem, while the second is a special case of Theorem 4.

Lemma 6. *Let us be given a forest F in K_N, which has its components $C_1, C_2, \ldots, C_m$ on $f_1, f_2, \ldots, f_m$ vertices. Then, the number of spanning trees T in K_N, such that F is contained by T, is*

$$f_1 f_2 \cdots f_m N^{N-2-\sum_i (f_i-1)}. \tag{9}$$

Proof. Recall Menon's Theorem (Problem 4.1 in [15]): the number of spanning trees in K_N with prescribed degrees $d_1, d_2, \ldots, d_N$ in vertices $1, 2, \ldots, N$ is the multinomial coefficient $\binom{N-2}{(d_1-1),\ldots,(d_N-1)}$. Contracting the components of F to single vertices, T contracts to a spanning tree T^* of $K_{N-\sum_i(f_i-1)}$. Let $v_1, \ldots, v_m$ denote the result of contraction of $C_1, C_2, \ldots, C_m$, and $u_1, u_2, \ldots, u_{N-\sum_i f_i}$ the vertices from $1, 2, \ldots, N$, not covered by F. By Menon's theorem, the number of H spanning trees of $K_{N-\sum_i(f_i-1)}$, with degree d_i in v_i and D_j in u_j, is

$$\binom{N-2-\sum_i(f_i-1)}{(d_1-1),\ldots,(d_m-1),(D_1-1),\ldots,(D_{N-\sum_i f_i}-1)}.$$

Note that every H spanning tree of $K_{N-\sum_i(f_i-1)}$ with degree d_i in v_i and D_j in u_j arises *precisely* $\prod_i f_i^{d_i}$ ways as a contraction T^* from some T spanning tree of K_N. Hence the number of spanning trees T containing F is

$$\sum \binom{N-2-\sum_i(f_i-1)}{(d_1-1),\ldots,(d_m-1),(D_1-1),\ldots,(D_{N-\sum_i f_i}-1)} \prod_i f_i^{d_i},$$

where the summation goes for all $d_1, \ldots, d_m$ and $D_1, \ldots, D_{N-\sum_i f_i}$ sequences. The multinomial theorem easily evaluates this summation to the required quantity. □

For the next lemma, we say two forests are in *strong conflict*, if they are not vertex disjoint.

Lemma 7. *Let $\mathscr{F}$ be a collection of forests in K_N. The strong conflict graph for the collection of canonical events $\{A^F \mid F \in \mathscr{F}\}$ is a negative dependency graph.*

Proof. To prove Lemma 7, we prove (5), where A_i is the set of spanning trees containing the forest F_i, and F_i is not in strong conflict with F_j for any $j \in S$. By inclusion–exclusion,

$$\Pr\left(\bigvee_{j\in S} A_j\right) = \sum_{R\subseteq S, |R|\geq 1} \Pr\left(\bigwedge_{j\in R} A_j\right)(-1)^{|R|-1}$$

and

$$\Pr\left(A_i \wedge \left(\bigvee_{j\in S} A_j\right)\right) = \sum_{R\subseteq S, |R|\geq 1} \Pr\left(A_i \wedge \left(\bigwedge_{j\in R} A_j\right)\right)(-1)^{|R|-1}.$$

Observe that the event $A_i \wedge (\wedge_{j\in R} A_j)$ consists of spanning trees that contain the forest F_i and $G_R = \cup_{j\in R} F_j$. The latter graph is either a forest or contains cycle. In the latter case, the corresponding event is impossible. Finally, we claim

$$\Pr(A_i)\Pr\left(\bigwedge_{j\in R} A_j\right) = \Pr\left(A_i \wedge \left(\bigwedge_{j\in R} A_j\right)\right)$$

either by G_R being impossible (and both sides are zero) or by F_i and G_R being vertex-disjoint forests, whose union is a forest again, having as components each and every component of F_i and G_R. Lemma 6 finishes the proof. □

Finally, to prove Theorem 4, let A_i denote again the set of spanning trees containing the forest F_i. We are going to prove (5). If F_i has no strong conflict with any F_j $(j \in S)$, then Lemma 7 already gives us the wanted result. Now suppose F_i does have strong conflict with some F_j $(j \in S)$. Define $F'_j = F_j \setminus F_i$ (we mean the difference of the edge sets). Let A'_j be the event corresponding to F'_j. We have

$$\begin{aligned}\Pr\left(A_i \wedge \left(\bigwedge_{j\in S} \overline{A_j}\right)\right) &= \Pr\left(A_i \wedge \left(\bigwedge_{j\in S} \overline{A'_j}\right)\right)\\ &= \Pr(A_i)\Pr\left(\bigwedge_{j\in S} \overline{A'_j}\right) && \text{(by Lemma 7)}\\ &\leq \Pr(A_i)\Pr\left(\bigwedge_{j\in S} \overline{A_j}\right).\end{aligned}$$

□

Note that in Lemma 7 we proved a negative dependency graph with equalities everywhere. This is in fact a negative dependency graph. It is likely that independence is lurking in the form of independent choice of entries in the Prüfer code or some other sequence encoding of trees.

There is one more interesting comment to make here. Fix any connected graph G and two of its edges e and f. In the uniform probability space of the spanning trees of G, the correlation inequality

$$\Pr(A^e)\Pr\left(A^f\right) \geq \Pr\left(A^e \wedge A^f\right) \tag{10}$$

holds [25]. This is the opposite of the inequality that we expect for (5)! There is no contradiction, however, as for $G = K_N$ and disjoint edges, (10) holds with identity, and for two edges sharing a single vertex, we have a conflict and we made no claim.

Change the underlying probability space of spanning trees to the uniform probability space of spanning forests of K_N, and let the canonical event associated with a forest be the set of all spanning forests containing it. Then neither *conflict* nor *strong conflict* of forests defines a negative dependency graph for their canonical events—this is in line of the conjecture of Kahn [13] that in every connected graph, (10) holds, where A^e is the set of spanning spanning forests containing edge e.

2.4 Upper Ideals in Distributive Lattices

Let X be an N-element set, and let Ω_N be the probability space consisting of all subsets of X and equipped with the uniform probability measure. For a fixed subset Y of X, define the *canonical event* A^Y to be the collection of all subsets of X that contain Y. In other words,

$$A^Y = \{Z \in \Omega_N \mid Y \subseteq Z\}.$$

Theorem 5. *Let $\mathscr{Y}$ be a collection of nonempty subsets of an N-element set. The graph with vertex set $\{A^Y \mid Y \in \mathscr{Y}\}$ is an edgeless negative dependency graph for the events A^Y.*

More generally, let Γ be a distributive lattice equipped with the uniform probability measure. For $Y \in \Gamma$, let

$$A^Y = \{Z \in \Gamma \mid Y \leq Z\}.$$

Theorem 6. *Let $\mathscr{Y}$ be a collection of elements of a distributive lattice Γ. The graph with vertex set $\{A^Y \mid Y \in \mathscr{Y}\}$ is an edgeless negative dependency graph for the events A^Y.*

Proof. Clearly Theorem 6 implies Theorem 5, if applied to the subset lattice. We have to show (5) for every $A_i = A^Y$ $(Y \in \mathscr{Y})$ and every $S \subseteq \Gamma \setminus \{Y\}$. Consider the indicator functions of the sets A^Y and $\vee_{U \in S} A^U$. These are increasing $\Gamma \to \mathbb{R}$ functions, to which the FKG inequality [10] applies, providing (5). Note that the FKG inequality follows from the even more general four functions theorem [1, 3]. The special case of (5) for the subset lattice already follows from [21].

2.5 Symmetric Events

We say that the events $A_1, A_2, \ldots, A_n$ are *symmetric* if the probability of any boolean expression of these events does not change if we substitute $A_{\pi(i)}$ to the place of A_i simultaneously for any permutation π of $[n]$. The following theorem was proved in [16]:

Theorem 7. *Assume that the events $A_1, A_2, \ldots, A_n$ are symmetric, and let p_i denote $\Pr(A_1 \wedge A_2 \wedge \ldots \wedge A_i)$ for $i = 1, 2, \ldots, n$, and let $p_0 = 1$. If the sequence p_i is logconvex, i.e., $p_k^2 \leq p_{k-1} p_{k+1}$ for $k = 1, 2, \ldots, n-1$, then these events have an empty negative dependency graph.*

3 Open Problems

3.1 Maximum Size Matchings in Graphs

The concept of canonical event and conflict, as defined in Sect. 2.1, can be extended in the case $k = 2$ for maximum size matchings in any graph G. Theorems 1 (for $k = 2$) and 3 can be interpreted that for the graphs $G = K_n$ and $K_{s,t}$, conflict of canonical events defines a negative dependency graph. Not every ambient graph will allow this result [17]. For example, for $G = C_6$, let e and f be any two opposite edges. Notice there are only two perfect matchings in C_6. We have that $\Pr\left(A^{\{e\}}\right) = \frac{1}{2}$, while $1 = \Pr\left(A^{\{e\}} \,\middle|\, \overline{A^{\{f\}}}\right) \not\leq \Pr\left(A^{\{e\}}\right)$. The 3-dimensional hypercube also fails this property. However, paths with even number of vertices have this property. Can we possibly classify the graphs that have this property?

3.2 Partition Lattice

The space of perfect matchings of K_N^k can be viewed as the space of partitions of an N-element set in which every block is of size k. Can we still find a negative dependency graph without this restriction on block sizes? To state this question more precisely, we will call a collection of disjoint subsets of an N-element set a *partial partition* and say that two partial partitions *conflict* whenever they have two classes neither disjoint nor identical, i.e., their union is not again a partial partition. (A partial partition may in fact fully partition the underlying set.) The ambient probability space is the space of all partitions of an N-element set (equipped with the uniform distribution) so that the canonical event A^M for a given partial partition M is the collection of all partitions extending M.

Conjecture 1. Let $\mathcal{M}$ be a collection of partial partitions of an N-element. The conflict graph for the collection of canonical events $\{A^M \mid M \in \mathcal{M}\}$ is a negative dependency graph.

Despite its apparent similarity to Theorem 1, the proof we gave cannot be applied when there are no restrictions on the block sizes. In particular, the necessary adaptation of Lemma 5, namely

$$\Pr_N\left(\bigwedge_{M \in \mathcal{M}} \overline{A_N^M}\right) \leq \Pr_{N+1}\left(\bigwedge_{M \in \mathcal{M}} \overline{A_{N+1}^M}\right),$$

may fail in some instances. For example, let $M_1 = \{\{1\},\{2\}\}$, $M_2 = \{\{1,\},\{3\}\}$, and $M_3 = \{\{2\},\{3\}\}$. One can compute by hand that $\Pr_3(\overline{A_3^{M_1}} \wedge \overline{A_3^{M_2}} \wedge \overline{A_3^{M_3}}) = \frac{4}{5}$, while $\Pr_4(\overline{A_4^{M_1}} \wedge \overline{A_4^{M_2}} \wedge \overline{A_4^{M_3}}) = \frac{11}{15}$. Theorem 6 is not going to help as the partition lattice is not distributive.

Let $M, M_1, \ldots, M_k$ be partial partitions of an N-element set such that M conflicts with none of the M_i (but M_i may conflict with M_j for $i \neq j$). The required $\Pr(A^M \mid \bigwedge_{i=1}^k \overline{A^{M_i}}) \leq \Pr(A^M)$ is equivalent to the inequality $\Pr(A^M)\Pr(\bigvee_{i=1}^k A^{M_i}) \leq \Pr(A^M \wedge (\bigvee_{i=1}^k A^{M_i}))$ (see (5)).

Let B_j denote the jth Bell number, which counts the number of partitions of a j-element set. The last inequality can be rewritten as

$$\left|A^M\right| \left|\bigcup_{i=1}^k A^{M_i}\right| \leq B_N \left|A^M \cap \left(\bigcup_{i=1}^k A^{M_i}\right)\right|. \tag{11}$$

$|A_M| = B_{N-|\cup M|}$, and the other two terms in (11) can be expressed by Bell numbers using inclusion–exclusion; however, the expression will depend heavily on which blocks the partial partitions have in common and whether they conflict. If we assume that all the partial partitions M_i have disjoint underlying sets and each cover exactly m elements, (11) turns into

$$B_{N-m} \sum_{i=1}^k (-1)^{i+1} \binom{k}{i} B_{N-i\cdot m} \leq B_N \sum_{i=1}^k (-1)^{i+1} \binom{k}{i} B_{N-(i+1)\cdot m}. \tag{12}$$

The inequality above has been verified asymptotically in N for small fixed values of m and k with Maple, using the modification of the Moser–Wyman formula for the Bell numbers found in [5]. [5] says that uniformly for $h = O(\ln n)$, as $n \to \infty$, $B_{n+h} = \frac{(n+h)!}{r^{n+h}} \frac{e^{e^r - 1}}{(2\pi B)^{1/2}} \left(1 + \frac{P_0 + hP_1 + h^2 P_2}{e^r} + \frac{Q_0 + hQ_1 + h^2 Q_2 + h^3 Q_3 + h^4 Q_4}{e^{2r}} + O(e^{-3r})\right)$, where $re^r = n$, $B = (r^2 + r)e^r$, P_i and Q_i are known rational functions of r. P_i and Q_i can be found explicitly in [4].

3.3 Permanent of Doubly Stochastic Matrices

Let $A = (a_{i,j})$ be an $n \times n$ doubly stochastic matrix with non-negative entries. For each $1 \leq i \leq n$, let X_i be independent random variables that select the element j from $\{1, \ldots, n\}$ with probability $a_{i,j}$. Define also B_i to be the event that $X_i = X_j$ for some $j \neq i$.

Conjecture 2. The collection of events $\{B_i \mid 1 \leq i \leq n\}$ are the vertices of an edgeless negative dependency graph.

This conjecture is relevant because of the continuing interest in lower bounds for the permanent. Computing the permanent is #P-hard [26] and is hard for the entire polynomial-time hierarchy [24]. Schrijver [20] was the first to give an interesting lower bound for the permanent in the form $\mathrm{per}(\tilde{A}) \geq \prod_{\substack{1\leq i\leq n \\ 1\leq j\leq n}}(1-a_{i,j})$, where $\tilde{A}$ is the matrix whose (i,j)th entry is $a_{i,j}(1-a_{i,j})$. Gurvits [12] has the current best lower bound, extending the ideas of [20]:

$$\mathrm{per}(A) \geq \prod_{\substack{1\leq i\leq n \\ 1\leq j\leq n}} (1-a_{i,j})^{1-a_{i,j}}. \tag{13}$$

Let us see what L4 gives, provided Conjecture 2 holds. Interpret X_i as selecting an entry j from row i of the doubly stochastic matrix. From this perspective, B_i is the event that, for some row $j \neq i$, the random variables X_i and X_j selected entries belonging to the same column for rows i and j. The product of n entries (one selected from each row) contributes to the permanent precisely when the chosen columns satisfy the event $\bigwedge_{i=1}^{n} \overline{B_i}$. Thus, $\mathrm{per}(A) = \Pr\left(\bigwedge_{i=1}^{n} \overline{B_i}\right) \geq \prod_{i=1}^{n}(1-\Pr(B_i)) = \prod_{i=1}^{n} \Pr\left(\overline{B_i}\right)$. Now, $\overline{B_i}$ is the event that, for all $k \neq i$, the value of X_i differs from the value of X_k. Letting $X_i = j$, the probability that $X_k \neq j$ is $1-a_{k,j}$ since the row sum is 1. Summing over j, we have

$$\Pr\left(\overline{B_i}\right) = \sum_{j=1}^{n} a_{i,j} \prod_{\substack{k=1 \\ k\neq i}}^{n} \left(1-a_{k,j}\right). \tag{14}$$

Finally, L4 would give the lower bound

$$\mathrm{per}(A) \geq \prod_{i=1}^{n} \sum_{j=1}^{n} a_{i,j} \prod_{\substack{k=1 \\ k\neq i}}^{n} \left(1-a_{k,j}\right). \tag{15}$$

It is interesting to compare our conjectured bound with the bound in [12]. [12] conjectures (13) makes the least fraction of the permanent of a doubly stochastic matrix on the matrix C, in which $c_{ii} = 1/2$ and for $i \neq j$ $c_{ij} = 1/(2n-2)$, with value $(\sqrt{2}+o(1))^{-n}$. Note that (13) evaluates to $(\sqrt{2e+o(1)})^{-n}$, while (15) evaluates to $(2\sqrt{e+o(1)})^{-n}$ on C, so our lower bound performs worse. However, (15) has terms more similar to the permanent than the terms in (13), possibly making easier to estimate the performance of the approximation.

Another interesting matrix to compare the bounds is $\frac{1}{n}J$, in which every entry is $1/n$. The famous van der Waerden conjecture stated that the permanent of non-negative doubly stochastic matrices is minimized on J, with $per(\frac{1}{n}J) = n!/n^n = (1+o(1))e^{-n}$. This conjecture was proved by Friedland [11] with $o(e^{-n})$ error term and exactly by Falikman [9] and Egorychev [6]. It is easy to see that both (13) and (15) evaluate $(1+o(1))e^{-n}$ on $\frac{1}{n}J$.

Evidence for the validity of Conjecture 2 is that it holds for $A = \frac{1}{n}J$ or A is a permutation matrix. For any fixed event B_i and any subset $S = \{k_1, k_2, \ldots, k_s\}$ of the vertices (with $i \notin S$), we have this generalization of (14):

$$\Pr\left(\bigwedge_{j\in S} \overline{B_j}\right) = \sum_{\substack{T\subseteq[n] \\ |T|=s}} \sum_{\substack{\pi:S\to T \\ injection}} b_{k_1\pi(k_1)} b_{k_2\pi(k_2)} \cdots b_{k_s\pi(k_s)} \prod_{\ell\notin S}\left(1-\sum_{t\in T} b_{\ell t}\right). \tag{16}$$

Using (16), for $A = \frac{1}{n}J$, the condition (4) boils down to

$$\Pr\left(\overline{B_i}\right) = \left(1 - \frac{1}{n}\right)^{n-1} \leq \left(1 - \frac{1}{n-s}\right)^{n-s-1}$$

$$= \frac{(n)_{s+1}n^{-s-1}(1 - \frac{s+1}{n})^{n-s-1}}{(n)_s n^{-s}(1 - \frac{s}{n})^{n-s}} = \Pr\left(\bigwedge_{j \in S \cup \{i\}} \overline{B_j} \,\middle|\, \bigwedge_{j \in S} \overline{B_j}\right).$$

Proving this inequality for arbitrary doubly stochastic A has so far eluded us. An alternative proof to Conjecture 2 with $A = \frac{1}{n}J$ is using Theorem 7 with B_i instead of A_i. Theorem 7 was designed for this in [16], although the context was estimating the number of injections.

Acknowledgment This research was supported in part by the NSF DMS contract 1000475. The third author acknowledges financial support from grant #FA9550-12-1-0405 from the U.S. Air Force Office of Scientific Research (AFOSR) and the Defense Advanced Research Projects Agency (DARPA). We thank Eva Czabarka for her useful suggestions to the manuscript.

References

1. Ahlswede, R., Daykin, D.E.: An inequality for the weights of two families of sets, their unions and intersections. Probab. Theor. Relat. Field. **43**(3), 183–185 (1978)
2. Albert, M., Frieze, A., Reed, B.: Multicolored Hamiltonian cycles. Electron. J. Combinator. **2**(#R10), 1–13 (1995)
3. Alon, N., Spencer, J.H.: The Probabilistic Method, 3rd edn. Wiley, New York (2008)
4. Canfield, E.R.: bellMoser.pdf, 6 pages manuscript, also see in http://www.math.sc.edu/~szekely/Aprilattemptformal.pdf
5. Canfield, E.R., Harper, L.H.: A simplified guide to large antichains in the partition lattice. Congressus Numerantium **100**, 81–88 (1994)
6. Egorychev, G.P.: The solution of van der Waerdens problem for permanents, Adv. Math. **42**, 299–305 (1981)
7. Erdős, P., Lovász, L.: Problems and results on 3-chromatic hypergraphs and some related questions. In: Hajnal, A. et. al. (eds.) Infinite and Finite Sets. Colloqium of the Mathematical Society János Bolyai, vol. 11, pp. 609–627. North Holland, Amsterdam (1975)
8. Erdős, P., Spencer, J.H.: Lopsided Lovász local lemma and latin transversals. Discrete Appl. Math. **30**, 151–154 (1991)
9. Falikman, D.I.: Proof of the van der Waerden conjecture on the permanent of a doubly stochastic matrix. Mat. Zametki **29**(6), 931–938 (1981) (in Russian)
10. Fortuin, C.M., Ginibre, J., Kasteleyn, P.N.: Correlation inequalities on some partially ordered sets. Commun. Math. Phys. **22**, 89–103 (1971)
11. Friedland, S.: A lower bound for the permanent of a doubly stochastic matrix. Ann. Math. **110**, 167–176 (1979)
12. Gurvits, L.: Unharnessing the power of Schrijver's permanental inequality. (2011). arXiv:1106.2844v5 [math.CO]
13. Kahn, J.: A normal law for matchings. Combinatorica **20**(3), 339–391 (2000)
14. Ku, C.Y.: Lovász local lemma. https://sites.google.com/site/kuchengyeaw/
15. Lovász, L.: Combinatorial Problems and Exercises. North-Holland, Amsterdam (1993)

16. Lu, L., Székely, L.A.: Using Lovász Local Lemma in the space of random injections. Electron. J. Combinator. **14**, R63 (2007)
17. Lu, L., Székely, L.A.: A new asymptotic enumeration technique: the Lovász local lemma. (2011) arXiv:0905.3983v3 [math.CO]
18. Mohr, A.: On negative dependency graphs in spaces of generalized random matchings (unpublished manuscript) (2011) http://austinmohr.com/work/negdep.pdf
19. Moser, R.A., Tardos, G.: A constructive proof of the general Lovász local lemma. Proceedings of the 41st annual ACM symposium on theory of computing (2009)
20. Schrijver, A.: Counting 1-factors in regular bipartite graphs. J. Comb. Theory, Series B **72**, 122–135 (1998)
21. Seymour, P.D.: On incomparable collections of sets, Mathematika **20**, 208–209 (1973)
22. Spencer, J.: Asymptotic lower bounds for Ramsey functions. Disc. Math. **20**, 69–76 (1977)
23. Spencer, J.H.: Ten lectures on the probabilistic method. Conference Board of the Mathematical Sciences, vol. 52, SIAM (1987)
24. Toda, S.: PP is as hard as the polynomial-time hierarchy. SIAM J. Comput. **20**, 865–877 (1991)
25. Tutte, W.T.: A problem on spanning trees. Quart. J. Math. Oxford **25**(2), 253–255 (1974)
26. Valiant, L.: The complexity of computing the permanent. Theretical Computer Science **8**(2), 189–201 (1979)

A Quantitative Open Mapping Theorem for Quasi-Pseudonormed Groups

Irina Mitrea, Marius Mitrea, and Elia Ziadé

Dedicated with friendship to Konstantin Oskolkov on the occasion of his 65th birthday.

Abstract We present a quantitative version of the classical Open Mapping Theorem in the context of groups equipped with topologies induced by quasi-pseudonorms, which are not necessarily group topologies.

1 Introduction

One of the cornerstones of modern functional analysis is the Open Mapping Theorem (OMT, for short), also known as the Banach-Schauder Theorem. This fundamental result states that *any continuous, surjective, linear operator between two Banach spaces is an open map*. More specifically, if X and Y are Banach spaces and if $T : X \to Y$ is a continuous, surjective, linear operator, then $T(U)$ is an open set in Y whenever U is an open set in X.

The proof of the classical OMT (cf., e.g., [8]) makes use of the Baire Category Theorem, and completeness of both X and Y is an essential hypothesis. For instance, the theorem may fail if either space is assumed to be just a normed space. Relevant examples, which are probably part of the folklore, are as follows. First, assume that $(Y, \|\cdot\|_Y)$ is a Banach space and pick an arbitrary linear functional Λ on Y. Define X to be the same vector space Y, this time equipped with the norm given by

---

I. Mitrea (✉)
Department of Mathematics, Temple University, Philadelphia, PA 19122, USA
e-mail: imitrea@temple.edu

M. Mitrea • E. Ziadé
Department of Mathematics, University of Missouri at Columbia, Columbia, MO 65211, USA
e-mail: mitream@missouri.edu; etzm7f@mail.missouri.edu

D. Bilyk et al. (eds.), *Recent Advances in Harmonic Analysis and Applications*, Springer Proceedings in Mathematics & Statistics 25, DOI 10.1007/978-1-4614-4565-4_22,

$\|x\|_X := \|x\|_Y + |\langle \Lambda, x\rangle|$. In this setting, consider T to be the identity operator. Then T is bijective and continuous since, trivially, $\|x\|_Y \leq \|x\|_X$. Now, if $(X, \|\cdot\|_X)$ were complete, this would entail, by the OMT, that T^{-1} is also bounded hence, ultimately, that $\|\cdot\|_X \approx \|\cdot\|_Y$. In turn, this would imply that Λ is continuous. However, there are Banach spaces (necessarily of infinite dimension) for which there exist unbounded linear functionals. This proves the necessity of the completeness of X in the context of the OMT. As for the necessity of Y being complete, a simple counterexample is provided by taking both X and Y to be the space of continuously differentiable functions on $[0,1]$, X equipped with the norm $\|f\|_X := \sup|f| + \sup|f'|$, Y equipped with the supremum norm, and T the identity operator.

This being said, the OMT continues to hold if both X and Y are F-spaces (see [4, Corollary 1.5, p. 10]). Recall that an F-space is a vector space furnished with a complete, additively invariant metric for which the scalar multiplication is separately continuous in each variable. It is therefore natural to search for other weaker contexts in which appropriate variants of the OMT are still valid.

The discovery in the early 1930s (cf. [1]) of the fact that a suitable version of OMT holds for complete metric groups (satisfying additional hypotheses) has eventually led to the development of a subbranch of functional analysis consisting of a collection of theorems aimed at identifying sufficient conditions ensuring that a surjective morphism (in a suitable algebraic/functional analytic framework) is open. Typically such theorems are still called "Open Mapping Theorems," and many important results of this sort are already known (cf., e.g., [2, 3, 5, 7], and the references therein).

As already mentioned, some of the most general formulations of the OMT are in the setting of complete metrizable topological groups. In this chapter, our goal is to further sharpen such results by weakening the hypotheses pertaining to the nature of the topologies considered on the groups in question. For the purpose of this introduction, a sample result (itself a consequence of a stronger theorem proved in the body of this chapter) is as follows:

Theorem 1. *Assume that G is a complete quasi-pseudonormed group and that S is a semi-topological group. Let $T \in \mathrm{Hom}(G,S)$ with closed graph be such that $\overline{T(U)}$ is a neighborhood of the neutral element in S for each neighborhood U of the neutral element in G. Then T is open.*

The reader is referred to Definition 10 for the notion of quasi-pseudonorm, and to Definition 5 for that of a semi-topological group. This result is strengthened in Theorem 7 which, in effect, constitutes the main result of this chapter. Theorem 7 may be naturally regarded as a quantitative version of the OMT, in a sense made precise in Comment 8. The proof of Theorem 7 makes essential use of the metrization theorem in the general setting of groupoids recently obtained in [6]. A summary of the key features of the work in [6] which are relevant for us here is presented in Theorem 6. Compared with the situation on metric groups, a quasi-norm may not necessarily be continuous in the topology it induces on the underlying group (cf. the discussion in Comment 4). Moreover, an artifact of direct repeated applications of a quasi-triangle inequality is that this typically yields multiplicative constants in the right-hand side which blow up exponentially as the number of iterations increases (see the discussion in Comment 5). On both accounts, Theorem 6 provides us with suitable surrogate alternatives.

Once established, Theorem 7 allows us to derive several corollaries which are more in line with the classical formulation of the OMT. As an example, we quote here the following:

Theorem 2. *Let G and S be two complete quasi-normed Abelian groups (equipped with the topologies induced by their respective quasi-norms). In addition, assume that S is torsion-free and divisible, and that either G is separable, or that*

$$G = \bigcup_{n\in\mathbb{N}} \{a^n : a \in V\} \tag{1}$$

for every symmetric neighborhood V of the neutral element in G and

$$\{a^n : a \in O\} \text{ is open in } S, \text{ for every open set } O \subseteq S. \tag{2}$$

Then any homomorphism of G onto S which is continuous is an open mapping.

This is a particular case of Theorems 9 and 10 in the body of this chapter. It is significant to note that the conditions stipulated on G and S are automatically satisfied in the case when these groups are the underlying additive groups of quasi-Banach spaces.

The organization of this chapter is as follows. In Sect. 2 we collect basic definitions and results of algebraical and topological nature, which are used throughout our present work. The concept of quasi-pseudonormed group is introduced and studied in Sect. 3 (cf. Definition 10 in particular). The main result of this chapter, Theorem 7 (which amounts to a quantitative OMT) is then formulated and proved in Sect. 4. In addition to the novel quantitative aspect of this result, it is perhaps remarkable that the groups considered in Theorem 7 are not assumed to be topological groups. This section also contains our main two corollaries of Theorem 7, namely Theorems 9 and 10, which may be viewed as more topologically flavored versions of the OMT.

In closing, the authors wish to thank the referee for his/her careful reading of the manuscript.

2 Algebraical and Topological Preliminaries

Let $\left(G, *, (\cdot)^{-1}, e_G\right)$ be a group, i.e., G is a set, $*$ is an associative binary operation on G, a^{-1} is the inverse of $a \in G$, and e_G is the neutral (or identity) element. In the subsequent discussion, we shall often abbreviate $\left(G, *, (\cdot)^{-1}, e_G\right)$ by $(G, *)$ and, occasionally, simply by G. Given two groups $(G, *)$, $(S, \circ)$, denote by $\mathrm{Hom}\,(G, S)$

the collection of all (group) homomorphisms from G to S. That is, $T \in \mathrm{Hom}(G,S)$ if and only if $T : G \to S$ satisfies

$$T(a*b) = (Ta) \circ (Tb), \qquad \forall a,b \in G. \tag{3}$$

In turn, (3) used with $a = b = e_G$ forces

$$Te_G = e_S, \qquad \forall T \in \mathrm{Hom}(G,S). \tag{4}$$

For each $T \in \mathrm{Hom}(G,S)$ we shall denote its graph by

$$\mathscr{G}_T := \{(a,Ta) : a \in G\} \subseteq G \times S. \tag{5}$$

Fix now a group $(G,*)$. For each $a \in G$ define

$$a^n := \underbrace{a*a*\cdots*a}_{n \text{ factors}}, \quad \forall n \in \mathbb{N}. \tag{6}$$

Also, for any two arbitrary subsets A, B of G introduce the notation

$$A*B := \{a*b : (a,b) \in A \times B\}, \qquad A^{-1} := \{a^{-1} : a \in A\}, \tag{7}$$

and call $A \subseteq G$ `symmetric` if $A = A^{-1}$. In particular, for $A \subseteq G$ and $a \in G$, we shall abbreviate

$$A*a := A*\{a\}, \quad a*A := \{a\}*A, \tag{8}$$

$$A^n := \underbrace{A*A*\cdots*A}_{n \text{ factors}}, \quad \forall n \in \mathbb{N}. \tag{9}$$

For each $A \subseteq G$ and each $n \in \mathbb{N}$, we shall also use the abbreviation

$$A^{(n)} := \{a^n : a \in A\}. \tag{10}$$

We stress that (10) is not to be confused with (9). Indeed, we have $A^{(n)} \subseteq A^n$ for every $A \subseteq G$, but, in general, the inclusion is strict.

We continue by recording a couple of definitions which are going to be relevant in Sect. 4.

Definition 1. Let $(G,*,(\cdot)^{-1},e_G)$ be a group. Define the `order` of a given element $a \in G$ as

$$\mathrm{ord}(a) := \inf\{n \in \mathbb{N} : a^n = e_G\}, \tag{11}$$

with the convention that $\inf \emptyset := +\infty$. An element $a \in G$ is said to be a `torsion element` if it has finite order. If the only torsion element in G is the identity element, then the group G is said to be `torsion-free`.

Definition 2. Call a group G `divisible` if for every $n \in \mathbb{N}$ and every $a \in G$ there exists $b \in G$ such that $b^n = a$.

Next, given a group $(G, *)$, for each $a \in G$, denote by s_a^R the right shift (or right translation) by a, i.e.,

$$s_a^R : G \longrightarrow G, \qquad s_a^R(x) := x * a, \quad \forall x \in G, \tag{12}$$

and by s_a^L the left shift (or left translation) by a, i.e.,

$$s_a^L : G \longrightarrow G, \qquad s_a^L(x) := a * x, \quad \forall x \in G. \tag{13}$$

Then, obviously,

$$s_{e_G}^R = s_{e_G}^L = \mathrm{id}_G, \text{ the identity mapping of } G, \tag{14}$$

$$s_a^R \circ s_b^R = s_{b*a}^R \text{ and } s_a^L \circ s_b^L = s_{a*b}^L, \quad \forall a, b \in G, \tag{15}$$

$$s_a^R, s_a^L \text{ are bijective, } \left(s_a^R\right)^{-1} = s_{a^{-1}}^R \text{ and } \left(s_a^L\right)^{-1} = s_{a^{-1}}^L, \quad \forall a \in G. \tag{16}$$

Also, the following intertwining identities hold

$$s_a^R \circ (\cdot)^{-1} = (\cdot)^{-1} \circ s_{a^{-1}}^L, \qquad s_a^L \circ (\cdot)^{-1} = (\cdot)^{-1} \circ s_{a^{-1}}^R, \qquad \forall a \in G. \tag{17}$$

We next describe the topology induced on a given group G by an arbitrary nonnegative (and possibly infinite) function ψ defined on G.

Definition 3. Given a group $(G, *, (\cdot)^{-1}, e_G)$ and a function $\psi : G \to [0, +\infty]$, define the right and left ψ-balls centered at $a \in G$ and with radius $r \in (0, +\infty)$, respectively, as

$$B_\psi^R(a, r) := \left\{x \in G : \psi(a * x^{-1}) < r\right\}, \tag{18}$$

$$B_\psi^L(a, r) := \left\{x \in G : \psi(x^{-1} * a) < r\right\}. \tag{19}$$

Then the `right-topology` τ_ψ^R induced by ψ on G is defined by

$$\tau_\psi^R := \left\{O \subseteq G : \forall a \in O\ \exists r \in (0, +\infty) \text{ such that } B_\psi^R(a, r) \subseteq O\right\}, \tag{20}$$

whereas the `left-topology` τ_ψ^L induced by ψ on G is defined by

$$\tau_\psi^L := \left\{O \subseteq G : \forall a \in O\ \exists r \in (0, +\infty) \text{ such that } B_\psi^L(a, r) \subseteq O\right\}. \tag{21}$$

Of particular interest is the case when a group is equipped with a topology which is compatible with the underlying algebraic structures, as described in the definition below.

Definition 4. A `topological group` is a group $(G, *, (\cdot)^{-1}, e_G)$ endowed with a topology τ on the set G with the property that the group operations $(\cdot)^{-1} : G \to G$ and $* : G \times G \to G$ are continuous functions (in the latter case, considering the product topology $\tau \times \tau$ on $G \times G$).

Alternatively, τ is called a group topology on G provided $(G, *, \tau)$ is a topological group.

Of course, given a group $(G, *)$ and a topology τ on G, in order for $(G, *, \tau)$ to be a topological group, it suffices to have

$$\big(G \times G, \tau \times \tau\big) \ni (x, y) \mapsto x * y^{-1} \in (G, \tau) \ \text{ continuous.} \tag{22}$$

Nonetheless, in the sequel, we shall work with groups equipped with topologies which satisfy weaker conditions than those required to render them topological groups. This is made precise in our next definition.

Definition 5. A `semi-topological group` is a group $\big(G, *, (\cdot)^{-1}, e_G\big)$ endowed with a topology τ on the set G with the property that the inverse operation as well as all right shifts are continuous. That is,

$$(\cdot)^{-1} : (G, \tau) \longrightarrow (G, \tau) \ \text{ and } \ s_a^R : (G, \tau) \longrightarrow (G, \tau), \quad \forall a \in G, \tag{23}$$

are continuous functions.

A few comments are in order.

Remark 1. (i) Obviously, any topological group is a semi-topological group.

(ii) Thanks to the intertwining identities from (17), the fact that the functions in (23) are continuous implies that the left shifts $s_a^L : (G, \tau) \to (G, \tau)$, $a \in G$, are continuous as well.

(iii) In any semi-topological group, the inverse operation as well as all right shifts and left shifts are in fact homeomorphisms (as seen from (16)).

At this point, we shall briefly digress for the purpose of introducing some useful notation. Given an arbitrary topological space (X, τ), denote by $\mathscr{N}(x; \tau)$ the family of neighborhoods of the point $x \in X$, relative to the topology τ. Also, for any $A \subseteq X$, we shall let $\mathrm{Int}(A; \tau)$ and $\mathrm{Clo}(A; \tau)$ denote, respectively, the interior and closure of the set A, relative to the topology τ. We shall make frequent use of the fact that if (X_1, τ_1) and (X_2, τ_2) are topological spaces and $f : X_1 \to X_2$ is a homeomorphism, then

$$f\big(\mathrm{Clo}(A; \tau_1)\big) = \mathrm{Clo}(f(A); \tau_2), \quad \forall A \subseteq X_1, \tag{24}$$

$$f\big(\mathrm{Int}(A; \tau_1)\big) = \mathrm{Int}(f(A); \tau_2), \quad \forall A \subseteq X_1, \tag{25}$$

$$f(U) \in \mathscr{N}\big(f(x); \tau_2\big), \quad \forall x \in X_1 \ \text{ and } \ \forall U \in \mathscr{N}(x; \tau_1). \tag{26}$$

Indeed, if $A \subseteq X_1$ is an arbitrary set, the continuity of the function f implies $f(\mathrm{Clo}(A;\tau_1)) \subseteq \mathrm{Clo}(f(A);\tau_2)$. Since f^{-1} is also continuous, we may now use the inclusion just established in order to obtain that $f^{-1}(\mathrm{Clo}(f(A);\tau_2)) \subseteq \mathrm{Clo}(A;\tau_1)$, by replacing f with its inverse f^{-1} and A with $f(A)$. The fact that f is bijective then implies $\mathrm{Clo}(f(A);\tau_2) \subseteq f(\mathrm{Clo}(A;\tau_1))$, finishing the proof of (24). Finally, (25) follows easily from (24) by taking complements, while (26) is a direct consequence of the fact that f is open.

Lemma 1. *Let $(G,*,(\cdot)^{-1},e_G,\tau)$ be a semi-topological group. Then*

$$U^{-1} \in \mathscr{N}(e_G;\tau) \text{ and } U*V \in \mathscr{N}(e_G;\tau), \quad \forall U,V \in \mathscr{N}(e_G;\tau), \tag{27}$$

$$\begin{aligned}\mathrm{Clo}(A;\tau) &= \bigcap_{V\in\mathscr{N}(e_G;\tau)} V*A = \bigcap_{V\in\mathscr{N}(e_G;\tau)} A*V \\ &= \bigcap_{V,W\in\mathscr{N}(e_G;\tau)} V*A*W, \quad \forall A \subseteq G,\end{aligned} \tag{28}$$

$$\mathrm{Clo}(A;\tau)*\mathrm{Clo}(B;\tau) \subseteq \mathrm{Clo}(A*B;\tau), \qquad \forall A,B \subseteq G. \tag{29}$$

Proof. Given any $U \in \mathscr{N}(e_G;\tau)$, part (*iii*) in Remark 1 together with (26) prove that $U^{-1} \in \mathscr{N}(e_G;\tau)$. For any sets $U,V \in \mathscr{N}(e_G;\tau)$, observe that $U*V \in \mathscr{N}(e_G;\tau)$ since $\mathscr{N}(e_G;\tau) \ni U \subseteq U*V$, given that $e_G \in V$. This proves (27). Next, if $A \subseteq G$ is arbitrary, then for each $a \in G$, we may write (by once again relying on part (*iii*) in Remark 1 and (26)) that

$$\begin{aligned}a \notin \mathrm{Clo}(A;\tau) &\Longleftrightarrow \exists V \in \mathscr{N}(e_G;\tau) \text{ such that } (V*a)\cap A = \emptyset \\ &\Longleftrightarrow \exists V \in \mathscr{N}(e_G;\tau) \text{ such that } a \notin V^{-1}*A.\end{aligned} \tag{30}$$

In concert with the first formula in (27) this proves the first equality in (28). The second equality in (28) is justified in a similar manner. Moreover, having established these two equalities, we may now make use of them and the second formula in (27) in order to write

$$\begin{aligned}\bigcap_{V,W\in\mathscr{N}(e_G;\tau)} V*A*W &= \bigcap_{V\in\mathscr{N}(e_G;\tau)} \Bigg(\bigcap_{W\in\mathscr{N}(e_G;\tau)} (V*A)*W \Bigg) \\ &= \bigcap_{V\in\mathscr{N}(e_G;\tau)} \mathrm{Clo}\,(V*A;\tau) \\ &= \bigcap_{V\in\mathscr{N}(e_G;\tau)} \Bigg(\bigcap_{W\in\mathscr{N}(e_G;\tau)} W*(V*A) \Bigg)\end{aligned}$$

$$= \bigcap_{V,W\in\mathcal{N}(e_G;\tau)} W*V*A$$

$$\subseteq \bigcap_{U\in\mathcal{N}(e_G;\tau)} U*A = \mathrm{Clo}\,(A;\tau). \tag{31}$$

On the other hand, since $e_G \in W$ for every $W \in \mathcal{N}(e_G;\tau)$, it follows that

$$\mathrm{Clo}\,(A;\tau) = \bigcap_{V\in\mathcal{N}(e_G;\tau)} V*A \subseteq \bigcap_{V,W\in\mathcal{N}(e_G;\tau)} V*A*W. \tag{32}$$

Now, the last equality in (28) is a consequence of (31)–(32). Finally, given two arbitrary sets $A,B \subseteq G$, we have

$$\mathrm{Clo}(A;\tau)*\mathrm{Clo}(B;\tau) = \left(\bigcap_{V\in\mathcal{N}(e_G;\tau)} V*A\right)*\left(\bigcap_{W\in\mathcal{N}(e_G;\tau)} B*W\right)$$

$$\subseteq \bigcap_{V,W\in\mathcal{N}(e_G;\tau)} V*A*B*W = \mathrm{Clo}(A*B;\tau), \tag{33}$$

by (28). This yields (29) and finishes the proof of the lemma. □

In our next lemma, we explore the extent to which the topology induced by a nonnegative function defined on a given group is compatible with the algebraic structure. Before stating this, we recall one piece of terminology. Given a topological space (X,τ) and a point $x \in X$, call a subset $\mathcal{U}$ of $\mathcal{N}(x;\tau)$ a `fundamental system of neighborhoods of` x (in the topology τ) if for every $V \in \mathcal{N}(x;\tau)$ there exists $U \in \mathcal{U}$ such that $U \subseteq V$.

Lemma 2. *Assume that $(G,*)$ is a group and that $\psi : G \to [0,+\infty]$ is an arbitrary function.*

(i) Both τ_ψ^R and τ_ψ^L are topologies on G.
(ii) For each $a \in G$, the shifts

$$s_a^R : (G,\tau_\psi^R) \longrightarrow (G,\tau_\psi^R), \qquad s_a^L : (G,\tau_\psi^L) \longrightarrow (G,\tau_\psi^L), \tag{34}$$

are homeomorphisms. Furthermore, for each $x \in G$ and $r \in (0,+\infty)$,

$$s_a^R\left(B_\psi^R(x,r)\right) = B_\psi^R(s_a^R(x),r) \quad \text{and} \quad s_a^L\left(B_\psi^L(x,r)\right) = B_\psi^L(s_a^L(x),r). \tag{35}$$

(iii) One has

$$a \in B_\psi^R(a,r)\ \forall a \in G,\ \forall r>0 \iff \psi(e_G)=0 \tag{36}$$

$$\iff a \in B_\psi^L(a,r)\ \forall a \in G,\ \forall r>0.$$

(iv) The mappings

$$(\cdot)^{-1} : (G, \tau_\psi^R) \longrightarrow (G, \tau_\psi^L), \qquad (\cdot)^{-1} : (G, \tau_\psi^L) \longrightarrow (G, \tau_\psi^R) \tag{37}$$

are homeomorphisms (which are inverse to one another) provided the function ψ has the property that

$$\begin{gathered} \forall \varepsilon > 0 \ \exists \delta > 0 \ \textit{such that} \\ \textit{if } x \in G \ \textit{satisfies } \ \psi(x) < \delta, \ \textit{then } \ \psi(x^{-1}) < \varepsilon. \end{gathered} \tag{38}$$

(v) Assume that the function ψ has the property that

$$\begin{gathered} \forall \varepsilon > 0 \ \exists \delta > 0 \ \textit{such that} \\ \psi(x * y) < \varepsilon \ \textit{if } x, y \in G \ \textit{satisfy } \ \psi(x) < \delta, \ \psi(y) < \delta. \end{gathered} \tag{39}$$

Then for every $A \subseteq G$, one has

$$\mathrm{Int}(A; \tau_\psi^R) = \{a \in A : \exists r \in (0, +\infty) \ \textit{such that } \ B_\psi^R(a, r) \subseteq A\}, \tag{40}$$

$$\mathrm{Int}(A; \tau_\psi^L) = \{a \in A : \exists r \in (0, +\infty) \ \textit{such that } \ B_\psi^L(a, r) \subseteq A\}. \tag{41}$$

Furthermore, if ψ satisfies (39) and $\psi(e_G) = 0$, then

$$\begin{gathered} \forall a \in G, \ \forall r > 0 \ \textit{there holds} \\ B_\psi^R(a, r) \in \mathscr{N}(a; \tau_\psi^R) \quad \textit{and} \quad B_\psi^L(a, r) \in \mathscr{N}(a; \tau_\psi^L). \end{gathered} \tag{42}$$

In fact, in this scenario, for each $a \in G$,

$$\left\{B_\psi^R(a, r)\right\}_{r>0} \textit{ is a fundamental system of neighborhoods of } a, \textit{ in } \tau_\psi^R, \tag{43}$$

and

$$\left\{B_\psi^L(a, r)\right\}_{r>0} \textit{ is a fundamental system of neighborhoods of } a, \textit{ in } \tau_\psi^L. \tag{44}$$

(vi) If the function ψ satisfies (38), (39), as well as

$$\psi^{-1}(\{0\}) = \{e_G\}, \tag{45}$$

then the topological spaces (G, τ_ψ^R) and (G, τ_ψ^L) are Hausdorff.

(vii) If the function ψ has the property that

$$\forall \varepsilon > 0 \ \exists \delta > 0 \ \textit{so that } \ \psi(y * x) < \varepsilon \ \textit{if } x, y \in G \ \textit{satisfy } \ \psi(x * y) < \delta, \tag{46}$$

then

$$\tau_\psi^R = \tau_\psi^L. \tag{47}$$

(viii) Suppose that (46) *holds and, in such a scenario, set* $\tau_\psi := \tau_\psi^R (= \tau_\psi^L)$*. In addition, assume that the function* ψ *satisfies* (38) *and* (39)*. Then*

$$\big(G, *, \tau_\psi\big) \ \text{is a topological group.} \tag{48}$$

Proof. It is clear from definitions that τ_ψ^R, τ_ψ^L are topologies on G. Also, the fact that the mappings in (34) are homeomorphisms follows from (16) and (20)–(21) after observing that (recall the convention made in (8))

$$B_\psi^R(a,r) * b = B_\psi^R(a*b,r), \qquad b * B_\psi^L(a,r) = B_\psi^L(b*a,r), \tag{49}$$

for all $a,b \in G$ and $r \in (0,+\infty)$. As a byproduct, the identities in (35) also follow. Also, the equivalences in (36) are clear from definitions. Going further, one can check without difficulty that (recall the piece of notation introduced in (7))

$$\Big(B_\psi^R(a,r)\Big)^{-1} = B^L_{\psi\circ(\cdot)^{-1}}(a^{-1},r) \ \text{ and } \ \Big(B_\psi^L(a,r)\Big)^{-1} = B^R_{\psi\circ(\cdot)^{-1}}(a^{-1},r) \tag{50}$$

for every $a \in G$ and $r \in (0,+\infty)$. In light of (50), condition (38) then becomes equivalent to the demand that

$$\begin{gathered} \forall \varepsilon > 0 \ \exists \delta > 0 \ \text{ such that for each } \ a \in G, \ \text{ one has} \\ B_\psi^L(a^{-1},\delta) \subseteq \Big(B_\psi^R(a,\varepsilon)\Big)^{-1} \ \text{ and } \ B_\psi^R(a^{-1},\delta) \subseteq \Big(B_\psi^L(a,\varepsilon)\Big)^{-1}. \end{gathered} \tag{51}$$

With this in hand, it readily follows that the maps in (37) are homeomorphisms if (38) holds. Moving on, consider (40) for some fixed $A \subseteq G$. In order to facilitate the presentation we shall temporarily use the notation

$$\widetilde{A} := \big\{a \in A : \exists r \in (0,+\infty) \ \text{ such that } \ B_\psi^R(a,r) \subseteq A\big\}. \tag{52}$$

In this regard, we make the claim that $\widetilde{A} \in \tau_\psi^R$. To justify this claim, pick some $a \in \widetilde{A}$. Then there exists $r > 0$ such that $B_\psi^R(a,r) \subseteq A$, and (39) guarantees the existence of some $\delta > 0$ with the property that

$$\psi(x*y) < r \ \text{ whenever } \ x,y \in G \ \text{ are such that } \psi(x) < \delta, \ \psi(y) < \delta. \tag{53}$$

Consider now two arbitrary elements, $b \in B_\psi^R(a,\delta)$ and $c \in B_\psi^R(b,\delta)$. Then we have $\psi(a*c^{-1}) = \psi\big((a*b^{-1})*(b*c^{-1})\big) < r$ by (53) and the fact that $\psi(a*b^{-1}) < \delta$ and $\psi(b*c^{-1}) < \delta$. This proves that $B_\psi^R(b,\delta) \subseteq B_\psi^R(a,r)$, hence $B_\psi^R(b,\delta) \subseteq A$, for every element $b \in B^R_\psi(a,\delta)$. Thus, ultimately, $B^R_\psi(a,\delta) \subseteq \widetilde{A}$, proving that $\widetilde{A} \in \tau^R_\psi$. Since by design $\widetilde{A} \subseteq A$, we may therefore conclude that $\widetilde{A} \subseteq \mathrm{Int}(A;\tau^R_\psi)$. In the converse direction, if $a \in \mathrm{Int}(A;\tau^R_\psi)$, then $a \in A$, and there exists $r > 0$ such that $B^R_\psi(a,r) \subseteq A$. Hence, based on this and the definition of $\widetilde{A}$, we obtain $\mathrm{Int}(A;\tau^R_\psi) \subseteq \widetilde{A}$. This finishes the proof of (40), and (41) is established similarly. Next, if ψ satisfies (39) and $\psi(e_G) = 0$, then, thanks to (40)–(41), we have

$$a \in \mathrm{Int}\big(B^R_\psi(a,r);\tau^R_\psi\big) \cap \mathrm{Int}\big(B^L_\psi(a,r);\tau^L_\psi\big), \qquad \forall\, a \in G,\ \forall\, r > 0, \tag{54}$$

from which (42) follows. In turn, (43)–(44) are immediate consequences of (42).

Suppose now that the function ψ satisfies (38), (39), as well as (45), and let $a,b \in G$ with $a \neq b$. Then $a * b^{-1} \neq e_G$ and, as such, $\psi(a * b^{-1}) > 0$. Pick next $\varepsilon \in \big(0, \psi(a * b^{-1})\big)$ and let $\delta > 0$ be associated with this ε as in (39). Also, making use of (38), select $\delta_1 \in (0,\delta)$ such that

$$\psi(x^{-1}) < \delta \ \text{ if } x \in G \text{ satisfies } \psi(x) < \delta_1. \tag{55}$$

We claim that

$$B^R_\psi(a,\delta_1) \cap B^R_\psi(b,\delta_1) = \emptyset. \tag{56}$$

To see this, reason by contradiction and assume that there exists an element $c \in B^R_\psi(a,\delta_1) \cap B^R_\psi(b,\delta_1)$. Then, since $\psi(b * c^{-1}) < \delta_1$, it follows from property (55) that $\psi(c * b^{-1}) < \delta$. With this in hand we have, thanks to our choice of δ and the fact that $\delta_1 < \delta$,

$$\psi(a * b^{-1}) = \psi\big((a * c^{-1}) * (c * b^{-1})\big) < \varepsilon. \tag{57}$$

This contradiction proves (56). Keeping in mind (42), we may then conclude from (56) that the topological space (G,τ^R_ψ) is Hausdorff. Moreover, a similar reasoning applies in the case of (G,τ^L_ψ).

Next, condition (46) is equivalent to the demand that

$$\begin{gathered} \forall\, \varepsilon > 0\ \exists\, \delta > 0 \ \text{ such that for each } \ a \in G, \ \text{ one has} \\ B^L_\psi(a,\delta) \subseteq B^R_\psi(a,\varepsilon) \ \text{ and } \ B^R_\psi(a,\delta) \subseteq B^L_\psi(a,\varepsilon), \end{gathered} \tag{58}$$

which, in turn, readily yields (47). Finally, if (46) and (38) hold, then from (47) and (37) we know that the mapping $(\cdot)^{-1} : \big(G,\tau_\psi\big) \to \big(G,\tau_\psi\big)$ is a homeomorphism. Thus, as far as (48) is concerned, there remains to show that

$$p : (G \times G, \tau_\psi \times \tau_\psi) \longrightarrow (G,\tau_\psi),\ p(x,y) := x * y,\ \forall\, x,y \in G, \ \text{ is continuous,} \tag{59}$$

i.e., that $p^{-1}(O) \in \tau_\psi \times \tau_\psi$ for each $O \in \tau_\psi$. Unraveling definitions, it is apparent that it suffices to check that for every $x_o, y_o \in G$ and any $r > 0$ there exists $\varepsilon > 0$ with the property that

$$B^R_\psi(x_o,\varepsilon) * B^R_\psi(y_o,\varepsilon) \subseteq B^R_\psi(x_o * y_o, r). \tag{60}$$

To this end, fix $x_o, y_o \in G$ along with $r > 0$ and, for some $\varepsilon > 0$ to be specified later, select two arbitrary elements $x \in B^R_\psi(x_o,\varepsilon)$ and $y \in B^R_\psi(y_o,\varepsilon)$. The goal is to specify ε such that we necessarily have $\psi\big(x_o * y_o * (x*y)^{-1}\big) < r$. The latter condition is equivalent to $\psi\big(x_o * y_o * y^{-1} * x^{-1}\big) < r$, and, granted (46), there exists $\delta_1 > 0$ such that this is true provided the inequality $\psi\big(x^{-1} * x_o * y_o * y^{-1}\big) < \delta_1$ holds. In turn, granted (39), there exists $\delta_2 > 0$ such this inequality holds if

$$\psi\big(x^{-1} * x_o\big) < \delta_2 \ \text{ and } \ \psi\big(y_o * y^{-1}\big) < \delta_2. \tag{61}$$

Note that $y \in B^R_\psi(y_o,\varepsilon)$ forces $\psi\big(y_o * y^{-1}\big) < \varepsilon$, so the second inequality in (61) is automatically satisfied if $\varepsilon < \delta_2$. Moreover, thanks to (46), there exists $\delta_3 > 0$ such that the first inequality in (61) holds provided $\psi\big(x_o * x^{-1}\big) < \delta_3$. However, given that $x \in B^R_\psi(x_o,\varepsilon)$, this last inequality is going to hold provided $\varepsilon < \delta_3$. All in all, (60) is verified if we choose $\varepsilon \in \big(0, \min\{\delta_2,\delta_3\}\big)$. This completes the proof of (48). □

We conclude this section by making a series of definitions which are going to be relevant in the next section.

Definition 6. Let $(G,*)$ be a group and suppose that $\psi : G \to [0,+\infty]$ is an arbitrary function. Then for each $\varepsilon > 0$ define the `right` ε`-enhancement` (with respect to ψ) of a given set $A \subseteq G$ as

$$[A]^R_{\psi,\varepsilon} := \bigcup_{a\in A} B^R_\psi(a,\varepsilon) \tag{62}$$

and define the `left` ε`-enhancement` (with respect to ψ) of a given set $A \subseteq G$ as

$$[A]^L_{\psi,\varepsilon} := \bigcup_{a\in A} B^L_\psi(a,\varepsilon). \tag{63}$$

Definition 7. Let $(G,*)$ be a group and suppose that τ is a topology on the set G. Also, suppose that a sequence $(a_i)_{i\in\mathbb{N}} \subseteq G$ and an element $a \in G$ have been given. Call the sequence $(a_i)_{i\in\mathbb{N}}$ `right-convergent to` a with respect to τ provided

$$\forall\, O \in \mathcal{N}(e_G;\tau)\ \exists\, n_O \in \mathbb{N} \ \text{such that}\ a_i * a^{-1} \in O,\ \forall\, i \in \mathbb{N} \ \text{with}\ i \geq n_O. \tag{64}$$

Also, call the sequence $(a_i)_{i\in\mathbb{N}}$ `left-convergent to` a with respect to τ provided

$$\forall\, O \in \mathcal{N}(e_G;\tau)\ \exists\, n_O \in \mathbb{N} \ \text{such that}\ a^{-1} * a_i \in O,\ \forall\, i \in \mathbb{N} \ \text{with}\ i \geq n_O. \tag{65}$$

Definition 8. Let $(G,*)$ be a group and suppose that τ is a topology on the set G. Call a sequence $(a_i)_{i\in\mathbb{N}} \subseteq G$ `right-Cauchy with respect to` τ provided

$$\begin{aligned} &\forall O \in \mathscr{N}(e_G;\tau)\ \exists n_O \in \mathbb{N} \text{ such that} \\ &a_i * a_j^{-1} \in O,\ \forall i,j \in \mathbb{N} \text{ with } \min\{i,j\} \geq n_O. \end{aligned} \tag{66}$$

Also, call a sequence $(a_i)_{i\in\mathbb{N}} \subseteq G$ `left-Cauchy with respect to` τ provided

$$\begin{aligned} &\forall O \in \mathscr{N}(e_G;\tau)\ \exists n_O \in \mathbb{N} \text{ such that} \\ &a_i^{-1} * a_j \in O,\ \forall i,j \in \mathbb{N} \text{ with } \min\{i,j\} \geq n_O. \end{aligned} \tag{67}$$

Definition 9. Let $(G,*)$ be a group and suppose that τ is a topology on the set G. Call G `right-complete with respect to` τ provided every sequence $(a_i)_{i\in\mathbb{N}} \subseteq G$ which is right-Cauchy with respect to τ is also right-convergent in the topology τ to some element $a \in G$.

Likewise, call G `left-complete with respect to` τ provided every sequence $(a_i)_{i\in\mathbb{N}} \subseteq G$ which is left-Cauchy with respect to τ is also left-convergent in the topology τ to some element $a \in G$.

3 Quasi-Pseudonormed Groups

Recall that, given a group $\big(G,*,(\cdot)^{-1},e_G\big)$, a function $\psi : G \to [0,+\infty)$ is called a `norm` on G provided

$$(i)\ \psi^{-1}(\{0\}) = \{e_G\}, \tag{68}$$

$$(ii)\ \psi(x^{-1}) = \psi(x), \quad \forall x \in G, \tag{69}$$

$$(iii)\ \psi(x*y) \leq \psi(x) + \psi(y), \quad \forall x,y \in G. \tag{70}$$

Moreover, a norm ψ on G is said to be `invariant` provided

$$\psi(x^{-1} * y * x) = \psi(y), \quad \forall x,y \in G. \tag{71}$$

When axiom (i) above is relaxed to $\psi(e_G) = 0$ while retaining (ii) – (iii) as stated above, the corresponding function ψ is typically called a `pseudonorm` (in the sense of Markov). In many situations of practical interest it is desirable to also weaken axioms (ii) and (iii) as well as allow the function ψ to eventually be infinite. We thus arrive at the notion of quasi-pseudonorm on a group, formally introduced below.

Definition 10. Let $\left(G,*,(\cdot)^{-1},e_G\right)$ be a group. Call a function $\psi : G \to [0,+\infty]$ a `quasi-pseudonorm` on G provided there exist $C_0, C_1 \in [1,+\infty)$ with the property that

$$\psi(e_G) = 0, \tag{72}$$

$$\psi(x^{-1}) \leq C_0\, \psi(x), \quad \forall x \in G, \tag{73}$$

$$\psi(x*y) \leq C_1\left(\psi(x)+\psi(y)\right), \quad \forall x,y \in G. \tag{74}$$

The pair (C_0, C_1) appearing in (73)–(74) will be referred to as the constants of the quasi-pseudonorm ψ.

A quasi-pseudonorm ψ is said to be `quasi-invariant` if there exists a constant $C_2 \in [1,+\infty)$ with the property that

$$\psi(x^{-1}*y*x) \leq C_2\, \psi(y), \quad \forall x,y \in G, \tag{75}$$

and call the quasi-pseudonorm ψ `finite` provided ψ actually takes values in $[0,+\infty)$.

Finally, a quasi-pseudonorm ψ on G is called a `quasi-norm` provided property (72) is strengthened to $\psi^{-1}(\{0\}) = \{e_G\}$, and a quasi-norm is `finite` or `quasi-invariant` provided it is so when viewed as a quasi-pseudonorm.

Comment 3. Compared with Definition 10, a slightly more economical (though ultimately equivalent) way of introducing the notion of quasi-pseudonorm on a group $\left(G,*,(\cdot)^{-1},e_G\right)$ is by stipulating that $\psi : G \to (-\infty,+\infty]$ is a function satisfying (72)–(74) for some constants $C_0, C_1 \in [1,+\infty)$. The fact that such a function is necessarily nonnegative is seen by writing, for each $x \in G$,

$$0 = \psi(e_G) = \psi(x*x^{-1}) \leq C_1\left(\psi(x)+\psi(x^{-1})\right) \leq C_1(1+C_0)\psi(x), \tag{76}$$

which forces $\psi(x) \geq 0$, as desired. ■

Conditions (73) and (74) in the definition of a quasi-pseudonorm may be naturally regarded as quantitative versions of the (topologically flavored) conditions (38) and (39), respectively, from Lemma 2. Likewise, the quasi-invariant condition (75) is a quantitative version of (46). These observations are at the core of the following result pertaining to the properties of quasi-pseudonorms on arbitrary groups.

Proposition 1. *Assume that $(G,*)$ is a group and that ψ is a quasi-pseudonorm on G.*

(i) The mappings

$$(\cdot)^{-1} : (G,\tau_\psi^R) \longrightarrow (G,\tau_\psi^L), \qquad (\cdot)^{-1} : (G,\tau_\psi^L) \longrightarrow (G,\tau_\psi^R), \tag{77}$$

are homeomorphisms (which are inverse to one another).

(ii) *For every $A \subseteq G$, one has*

$$\mathrm{Int}\big(A;\tau_\psi^R\big) = \big\{a \in A : \exists\, r \in (0,+\infty) \ \text{such that} \ B_\psi^R(a,r) \subseteq A\big\}, \tag{78}$$

$$\mathrm{Int}\big(A;\tau_\psi^L\big) = \big\{a \in A : \exists\, r \in (0,+\infty) \ \text{such that} \ B_\psi^L(a,r) \subseteq A\big\}. \tag{79}$$

Furthermore,

$$B_\psi^R(a,r) \in \mathcal{N}\big(a;\tau_\psi^R\big) \quad \text{and} \quad B_\psi^L(a,r) \in \mathcal{N}\big(a;\tau_\psi^L\big), \quad \forall a \in G,\ \forall r > 0. \tag{80}$$

In fact, for every $a \in G$ and every $r \in (0,+\infty)$,

$$\bigcup_{C>C_1} B_\psi^R(a,r/C) \subseteq \mathrm{Int}\big(B_\psi^R(a,r);\tau_\psi^R\big), \tag{81}$$

$$\bigcup_{C>C_1} B_\psi^L(a,r/C) \subseteq \mathrm{Int}\big(B_\psi^L(a,r);\tau_\psi^L\big). \tag{82}$$

(iii) *For each $a \in G$,*

$$\Big\{B_\psi^R(a,r)\Big\}_{r>0} \ \text{is a fundamental system of neighborhoods of } a, \text{ in } \tau_\psi^R, \tag{83}$$

$$\Big\{B_\psi^L(a,r)\Big\}_{r>0} \ \text{is a fundamental system of neighborhoods of } a, \text{ in } \tau_\psi^L, \tag{84}$$

where $C_1 \in [1,+\infty)$ is as in (74).

(iv) *A sequence $(a_i)_{i\in\mathbb{N}} \subseteq G$ is right-convergent to $a \in G$ with respect to τ_ψ^R if and only if for every $\varepsilon \in (0,+\infty)$ there exists $n_\varepsilon \in \mathbb{N}$ such that $a_i \in B_\psi^R(a,\varepsilon)$ whenever $i \in \mathbb{N}$ satisfies $i \geq n_\varepsilon$. Furthermore, a similar characterization of left-convergence with respect to the topology τ_ψ^L holds.*

(v) *A sequence $(a_i)_{i\in\mathbb{N}} \subseteq G$ is right-Cauchy with respect to τ_ψ^R if and only if it is left-Cauchy with respect to τ_ψ^L. Moreover, either of these conditions is equivalent to the demand that for every $\varepsilon \in (0,+\infty)$ there exists $n_\varepsilon \in \mathbb{N}$ such that $\psi(a_i * a_j^{-1}) < \varepsilon$ whenever $i, j \in \mathbb{N}$ satisfy $i \geq j \geq n_\varepsilon$.*

(vi) *One has*

$$G \text{ is right-complete with respect to } \tau_\psi^R \iff G \text{ is left-complete with respect to } \tau_\psi^L. \tag{85}$$

(vii) *If the quasi-pseudonorm ψ is actually quasi-invariant, then $\tau_\psi^R = \tau_\psi^L =: \psi_\tau$. Moreover, in such a scenario,*

$$\big(G,*,\tau_\psi\big) \ \text{is a topological group}. \tag{86}$$

Proof. With the exception of (81)–(82), all other properties follow directly from Lemma 2 and the comments made just prior to the statement of the proposition. To justify (81), suppose that $a \in G$ and $r \in (0,+\infty)$ are given. Pick a number $C > C_1$ along with some element $b \in B^R_\psi(a,r/C)$. In particular, $\psi(a * b^{-1}) < r/C$. Now, if $\varepsilon \in \big(0, r(C_1^{-1} - C^{-1})\big)$ and $c \in B^R_\psi(b,\varepsilon)$, it follows that

$$\psi(a * c^{-1}) = \psi\big((a * b^{-1}) * (b * c^{-1})\big) \leq C_1\big(\psi(a * b^{-1}) + \psi(b * c^{-1})\big)$$
$$< C_1\big(r/C + \varepsilon\big) < r. \tag{87}$$

From this we deduce that $B^R_\psi(b,\varepsilon) \subseteq B^R_\psi(a,r)$ and, further, $b \in \operatorname{Int}\big(B^R_\psi(a,r);\tau^R_\psi\big)$ by (78). Hence, ultimately, $B^R_\psi(a,r/C) \subseteq \operatorname{Int}\big(B^R_\psi(a,r);\tau^R_\psi\big)$, proving (81). Formula (82) is then established in a similar manner, and this completes the proof of the proposition. □

Remark 2. By a quasi-pseudonormed group we shall always understand a group equipped with a quasi-pseudonorm. Also, in light of parts (v) and (vi) in Proposition 1, given a quasi-pseudonormed group, we agree to drop the adjectives left/right when referring to Cauchy sequences and completeness.

Comment 4. Let $(G,*)$ be a group. As opposed to the case of a genuine pseudonorm on G, a mere quasi-pseudonorm ψ on G *may not be continuous* as a function from (G,τ^R_ψ) into $[0,+\infty]$. For example, if $(G,*) := (\mathbb{R},+)$ and

$$\psi(x) := \begin{cases} |x| & \text{if } x \in \mathbb{Q}, \\ 2|x| & \text{if } x \in \mathbb{R} \setminus \mathbb{Q}, \end{cases} \qquad \forall x \in \mathbb{R}, \tag{88}$$

then τ^R_ψ is just the ordinary topology on the real line and $\psi : \mathbb{R} \to [0,+\infty)$ is a quasi-norm which is discontinuous at every point except at the origin. The latter is no accident. Indeed, as one may readily verify from definitions, any quasi-pseudonorm ψ on an arbitrary group G is continuous (both in the topology τ^R_ψ and τ^L_ψ) at the neutral element $e_G \in G$. ■

Comment 5. Assume that ψ is a quasi-pseudonorm on a group $(G,*)$ with constants (C_0,C_1). Then repeated applications of (74) give that, for each $x_1,\dots,x_N \in G$,

$$\psi(x_1 * \cdots * x_N) \leq C_1\psi(x_1) + C_1^2\psi(x_2) + \cdots + C_1^{N-1}\psi(x_{N-1}) + C_1^{N-1}\psi(x_N). \tag{89}$$

Note that, with the exception of the situation when $C_1 = 1$ (as in the case of a pseudonorm), the largest coefficient in the right-hand side of (89) increases exponentially with N. ■

The discussion in Comments 4–5 exposes some of the most significant differences between pseudonorms and quasi-pseudonorms on groups. The aforementioned shortcomings of quasi-pseudonorms cause significant problems in applications; hence, the case of quasi-pseudonormed groups is more subtle than that of, say, metric and pseudometric groups.

A key technical tool in the proof of the quantitative version of the OMT (formulated in Theorem 7), which is brought into play specifically to address the deficiencies inherent to quasi-pseudonorms, is contained in Theorem 6. Before stating it, we remind the reader that a subset of a topological space is called `nowhere dense` if the interior of its closure is empty. Also, a topological space is said to be of `second Baire category` provided it may not be written as the union of countably many nowhere dense subsets.

Theorem 6. *Let $(G,*)$ be a semigroup and assume that $\psi : G \to [0,+\infty]$ is a quasi-subadditive function, i.e., there exists a constant $\kappa \in [1,+\infty)$ such that*

$$\psi(a*b) \leq \kappa \max\{\psi(a),\psi(b)\}, \quad \text{for all } a,b \in G. \tag{90}$$

Fix a number

$$\beta \in \left(0,(\log_2 \kappa)^{-1}\right]. \tag{91}$$

Then, for each integer $N \in \mathbb{N}$, the function ψ satisfies

$$\psi(a_1 * \cdots * a_N) \leq \kappa^2 \left\{ \sum_{i=1}^{N} \psi(a_i)^{\beta} \right\}^{\frac{1}{\beta}}, \qquad \text{for all } a_1,\ldots,a_N \in G. \tag{92}$$

In particular, for every sequence $(a_i)_{i\in\mathbb{N}} \subseteq G$, one has

$$\sup_{N\in\mathbb{N}} \psi(a_1 * \cdots * a_N) \leq \kappa^2 \left\{ \sum_{i=1}^{\infty} \psi(a_i)^{\beta} \right\}^{\frac{1}{\beta}}. \tag{93}$$

While, in general, the function ψ may not be continuous when G is equipped with either the topology τ_ψ^R or the topology τ_ψ^L, a closely related property holds under additional assumptions. Specifically, assume that actually $(G,*)$ is a group and ψ takes finite values (i.e., $\psi : G \to [0,+\infty)$) and is quasi-symmetric, in the sense that there exists $C_0 \in [1+\infty)$ such that (73) *holds. In such a scenario, for any sequence $(a_n)_{n\in\mathbb{N}} \subseteq G$ which converges to some $a \in G$, either in the topology τ_ψ^R or in the topology τ_ψ^L, one has*

$$\kappa^{-2} C_0^{-1} \psi(a) \leq \liminf_{n\to\infty} \psi(a_n) \leq \limsup_{n\to\infty} \psi(a_n) \leq \kappa^2 C_0\, \psi(a). \tag{94}$$

If, in addition to the conditions just described above, one also has $\psi(e_G) = 0$, then

$$G \text{ complete with respect to } \tau_\psi^R \implies (G,\tau_\psi^R) \text{ is of second Baire category}, \tag{95}$$

$$G \text{ complete with respect to } \tau_\psi^L \implies (G,\tau_\psi^L) \text{ is of second Baire category}. \tag{96}$$

For a proof, the interested reader is referred to [6], where a more general result of this nature (formulated in the setting of groupoids) may be found. Here, we only wish to note that all quantitative aspects of the assertions formulated in the above theorem are sharp.

4 A Quantitative Open Mapping Theorem

The main result in this section is Theorem 7. As a preamble, in the lemma below, we isolate a useful technical ingredient.

Lemma 3. *Assume that* $(G,*)$ *is a group and that* $\psi : G \to [0,+\infty]$ *is an arbitrary nonnegative function. Also, suppose that* $(S,\circ,\tau_S)$ *is a semi-topological group and consider* $T \in \mathrm{Hom}(G,S)$ *with the property that there exists* $r_0 \in (0,+\infty)$ *such that*

$$\mathrm{Clo}\Big(T\big(B^R_\psi(e_G,r_0)\big);\tau_S\Big) \in \mathscr{N}(e_S;\tau_S). \tag{97}$$

Then for every $A \subseteq G$ *one has*

$$\mathrm{Clo}\big(T(A);\tau_S\big) \subseteq \bigcup_{a\in A} \mathrm{Clo}\Big(T\big(B^R_\psi(a,r_0)\big);\tau_S\Big). \tag{98}$$

Proof. From (97) it follows that there exists $V \in \tau_S$ such that

$$e_S \in V \subseteq \mathrm{Clo}\Big(T\big(B^R_\psi(e_G,r_0)\big);\tau_S\Big). \tag{99}$$

Also, since the mapping

$$(\cdot)^{-1} : (S,\tau_S) \longrightarrow (S,\tau_S) \tag{100}$$

is a homeomorphism (cf. part (*iii*) in Remark 1) we obtain $e_S \in V^{-1} \in \tau_S$. Hence, if we set $W := V \cap V^{-1}$, then

$$W \in \tau_S, \quad e_S \in W \subseteq V, \quad \text{and} \quad W^{-1} = W. \tag{101}$$

In particular, $W \in \mathscr{N}(e_S;\tau_S)$. Next, consider a set $A \subseteq G$ and fix an arbitrary element

$$y \in \mathrm{Clo}\big(T(A);\tau_S\big). \tag{102}$$

Since Definition 5 and assumptions ensure that $s^R_y : (S,\tau_S) \to (S,\tau_S)$ is a homeomorphism, it follows that $W \circ y = s^R_y(W) \in \mathscr{N}(y;\tau_S)$ (cf. (26)). Together with (102), the latter condition entails $(W \circ y) \cap T(A) \neq \emptyset$. Thus, there exists $a \in A$ such that $Ta \in W \circ y$ which, in turn, implies $(Ta)^{-1} \in y^{-1} \circ W^{-1} = y^{-1} \circ W$. Hence,

$$
\begin{aligned}
y \in W \circ Ta \subseteq V \circ Ta \subseteq \mathrm{Clo}\Big(T\big(B^R_\psi(e_G, r_0)\big); \tau_S\Big) \circ Ta \\
&= \mathrm{Clo}\Big(T\big(B^R_\psi(e_G, r_0)\big) \circ Ta; \tau_S\Big) \\
&= \mathrm{Clo}\Big(T\big(B^R_\psi(e_G, r_0) * a\big); \tau_S\Big) \\
&= \mathrm{Clo}\Big(T\big(B^R_\psi(a, r_0)\big); \tau_S\Big).
\end{aligned} \tag{103}
$$

The first inclusion in (103) is a consequence of (101), while the second inclusion makes use of (99). Next, the first equality in (103) is a consequence of the fact that all right shifts in a semi-topological group are homeomorphisms (cf. Definition 5). Finally, the second equality in (103) holds since $T \in \mathrm{Hom}(G,H)$, while the last equality is a consequence of (49).

Given that y in (102) has been arbitrarily chosen, (98) follows. □

We are now in a position to formulate and prove the main result in this chapter.

Theorem 7 (Quantitative OMT). *Let $(G,*)$ be a group and assume that ψ is a finite quasi-pseudonorm on G, with constants (C_0,C_1), such that $(G,*)$ is complete with respect to τ^R_ψ. Consider a semi-topological group (S,τ_S) and suppose that $T \in \mathrm{Hom}(G,S)$ has the property that*

$$\mathrm{Clo}\Big(T\big(B^R_\psi(e_G, r)\big); \tau_S\Big) \in \mathcal{N}(e_S; \tau_S), \qquad \forall r \in (0,+\infty), \tag{104}$$

and that its graph satisfies

$$\mathcal{G}_T \text{ is a closed subset of } \big(G \times S, \tau^R_\psi \times \tau_S\big). \tag{105}$$

Then for each $A \subseteq G$, one has

$$\mathrm{Clo}\big(T(A); \tau_S\big) \subseteq \bigcap_{\varepsilon>0} T\big([A]^R_{\psi,\varepsilon}\big). \tag{106}$$

In particular,

$$\mathrm{Clo}\Big(T\big(B^R_\psi(a, r)\big); \tau_S\Big) \subseteq \bigcap_{C>C_1} T\big(B^R_\psi(a, Cr)\big), \quad \forall a \in G, \ \forall r \in (0,+\infty), \tag{107}$$

and, as a corollary, the map

$$T : (G, \tau^R_\psi) \longrightarrow (S, \tau_S) \quad \text{is open.} \tag{108}$$

Finally, if the mapping $T : (G, \tau^R_\psi) \to (S, \tau_S)$ is continuous, then one actually has equality in (106), *i.e.,*

$$\mathrm{Clo}\big(T(A); \tau_S\big) = \bigcap_{\varepsilon>0} T\big([A]^R_{\psi,\varepsilon}\big). \tag{109}$$

Proof. Fix a set $A \subseteq G$ and choose an arbitrary number $\varepsilon \in (0,+\infty)$. Pick a number $\beta \in \big(0,(1+\log_2 C_1)^{-1}\big]$ along with a numerical sequence $(r_i)_{i\in\mathbb{N}} \subseteq (0,+\infty)$ such that

$$16C_1^4C_0^2\Big\{\sum_{i\in\mathbb{N}} r_i^\beta\Big\}^{1/\beta} < \varepsilon. \tag{110}$$

Finally, select an arbitrary element $y_0 \in \mathrm{Clo}\big(T(A);\tau_S\big)$. Then, by Lemma 3 and (104), we have $y_0 \in \bigcup_{a\in A} \mathrm{Clo}\Big(T\big(B^R_\psi(a,r_1)\big);\tau_S\Big)$. Hence, there exists $a_1 \in A$ such that

$$y_0 \in \mathrm{Clo}\Big(T\big(B^R_\psi(a_1,r_1)\big);\tau_S\Big) \subseteq \bigcup_{a\in B^R_\psi(a_1,r_1)} \mathrm{Clo}\Big(T\big(B^R_\psi(a,r_2)\big);\tau_S\Big), \tag{111}$$

where the inclusion in (111) is a consequence of Lemma 3 and (104). Inductively, this procedure yields a sequence $(a_i)_{i\in\mathbb{N}} \subseteq G$ such that

$$\begin{gathered} a_1 \in A \quad \text{and} \quad a_{i+1} \in B^R_\psi(a_i,r_i) \;\text{ for each }\; i \in \mathbb{N}, \\ \text{with the property that }\; y_0 \in \mathrm{Clo}\Big(T\big(B^R_\psi(a_i,r_i)\big);\tau_S\Big), \quad \forall\, i \in \mathbb{N}. \end{gathered} \tag{112}$$

Note that if $i,k \in \mathbb{N}$, then, based on (92) in Theorem 6 used here with $\kappa := 2C_1$, we may estimate

$$\begin{aligned} \psi(a_i * a_{i+k}^{-1}) &= \psi\big((a_i * a_{i+1}^{-1}) * (a_{i+1} * a_{i+2}^{-1}) * \cdots * (a_{i+k-1} * a_{i+k}^{-1})\big) \\ &\le 4C_1^2\Big\{\sum_{j=i}^{i+k-1} \psi(a_j * a_{j+1}^{-1})^\beta\Big\}^{1/\beta} \le 4C_1^2\Big\{\sum_{j=i}^{\infty} r_j^\beta\Big\}^{1/\beta}. \end{aligned} \tag{113}$$

Since the last expression above converges to zero as $i \to \infty$, this implies that

$$(a_i)_{i\in\mathbb{N}} \;\text{ is a Cauchy sequence in } (G,*) \text{ relative to } \tau^R_\psi. \tag{114}$$

Recall that $(G,*)$ is assumed to be complete with respect to τ^R_ψ and, as such, there exists $a \in G$ with the property that

$$a_i \longrightarrow a \;\text{ in }\; \tau^R_\psi, \;\text{ as }\; i \to \infty. \tag{115}$$

Upon noting that the map $s^R_{a_1^{-1}} : (G,\tau^R_\psi) \to (G,\tau^R_\psi)$ is continuous (cf. part *(ii)* in Lemma 2), we further obtain from (115) that

$$a_i * a_1^{-1} \longrightarrow a * a_1^{-1} \;\text{ in }\; \tau^R_\psi, \;\text{ as }\; i \to \infty. \tag{116}$$

Also, generally speaking,

$$\left.\begin{array}{r}(x_i)_{i\in\mathbb{N}}\subseteq G,\ x\in G \text{ such that}\\ x_i\longrightarrow x \text{ in } \tau_\psi^R \text{ as } i\to\infty\end{array}\right\}\Longrightarrow \psi(x)\leq 4C_1^2C_0\limsup_{i\to\infty}\psi(x_i). \tag{117}$$

This follows from (94) in Theorem 6, used here with $\kappa := 2C_1$. Now, by (113) in which we first set $i := 1$ and then take $k := i-1$, we have

$$\psi(a_i * a_1^{-1}) \leq 4C_1^2\left\{\sum_{j=1}^{i-1} r_j^\beta\right\}^{1/\beta}, \quad \text{for } i=2,3,\dots. \tag{118}$$

Keeping in mind (73) and combining (116), (117), (118), and (110), we arrive at the conclusion that

$$\psi\big(a_1 * a^{-1}\big) \leq C_0\,\psi\big(a * a_1^{-1}\big) \leq 16C_1^4C_0^2\left\{\sum_{j=1}^{\infty} r_j^\beta\right\}^{1/\beta} < \varepsilon. \tag{119}$$

Thus, ultimately,

$$a\in B_\psi^R(a_1,\varepsilon) \text{ for some } a_1\in A. \tag{120}$$

We now claim that

$$(a,y_0)\in \mathrm{Clo}\big(\mathscr{G}_T;\tau_\psi^R\times\tau_S\big). \tag{121}$$

To justify the claim above, select two arbitrary neighborhoods $V\in\mathscr{N}(a;\tau_\psi^R)$ and $W\in\mathscr{N}(y_0;\tau_S)$. Then, thanks to (83), there exists $r\in(0,+\infty)$ such that

$$B_\psi^R(a,r)\subseteq V. \tag{122}$$

On the other hand, from (80) we know that $B_\psi^R\big(a,r/(2C_1)\big)\in\mathscr{N}\big(a;\tau_\psi^R\big)$. This, in concert with (110) and (115), implies that there exists $i_0\in\mathbb{N}$ such that

$$r_i < r/(2C_1), \quad \forall i\in\mathbb{N} \text{ with } i\geq i_0, \tag{123}$$

$$a_i\in B_\psi^R\big(a,r/(2C_1)\big), \quad \forall i\in\mathbb{N} \text{ with } i\geq i_0. \tag{124}$$

In particular, if $x\in B_\psi^R(a_i,r_i)$, then $\psi(a_i*x^{-1})<r_i$; hence,

$$\begin{aligned}\psi(a*x^{-1}) &= \psi\big((a*a_i^{-1})*(a_i*x^{-1})\big)\leq C_1\big(\psi(a*a_i^{-1})+\psi(a_i*x^{-1})\big)\\ &< C_1\big(r/(2C_1)+r_i\big)<r.\end{aligned} \tag{125}$$

This analysis proves that there exists $i_0 \in \mathbb{N}$ such that

$$B_\psi^R(a_i, r_i) \subseteq B_\psi^R(a, r), \quad \forall i \in \mathbb{N} \text{ with } i \geq i_0. \tag{126}$$

Now (122) and (126) yield $B_\psi^R(a_i, r_i) \subseteq V$ whenever $i \in \mathbb{N}$ satisfies $i \geq i_0$; hence,

$$T\big(B_\psi^R(a_i, r_i)\big) \subseteq T(V), \quad \forall i \in \mathbb{N} \text{ with } i \geq i_0. \tag{127}$$

On the other hand, (112) ensures that $W \cap T\big(B_\psi^R(a_i, r_i)\big) \neq \emptyset$ for every $i \in \mathbb{N}$ which, when combined with (127), gives that

$$W \cap T(V) \neq \emptyset. \tag{128}$$

In turn, (128) yields the existence of an element $v \in V$ with the property that $Tv \in W$. This entails $(v, Tv) \in \mathscr{G}_T \cap (V \times W)$; therefore, $\mathscr{G}_T \cap (V \times W) \neq \emptyset$. This shows that (121) holds, since $V \in \mathscr{N}(a; \tau_\psi^R)$ and $W \in \mathscr{N}(y_0; \tau_S)$ have been arbitrarily chosen. Having established (121) and since, by assumption, $\mathscr{G}_T$ is closed in $\big(G \times S, \tau_\psi^R \times \tau_S\big)$, we may conclude that $(a, y_0) \in \mathscr{G}_T$, which forces

$$y_0 = Ta. \tag{129}$$

Now, (120) implies that $a \in [A]_{\psi,\varepsilon}^R$, hence using (129) yields $y_0 \in T\big([A]_{\psi,\varepsilon}^R\big)$. Given that $y_0 \in \mathrm{Clo}\big(T(A); \tau_S\big)$ and $\varepsilon > 0$ have been arbitrarily chosen, it follows that (106) holds.

Moving on, consider $a \in G$, $r \in (0, +\infty)$, $\varepsilon > 0$, and recall that

$$\big[B_\psi^R(a, r)\big]_{\psi,\varepsilon}^R = \bigcup_{b \in B_\psi^R(a,r)} B_\psi^R(b, \varepsilon). \tag{130}$$

Now, for each $b \in B_\psi^R(a, r)$ and each $c \in B_\psi^R(b, \varepsilon)$, we may write

$$\psi(a * c^{-1}) \leq C_1\big(\psi(a * b^{-1}) + \psi(b * c^{-1})\big) < C_1(r + \varepsilon), \tag{131}$$

which shows that

$$\big[B_\psi^R(a, r)\big]_{\psi,\varepsilon}^R \subseteq B_\psi^R\big(a, C_1(r + \varepsilon)\big). \tag{132}$$

If we now apply (106) with $A := B_\psi^R(a, r)$ and use (132), we obtain

$$\begin{aligned} \mathrm{Clo}\Big(T\big(B_\psi^R(a, r)\big); \tau_S\Big) &\subseteq \bigcap_{\varepsilon > 0} T\big(\big[B_\psi^R(a, r)\big]_{\psi,\varepsilon}^R\big) \\ &\subseteq \bigcap_{\varepsilon > 0} T\big(B_\psi^R(a, C_1(r + \varepsilon))\big) \\ &= \bigcap_{C > C_1} T\big(B_\psi^R(a, Cr)\big), \end{aligned} \tag{133}$$

proving (107). As a corollary of (107) we also see that

$$\mathrm{Clo}\Big(T\big(B^R_\psi(a,r)\big);\tau_S\Big) \subseteq T\big(B^R_\psi(a,2C_1 r)\big), \quad \forall a \in G,\ \forall r \in (0,+\infty). \tag{134}$$

At the same time,

$$\mathrm{Clo}\Big(T\big(B^R_\psi(a,r)\big);\tau_S\Big) = \mathrm{Clo}\Big(T\big(B^R_\psi(e_G,r)\big);\tau_S\Big) \circ Ta \in \mathscr{N}(Ta;\tau_S), \tag{135}$$

by (104) and the fact that shifts are continuous on (S,τ_S). When combined with (134), this gives that

$$T\big(B^R_\psi(a,r)\big) \in \mathscr{N}(Ta;\tau_S), \quad \forall a \in G,\ \forall r \in (0,+\infty). \tag{136}$$

Now, consider some $\mathscr{O} \in \tau^R_\psi$ and select an arbitrary $b \in T(\mathscr{O})$. Then there exists an element $a \in \mathscr{O}$ such that $b = Ta$; hence, there exists $r \in (0,+\infty)$ with the property that $B^R_\psi(a,r) \subseteq \mathscr{O}$. Consequently, $T\big(B^R_\psi(a,r)\big) \subseteq T(\mathscr{O})$. Since by (136) the set $T\big(B^R_\psi(a,r)\big)$ is a neighborhood of Ta in τ_S, it follows that $T(\mathscr{O}) \in \mathscr{N}(b;\tau_S)$ for every $b \in T(\mathscr{O})$. Therefore, $T(\mathscr{O}) \in \tau_S$, proving that T is open.

At this stage in the proof, there remains to show that (109) holds whenever T is continuous as a mapping from (G,τ^R_ψ) into (S,τ_S). To this end, we note that once an element $b \in \bigcap_{\varepsilon>0} T\big([A]^R_{\psi,\varepsilon}\big)$ has been arbitrarily chosen, for every $n \in \mathbb{N}$, there exists an element $x_n \in [A]^R_{\psi,1/n}$ such that $b = Tx_n$. Having constructed such a sequence $(x_n)_{n\in\mathbb{N}} \subseteq G$, for each $n \in \mathbb{N}$, it is then possible to select $y_n \in A$ with the property that $x_n \in B^R_\psi(y_n,1/n)$. Hence, $\psi(y_n * x_n^{-1}) < 1/n$ for each $n \in \mathbb{N}$ which, in light of part (*iv*) in Proposition 1, proves that

$$y_n * x_n^{-1} \longrightarrow e_G \ \text{ in } \ \tau^R_\psi, \ \text{ as } \ n \to \infty. \tag{137}$$

Granted that $T : (G,\tau^R_\psi) \to (S,\tau_S)$ is continuous, we deduce from (137), and our choice of the x_n's that

$$(Ty_n) \circ b^{-1} = T\big(y_n * x_n^{-1}\big) \longrightarrow e_S \ \text{ in } \ \tau_S, \ \text{ as } \ n \to \infty. \tag{138}$$

Upon recalling from Definition 5 that the map $s^R_b : (S,\tau_S) \to (S,\tau_S)$ is continuous, this further implies that

$$Ty_n \longrightarrow b \ \text{ in } \ \tau_S, \ \text{ as } \ n \to \infty. \tag{139}$$

Having proved (139), the fact that $y_n \in A$ for every $n \in \mathbb{N}$ entails $b \in \mathrm{Clo}\big(T(A);\tau_S\big)$. Hence, ultimately,

$$\bigcap_{\varepsilon>0} T\big([A]^R_{\psi,\varepsilon}\big) \subseteq \mathrm{Clo}\big(T(A);\tau_S\big) \tag{140}$$

which, together with (106), establishes (109) and completes the proof of the theorem. □

Of course, a version of Theorem 7 emphasizing the left-topology induced by the quasi-pseudonorm ψ is also valid. We leave the formulation of such a result to the interested reader.

Comment 8. The quantitative aspect of Theorem 7 is most apparent by considering the case when S is also a quasi-pseudonormed group. Concretely, assume that the function $\varphi : S \to [0,+\infty]$ is a quasi-pseudonorm with the property that τ_S is the right-topology induced by φ on S. Condition (104) then is equivalent with the demand that there exists a function $\eta : (0,+\infty) \to (0,+\infty)$ with the property that

$$B^R_\varphi(e_S,\eta(r)) \subseteq \mathrm{Clo}\Big(T\big(B^R_\psi(e_G,r)\big);\tau_S\Big), \qquad \forall r \in (0,+\infty). \tag{141}$$

In conjunction with (107), this yields

$$B^R_\varphi(Ta,\eta(r)) \subseteq \bigcap_{C>C_1} T\big(B^R_\psi(a,Cr)\big), \qquad \forall a \in G,\ \forall r \in (0,+\infty), \tag{142}$$

which provides concrete information about the size of the neighborhood of Ta one expects to be contained in the image of a ball $B^R_\psi(a,R)$ under the mapping T. ■

In the last part of this section we shall discuss two corollaries of the quantitative OMT formulated in Theorem 7 which are more akin to the classical formulation of this result. In preparation, we first prove an auxiliary lemma.

Lemma 4. *Let $\big(G,*,(\cdot)^{-1},e_G\big)$ be a group and assume that τ is a topology on the set G with the property that the mapping*

$$\big(G\times G,\tau\times\tau\big) \ni (x,y) \mapsto x*y \in (G,\tau) \ \text{ is continuous.} \tag{143}$$

Then for every set $A \subseteq G$, one has

$$\mathrm{Int}(A;\tau) \neq \emptyset \Longrightarrow A*A^{-1} \in \mathcal{N}(e_G;\tau). \tag{144}$$

Proof. Indeed, if $A \subseteq G$ is such that $\mathrm{Int}(A;\tau) \neq \emptyset$, then there exists a nonempty set $O \in \tau$ with the property that $O \subseteq A$. Consequently, on the one hand, we have

$$e_G \in O*O^{-1} \subseteq A*A^{-1}. \tag{145}$$

On the other hand, the continuity of the function (143) forces all right shifts on G to be continuous, hence ultimately, homeomorphisms of (G,τ) by (16). Given this, keeping in mind that $O \in \tau$ and writing

$$O*O^{-1} = \bigcup_{a\in O} s^R_{a^{-1}}(O), \tag{146}$$

we deduce that $O*O^{-1} \in \tau$. Together with property (145) this proves that, as desired, $A*A^{-1} \in \mathcal{N}(e_G;\tau)$. □

Here is the first consequence of Theorem 7 alluded to above. To state it, recall the convention made in (10).

Theorem 9 (Topological OMT: Version 1). *Let G be a group equipped with a finite quasi-invariant quasi-pseudonorm inducing a right-topology τ_G with respect to which G is complete, and such that*

$$G = \bigcup_{n\in\mathbb{N}} V^{(n)}, \quad \textit{for every symmetric neighborhood } V \textit{ of } e_G. \tag{147}$$

Also, suppose that S is an Abelian, torsion-free, divisible group, endowed with a finite quasi-norm, inducing a topology τ_S on S with the property that

$$\mathscr{O}^{(n)} \in \tau_S, \quad \textit{for every } \mathscr{O} \in \tau_S \textit{ and every } n \in \mathbb{N}. \tag{148}$$

Then any $T \in \operatorname{Hom}(G,S)$ which is continuous and surjective is an open mapping.

Proof. Let $G = \big(G, *, (\cdot)^{-1}, e_G\big)$ and $S = \big(S, \circ, (\cdot)^{-1}, e_S\big)$ be as in the statement of the theorem. Also, denote by τ_G and τ_S the right-topologies induced on G and S by their respective quasi-pseudonorms. Finally, consider $T \in \operatorname{Hom}(G,S)$ which is continuous and surjective, and fix an arbitrary $U \in \mathscr{N}(e_G;\tau_G)$. Then $(G, *, \tau_S)$ becomes a topological group, as seen from part (*viii*) in Lemma 2. As such, from the continuity of the function (cf. (22))

$$\big(G \times G, \tau_G \times \tau_G\big) \ni (x,y) \mapsto x * y^{-1} \in (G, \tau_G), \tag{149}$$

at (e_G, e_G) it follows that there exists $V_0 \in \mathscr{N}(e_G;\tau_G)$ such that

$$V_0 * V_0^{-1} \subseteq U. \tag{150}$$

Thus, if we set $V := V_0 \cap V_0^{-1}$, it follows that

$$V \in \mathscr{N}(e_G;\tau_G), \ V \text{ is symmetric, and } V^2 \subseteq U. \tag{151}$$

At the same time, (147) and the fact that $T \in \operatorname{Hom}(G,S)$ is surjective yield

$$S = T(G) = T\Big(\bigcup_{n\in\mathbb{N}} V^{(n)}\Big) = \bigcup_{n\in\mathbb{N}} T\big(V^{(n)}\big) = \bigcup_{n\in\mathbb{N}} \big(T(V)\big)^{(n)}. \tag{152}$$

In concert with (95), this ensures that there exists $n \in \mathbb{N}$ with the property that

$$\operatorname{Int}\Big(\operatorname{Clo}\big((T(V))^{(n)};\tau_S\big);\tau_S\Big) \neq \emptyset. \tag{153}$$

In particular, there exists $\mathscr{O} \in \tau_S$ such that

$$\emptyset \neq \mathscr{O} \subseteq \operatorname{Clo}\big((T(V))^{(n)};\tau_S\big). \tag{154}$$

Consider now

$$\phi : S \longrightarrow S, \quad \phi(x) := x^n, \quad \forall x \in S. \tag{155}$$

The fact that $(S, \circ, \tau_S)$ is a topological group implies that ϕ is continuous, while the fact that the group $(S, \circ)$ is Abelian, torsion-free, and divisible forces ϕ to be a bijection. Finally, (148) ensures that ϕ is open. As a result,

$$\phi : (S, \tau_S) \longrightarrow (S, \tau_S) \text{ is a homeomorphism.} \tag{156}$$

Hence, given that $\phi(A) = A^{(n)}$ for any $A \subseteq G$, we have

$$\mathrm{Clo}\big((T(V))^{(n)}; \tau_S\big) = \Big(\mathrm{Clo}\big(T(V); \tau_S\big)\Big)^{(n)} \tag{157}$$

by (156) and (24). Consequently, on the one hand,

$$\phi^{-1}(\mathscr{O}) \subseteq \phi^{-1}\Big[\Big(\mathrm{Clo}\big(T(V); \tau_S\big)\Big)^{(n)}\Big] = \mathrm{Clo}\big(T(V); \tau_S\big), \tag{158}$$

by (157), (154), and the fact that ϕ is bijective. On the other hand, $\emptyset \neq \phi^{-1}(\mathscr{O}) \in \tau_S$ thanks to (156). Granted this and keeping (158) in mind, we arrive at the conclusion that

$$\mathrm{Int}\left(\mathrm{Clo}\big(T(V); \tau_S\big); \tau_S\right) \neq \emptyset. \tag{159}$$

Together with Lemma 4 and the nature of τ_S, this allows us to conclude that

$$\mathrm{Clo}\big(T(V); \tau_S\big) \circ \Big(\mathrm{Clo}\big(T(V); \tau_S\big)\Big)^{-1} \in \mathscr{N}(e_S; \tau_S). \tag{160}$$

At the same time,

$$\begin{aligned} \mathrm{Clo}\big(T(V); \tau_S\big) \circ \Big(\mathrm{Clo}\big(T(V); \tau_S\big)\Big)^{-1} &= \mathrm{Clo}\big(T(V); \tau_S\big) \circ \mathrm{Clo}\big((T(V))^{-1}; \tau_S\big) \\ &\subseteq \mathrm{Clo}\big(T(V) \circ (T(V))^{-1}; \tau_S\big) \\ &= \mathrm{Clo}\big(T(V \circ V^{-1}); \tau_S\big) \\ &\subseteq \mathrm{Clo}\big(T(U); \tau_S\big), \end{aligned} \tag{161}$$

by (29), (151), the fact that $(\cdot)^{-1} : (S, \tau_S) \to (S, \tau_S)$ is a homeomorphism, and (24). In combination with (160), (161) proves that

$$\mathrm{Clo}\big(T(U); \tau_S\big) \in \mathscr{N}(e_S; \tau_S), \qquad \forall U \in \mathscr{N}(e_G; \tau_G). \tag{162}$$

With this in hand, Theorem 7 applies and gives that T is an open mapping, as soon as we check that $\mathscr{G}_T$, the graph of T, is a closed subset of $\big(G \times S, \tau_G \times \tau_S\big)$. This, however, follows by noting that

$$\mathscr{G}_T = F^{-1}\big(\{e_S\}\big), \tag{163}$$

where

$$F : G \times S \longrightarrow S, \quad F(x,y) := (Tx) \circ y^{-1}, \quad \forall (x,y) \in G \times S. \tag{164}$$

Now, the fact that $T : (G, \tau_G) \to (S, \tau_S)$ is continuous and that $(S, \circ, \tau_S)$ is a topological group (cf. part (vii) in Proposition 1) implies that $F : \big(G \times S, \tau_G \times \tau_S\big) \to (S, \tau_S)$ is continuous. Also, since S is a quasi-normed group, it follows from part (vi) in Lemma 2 that the singleton $\{e_S\}$ is closed in (S, τ_S). Keeping this in mind, we then conclude from (163) that $\mathscr{G}_T$ is a closed subset of $\big(G \times S, \tau_G \times \tau_S\big)$, thus finishing the proof of the theorem. □

We conclude by presenting a second consequence of Theorem 7.

Theorem 10 (Topological Open Mapping Theorem: Version 2). *Let G be a group equipped with a finite quasi-invariant quasi-pseudonorm inducing a topology τ_G with respect to which G is complete, and such that*

$$\textit{the topological space } (G, \tau_G) \textit{ is separable.} \tag{165}$$

Also, suppose that S is a group equipped with a finite quasi-invariant quasi-norm, inducing a topology with respect to which S is complete.

Then any $T \in \mathrm{Hom}\,(G,S)$ which is continuous and surjective is an open mapping.

Proof. Consider $T \in \mathrm{Hom}\,(G,S)$ which is continuous and surjective, and fix an arbitrary $U \in \mathscr{N}(e_G; \tau_G)$. As in (151), select $V \in \mathscr{N}(e_G; \tau_G)$ such that $V * V^{-1} \subseteq U$. To proceed, let $A := \{a_n\}_{n \in \mathbb{N}}$ be a countable dense subset of the topological space (G, τ_G). Then

$$G = \mathrm{Clo}(A; \tau_G) = A * V \tag{166}$$

by (28). Hence, since T is surjective, we may write

$$S = T(G) = \bigcup_{n \in \mathbb{N}} (Ta_n) \circ T(V). \tag{167}$$

On the other hand, since from (95) we know that (S, τ_S) is of second Baire category, there exists $n \in \mathbb{N}$ such that

$$\mathrm{Int}\Big(\mathrm{Clo}\big((Ta_n) \circ T(V); \tau_S\big); \tau_S\Big) \neq \emptyset. \tag{168}$$

However, from part (ii) in Lemma 2 and (25), we deduce that

$$\operatorname{Int}\Big(\operatorname{Clo}\big((Ta_n)\circ T(V);\tau_S\big);\tau_S\Big) = (Ta_n)\circ\operatorname{Int}\Big(\operatorname{Clo}\big(T(V);\tau_S\big)\Big); \tag{169}$$

hence, further,

$$\operatorname{Int}\Big(\operatorname{Clo}\big(T(V);\tau_S\big)\Big) \neq \emptyset. \tag{170}$$

With this in hand, the rest of the proof now proceeds as in the case of Theorem 9 (compare with (159)). □

References

1. Banach, S.: Metrische gruppen. Studia Math. **3**, 101–113 (1931)
2. Brown, L.G.: Note on the open mapping theorem. Pac. J. Math. **38**(1), 25–28 (1971)
3. Husain, T.: Introduction to Topological Groups. W.B. Saunders, Philadelphia (1966)
4. Kalton, N.J., Peck, N.T., Roberts, J.W.: An F-space sampler, London Mathematical Society Lecture Notes Series, vol. 89. Cambridge University Press, Cambridge (1984)
5. Kelley, J.L.: General Topology. van Nostrand, Toronto, New York, London (1955)
6. Mitrea, D., Mitrea, I., Mitrea, M., Monniaux, S.: Groupoid Metrization Theory with Applications to Analysis on Quasi-Metric Spaces and Functional Analysis, to appear in Applied and Numerical Harmonic Analysis, Birkhäuser, 481 (2012)
7. Pettis, B.: On continuity and openness of homomorphisms in topological groups. Ann. Math. **54**, 293–308 (1950)
8. Rudin, W.: Functional Analysis. McGraw-Hill, New York (1973)

The Buckley Dyadic Square Function

Fedor Nazarov and Michael Wilson

Abstract We explore the relationships and differences between the "standard" dyadic square function $S(f)$ and a pointwise smaller square function $S_b(f)$ due to Stephen Buckley. In dimension one, $S(f) = S_b(f)$, but in higher dimensions, $S_b(f)$ can vanish on a cube in which $S(f)$ is everywhere positive, and uniform boundedness of $S_b(f)$ does not imply the same of $S(f)$. However, uniform boundedness of $S_b(f)$ implies local exponential-square integrability of both $S(f)$ and f. The second implication is surprising because the result of [2] requires $S(f) \in L^\infty$ to infer that f is in the local exponential L^2 class.

AMS Subject Classification (2000): 42B25.

1 Introduction

The dyadic square function is a familiar tool in probability and analysis, and many well known theorems are attached to its name. Among these is a result from [2], which states that if the dyadic square function of a function f belongs to L^∞, then f is locally exponentially square integrable.

What seems to be less well-known (this might be an understatement) is that there are *two* "dyadic square functions" abroad in the world; and, although they are closely related, they are not equivalent.

Unfortunately, explaining all of these requires some definitions.

F. Nazarov
Department of Mathematics, University of Wisconsin, Madison, WI 53706, USA

M. Wilson (✉)
Department of Mathematics, University of Vermont, Burlington, VT 05405, USA
e-mail: jmwilson@uvm.edu

D. Bilyk et al. (eds.), *Recent Advances in Harmonic Analysis and Applications*, Springer Proceedings in Mathematics & Statistics 25, DOI 10.1007/978-1-4614-4565-4_23,

Let $\mathscr{D}$ denote the family of dyadic cubes in $\mathbf{R}^d$, where a cube is called dyadic if it has the form $[j_1 2^k, (j_1+1)2^k) \times [j_1 2^k, (j_1+1)2^k) \times \cdots \times [j_d 2^k, (j_d+1)2^k)$ for some integers $j_1, j_2, \ldots, j_d$, and k. (We will only be considering dyadic cubes in this paper.) The number 2^k is called Q's sidelength, which we denote by $\ell(Q)$. Every $Q \in \mathscr{D}$ can be divided into 2^d congruent dyadic subcubes ("successors") by bisecting each of Q's component intervals $[j_l 2^k, (j_l+1)2^k)$; we call the set of Q's successors $N(Q)$. (The collection of *all* dyadic subcubes of a dyadic cube Q—including Q itself—will be denoted by $\mathscr{D}(Q)$.) Every $Q \in \mathscr{D}$ is the successor of a unique dyadic cube, denoted by $\tilde{Q}$, which we will call Q's dyadic double (sometimes called Q's "father" or "predecessor"). If $f \in L^1_{\text{loc}}(\mathbf{R}^d)$ and $Q \in \mathscr{D}$, f_Q will mean f's average over Q:

$$f_Q \equiv \frac{1}{|Q|} \int_Q f(x)\,\mathrm{d}x,$$

where $|Q|$ means Q's Lebesgue measure.

For k an integer and $f \in L^1_{\text{loc}}(\mathbf{R}^d)$, we define

$$f_{(k)}(x) \equiv \sum_{Q:\ell(Q)=2^k} f_Q \chi_Q(x).$$

That is, we have replaced f by its averages over dyadic cubes of sidelength 2^k. If $\ell(Q) = 2^k$, we define

$$a_Q(f)(x) \equiv (f_{(k-1)} - f_{(k)})\chi_Q(x).$$

The *standard dyadic square function* $S(f)$ is defined by the formula:

$$S(f)(x) \equiv \left(\sum_{Q \in \mathscr{D}} \frac{\|a_Q(f)\|_2^2}{|Q|} \chi_Q(x) \right)^{1/2}.$$

This is the one treated in [2], where one can find the following theorem (Theorem 3.1, p. 236):

Theorem 1. *There exist positive constants $C_1(d)$ and $C_2(d)$ such that if $f \in L^1_{\text{loc}}(\mathbf{R}^d)$ and $\|S(f)\|_\infty \leq 1$, then, for all dyadic cubes, Q and $\lambda > 0$,*

$$|\{x \in Q : |f - f_Q| > \lambda\}| \leq C_1(d)|Q| \exp(-C_2(d)\lambda^2). \tag{1}$$

Furthermore, when $d = 1$, we may take $C_1(1) = 2$ and $C_2(1) = 1/2$.

Remark. This is not exactly how Theorem 3.1 is stated in [2], but it is an immediate consequence of it (see the paragraph preceding Theorem 3.1). The central limit theorem implies that the bound (1) is essentially sharp.

The *Buckley dyadic square function* $S_b(f)$ [1] is defined by the formula

$$S_b(f)(x) \equiv \left(\sum_k |f_{(k-1)}(x) - f_{(k)}(x)|^2 \right)^{1/2}. \tag{2}$$

The Buckley square function can also be written as

$$\left(\sum_{Q\in\mathscr{D}} |f_Q - f_{\tilde{Q}}|^2 \chi_Q(x) \right)^{1/2} \tag{3}$$

or

$$\left(\sum_{Q\in\mathscr{D}} |a_Q(f)(x)|^2 \right)^{1/2}. \tag{4}$$

The reader should satisfy himself that (2)–(4) are all equal, even though a "χ_Q" might seem to be missing in (4) (it is only hidden).

A little thought shows that the standard square function $S(f)$ is pointwise comparable to

$$\left(\sum_{Q\in\mathscr{D}} \|a_Q(f)\|_\infty^2 \chi_Q(x) \right)^{1/2},$$

which is the definition used in Theorem 3.1 from [2].

When $d = 1$, $S(f) = S_b(f)$. When $d > 1$, the two square functions are not pointwise comparable. It is easy to see that $S_b(f) \leq C_d S(f)$. But now let $d = 2$, and subdivide $[0,1)^2$ into 4 congruent dyadic subsquares. Set f equal to 1 on one of the subsquares, -1 on another, 0 on the remaining two, and equal to 0 outside $[0,1)^2$. Then $S(f) \equiv 1/\sqrt{2}$ on $[0,1)^2$, but $S_b(f)$ equals 0 on the two "zeroed" subsquares of $[0,1)^2$.

We can iterate this construction to show that even *uniform boundedness* of $S_b(f)$ does not imply uniform boundedness of $S(f)$. Consider the function f we just constructed. For definiteness, let us say that $[0,1/2)^2$—the "southwestern" subsquare—was one of the zeroed ones. For $k = 0, 1, 2, \ldots$, define

$$\psi_k(x) \equiv f(2^k x),$$

and set

$$h(x) \equiv \sum_0^\infty \psi_k(x).$$

The functions ψ_k have disjoint supports: the support of ψ_k is contained in the square $[0,1/2^k)^2$, but ψ_k is identically zero on $[0,1/2^{k+1})^2$, which contains the support of ψ_{k+1}. If $x \in \mathbf{R}^2$, then $|h_Q - h_{\tilde{Q}}|$ can be nonzero for at most one Q containing x, and if it is nonzero, $|h_Q - h_{\tilde{Q}}|$ has to equal 1. Therefore, $\|S_b(h)\|_\infty = 1$. However, if $x \in [0,1/2^k)^2$,

$$(S(h)(x))^2 \geq \frac{k+1}{2} \to \infty.$$

In spite of this, the Buckley square function controls $S(f)$ and f itself in a fairly strong manner. To state our first theorem we need to recall two familiar definitions. A *weight* is any nonnegative function in $L^1_{\mathrm{loc}}(\mathbf{R}^d)$. If g is any locally integrable function, the dyadic Hardy–Littlewood maximal function of g, $M_d(g)$, is defined by

$$M_d(g)(x) \equiv \sup_{\substack{Q \in \mathscr{D} \\ x \in Q}} \frac{1}{|Q|} \int_Q |g(t)|\,\mathrm{d}t.$$

We prove the following theorem in Sect. 2.

Theorem 2. *For every* $1 < p \leq 2$ *there is a constant* $C(p,d)$ *such that if* $f \in L^1_{\mathrm{loc}}(\mathbf{R}^d)$ *and* v *is any weight, then*

$$\int_{\mathbf{R}^d} (S(f))^p\, v\,\mathrm{d}x \leq C(p,d) \int_{\mathbf{R}^d} (S_b(f))^p M_d(v)\,\mathrm{d}x.$$

In Sect. 2 we also show how Theorem 2 implies that, if $S_b(f) \in L^\infty$, then $S(f)$ is, in a precise sense, locally exponentially square integrable. The example we gave a few paragraphs ago (we give the details in Sect. 2) shows that this bound is essentially best possible.

In Sects. 3 and 4 we prove:

Theorem 3. *There exist positive constants* $C_1(d)$ *and* $C_2(d)$ *such that if* $f \in L^1_{\mathrm{loc}}(\mathbf{R}^d)$ *and* $\|S_b(f)\|_\infty \leq 1$, *then, for all dyadic cubes,* Q *and* $\lambda > 0$,

$$|\{x \in Q : |f - f_Q| > \lambda\}| \leq C_1(d)|Q| \exp(-C_2(d)\lambda^2).$$

In light of Theorems 1 and 2, and the fact that their exponential-square estimates cannot be improved, Theorem 3 comes as something of a surprise.

We give two proofs of Theorem 3. The first proof (which came second and appears in Sect. 3) has a Bellman-function flavor. Not surprisingly, it gives the sharper constants. The second proof (which came first and appears in Sect. 4) is more combinatorial. It uses something called "the crot lemma," which might be of independent interest.

Notation. When $E \subset \mathbf{R}^d$ is a measurable set, $|E|$ means E's Lebesgue measure. When we write $A \sim B$ for two nonnegative quantities A and B, we mean that there are positive constants c and C such that $cA \leq B \leq CA$, where the constants do not depend on the parameters involved so as to make the inequalities trivial. We will use C to mean a positive constant that might vary from place to place but not in a way that makes the inequalities in which it occurs into trivialities.

2 Proof of Theorem 2

The following function will simplify our discussion.

Definition 1. The "look-ahead" Buckley square function $S_a(f)$ is defined by the formula

$$S_a(f)(x) \equiv \left(\sum_{Q \in \mathscr{D}} |f_Q - f_{\tilde{Q}}|^2 \chi_{\tilde{Q}}(x) \right)^{1/2}.$$

The only difference between $S_a(f)$ and $S_b(f)$ (as defined by (3)) is that χ_Q has been replaced with $\chi_{\tilde{Q}}$. The reader should make sure that he sees this.

The functions $S(f)$ and $S_a(f)$ are pointwise comparable because

$$\frac{\|a_Q(f)\|_2^2}{|Q|} \sim \|a_Q(f)\|_\infty^2 \sim \sum_{Q' \in N(Q)} |f_{Q'} - f_Q|^2.$$

Therefore, it suffices to prove Theorem 2 with $S_a(f)$ in place of $S(f)$.

We will prove Theorem 2 in two stages. First we will prove it for $p = 2$, which is quite easy. Then we will adapt an argument due to Chanillo and Wheeden [3] to show that for any $\lambda > 0$, all f and all weights v,

$$v(\{x\colon S_a(f) > \lambda\}) \leq \frac{C}{\lambda} \int_{\mathbf{R}^d} S_b(f) M_d(v)\, dx,$$

for a constant C that only depends on the dimension; and we are using $v(E)$ to mean $\int_E v\, dx$, the v-measure of a set E. Theorem 2 will follow by interpolation.

Step one: L^2 bound

$$\begin{aligned}
\int_{\mathbf{R}^d} (S_a(f))^2 v\, dx &= \sum_Q |f_Q - f_{\tilde{Q}}|^2 v(\tilde{Q}) \\
&\leq C \sum_Q |f_Q - f_{\tilde{Q}}|^2 \int_Q M_d(v)\, dx \\
&= \int_{\mathbf{R}^d} (S_b(f))^2 M_d(v)\, dx,
\end{aligned}$$

where the important inequality to observe is

$$v(\tilde{Q}) \leq C \int_Q M_d(v)\, dx.$$

□

Step two: The weak$(1,1)$ bound

For R a large positive number, let us temporarily define

$$S_{a,R}(f)(x) = \left(\sum_{Q \in \mathscr{D}: \ell(Q) \leq R} |f_Q - f_{\tilde{Q}}|^2 \chi_{\tilde{Q}}(x) \right)^{1/2}$$

$$S_{b,R}(f)(x) = \left(\sum_{Q \in \mathscr{D}: \ell(Q) \leq R} |f_Q - f_{\tilde{Q}}|^2 \chi_Q(x) \right)^{1/2}.$$

It will suffice to show the corresponding weak-type inequality between $S_{a,R}(f)$ and $S_{b,R}(f)$, with a constant independent of R.

For $\lambda > 0$, let $\{Q_i\} \subset \mathscr{D}$ be the family of maximal dyadic cubes such that

$$\sum_{\substack{Q \in \mathscr{D}: \ell(Q) \leq R \\ Q_i \subset Q}} |f_Q - f_{\tilde{Q}}|^2 > \lambda^2.$$

(Our restriction to sidelengths less than or equal to R ensures that such maximal cubes exist.) Consider the family of cubes $\{\tilde{Q}_i\}$, the predecessors of the Q_i's. Denote the maximal members of this family by $\{I_j\}$. Each I_j contains some Q_i, and $\cup Q_i$ is the set where $S_{b,R}(f) > \lambda$. Our observation in the proof of Theorem 2 shows that, for each j,

$$v(I_j) \leq C \sum_{Q_i \subset I_j} \int_{Q_i} M_d(v)\, \mathrm{d}x;$$

and therefore,

$$\sum_j v(I_j) \leq C \int_{\{x:\ S_{b,R}(f) > \lambda\}} M_d(v)\, \mathrm{d}x,$$

which, by Chebyshev's Inequality, is less than or equal to

$$\frac{C}{\lambda} \int_{\mathbf{R}^d} S_{b,R}(f) M_d(v)\, \mathrm{d}x.$$

Thus, the weak-type inequality will follow if we can show

$$v(\{x \notin \cup I_j :\ S_{a,R}(f) > \lambda\}) \leq \frac{C}{\lambda} \int_{\mathbf{R}^d} S_{b,R}(f) M_d(v)\, \mathrm{d}x.$$

Write $f = g + b$, where

$$g(x) = \begin{cases} f_{I_j} & \text{if } x \in I_j; \\ f(x) & \text{if } x \notin \cup I_j. \end{cases}$$

If Q is any cube that is not *strictly* contained in some I_j, then $\int_Q b\,\mathrm{d}x = 0$. This is because such a cube Q must either (a) *be* an I_j, (b) be disjoint from $\cup I_j$, or (c) strictly contain some collection of I_j's. Since b's support is contained in $\cup I_j$ and $\int_{I_j} b\,\mathrm{d}x = 0$ for all j, the integral $\int_Q b\,\mathrm{d}x$ will be 0 in all three cases. But, if $x \notin \cup I_j$ and $x \in \tilde{Q}$, none of $\tilde{Q}$'s immediate dyadic subcubes can be strictly contained in an I_j (because then $\tilde{Q}$ would be contained in I_j, forcing $x \in I_j$). Therefore, if $x \notin \cup I_j$, $S_{a,R}(b)(x) = 0$. Since $S_{a,R}(f) \leq S_{a,R}(g) + S_{a,R}(b)$, our weak-type inequality now reduces to showing

$$v(\{x \notin \cup I_j :\ S_{a,R}(g) > \lambda\}) \leq \frac{C}{\lambda} \int_{\mathbf{R}^d} S_{b,R}(f) M_d(v)\,\mathrm{d}x.$$

We will actually show something stronger, namely

$$v\left(\{x \in \mathbf{R}^d :\ S_{a,R}(g) > \lambda\}\right) \leq \frac{C}{\lambda} \int_{\mathbf{R}^d} S_{b,R}(f) M_d(v)\,\mathrm{d}x.$$

By the way we constructed g, we have $S_{b,R}(g) \leq \lambda$ everywhere. It is easy to see that $S_{b,R}(g) \leq S_{b,R}(f)$ everywhere. (The sum that defines $S_{b,R}(g)$ is the same as the one defining $S_{b,R}(f)$, but with possibly fewer terms.) Because of our L^2 inequality ("step one") and our estimates on $S_{b,R}(g)$, we can write

$$\begin{aligned}\int_{\mathbf{R}^d} (S_{a,R}(g))^2 v\,\mathrm{d}x &\leq C \int_{\mathbf{R}^d} (S_{b,R}(g))^2 M_d(v)\,\mathrm{d}x \\ &\leq C\lambda \int_{\mathbf{R}^d} (S_{b,R}(g)) M_d(v)\,\mathrm{d}x \\ &\leq C\lambda \int_{\mathbf{R}^d} (S_{b,R}(f)) M_d(v)\,\mathrm{d}x.\end{aligned}$$

Therefore,

$$\int_{\mathbf{R}^d} (S_{a,R}(g))^2 v\,\mathrm{d}x \leq C\lambda \int_{\mathbf{R}^d} S_{b,R}(f) M_d(v)\,\mathrm{d}x.$$

The weak-type bound follows by Chebyshev's Inequality (dividing both sides by λ^2). Theorem 2 is proved.

The $p = 2$ case of Theorem 2 yields an exponential-square estimate between $S(f)$ and $S_b(f)$.

Corollary 1. *There are positive constants c_1 and c_2, depending only on d, such that, if $\|S_b(f)\|_\infty \leq 1$ then, for all dyadic cubes, Q and all $\lambda > 0$,*

$$|\{x \in Q : S(f - f_Q)(x) > \lambda\}| \leq c_1 |Q| \exp(-c_2 \lambda^2).$$

Proof. By replacing f with $(f - f_Q)\chi_Q$, we may, without loss of generality, assume that $S_b(f)$ is supported in Q. Set $E_\lambda \equiv \{x \in Q :\ S(f)(x) > \lambda\}$, and define $v = \chi_{E_\lambda}$. Theorem 2 and Chebyshev's Inequality imply

$$\lambda^2 |E_\lambda| \leq C \int_Q M_d(\chi_{E_\lambda})(x)\,\mathrm{d}x. \tag{5}$$

If $|E|_\lambda = 0$, we are done. Otherwise, the right-hand side of (5) is less than or equal to a constant times

$$\int_Q v(x)\log(1+v(x)/v_Q)\,dx \le C|E_\lambda|\log(1+|Q|/|E_\lambda|).$$

(see [4], p. 23 or [5], p. 17). Therefore,

$$\lambda^2 \le C\log(1+|Q|/|E_\lambda|);$$

which, after some algebraic unwinding, gives our result.

In the counterexample given in the introduction, showing that $S(f)$ can be unbounded even if $S_b \in L^\infty$, the set on which $S(f) > \sqrt{k}$ (that is where $(S(f))^2 > k$) has measure $\sim 2^{-2k}$. Therefore, beyond sharpening the constants c_1 and c_2, Corollary 1's exponential-square estimate cannot be improved.

3 The Exponential-Square Estimate for f: "Bellman" Proof

Without loss of generality we assume that f is real-valued, supported in $Q_0 \equiv [0,1)^d$ and that $\int f\,dx = 0$. By an approximation argument, we may also assume that f is constant on dyadic cubes of sidelength 2^{-N}, for some large positive N, so long as we obtain a bound on $|\{x \in Q_0 : |f(x)| > \lambda\}|$ that is independent of N.

Lemma 1. *If* $0 < c < 1/2$, $t^2 \le 1/2$, *and* $\xi \in \mathbf{R}$, *then*

$$c(1-t^2)(\xi+t)^2 - t^2 \le c\xi^2 + 2c\xi t - \frac{c}{2}\xi^2 t^2. \tag{6}$$

Proof. In inequality (6), the difference of the right- and left-hand sides equals

$$\begin{aligned}
&c\xi^2 + 2c\xi t - \frac{c}{2}\xi^2 t^2 - c\xi^2 - 2\xi ct - ct^2 + c\xi^2 t^2 + 2\xi ct^3 + ct^4 + t^2\\
&= \frac{c}{2}\xi^2 t^2 + 2c\xi t^3 + ct^4 + t^2 - ct^2\\
&= \frac{ct^2}{2}(\xi^2 + 4\xi t + 4t^2) - ct^4 + t^2 - ct^2\\
&= \frac{ct^2}{2}(\xi + 2t)^2 + ct^2 - ct^4 + t^2 - ct^2\\
&= \frac{ct^2}{2}(\xi + 2t)^2 + ct^2(1-t^2) + t^2(1-2c) \ge 0.
\end{aligned}$$

Lemma 2. *Let m be a positive integer. There is a positive constant $c(m)$ such that, for all $0 < c \le c(m)$ and all finite sequences $\{x_i\}_1^m \subset \mathbf{R}$ satisfying $\sum_1^m x_i = 0$, we have*

$$\frac{1}{m}\sum_1^m \exp\left(2cx_i - \frac{c}{2}x_i^2\right) \le 1.$$

Proof. Write $\phi(s) = \exp(2cs - \frac{c}{2}s^2) \equiv \exp(P(s))$. Then $\phi''(s) = ((P'(s))^2 + P''(s))\phi(s)$. But

$$(P'(s))^2 + P''(s) = c(c(2-s)^2 - 1).$$

If all of the x_i satisfy $c(2-x_i)^2 - 1 \le 0$, we get the result by the "concave" form of Jensen's inequality. If that is not the case, then some x_i lies outside the set $\{s \in \mathbf{R} : c(2-s)^2 \le 1\} = \{s \in \mathbf{R} : |2-s| \le c^{-1/2}\}$, which is the interval $[2-c^{-1/2}, 2+c^{-1/2}]$. The supremum of P off this set is attained at the endpoints of the interval, where P equals

$$2c\left(2 \pm \frac{1}{\sqrt{c}}\right) - \frac{c}{2}\left(2 \pm \frac{1}{\sqrt{c}}\right)^2 = 2c - \frac{1}{2}.$$

The global maximum of P is attained at $s = 2$, where $P = 2c$. Therefore, the worst that can happen in this case is we get one term no bigger than $\exp(2c - 1/2)$, and $m-1$ terms no bigger than $\exp(2c)$, giving us

$$\frac{1}{m}\sum_1^m \exp\left(2cx_i - \frac{c}{2}x_i^2\right) \le \frac{e^{2c}}{m}\left(e^{-1/2} + (m-1)\right),$$

which will be ≤ 1 so long as

$$e^{2c} \le \frac{m}{e^{-1/2} + (m-1)},$$

or $c \le (1/2)\log(\frac{m}{e^{-1/2}+(m-1)}) \equiv c_0(m)$.

We observe, because it will be important later, that $c_0(m)$ is always $< 1/2$.

Lemma 3. *Let Q_0 be the unit cube in $\mathbf{R}^d$, $f : Q_0 \mapsto \mathbf{R}$ integrable, $\int_{Q_0} f\,dx = 0$, and suppose that $(S_b(f))^2 \le 1/2$ everywhere. For $k \ge 1$ define*

$$S_{(k)}(f)(x) \equiv \sum_{j=1}^k \left(f_{(j)}(x) - f_{(j-1)}(x)\right)^2$$

and

$$P_{(k)}(f)(x) \equiv \prod_{j=1}^k \left[1 - (f_{(j)}(x) - f_{(j-1)}(x))^2\right].$$

Then the sequence of numbers

$$\int_{Q_0} \exp(c_0(2^d) f_{(k)}^2(x) P_{(k)}(f)(x)) \exp(-S_{(k)}(f)(x))\,dx$$

is nonincreasing.

Proof. We observe that $S_{(k)}(f) \leq 1/2$ and $1/2 \leq P_{(k)}(f) \leq 1$ everywhere. Suppose $Q \subset Q_0$ is dyadic, has sidelength 2^{-k}, and has 2^d dyadic successors $\{Q_j\}_1^{2^d}$. On Q, the functions $f_{(k)}$, $P_{(k)}(f)$, and $S_{(k)}(f)$ have some constant values: call them ξ, p, and s. On each Q_j, $f_{(k+1)}$ equals $\xi + t_j$, $P_{(k+1)}(f)$ equals $p(1-t_j^2)$, and $S_{(k+1)}(f)$ equals $s + t_j^2$, where $t_j^2 \leq 1/2$ for each j and $\sum_1^{2^d} t_j = 0$. Now put $c = c_0(2^d)p < 1/2$ and, for each j, apply Lemma 1 to ξ and $t = t_j$, to get, on each Q_j,

$$\begin{aligned}
c_0(2^d)p(1-t_j)^2(\xi+t_j)^2 - (s+t_j^2) &\leq c_0(2^d)p\xi^2 + 2c_0(2^d)p\xi t_j - \frac{c_0(2^d)p}{2}\xi^2 t_j^2 - s \\
&= \left(c_0(2^d)p\xi^2 - s\right) + \left(2c_0(2^d)p\xi t_j - \frac{c_0(2^d)p}{2}\xi^2 t_j^2\right) \\
&\equiv (I) + (II)_j,
\end{aligned}$$

where $(II)_j$ depends on j and (I) does not (i.e., is constant on all of Q). Quantity (I) is the value that $c_0(2^d)f_{(k)}^2 P_{(k)}(f)^2 - S_{(k)}(f)$ has on all of Q. Therefore,

$$\begin{aligned}
&\frac{1}{|Q|}\int_Q \exp(c_0(2^d)f_{(k+1)}^2(x)P_{(k+1)}(f)(x))\exp(-S_{(k+1)}(f)(x))\,dx \\
&\quad = \exp\left(c_0(2^d)f_{(k)}^2 P_{(k)}(f)^2 - S_{(k)}(f)\right)\left(2^{-d}\sum_1^{2^d}\exp((II)_j)\right).
\end{aligned}$$

We can write $(II)_j$ as $2cx_j - \frac{c}{2}x_j^2$ by setting $c = c_0(2^d)p$ as before and putting $x_j = \xi t_j$. Lemma 2 implies that

$$2^{-d}\sum_1^{2^d}\exp((II)_j) \leq 1,$$

proving Lemma 3.

The exponential-square estimate follows from the fact that $c_0(2^d)P_{(k)}(f)$ is strictly bounded below, independent of f. Theorem 3 is proved.

4 The Exponential-Square Estimate for f: Combinatorial Proof

The basic idea of the proof is simple. We are given that $S_b(f) \leq 1$ everywhere. We will decompose f into a sum

$$f = \sum_1^{2^d-1} f_j,$$

where each f_j is also supported in Q_0 and satisfies $\int f_j \, dx = 0$. For each f_j we will construct a function $\phi_j : [0,1) \mapsto \mathbf{R}$ such that

(a) for all $\lambda > 0$,

$$\left|\{x \in Q_0 : \ |f_j(x)| > \lambda\}\right| = \left|\{t \in [0,1) : \ |\phi_j(t)| > \lambda\}\right|, \textit{ and}$$

(b) $\|S(\phi_j)\|_\infty \leq 1$, where here $S(\cdot)$ means the one-dimensional dyadic square function. Then Theorem 3.1 from [2] will give us

$$\left|\{t \in [0,1) : \ |\phi_j(t)| > \lambda\}\right| \leq 2\exp(-\lambda^2/2)$$

and therefore

$$\left|\{x \in Q_0 : \ |f_j(x)| > \lambda\}\right| = 2\exp(-\lambda^2/2),$$

for each j. The exponential-square estimate for f will follow from summing (at the price of some unpleasant dimensional constants, but that is to be expected).

The decomposition and the square function bounds for the ϕ_j's depend on a lemma, and that lemma requires a definition.

Definition 2. If $n \geq 2$, a *crot* is a function $\psi : \{1,2,3,\dots,n\} \mapsto \mathbf{R}$ such that $\Sigma_1^n \psi(k) = 0$ and $\psi(k)$ is nonzero for at most two values of k.

Remark. In other words, a crot equals $+a$ at one point, $-a$ at another point, and is zero everywhere else. The number a can equal 0.

Lemma 4. *Suppose that $n \geq 2$ and $g : \{1,2,3,\dots,n\} \mapsto \mathbf{R}$ satisfies $\Sigma_1^n g(k) = 0$. There exist crots $\psi_j : \{1,2,3,\dots,n\} \mapsto \mathbf{R}$ $(1 \leq j \leq n-1)$ such that*

$$g = \sum_{j=1}^{n-1} \psi_j,$$

and, for every $k \in \{1,2,3,\dots,n\}$,

$$\sum_{j=1}^{n-1} |\psi_j(k)| = |g(k)|. \tag{7}$$

Remark. Since "$\geq$" comes for free in the last inequality, the content of (7) is the "$\leq$" direction. That happens to be the direction we want, but the proof of Theorem 3 will not use the full force of the inequality. We will only need

$$\max_{1 \leq j \leq n-1} |\psi_j(k)| \leq |g(k)| \tag{8}$$

for all $k \in \{1,2,3,\dots,n\}$.

Remark. We shall refer to Lemma 4 as the Crot Lemma.

Proof of Theorem 4. We do induction on n. By renumbering we can assume that $g(n-1)$ and $g(n)$ have opposite signs (they might both be zero) and $|g(n)| = \min_{1\le k\le n}|g(k)|$. We decompose g into h_1+h_2, where

$$h_1(k) = \begin{cases} g(n) & \text{if } k=n; \\ -g(n) & \text{if } k=n-1; \\ 0 & \text{otherwise;} \end{cases}$$

and

$$h_2(k) = \begin{cases} 0 & \text{if } k=n; \\ g(n-1)+g(n) & \text{if } k=n-1; \\ g(k) & \text{if } k<n-1. \end{cases}$$

The function h_1 is a crot. Our strictures on $g(n)$ and $g(n-1)$ (their opposite signs and the minimality of $|g(n)|$) imply that $|g(k)| = |h_1(k)|+|h_2(k)|$ for all k. We notice that $\sum_1^{n-1} h_2(k) = \sum_1^n g(k) = 0$. By induction, the restriction of h_2 to $\{1,2,3,\dots,n-1\}$ can be decomposed into $n-2$ crots ψ_j satisfying

$$\sum_{j=1}^{n-2} |\psi_j(k)| = |h_2(k)|$$

for all $k\in\{1,2,3,\dots,n-1\}$. Extend each ψ_j to all of $\{1,2,3,\dots,n\}$ by setting $\psi_j(n)=0$. Then $\{h_1, \psi_1, \psi_2, \dots, \psi_{n-2}\}$ is the desired family of crots. □

Remark. Notice that the preceding construction works even if $g\equiv 0$.

Remark. With obvious modifications, the Crot Lemma applies to any real-valued function g defined on a set S with n elements, so long as

$$\sum_{s\in S} g(s) = 0.$$

Remark. According to the website *grammar.about.com*, a "crot" is "A verbal bit or fragment used as an autonomous unit to create an effect of abruptness and rapid transition." We encountered the word in a book by John Gardner (the author of *Grendel*, not the thriller writer).

If $Q\subset Q_0$ is a dyadic cube then

$$a_Q(f) = \sum_{Q'\in N(Q)} c_{Q'}\chi_{Q'},$$

for some constants $c_{Q'}$. (Recall that $N(Q)$ is the family of Q's immediate dyadic successors.) The constants satisfy $\sum_{Q' \in N(Q)} c_{Q'} = 0$, and the set $N(Q)$ has 2^d elements. By the Crot Lemma applied to the set $N(Q)$ and the function $g : N(Q) \mapsto \mathbf{R}$ defined by $g(Q') \equiv c_{Q'}$, we can find constants $\{\alpha_j\}_1^{2^d-1}$ and pairs of cubes $\{A_{Q,j}, B_{Q,j}\}_1^{2^d-1}$ such that (a) for each j, $A_{Q,j}$ and $B_{Q,j}$ both belong to $N(Q)$; (b) for each j, $A_{Q,j} \neq B_{Q,j}$; (c) for all $x \in Q$,

$$a_Q(f)(x) = \sum_{j=1}^{2^d-1} \alpha_{j,Q} \left(\chi_{A_{Q,j}}(x) - \chi_{B_{Q,j}}(x) \right);$$

and (d) for all $x \in Q$

$$\sum_{j=1}^{2^d-1} |\alpha_{j,Q}| \left(\chi_{A_{Q,j}}(x) + \chi_{B_{Q,j}}(x) \right) = |a_Q(f)(x)|.$$

We are now ready to define the f_j's. They are

$$f_j(x) = \sum_{Q \in \mathscr{D}(Q_0)} \alpha_{j,Q} \left(\chi_{A_{Q,j}}(x) - \chi_{B_{Q,j}}(x) \right), \tag{9}$$

where, as defined in the introduction, we are using $\mathscr{D}(Q_0)$ to mean the collection of all of Q_0's dyadic subcubes (including Q_0 itself). The reader should note that the sum (9) is finite (i.e., has only finitely many nonzero terms) because f is constant on sufficiently small dyadic cubes. The reader should also note that, for each j, $\alpha_{j,Q}$, $A_{Q,j}$, and $B_{Q,j}$ are defined for *all* of Q_0's dyadic subcubes Q.

Each f_j is a linear sum of things that look like "Haar functions," equaling $+\alpha_{j,Q}$ on one cube in $N(Q)$, $-\alpha_{j,Q}$ on another, and zero on the rest. Because of the Crot Lemma,

$$|\alpha_{j,Q} \left(\chi_{A_{Q,j}}(x) - \chi_{B_{Q,j}}(x) \right)| \leq |a_Q(f)(x)|$$

pointwise. Therefore, for all $x \in Q_0$,

$$\begin{aligned} \sum_{Q \in \mathscr{D}(Q_0)} \left| \alpha_{j,Q} \left(\chi_{A_{Q,j}}(x) - \chi_{B_{Q,j}}(x) \right) \right|^2 &\leq \sum_{Q \in \mathscr{D}(Q_0)} |a_Q(f)(x)|^2 \\ &\leq (S_b(f)(x))^2 \\ &\leq 1 \end{aligned} \tag{10}$$

because of our assumption on $S_b(f)$. Looking a little harder, we can see that

$$\begin{aligned} \sum_{Q \in \mathscr{D}(Q_0)} \left| \alpha_{j,Q} \left(\chi_{A_{Q,j}}(x) - \chi_{B_{Q,j}}(x) \right) \right|^2 &= \sum_{Q \in \mathscr{D}(Q_0)} |\alpha_{j,Q}|^2 \left(\chi_{A_{Q,j}}(x) + \chi_{B_{Q,j}}(x) \right) \\ &= (S_b(f_j)(x))^2 \end{aligned} \tag{11}$$

everywhere in Q_0. The reader should be sure that he understands this; it will become important very soon.

Now we are ready to define the ϕ_j's. Since j will remain fixed until the very end of the proof, we will temporarily suppress it; until we say otherwise, ϕ means ϕ_j, f means f_j, α_Q means $\alpha_{j,Q}$, A_Q means $A_{Q,j}$, etc.

We will define ϕ by means of an identification between the dyadic subcubes of Q_0 and certain of the dyadic subintervals of $[0,1)$. Recall that we defined $\mathscr{D}(Q)$ to mean the collection of Q's dyadic subcubes (including Q itself). We will use $\mathscr{I}_d([0,1))$ to mean the 2^d-adic subintervals of $[0,1)$ (including $[0,1)$ itself). In other words, $J \in \mathscr{I}_d([0,1))$ if and only if $J \subset [0,1)$ and $J = [j2^{nd}, (j+1)2^{nd})$ for some integer j and some (necessarily nonpositive) integer n.

We define our identification $\psi : \mathscr{D}(Q_0) \mapsto \mathscr{I}_d([0,1))$ *inductively*.

Step one: $\psi(Q_0) = [0,1)$.

Inductive step: Suppose that $\psi(Q) = I \in \mathscr{I}_d([0,1))$. We can write

$$N(Q) = \{A_Q, B_Q\} \cup R(Q),$$

where $R(Q) \equiv N(Q) \setminus \{A_Q, B_Q\}$ (the "remainder" of $N(Q)$) has $2^d - 2$ elements. We can also write

$$I = \cup_1^{2^d} I_l,$$

a disjoint union of 2^d-adic subintervals, where the measure (length) of each I_l equals $2^{-d}|I|$ and the I_ls are ordered from left to right: I_1 is leftmost in I, I_2 lies immediately to the right of I_1, and so on.

We now define $\psi(A_Q) = I_1$, $\psi(B_Q) = I_2$, and we let ψ assign the elements of $R(Q)$ to $\{I_3, I_4, \dots, I_{2^d}\}$ in any convenient, one-to-one fashion.

We recall that the function f is defined by

$$f(x) \equiv \sum_{Q \in \mathscr{D}(Q_0)} \alpha_Q \left(\chi_{A_Q}(x) - \chi_{B_Q}(x)\right).$$

The function ϕ is given by

$$\phi(t) \equiv \sum_{Q \in \mathscr{D}(Q_0)} \alpha_Q \left(\chi_{\psi(A_Q)}(t) - \chi_{\psi(B_Q)}(t)\right).$$

It is easy to see that $\psi : \mathscr{D}(Q_0) \mapsto \mathscr{I}_d([0,1))$ sets up a one-to-one correspondence between $\mathscr{D}(Q_0)$ and $\mathscr{I}_d([0,1))$ that preserves the set relations inclusion and disjointness: for all Q and Q' in $\mathscr{D}(Q_0)$, $Q \subset Q'$ if and only if $\psi(Q) \subset \psi(Q')$, and $Q \cap Q' = \emptyset$ if and only if $\psi(Q) \cap \psi(Q') = \emptyset$. The reader should notice that the sets A_Q and B_Q are, in a manner, joined at the hip. If $Q' \in \mathscr{D}(Q_0)$ and $A_Q \subset Q'$ (or, by symmetry, if $B_Q \subset Q'$), then we have two possibilities: $Q = Q'$ or $A_Q \cup B_Q$ is entirely contained in one of $A_{Q'}$ or $B_{Q'}$. The mapping preserves this property in the images of the A_Q's and B_Q's. The mapping also preserves measure: for all $Q \in \mathscr{D}(Q_0)$, $|Q| = |\psi(Q)|$. Because we have set things up so that, for every $Q \in \mathscr{D}(Q_0)$, $\psi(A_Q)$ and $\psi(B_Q)$ are the left and right halves of a dyadic interval I, we have that

$$(S(\phi)(t))^2 = \sum_{Q : t \in \psi(A_Q) \cup \psi(B_Q)} |\alpha_Q|^2.$$

We claim that this is less than or equal to 1 everywhere. Suppose $S(\phi)(t) > 0$, and let $Q^* \in \mathscr{D}(Q_0)$ be the minimal cube such that $t \in \psi(A_{Q^*}) \cup \psi(B_{Q^*})$ and $\alpha_{Q^*} \neq 0$. Such a cube exists because (9) has only finitely many nonzero terms. For this t, $(S(\phi)(t))^2$ equals

$$\sum_{\substack{Q \in \mathscr{D}(Q_0) \\ \psi(A_{Q^*}) \cup \psi(B_{Q^*}) \subset \psi(A_Q) \cup \psi(B_Q)}} |\alpha_Q|^2.$$

But the set-inclusion properties of ψ imply that this last sum also equals

$$\sum_{\substack{Q \in \mathscr{D}(Q_0) \\ A_{Q^*} \cup B_{Q^*} \subset A_Q \cup B_Q}} |\alpha_Q|^2,$$

which we know to be less than or equal to 1, from (10) and (11).

Because of Theorem 3.1 from [2], we now know that, for all $\lambda > 0$,

$$|\{t \in [0,1): \ |\phi(t)| > \lambda\}| \leq 2\exp(-\lambda^2/2).$$

In order to get the corresponding estimate for f, we need to show that f and ϕ have the same distribution functions. The easiest way to see this fact is to first observe that f and ϕ are, for all intents and purposes, defined on discrete probability spaces. This is because f is constant on dyadic subcubes of sidelength 2^{-N}, and ϕ is constant on dyadic intervals of length 2^{-Nd}. To make things precise, we will call these two spaces (Ω_1, P_1) and (Ω_2, P_2), where

$$\Omega_1 \equiv \{Q \in \mathscr{D}(Q_0): \ \ell(Q) = 2^{-N}\}$$

$$\Omega_2 \equiv \{I \in \mathscr{I}_d([0,1))): \ |I| = 2^{-Nd}\},$$

$P_1(Q) = 2^{-Nd}$ for all $Q \in \Omega_1$, and $P_2(I) = 2^{-Nd}$ for all $I \in \Omega_2$. For $Q \in \Omega_1$ we define $\tilde{f}: \Omega_1 \mapsto \mathbf{R}$ by

$$\tilde{f}(Q) \equiv f_Q = \sum_{Q' \in \mathscr{D}(Q_0)} \alpha_{Q'} \left(\chi_{A_{Q'}}(x_Q) - \chi_{B_{Q'}}(x_Q)\right),$$

where x_Q means the center of Q; and for $I \in \Omega_2$, $I = \psi(Q)$, we define $\tilde{\phi}: \Omega_2 \mapsto \mathbf{R}$ by

$$\tilde{\phi}(I) \equiv \phi_I = \sum_{Q' \in \mathscr{D}(Q_0)} \alpha_{Q'} \left(\chi_{\psi(A_{Q'})}(t_I) - \chi_{\psi(B_{Q'})}(t_I)\right),$$

where t_I means the center of I. It is clear that f and $\tilde{f}$ have the same distribution functions on their respective probability spaces (Q_0 with Lebesgue measure vs. (Ω_1, P_1)) as do ϕ and $\tilde{\phi}$ on theirs ($[0,1)$ with Lebesgue measure vs. (Ω_2, P_2)). The function ψ induces, in the natural sense, a measure (probability) preserving bijection between (Ω_1, P_1) and (Ω_2, P_2), and it is trivial that $\tilde{f}(Q) = \tilde{\phi}(\psi(Q))$ for all $Q \in \Omega_1$. Therefore, $\tilde{f}$ and $\tilde{\phi}$—hence f and ϕ—have the same distribution functions, meaning, for all $\lambda > 0$,

$$|\{x \in Q_0: \ |f(x)| > \lambda\}| \leq 2\exp(-\lambda^2/2).$$

We will now bring back the j's. Summing the exponential-square estimates for each f_j yields, for our *original* function f,

$$|\{x \in Q_0 : |f(x)| > \lambda\}| \leq 2(2^d - 1)\exp\left(-\frac{\lambda^2}{2(2^d-1)^2}\right).$$

Theorem 3 is proved again.

References

1. Buckley, S.M.: Summation conditions on weights. Michigan Math. Journal **40**, 153–170 (1993)
2. Chang, S.Y.A., Wilson, J.M., Wolff, T.H.: Some weighted norm inequalities concerning the Schroedinger operators. Comm. Math. Helv. **60**, 217–246 (1985)
3. Chanillo, S., Wheeden, R.L.: Some weighted norm inequalities for the area integral. Indiana U. Math. Jour. **36**, 277–294 (1987)
4. Stein, E.M.: Singular Integrals and Differentiability Properties of Functions. Princeton University Press, Princeton (1970)
5. Wilson, M.: Weighted Littlewood–Paley theory and exponential-square integrability. Springer Lecture Notes in Mathematics, vol. 1924. Springer, New York (2007)

L_∞-Bounds for the L_2-Projection onto Linear Spline Spaces

Peter Oswald

Dedicated to Kostja Oskolkov

Abstract In the univariate case, the L_2-orthogonal projection P_V onto a spline space V of degree k is bounded as an operator in L_∞ by a constant $C(k)$ depending on the degree k but independent of the knot sequence. In the case of linear spline spaces the sharp bound is

$$||P_V||_{L_\infty \to L_\infty} < 3,$$

as established by Ciesielski, Oskolkov, and the author. As was shown more recently, the L_2-orthogonal projection P_V onto spaces $V = V(\mathcal{T})$ of linear splines over triangulations $\mathcal{T}$ of a bounded polygonal domain in $\mathbb{R}^2$ cannot be bounded in L_∞ by a constant that is independent of the underlying triangulation. Similar counterexamples show this for higher dimensions as well. In this note we state a new geometric condition on families of triangulations under which uniform boundedness of $||P_V||_{L_\infty \to L_\infty}$ can be guaranteed. It covers certain families of triangular meshes of practical interest, such as Shishkin and Bakhvalov meshes. On the other hand, we show that even for type-I triangulations of a rectangular domain uniform boundedness of P_V in L_∞ cannot be established.

Mathematics Subject Classification (2000): 65N30, 41A15.

P. Oswald (✉)
Jacobs University Bremen, College Ring 1, D-28759 Bremen, Germany
e-mail: p.oswald@jacobs-university.de

D. Bilyk et al. (eds.), *Recent Advances in Harmonic Analysis and Applications*, Springer Proceedings in Mathematics & Statistics 25, DOI 10.1007/978-1-4614-4565-4_24,

1 Introduction

The study of the boundedness of projections onto spline spaces V with respect to various norms has been motivated by applications to the numerical analysis of the finite element method and by investigations on spline basis in function spaces. A particular problem is to find bounds for L_2-orthogonal projections P_V in the L_∞-norm. In the univariate case, A. Shadrin [14] has proved that

$$\|P_V\|_{L_\infty \to L_\infty} \leq C(k) < \infty \tag{1}$$

for any space of splines of degree k, answering a longstanding conjecture by C. de Boor (see, e.g., [2]). As far as we know, only for $k = 1$ the exact value of $C(k)$ is known: For linear spline spaces V and any knot sequence, we have $\|P_V\|_{L_\infty \to L_\infty} < C(1) := 3$ by a result of Z. Ciesielski [3], and the sharpness of this bound was independently shown by K. I. Oskolkov [8] and the author [9].

In higher dimensions, boundedness of projections onto spline and finite element spaces was studied under various geometric conditions on the underlying partitions. We refer to [1, 4, 5] for conditions based on maximum/minimum angle and other regularity assumptions. It was folklore that a bound (1) cannot hold in $\mathbb{R}^d$, $d > 1$, but formal proof was given only recently. In [10], we constructed a sequence of triangulations $\mathcal{T}_J$ of a square into $O(J)$ triangles such that the L_2-orthogonal projection operator $P_V : L_2(\Omega) \to V$ onto the linear spline space $V = V(\mathcal{T})$ defined over a finite triangulation $\mathcal{T}$ of a bounded polygonal domain $\Omega \in \mathbb{R}^2$ satisfies the lower bound

$$\|P_{V(\mathcal{T}_J)}\|_{L_\infty \to L_\infty} \geq J, \quad J \geq 1. \tag{2}$$

We also mentioned in [10] unpublished work by A. A. Privalov who (according to A. Shadrin and Yu. N. Subbotin) stated examples of a similar flavor based on type-I triangulations of a rectangle in a conference talk around 1984. At the time of the preparation of [10], we were not able to verify this information, and claimed that the methods leading to (2) cannot be used to give similar lower bounds for tensor-product type-I triangulations. In Sect. 3, we correct this wrong statement by showing the same result for a sequence of type-I triangulations $\mathcal{T}_J$ partitioning a square into $2J^2$ triangles.

The reason for returning to this question was, however, a different one: Triangulations obtained by bisecting the rectangles of a tensor-product rectangular partition are popular in certain finite element applications, e.g., as Shishkin- or Bakhvalov-type meshes for boundary layer treatment in convection-dominated second-order elliptic problems (see [11] for some survey and further references), and having L_∞-bounds for P_V for such triangulations could be helpful in the analysis of these problems, compare [7, 13]. In Sect. 2, Theorem 1, we provide a uniform upper bound in terms of certain geometric parameters (such as maximal valence of vertices and growth of local triangle area ratios) that is applicable to these classes of meshes. Using these techniques, we establish that for triangulations $\mathcal{T}$ obtained from an arbitrary tensor-product partition $\mathcal{P}$ of a rectangle by bisecting the rectangular cells in $\mathcal{P}$, the bisection pattern can always be chosen such that $\|P_{V(\mathcal{T})}\|_{L_\infty \to L_\infty}$ remains uniformly bounded, independently of $\mathcal{P}$ (see Theorem 2 in Sect. 3). The above mentioned counterexample shows that such a result cannot hold for all bisection patterns.

The author is indebted to H.-G. Roos for attracting his attention to Shishkin meshes which triggered work on this note. He also acknowledges the helpful information exchange with A. Shadrin and Yu. N. Subbotin about the unaccessible earlier work by A. A. Privalov.

2 Sufficient Geometric Conditions

We recall some simple facts and notation already used in [10]. As usual, we parameterize $g \in V(\mathcal{T})$ by its nodal values $x_P = g(P)$ at the vertices $P \in \mathcal{V}_\mathcal{T}$ of $\mathcal{T}$, i.e.,

$$g = \sum_{P \in \mathcal{V}_\mathcal{T}} x_P \phi_P, \tag{3}$$

where $\{\phi_P \in V(\mathcal{T})\}$ denotes the standard nodal basis in $V(\mathcal{T})$. The support set $\Omega_P := \operatorname{supp} \phi_P$ of the nodal basis function ϕ_P corresponds to the 1-ring neighborhood of P in $\mathcal{T}$; its area is denoted by A_P. We denote by $\mathcal{V}_P = \{Q \in \Omega_P \cap \mathcal{V}_\mathcal{T} : Q \neq P\}$ the set of all neighboring vertices to P, by $K_P = \#\mathcal{V}_P$ the valence of P in $\mathcal{T}$, and by A_{PQ} the sum of the areas of the triangles attached to the edge PQ, $Q \in \mathcal{V}_P$. The following lemma was proved in [10].

Lemma 1. *We have*

$$\|P_{V(\mathcal{T})}\|_{L_\infty \to L_\infty} \asymp \|A^{-1}\|_\infty, \tag{4}$$

with constants independent of $\mathcal{T}$, where the matrix A is the diagonally scaled Gram matrix of the nodal basis, with entries a_{PQ} given by the formula

$$a_{PQ} = \begin{cases} 1, & Q = P, \\ \frac{(\phi_Q, \phi_P)}{(\phi_P, \phi_P)} = \frac{A_{PQ}}{2A_P} > 0, & Q \in \mathcal{V}_P, \\ 0, & \text{otherwise}, \end{cases} \quad . \tag{5}$$

In particular,

$$1 = \sum_{Q \neq P} a_{PQ} = \sum_{Q \in \mathcal{V}_P} a_{PQ}, \qquad \forall\, P \in \mathcal{V}_\mathcal{T}. \tag{6}$$

We will use this structure of A to give uniform bounds for $\|P_{V(\mathcal{T})}\|_{L_\infty \to L_\infty}$ in terms of the *maximal valence* $K_{\max}(\mathcal{T}) := \max_{P \in \mathcal{V}} K_P$ and another, a bit unusual geometric characteristics $L_{\max}(\mathcal{T}, r) := \max_{P \in \mathcal{V}} L_P(r)$ of $\mathcal{T}$ called *depth of local area growth* with ratio $r \geq 1$. To define it, we denote by Δ'_Q the area-largest triangle attached to $Q \in \mathcal{V}$ and by A'_Q its area (if there are several ones, we consider all of them one-by-one). The quantity $L_P(r)$ equals the length L of the *shortest* chain of vertices $P_0 = P, P_1, \ldots, P_L$ such that (a) the vertex $P_k \neq P_{k-1}$ belongs to $\Delta'_{P_{k-1}}$ for $k = 1, \ldots, L$, and (b) $A'_{P_L} \leq rA'_{P_{L-1}}$ (as a consequence of the requirement that this chain is the shortest among all chains with properties (a) and (b), we automatically have $A'_{P_k} \geq rA'_{P_{k-1}}$ for $k = 1, \ldots, L-1$). Obviously, $L_P(r)$ (and thus $L_{\max}(\mathcal{T}, r)$) are bounded by the number of triangles in $\mathcal{T}$. The counterexamples from [10] provide evidence that $L_{\max}(\mathcal{T}, r)$ can grow proportionally with the number of triangles.

Theorem 1. *Given positive integers K_0, L_0, and a real number $r_0 > 1$, there is a finite constant $C(K_0, L_0, r_0)$ such that for any triangulation $\mathcal{T}$ satisfying $K_{\max}(\mathcal{T}) \leq K_0$ and $L_{\max}(\mathcal{T}, r) \leq L_0$ for some $r \leq r_0$, we have the bound*

$$\|P_{V(\mathcal{T})}\|_{L_\infty \to L_\infty} \leq C(K_0, L_0, r_0).$$

Proof. According to (4), it is enough to show that the solution of the system $Ax = b$ satisfies

$$\|x\|_\infty := \max_{P \in \mathcal{V}_\mathcal{T}} |x_P| \leq C(K_0, L_0, r_0) \tag{7}$$

for any right-hand side b with $\|b\|_\infty \leq 1$. Denote $M = \|x\|_\infty$, and find the vertex P with $|x_P| = M$. W.l.o.g., let $x_P = M$. Since by assumption $L_P(r) \leq L_0$, we can find a sequence of points $P_0 = P, P_1, \ldots, P_L$, $L \leq L_0$, with properties (a) and (b) stated above. Now, set $a_k := M - (-1)^k x_{P_k}$. Obviously, $a_0 = 0$, and $a_k \geq 0$ because $M \pm x_Q \geq 0$ for all Q (this latter fact will be used without further mentioning). By induction in $k = 1, \ldots, L$, we show an upper bound for a_k for $k \leq L$. Indeed, by (a) $P_k \neq P_{k-1}$ is one of the vertices of $\Delta'_{P_{k-1}}$ and belongs to $\mathcal{V}_{P_{k-1}}$. Since $\Delta'_{P_{k-1}}$ is the largest in area triangle among the $K_{P_{k-1}} \leq K_{\max}(\mathcal{T}) \leq K_0$ triangles attached to P_{k-1}, it follows from the definition (5) that $a_{P_{k-1}P_k} \geq 1/(2K_0)$ and

$$\begin{aligned}
a_k &\leq 2K_0 a_{P_{k-1}P_k}(M + (-1)^k x_{P_k}) \leq 2K_0 \left(\sum_{Q \in \mathcal{V}_{P_{k-1}}} a_{P_{k-1}Q}(M + (-1)^k x_Q) \right) \\
&= 2K_0 \left(M + (-1)^k \sum_{Q \in \mathcal{V}_{P_{k-1}}} a_{P_{k-1}Q} x_Q \right) \\
&= 2K_0(M + (-1)^k (b_{P_{k-1}} - x_{P_{k-1}})) \leq 2K_0(a_{k-1} + 1),
\end{aligned}$$

and this gives by induction

$$\max_{k=0,\ldots,L} a_k \leq (2K_0)^L + \ldots + (2K_0)^2 + 2K_0 = 2K_0 \frac{(2K_0)^L - 1}{2K_0 - 1} \leq 2(2K_0)^{L_0}.$$

Now consider $\Delta'_{P_{L-1}}$. By construction, it has vertices P_{L-1}, P_L, and a third one which we denote by R. Let $a_R = M-(-1)^L x_R$, and observe that by the same reasoning as above for $k = L-1$, we also have $0 \le a_R \le 2(2K_0)^{L_0}$, i.e., substituting this information into the equation corresponding to P_k, we have

$$\begin{aligned}(-1)^L b_{P_L} &= (-1)^L x_{P_L} + a_{P_L R}(-1)^L x_R + \sum_{Q\in\mathcal{V}_{P_L}, Q\neq R} a_{P_L Q}(-1)^L x_Q \\ &\ge (1+a_{P_L R})(M-2(2K_0)^{L_0}) - M \sum_{Q\in\mathcal{V}_{P_L}, Q\neq R} a_{P_L Q} \\ &\ge (1+a_{P_L R})(M-2(2K_0)^{L_0}) - (1-a_{P_L R})M \\ &\ge 2a_{P_L R}M - 4(2K_0)^{L_0}.\end{aligned}$$

This gives

$$M = \|x\|_\infty \le \frac{4(2K_0)^{L_0}+1}{2a_{P_L R}}.$$

It remains to observe that by property (b) of the chain $A'_{P_L} \le rA'_{P_{L-1}}$, and thus

$$a_{P_L R} \ge \frac{A'_{P_{L-1}}}{2A_{P_L}} \ge \frac{A'_{P_{L-1}}}{2K_{P_L}A'_{P_L}} \ge \frac{1}{2K_0 r_0}.$$

Putting the last two estimates together, we arrive at (7). This concludes the proof of Theorem 1.

Remark 1. The same bound holds if we consider the L_2-projection onto subspaces of $V \subset V(\mathcal{T})$ generated by any subset of nodal basis functions (e.g., subspaces of linear finite elements over $\mathcal{T}$ satisfying homogeneous Dirichlet boundary conditions). What changes is that equality in (6) is replaced by inequality if $\mathcal{V}_P$ contains an eliminated node. The modifications in the proof are minimal and left to the reader.

Remark 2. Here are a couple of simpler conditions on $\mathcal{T}$ that imply bounds on $L_{\max}(\mathcal{T},r)$. *Locally area-quasiuniform triangulations* are characterized by a uniform bound $\le r_0$ for the ratios of the areas of any two triangles sharing at least one vertex. According to our definition of the depth of local area growth, this implies $L_{\max}(\mathcal{T},r_0) = 1$. *Regular triangulations* (a standard assumption in finite element applications) for which the ratio of radii of circumscribed to inscribed circle is bounded by γ for any triangle in $\mathcal{T}$ are obviously locally area-quasiuniform for some properly chosen r_0 (and satisfy in addition $K(\mathcal{T}) \le K_0$ for some K_0 depending on γ only).

Another example is as follows. Suppose that the triangles in $\mathcal{T}$ can be sorted into L_0 classes where within each class any two triangles (not only those touching each other!) have an area ratio $\le r_0$. Then necessarily $L_{\max}(\mathcal{T},r_0) \le L_0$. Indeed,

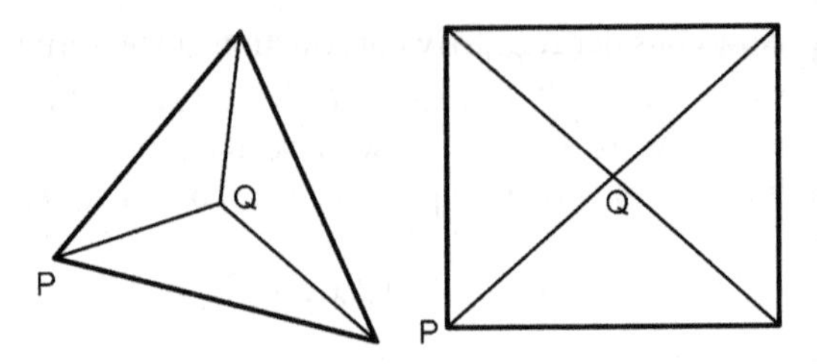

Fig. 1 Clough-Tocher split of a triangle respectively criss-cross split of a rectangle

choose $P \in \mathcal{V}$ such that $L_P(r_0) = L_{\max}(\mathcal{T}, r_0)$, and consider the associated shortest chain $P_0 = P, P_1, \ldots, P_{L_P(r_0)}$ satisfying conditions (a) and (b). Since $A'_{P_k} > r_0 A'_{P_{k-1}}$ for $k = 1, \ldots, L_P(r_0) - 1$, the associated triangles Δ'_{P_k}, $k = 0, 1, \ldots, L_P(r_0) - 1$, must belong to different classes. Thus, $L_{\max}(\mathcal{T}, r_0) = L_P(r_0) \le L_0$. Since Shishkin-type meshes are composed of a finite number of quasiuniform meshes, this assures that for those uniform L_∞-boundedness of $P_{\mathcal{V}(\mathcal{T})}$ holds as long as the number of different boundary/interior layers is fixed (often $L_{\max}(\mathcal{T}, r_0)$ is much smaller than the crude upper bound given by the number of layers). It is not hard to show that also Bakhvalov-type meshes [11, 12] have uniformly bounded $L_{\max}(\mathcal{T}, r)$ if $r > 1$ is properly chosen, and that layer-adapted partitions on general domains [6] can be treated as well.

Last but not least, for any $\mathcal{T}$ we can obtain a refined triangulation $\tilde{\mathcal{T}}$ by inserting one new vertex into each triangle at its barycenter and subdividing it into 3 triangles of equal area (Clough-Tocher split). Then

$$L_{\max}(\tilde{\mathcal{T}}, r) \le 2, \qquad r > 1.$$

Similarly, if we take any tensor-product partition $\mathcal{P}$ of a rectangle and subdivide each of the rectangular cells in $\mathcal{P}$ into 4 triangles of equal area by inserting its two diagonals, then the resulting crisscross triangulation $\mathcal{T}_{\mathcal{P}}$ also satisfies

$$L_{\max}(\mathcal{T}_{\mathcal{P}}, r) \le 2, \qquad r \ge 1,$$

while $K_{\max}(\mathcal{T}_{\mathcal{P}}) \le K_0 := 8$ holds automatically. Indeed, in both cases there are only two types of vertices: old vertices from $\mathcal{T}$ resp. $\mathcal{P}$ and newly inserted vertices (one per cell) that lead to $\tilde{\mathcal{T}}$ resp. $\mathcal{T}_{\mathcal{P}}$. By construction, for each old vertex P, there are at least two neighboring area-largest triangles Δ'_P which share one new vertex Q, see Fig. 1. Obviously, all triangles attached to Q have the same area; thus, $A'_Q = A'_P \le rA'_P$ for any $r \ge 1$, and thus $L_P(r) = 1$. If we start at a new vertex Q and for any of the old vertices P of the cell containing Q, we have $A'_P \le rA'_Q$, then again $L_Q(r) = 1$. Otherwise, we have $L_Q(r) = L_P(r) + 1 = 2$ by the already established fact for old vertices P. Note that with appropriately chosen $r > 1$, the result also holds if the cells in a triangulation or quadrilateral partition are subdivided into 3 or 4 triangles, respectively, of approximately the same area.

Remark 3. The above proof heavily relies on the special properties of the diagonally scaled Gram matrices A which are nonnegative and weakly diagonally dominant, see the formulas (5) and (6). These properties do not carry over to higher dimensions nor to most of the other finite element and spline spaces in use.

3 Triangulations Obtained from Tensor-Product Rectangular Partitions

This section deals with linear spline spaces V on triangulations of rectangles (resp. of domains composed of a few axis-parallel rectangles such as an L-shaped domain) obtained from a tensor-product rectangular partition by bisecting each of its rectangular cells. The major observation is that the uniform boundedness of P_V crucially depends on the particular bisection pattern.

Our first result concerns the existence of tensor-product rectangular partitions $\mathcal{P}_J$ of a rectangle into $\mathrm{O}(J^2)$ rectangles such that for the associated type-II triangulations $\mathcal{T}_J$, the lower bound (2) holds. Such an example was, according to recollections by A. Shadrin and Yu. N. Subbotin, first given by A. A. Privalov in a conference talk around 1984, but no published record seems available. How $\mathcal{T}_J$ looks like is depicted in Fig. 2.

Theorem 2. *There exist rectangular partitions $\mathcal{P}_J$ and associated type-II triangulations $\mathcal{T}_J$ of a rectangular domain into $\mathrm{O}(J^2)$ rectangles and triangles, respectively, such that*

$$\|P_{V(\mathcal{T}_J)}\|_{L_\infty \to L_\infty} \geq J, \qquad J \geq 1.$$

Proof. The argument is similar to the one used in [10]. We temporarily fix J, and look at tensor products of the same univariate partition $0 = t_0 < t_1 < \ldots < t_J < t_{J+1} = 1$ to generate a family of rectangular partitions $\mathcal{P}$ and of the associated type-II triangulations $\mathcal{T}$ with $\mathrm{O}(J^2)$ rectangles and triangles, respectively. If we let the local mesh ratios $(t_{j-1} - t_j)/(t_j - t_{j+1})$, $j = 1, \ldots, J$, simultaneously tend to ∞, then the scaled Gram matrices A associated with $P_{V(\mathcal{T})}$ converge to an invertible limit matrix A_J satisfying the bound

$$\|A_J^{-1}\|_\infty \geq \frac{3}{2} J, \qquad J \geq 1. \tag{8}$$

This shows the claim of Theorem 2 since by continuity we can choose from the considered family of triangulations a type-II triangulation $\mathcal{T} =: \mathcal{T}_J$ into $2(J+1)^2$ triangles with $A \approx A_J$ and $A^{-1} \approx A_J^{-1}$ such that $\|A^{-1}\|_\infty \geq J$ (the constant $c > 0$ coming from (4) can easily be compensated for by choosing a slightly larger $J' \leq CJ$ in this construction).

To establish (8), we use pairs (i,j), $i,j = 0, \ldots, J+1$ to index the entries associated with the "virtual" vertex that corresponds to the limit of $P_{i,j} = (t_i, t_j)$ (see Fig. 1) of the column vectors b and x in the linear system $A_J x = b$ (note that in

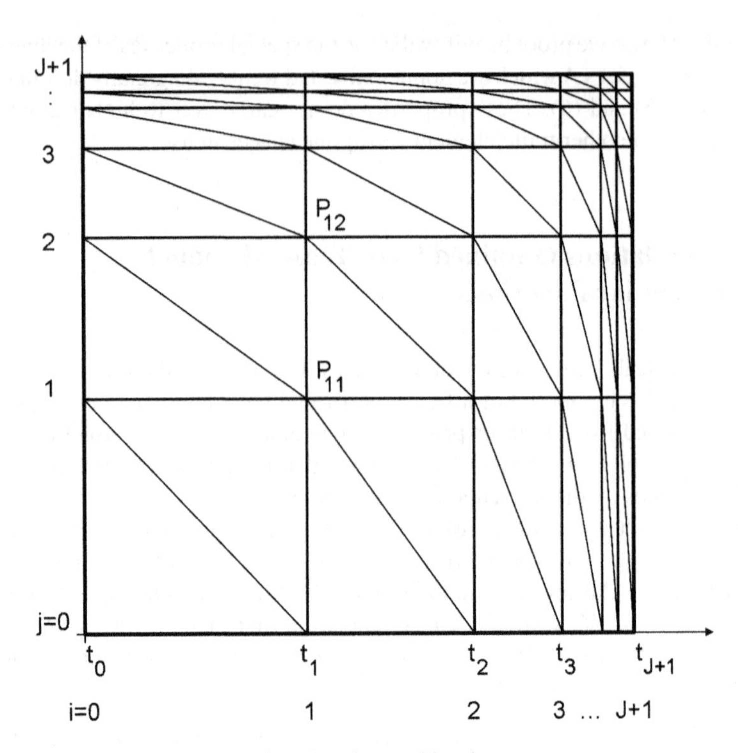

Fig. 2 Sketch of a type-II triangulation $\mathcal{T}_J$ under consideration

the "physical world" all these points P_{ij} collapse into one of the corners of the unit square in the limit). For the ease of proof, we use a modified checkerboard pattern for the right-hand side b given by

$$b_{i,j} = 0, \quad \min(i,j) \leq 1, \qquad b_{i,j} = (-1)^{i+j}, \quad \min(i,j) > 1.$$

This choice corresponds to an extremal function f on the square that takes value zero in the strips of rectangles attached to the axes and value $(-1)^{i+j}$ in the remaining rectangles with P_{ij} as lower-left corner, $\min(i,j) \geq 1$.

Because of symmetry, the solution vector x satisfies $x_{i,j} = x_{j,i}$. Using the geometric definition (6) of the matrix elements, the limit system $A_J x = b$ can be explicitly given and solved in a simple recursive fashion since it is essentially in lower triangular form if the variables are ordered properly. We use the above symmetry to reduce its size; the order we write down the system corresponds to marching in the rectangular mesh as follows: Start from the bottom-left square, move one-by-one to the right end in the first row of rectangles, then switch to the "diagonal" square in the second row, move to the right end in the second row, switch to the "diagonal" square in the third row, and so on. The three equations in $Ax = b$ associated with the bottom-left square are

$$x_{00}+x_{10}=0, \qquad \frac{3}{2}x_{10}+\frac{1}{4}(x_{00}+x_{11})=0, \qquad x_{11}+x_{10}=0,$$

from which we get $x_{ij}=0$ for $i,j\le 1$. We proceed with the remaining equations corresponding to the first row of rectangles:

$$\begin{aligned} x_{i0}+\frac{1}{4}x_{i1} &= -\frac{1}{4}x_{i-1,0}-\frac{1}{2}x_{i-1,1},\\ \frac{1}{2}x_{i0}+x_{i1} &= -\frac{1}{2}x_{i-1,1}, \end{aligned}$$

which by recursion in $i=2,\dots,J+1$ gives $x_{i,j}=0$ for all i,j with $\min(i,j)\le 1$.

The remaining equations (one for each $(x_{i,j})$ with $\min(i,j)>1$) can be written in the form

$$x_{ij}=(-1)^{i+j}-\frac{1}{2}(x_{i-1,j}+x_{i,j-1}) \quad\Longleftrightarrow\quad y_{ij}=\frac{1}{2}(y_{i-1,j}+y_{i,j-1})+1,$$

where $y_{ij}=(-1)^{i+j}x_{ij}$. We start with $j=2$ and obtain by recursion in $i=2,\dots,J+1$ that $y_{22}=1$, $y_{32}=3/2$ and generally $y_{i2}\ge 3/2$ for $i\ge 3$. Now, switch to $j=3$. Looking at the equation for $i=3$, we get $y_{33}=y_{32}+1=5/2$, and for $i=4$ we obtain $y_{43}=\frac{1}{2}(y_{33}+y_{42})+1\ge 3$ and more generally $y_{i3}\ge 3$ for all $i=4,\dots,J+1$. By induction in j, it is easy to see that this pattern generalizes to

$$y_{jj}\ge (j-1)\frac{3}{2}-\frac{1}{2}, \qquad y_{ij}\ge (j-1)\frac{3}{2}, \quad i=j+1,\dots,J+1,$$

for all $j=2,\dots,J+1$. Thus, taking $j=J+1$, we get

$$\|A_J^{-1}\|_\infty \ge \|x\|_\infty \ge |y_{J+1,J+1}| \ge \frac{3}{2}J$$

as was claimed above. The statement is proved.

Remark 4. It is not hard to show that $\|A_J^{-1}\|_\infty \asymp J$ as $J\to\infty$ in the above example. Thus, as to the asymptotic growth of the L_∞-norm of $P_{V(\mathcal{T})}$ with respect to the size of $\mathcal{T}$, the above example is less effective than the one provided in [10] since the latter provides the same lower estimate for triangulations into $\mathrm{O}(J)$ triangles. Recall that we conjectured in [10] that a matching upper bound of

$$\|P_{V(\mathcal{T})}\|_{L_\infty\to L_\infty}=\mathrm{O}(\#\mathcal{T})$$

is to be expected. This conjecture remains open.

Remark 5. The above counterexample should be contrasted with the situation for L_2-orthogonal projections onto the space of bilinear Q1 elements over tensor-product rectangular partitions $\mathcal{P}$, where a uniform boundedness result holds, with no restrictions on the univariate partitions that generate $\mathcal{P}$. The proof is an easy exercise on tensor-product approximation. Also note that if we allow crisscrossing of all rectangles into 4 (rather than bisecting into 2) equal-area triangles, then, according to the last example of Remark 2, Theorem 1 yields uniform boundedness of $P_{V(\mathcal{T}_\mathcal{P})}$ independently of $\mathcal{P}$.

In other words, cutting each rectangle in $\mathcal{P}$ into two triangles and switching from rectangular Q1 element to triangular P1 element spaces make a difference. A natural question is whether this effect depends on how we bisect the rectangles. For example, if we flip all bisection lines in the above example and consider the limit case of the resulting type-I triangulations, then $\|A_J^{-1}\|_\infty$ is uniformly bounded (this is a difficulty in coming up with the counterexamples and partly explains our misconception concerning tensor-product meshes that entered [10]). Since for both the type-I and the type-II triangulations the depth of local area growth with ratio r equals $J+1$ in the limit (to this end, consider the upper-right corner of the rectangle as P, and check the definition of $L_P(r)$ as $\min_{j=1,\dots,J}(t_{j-1}-t_j)/(t_j-t_{j+1}) \to \infty$), we also see that the boundedness of $L_{\max}(\mathcal{T},r)$ for some $r \geq 1$ required in Theorem 1 is only sufficient but not necessary for getting L_∞-bounds. A similar comment is true for the assumption on $K_{\max}(\mathcal{T})$ in Theorem 1, as can be seen from examining the case of 1-ring triangulations around a single interior vertex of high valence.

In light of the previous Remark 5, it is natural to ask if for any tensor-product rectangular partition $\mathcal{P}$ of a rectangle, there exists a bisection pattern such that $\|P_{V(\mathcal{T})}\|_{L_\infty \to L_\infty}$ is bounded from above by a universal constant. The answer is indeed yes if the following bisection algorithm is adopted: For any ordering of the vertices in $\mathcal{P}$, loop through the vertices. If P is the current vertex, consider its neighboring rectangles, and mark the area-largest ones among them. Marked rectangles that are not yet bisected should be bisected using the diagonal emanating from P. For marked but already bisected rectangles, no action is necessary. Then proceed with the next vertex in the list. After all vertices are processed, a few rectangles may be left undivided. Those can be bisected by choosing a diagonal randomly. This algorithm is illustrated in Fig. 3.

Theorem 3. *For any tensor-product partition $\mathcal{P}$ of a rectangular domain, the above bisection algorithm yields a triangulation $\mathcal{T}$ such that the norms of the associated projectors $P_{V(\mathcal{T})}$ remain uniformly bounded. In other words,*

$$\sup_{\mathcal{P}} \|P_{V(\mathcal{T})}\|_{L_\infty \to L_\infty} < \infty.$$

Proof. Assume that x is the solution with maximal max-norm $M = \|x\|_\infty$ of the system $Ax = b$ with right-hand side b of unit norm $\|b\|_\infty = 1$. Without loss of generality, let P be the vertex of $\mathcal{T}$ for which $x_P = M$. Since the vertex set of $\mathcal{T}$ coincides with the vertex set of $\mathcal{P}$, this P has between one to four neighboring rectangles, the area-largest one of which is denoted by R'_P and its area by A'_P (if there are several area-largest rectangles attached to P, we pick one of them). Our algorithm guarantees that R'_P or has been bisected with the diagonal emanating

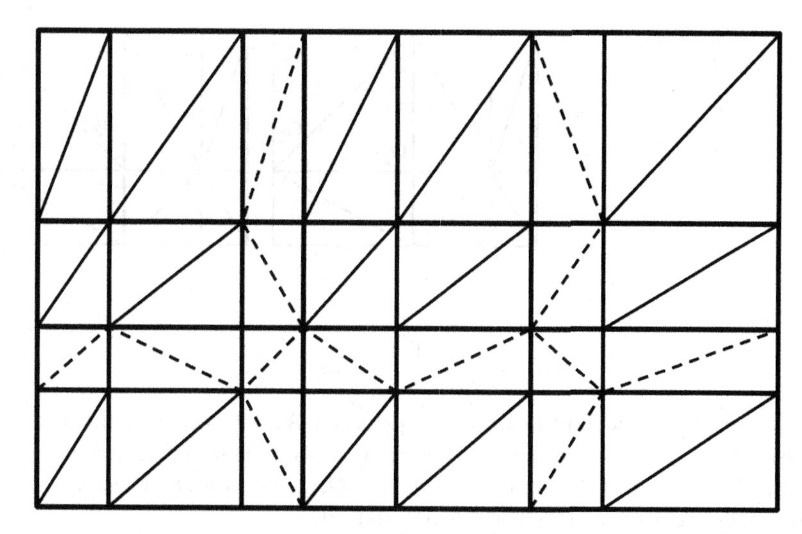

Fig. 3 Bisection algorithm for a sample $\mathcal{P}$. The ordering of vertices is row-by-row from *left* to *right*, starting with the *bottom row*. *Dashed bisection lines* indicate *rectangles* that have been randomly bisected after the loop over all vertices was finished

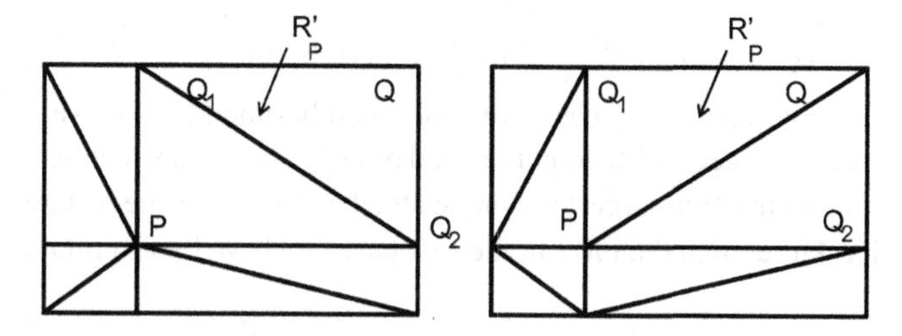

Fig. 4 Two possible situations of bisection of R'_P

from P (e.g., when P was processed in the loop of the algorithm) or with the other diagonal when one of the two vertices of this rectangle neighboring with P was processed (and led to bisecting R'_P) earlier in the algorithm. The two possibilities are shown in Fig. 4: the situation on the left occurs if one of the neighbor vertices Q of P in R'_P has triggered the bisection of R_P; the situation on the right occurs if the examination of either P or the vertex Q opposite to P led to the bisection.

We first give the complete argument for the situation depicted in Fig. 4 on the right; the situation on the left is similar and even simpler to deal with, details are given at the end of the proof. Since R_P is the area-largest rectangle attached to P, the coefficients a_{PS}, $S = Q, Q_1, Q_2$ are safely bounded away from zero. The following crude estimates suffice. We have $A_P \leq 4R'_P$, and

$$A_{PQ_1}, A_{PQ_2} \geq A'_P/2, \qquad A_{PQ} = A'_P,$$

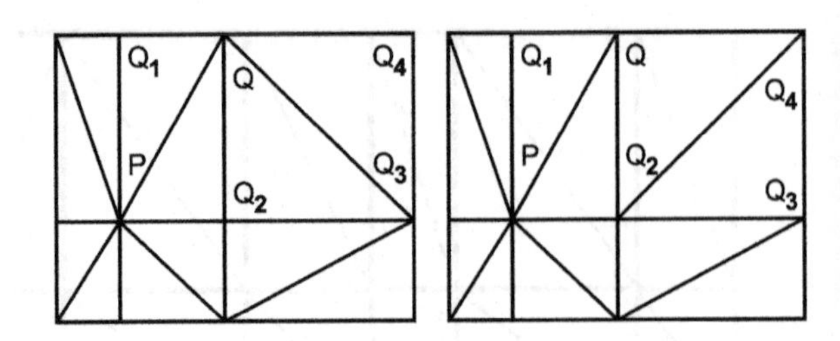

Fig. 5 Two bisection situations for R'_{Q2}

which yields

$$a_{PQ_1}, a_{PQ_2} \geq \frac{1}{16}, \qquad a_{PQ} \geq \frac{1}{8}.$$

As in the proof of Theorem 1, from the equation in $Ax = b$ corresponding to P and the assumptions $\|b\|_\infty = 1$ and $M = x_P = \|x\|_\infty$, we obtain upper bounds for the quantities $y_S := x_S + M \geq 0$ for the vertices $S = Q, Q_1, Q_2$ of R'_P (suggesting that the corresponding entries x_S are close to $-M$). Indeed,

$$1 \geq b_P = x_P + \sum_{S \neq P} a_{PS} x_S = \sum_{S \neq P} a_{PS} y_S \geq a_{PS} y_S, \quad S \neq P$$

implies

$$y_Q \leq 1/a_{PQ} \leq 8, \qquad y_{Q_i} \leq 1/a_{PQ_i} \leq 16, \quad i = 1, 2. \tag{9}$$

Now consider without loss of generality the neighborhood of Q_2 (we could have equally chosen to proceed with neighborhood of Q_1). There are two cases: Or $R'_{Q_2} = R'_P$ (i.e., R'_P is also the area-largest rectangle attached to Q_2), or the rectangle $R'_{Q_2} := QQ_2Q_3Q_4$ has larger area than R'_P. In the first case, we have by a similar reasoning

$$A_{Q_2Q}, A_{Q_2P} \geq A'_P/2, \qquad A_{Q_2} \leq 7A'_P/2,$$

and thus

$$a_{Q_2Q}, a_{Q_2P} \geq \frac{1}{14}.$$

Using (9) and inspecting the equation in $Ax = b$ corresponding to Q_2 results in

$$\begin{aligned} -1 \leq b_{Q_2} &= x_{Q_2} + \sum_{S \neq Q_2} a_{Q_2S} x_S = y_{Q_2} + \sum_{S \neq Q_2} a_{Q_2S}(x_S - M) \\ &\leq 16 + a_{Q_2Q}(x_Q - M) \leq 16 + (y_Q - 2M)/14 \leq 16 + 4/7 - M/7, \end{aligned}$$

since $a_{Q_2S} \geq 0$ and $x_S - M \leq 0$ for all $S \neq Q_2$. Consequently, we arrive at the crude upper bound $M \leq 123$.

In the second case, it depends on how R'_{Q_2} is bisected. The two situations are depicted in Fig. 5. In either case, we find as above that

$$A_{Q_2} \leq 4A'_{Q_2}, \quad A_{Q_2Q} \geq A'_{Q_2}/2 \quad \Longrightarrow \quad a_{Q_2Q} \geq \frac{1}{16}.$$

Thus, as above,

$$-1 \le b_{Q_2} \le y_{Q_2} + a_{Q_2Q}(x_Q - M) \le 16 + (y_Q - 2M)/16 \le 16 + 1/2 - M/8,$$

and $M \le 140$. This concludes the proof of Theorem 3 for the situation shown in Fig. 4 on the right.

The remaining situation on the left in Fig. 4 is even easier to deal with. In this case, the bisection of R'_P has been triggered by either Q_1 or Q_2, without loss of generality assume that it was Q_2. Thus, R'_P is the area-largest among the rectangles attached to both P and Q_2. As above, we obtain in a first step

$$A_P \le 7A'_P/2, \quad A_{PQ_1}, A_{PQ_2} \ge A'_P/2 \quad \Longrightarrow \quad a_{PQ_1}, a_{PQ_2} \ge \frac{1}{14},$$

and consequently

$$y_{Q_1}, y_{Q_2} \le 14.$$

In the second step, we examine the equation for Q_2 and get

$$A_{Q_2} \le 4A'_P, \quad A_{Q_2Q_1} = A'_P \quad \Longrightarrow \quad a_{Q_2Q_1} \ge \frac{1}{8},$$

which leads to

$$-1 \le b_{Q_2} \le y_{Q_2} + a_{Q_2Q_1}(x_{Q_1} - M) \le 14 + (y_{Q_1} - 2M)/8 \le 14 + 7/4 - M/4,$$

and eventually to $M \le 67$. The proof of Theorem 3 is complete.

Remark 6. The above obtained numerical upper bound of

$$\|P_{V(\mathcal{T})}\|_\infty \asymp \|A^{-1}\|_\infty \le 140$$

is certainly too pessimistic. Note that for the case of bilinear finite element spaces over tensor-product partitions $\mathcal{P}$ one has the sharp upper bound of

$$\|P_{V(\mathcal{P})}\|_{L_\infty \to L_\infty} < 9,$$

which follows from Z. Ciesielski result [3] for the univariate case.

Remark 7. If one applies the bisection algorithm underlying Theorem 3 to the family of $\mathcal{P}$ used for establishing Theorem 2 (recall that the counterexamples require the partitions to collapse into the upper-right corner of the rectangle depicted in Fig. 2), then one ends up with essentially a type-I triangulation. To see this, start the loop through the vertex set of $\mathcal{P}$ from the upper-right corner and move row-by-row from right to left. This confirms the observation reported in Remark 5 for this type of $\mathcal{P}$.

Remark 8. The result of Theorem 3 applies to the two-dimensional examples of Shishkin- and Bakhvalov-type meshes for the treatment of elliptic and parabolic boundary layers in diffusion–convection problems considered in [12, 2.4.1–2] and can be carried over to the more general h-FEM meshes discussed in [6, Section 2.6]. Note the difference with applying Theorem 1 to these families of meshes: While Theorem 1 guarantees a bound independent of the bisection pattern but formally depending on the number of layers with different mesh-sizes, Theorem 3 gives a bound for a specific bisection pattern, independently of the number of layers with different mesh-sizes!

References

1. Babuska, I., Aziz, A.K.: On the angle condition in the finite element method. SIAM J. Numer. Anal. **13**, 214–226 (1976)
2. de Boor, C.: On a max-norm bound for the least-squares spline approximant. In: Ciesielski, Z. (ed.) *Approximation and Function Spaces* (Gdansk, 1979), pp. 163–175. North-Holland, Amsterdam (1981)
3. Ciesielski, Z.: Properties of the orthonormal Franklin system. Studia Math. **23**, 141–157 (1963)
4. Desloux, J.: On finite element matrices. SIAM J. Numer. Anal. **9**(2), 260–265 (1972)
5. Douglas, J. Jr., Dupont, T., Wahlbin, L.: The stability in L^q of the L^2-projection into finite element function spaces. Numer. Math. **23**, 193–197 (1975)
6. Melenk, J.: hp-finite element method for singular perturbations. Lecture Notes in Mathematics, vol. 1796. Springer, Berlin (2002)
7. Kaland, L., Roos, H.-G.: Parabolic singularly perturbed problems with exponential layers: robust discretizations using finite elements in space on Shishkin meshes. Int. J. Numer. Anal. Model. **7**(3), 593–606 (2010)
8. Oskolkov, K.I.: The upper bound of the norms of orthogonal projections onto subspaces of polygonals. In: Approximation Theory, Proc. VIth Semester Stefan Banach International Mathematical Center, Warsaw, 1975, Banach Center Publications, vol. 4, pp. 177–183. PWN, Warsaw (1979)
9. Oswald, P.: On the C norm of orthoprojections onto subspaces of polygons. Matem. Zametki **21**(4), 495–502 (1977) (in Russian)
10. Oswald, P.: A counterexample concerning the L_2-projector onto linear spline spaces, Math. Comp. **77**, 221–226 (2008)
11. Roos, H.-G.: Stabilized FEM for convection-diffusion problems on layer-adapted meshes. J. Comput. Math. **27**, 266–279 (2009)
12. Roos, H.-G., Stynes, M., Tobiska, L.: Robust numerical methods for singularly perturbed differential equations. Springer Series in Computational Mathematics, vol. 24, 2nd edn. Springer, Berlin (2008)
13. Schopf, M., Roos, H.-G.: Convergence and stability in balanced norms of finite element mathods on Shishkin meshes for reaction-diffusion problems (submitted).
14. Shadrin, A. Yu. The L_∞-norm of the L_2-spline projector is bounded independently of the knot sequence: a proof of de Boor's conjecture, Acta Math. **187** (2001), 59–137.

Fast Implementation of ℓ^1-Greedy Algorithm

Alexander Petukhov and Inna Kozlov

Abstract We present an algorithm for finding sparse solutions of the system of linear equations $A\mathbf{x} = \mathbf{b}$ with the rectangular matrix A of size $n \times N$, where $n < N$. The algorithm basic constructive block is one iteration of the standard interior-point linear programming algorithm. To find the sparse representation we modify (reweight) each iteration in the spirit of Petukhov (Fast implementation of orthogonal greedy algorithm for tight wavelet frames. Signal Process. **86**, 471–479 (2006)). However, the weights are selected according to the ℓ^1-greedy strategy developed in Kozlov and Petukhov (Sparse solutions for underdetermined systems of linear equations. In: Freeden, W., Nashed, M.Z., Sonar, T. (eds.) Handbook of Geomathematics, pp. 1243–1259. Springer, Berlin (2010)). Our algorithm combines computational complexity close to plain ℓ^1-minimization with the significantly higher efficiency of the sparse representations recovery than the reweighted ℓ^1-minimization (Candes et al.: Enhancing sparsity by reweighted ℓ_1 minimization. J. Fourier Anal. Appl. **14**, 877–905 (2008) (special issue on sparsity)), approaching the capacity of the ℓ^1-greedy algorithm.

1 Introduction

The problem of finding sparse solutions of a system of linear equations

$$A\mathbf{x} = \mathbf{b}, \quad \mathbf{x} \in \mathbb{R}^N, \mathbf{b} \in \mathbb{R}^n, N > n, \tag{1}$$

A. Petukhov (✉)
University of Georgia, Athens, GA 30602, USA
e-mail: petukhov@math.uga.edu

I. Kozlov
Algosoft Tech, Bogart, GA, USA
e-mail: kozlovinna@yahoo.com

D. Bilyk et al. (eds.), *Recent Advances in Harmonic Analysis and Applications*, Springer Proceedings in Mathematics & Statistics 25, DOI 10.1007/978-1-4614-4565-4_25,

is interesting in many information theory-related contexts. It is tractable as reconstruction of the sparse data vector $\mathbf{x}$ compressed with the linear operator A. In channel encoding context, the vector $\mathbf{b}$ can be viewed as a syndrome measured by the receiver. Then the vector $\mathbf{x}$ is the vector of transmission errors. Since the problem cannot be solved in polynomial time (cf. [11]), it also has connection to cryptographic encoding. Within this chapter, we say that a vector is *sparse* if

$$k := \#\{i \mid x_i \neq 0, 1 \leq i \leq N\} < n,$$

i.e., it has the restricted number of nonzero entries. This number is called also the Hamming weight of the vector $\mathbf{x}$. For a randomly selected matrix A and a vector $\mathbf{b}$ of system (1), if a sparse solution exists, it is unique almost for sure.

The recent compressed sensing (compressive sampling) studies gave a new push to sparse solutions theory and algorithms. It turned out that for a reasonable value of the sparsity, the vector $\mathbf{x}$ can be recovered precisely using linear programming algorithm (LPA) for finding a solution to (1) with minimal ℓ^1-norm ([2, 6, 13]). While, in practice, the number k characterizing the sparsity of the vectors $\mathbf{x}$ which can be recovered with ℓ^1-minimization is far from the magic number $n-1$, this case is well studied and reliable tool for solving systems (1) for many applied problems.

Orthogonal greedy algorithm (OGA) is a very serious competitor of LPA. If it is appropriately implemented ([7,9,12]), it outperforms LPA in both the computational complexity and in the ability to recover sparse representations with the greater sparsity number k. In some papers (e.g., [7]), this modification of OGA is called stagewise orthogonal matching pursuit (StOMP = StOGA).

LPA found a nice turn in the reweighted ℓ^1-minimization algorithm (RLA) ([3]). Similar ideas were used also in [4, 8], and [10] for the design of algorithms with non-convex optimization. The number k was significantly moved up. However, the computational complexity also has grown up. Further noticeable growth of the k at cost of growing computing complexity was reached in ℓ^1-greedy algorithm (LGA) ([9]). LGA is based on a combination of ideas of OGA and RLA.

In this chapter, we present the algorithm with computational complexity comparable with the regular LPA, whereas its recovery ability outperforms RLA and is very close to LGA.

2 Basic Algorithms

2.1 ℓ^1-Norm Minimization

LPA will be a starting point for our algorithm. To reduce the problem (1) to linear programming, we have to formulate it in the weaker form:

$$\|\mathbf{x}\|_1 := \sum |x_i| \to \min, \quad \text{subject to} \quad A\mathbf{x} = \mathbf{b}. \tag{2}$$

As we mentioned above, if a sparse solution with a moderate number of nonzero entries exists, it can be found by solving (2).

There are many competing algorithms allowing to solve the problem above. We are interested, in particular, in the primal-dual interior-point algorithm as it is described in Sect. 11.7 in [1]. Our preference is just due to the fact that we will take the general iteration step of that algorithm as a brick for constructing our fast modification of the LGA. More details of its implementation will be discussed in Sect. 3.

2.2 Reweighted ℓ^1-Minimization

In [3], Candès, Wakin, and Boyd developed the RLA. Instead of computing a solution with the minimum of the ℓ^1-norm, they suggested to iteratively recompute it a few times with the weighted norm

$$\|\mathbf{x}\|_{w_{j,1}} = \sum w_{j,i}|x_i|. \tag{3}$$

The weight $w_{j,i}$ is defined by the previous iteration $\mathbf{x}_{j-1}$:

$$w_{j,i} = (\varepsilon + |x_{j-1,i}|)^{-1}. \tag{4}$$

When $\varepsilon \to 0$ the functional $\sum |x_i|/(\varepsilon + |x_i|)$ becomes "close" to the Hamming weight. Numerical experiments showed that the recovery capacity of the reweighted algorithm is significantly higher than for the regular ℓ^1-optimization. This improvement in the numerical experiments was reached at cost of about eightfold increasing the computing time.

Foucart and Lai [8] considered the reweighted ℓ^1-minimization with the weight

$$w_{j,i} = (\varepsilon + |x_{j-1,i}|)^{q-1}, \quad 0 < q \leq 1. \tag{5}$$

Such algorithm gave some improvement of the recovery with about ten times higher than in RLA computational complexity.

The reweighted ℓ^1-minimization with the same weight (5) was considered also in [4].

The reweighted ℓ^2-minimization with the detailed analysis of the convergence was considered in [5].

2.3 ℓ^1-Greedy Algorithm

The LGA introduced in [9] has two main distinctions with RLA. It uses a variable weight function which changes with algorithm iterations. The weight is a piecewise constant function.

That algorithm was based on a simple observation that RLA algorithm is designed in the "greedy" style for the ℓ^1-norm. The weight in RLA reduces the influence of the big entries (found on the previous iteration) on the current step of the optimization procedure.

This idea is the keystone of OGA (and more general greedy algorithms), the main competitor of ℓ^1-minimization in hunting for sparse solutions/representations. Our approach was to use this idea in its absolute form, by exempting already found large coefficients from consideration. However, since there is no notion of the orthogonal projection for ℓ^1-norm, we emulate this exemption by setting a very little weight for large coefficients.

In spite of its simplicity, LGA outperforms other algorithms based on the reweighted ℓ^1-minimization.

ℓ^1-Greedy Algorithm

1. *Initialization*

   a. Take $w_{0,i} \equiv 1$, $i = 1,\ldots,n$.
   b. Find a solution $\mathbf{x}_0$ to (1) providing the minimum for (3).
   c. Set $M_0 := \max\{x_{0,i}\}$ and $j := 1$.

2. *General Step*

   a. Set $M_j := \alpha M_{j-1}$ and the weights

$$w_{j,i} := \begin{cases} \varepsilon, & |x_{j-1,i}| > M_j, \\ 1, & |x_{j-1,i}| \le M_j. \end{cases}$$

   b. Find a solution $\mathbf{x}_j$ to (1) providing the minimum for (3).
   c. Set $j := j+1$.
   d. If stopping criterion is not satisfied, go to 2 (a).

While in our modeling we used a number of iterations as a simple stopping criterion, probably the cardinality of the set of large coefficients or the behavior of the ℓ^p-quasinorm ($p < 1$) of the solution can be used for this purpose.

2.4 Fast Orthogonal Greedy Algorithm

In [12], the first author designed the algorithm which was in main features a traditional OGA with a few distinctions allowing the reduction of the computational cost dramatically. The two main principles of a general iteration step of that algorithm are:

(1) Pick up all significant coefficients (not only the biggest one) above a threshold.
(2) Do not compute precise intermediate orthogonal projection. Just make one step toward it.

The implementation of OGA based on those principles has complexity comparable with computing one orthogonal projection. That algorithm gives an opportunity to solve large-scale problems of sparse representation efficiently. OGA is especially efficient for some special important problems when the data decomposition and reconstruction algorithms in redundant systems have low complexity. Say, sparse approximations of a few megapixel images with redundant wavelet frames can be easily found even on a low-end laptop.

3 Fast ℓ^1-Greedy Algorithm

We take the primal-dual interior-point algorithm as it is described in [1] as a basic building block.

The problem (2) has an equivalent reformulation in terms of classical linear programming settings:

$$\sum_{i=1}^{N} u_i \to \min, \text{ subject to (1) and} x_i - u_i \leq 0,\ -x_i - u_i \leq 0. \tag{6}$$

If we introduce the vectors

$$\mathbf{c}_0 := \begin{pmatrix} \mathbf{0}_N \\ \mathbf{1}_N \end{pmatrix} \in \mathbb{R}^{2N}, \quad \mathbf{z} := \begin{pmatrix} \mathbf{x} \\ \mathbf{u} \end{pmatrix} \in \mathbb{R}^{2N}$$

and matrices

$$A_0 := (A \quad 0_{n\times N}) \in \mathbb{R}^{n\times 2N}$$

and

$$C = \begin{pmatrix} \mathrm{Id}_N & -\mathrm{Id}_N \\ -\mathrm{Id}_N & -\mathrm{Id}_N \end{pmatrix} \in \mathbb{R}^{2N\times 2N},$$

where Id_N is the identity matrix of size $N\times N$, linear program problem (6) can be rewritten as

$$\langle \mathbf{c}_0, \mathbf{z}\rangle \to \min \text{ subject to}$$
$$A_0\mathbf{z} = \mathbf{b}, \quad C\mathbf{z} \leq \mathbf{0}.$$

Then the solution to (6) has to satisfy the Karush–Kuhn–Tucker conditions:

$$\begin{aligned} r_{\mathrm{dual}} &:= \mathbf{c}_0 + A_0^T\mathbf{v}^* + C^T\boldsymbol{\lambda}^* = \mathbf{0}; \\ r_{\mathrm{cent}} &:= \Lambda C\mathbf{z}^* = \mathbf{0}, \quad \text{where } \Lambda := \mathrm{diag}(\boldsymbol{\lambda}^*); \\ & C\mathbf{z}^* \leq 0, \quad \boldsymbol{\lambda}^* \geq 0; \\ r_{\mathrm{pri}} &:= A_0\mathbf{z}^* - \mathbf{b} = \mathbf{0}, \end{aligned} \tag{7}$$

where $\mathbf{v}^* \in \mathbb{R}^n$ and $\boldsymbol{\lambda}^* \in \mathbb{R}^{2N}$ are the optimal Lagrange multipliers.

While, in such dual settings, we have a problem nonlinear with respect to the combined set of primal and dual variables $(\mathbf{z},\mathbf{v},\lambda)$, it can be solved with the Newton method. Due to the iterative nature of the Newton method, the replacement of (7) with

$$r_{\text{cent}} := \Lambda C\mathbf{z}^* + (1/\tau)\mathbf{1} = \mathbf{0}, \quad \tau > 0,$$

baring in mind to provide $\tau \to +\infty$ through the Newton iterations, looks reasonable for better convergence. Thus, to solve an LP problem, the equation

$$r_\tau(\mathbf{z},\mathbf{v},\lambda) := \begin{pmatrix} r_{\text{dual}} \\ r_{\text{cent}} \\ r_{\text{pri}} \end{pmatrix} = \mathbf{0} \tag{8}$$

has to be solved. Assuming that we know close but not precise (at least, because τ was refined), we need to find vectors $\Delta\mathbf{z}$, $\Delta\mathbf{v}$, and $\Delta\lambda$ such that

$$r_\tau(\mathbf{z}+\Delta\mathbf{z},\mathbf{v}+\Delta\mathbf{v},\lambda+\Delta\lambda) = \mathbf{0}.$$

The next step is a linearization of the function

$$\begin{aligned}\mathbf{0} &= r_\tau(\mathbf{z}+\Delta\mathbf{z},\mathbf{v}+\Delta\mathbf{v},\lambda+\Delta\lambda)\\ &\approx r_\tau(\mathbf{z},\mathbf{v},\lambda) + J_\tau(\mathbf{z},\mathbf{v},\lambda)\begin{pmatrix}\Delta\mathbf{z}\\ \Delta\mathbf{v}\\ \Delta\lambda\end{pmatrix}.\end{aligned}$$

Thus, we have a system of linear equations:

$$\begin{pmatrix} 0 & A_0^T & C^T \\ \Lambda C & 0 & \operatorname{diag}(C\mathbf{z}) \\ A_0 & 0 & 0 \end{pmatrix}\begin{pmatrix}\Delta\mathbf{z}\\ \Delta\mathbf{v}\\ \Delta\lambda\end{pmatrix} \approx -\begin{pmatrix} \mathbf{c}_0 + A_0^T\mathbf{v} + C^T\lambda \\ \Lambda C\mathbf{z} + \frac{1}{\tau}\mathbf{1} \\ A_0\mathbf{z} - \mathbf{b} \end{pmatrix}.$$

Resolving the second equation with respect to $\Delta\lambda$, we have

$$\Delta\lambda = -\operatorname{diag}^{-1}(C\mathbf{z})\left(\Lambda C\Delta\mathbf{z} + \frac{1}{\tau}\mathbf{1}\right) - \lambda.$$

When we substitute the computed value $\Delta\lambda$, the first and the third equations turn into

$$\begin{pmatrix} -C^T\operatorname{diag}^{-1}(C\mathbf{z})\Lambda C & A_0^T \\ A_0 & 0 \end{pmatrix}\begin{pmatrix}\Delta\mathbf{z}\\ \Delta\mathbf{v}\end{pmatrix} = \begin{pmatrix} \frac{1}{\tau}C^T\operatorname{diag}^{-1}(Cz)\mathbf{1} - \mathbf{c}_0 - A_0^T\mathbf{v} \\ \mathbf{b} - A_0\mathbf{z} \end{pmatrix}. \tag{9}$$

Now, taking into account that $C^{-1} = (1/2)C$ and $C^T = C$, we resolve the first equation for $\Delta\mathbf{z}$:

$$\Delta\mathbf{z} = -\frac{1}{2\tau}C\lambda^{-1} - \frac{1}{4}C\Lambda^{-1}\operatorname{diag}(C\mathbf{z})\mathbf{1} + \frac{1}{4}C\Lambda^{-1}\operatorname{diag}(C\mathbf{z})CA_0^T(\mathbf{v}+\Delta\mathbf{v}).$$

Substituting $\Delta\mathbf{z}$ in the second equation of (9), we have

$$A_0C\Lambda^{-1}\operatorname{diag}(C\mathbf{z})CA_0^T(\mathbf{v}+\Delta\mathbf{v}) = \frac{2}{\tau}A_0C\lambda^{-1} + A_0C\Lambda^{-1}\operatorname{diag}(C\mathbf{z})\mathbf{1} + 4(\mathbf{b}-A_0\mathbf{z}).$$

If we introduce the notation $\Lambda_1 := \operatorname{diag}\{\lambda_1,\dots,\lambda_N\}$, $\Lambda_2 := \operatorname{diag}\{\lambda_{N+1},\dots,\lambda_{2N}\}$, $Z_1 := \operatorname{diag}\{x_1-u_1,\dots,x_N-u_N\}$, $Z_2 := \operatorname{diag}\{-x_1-u_1,\dots,-x_N-u_N\}$, then the last system of linear equations can be rewritten as

$$\begin{aligned} &A\left(\Lambda_1^{-1}Z_1+\Lambda_2^{-1}Z_2\right)A^T(\mathbf{v}+\Delta\mathbf{v}) \\ &= \tfrac{2}{\tau}A\left(\Lambda_1^{-1}-\Lambda_2^{-1}\right)\mathbf{1} + A(\Lambda_1^{-1}Z_1-\Lambda_2^{-1}Z_2)\,\mathbf{1} + 4(\mathbf{b}-A\mathbf{x}). \end{aligned}$$

From the last system we can find $\mathbf{v}+\Delta\mathbf{v}$ and then consequently $\Delta\mathbf{v}$, $\Delta\mathbf{x}$, $\Delta\mathbf{u}$, and $\Delta\lambda$. Then the vector can be scaled to keep the intermediate values of the parameters in the "interior." In this and some other steps in our practical implementation, we follow the main steps of the routines from MATLAB package "ℓ^1-MAGIC" (http://www-stat.stanford.edu/~candes/l1magic/) developed by E. Candès and J. Romberg. In particular, we use for the initialization the minimum ℓ^2-norm solution of (1) as $\mathbf{x}_0$, $u_{0,i} := 0.95|x_{0,i}| + 0.1m$, where $m := \max_i |x_{0,i}|$,

$$\lambda_{0,i} := \begin{cases} 1/(x_{0,i}-u_{0,i}),\ i=1,\dots,N; \\ -1/(x_{0,i-N}+u_{0,i-N}),\ i=N+1,\dots,2N; \end{cases}$$

$\mathbf{v}_0 = -A(\Lambda_{0,1}-\Lambda_{0,2})\mathbf{1}$.

We use 2-stage algorithm:

1. Adaptive greedy part
   (a) Initialize $\mathbf{x}_0$, $\mathbf{u}_0$, $\mathbf{v}_0$, λ_0, $\mathbf{W}_0 := \mathbf{1} \in \mathbb{R}^N$, $M_0 := \|\mathbf{x}_0\|_\infty$, $A_0 = A$.
   (b) For current $\mathbf{x}_j$, $\mathbf{u}_j$, $\mathbf{v}_j$, and λ_j find next interior point $\mathbf{x}_{j+1} = \mathbf{x}_j + \Delta\mathbf{x}$, $\mathbf{u}_{j+1} = \mathbf{u}_j + \Delta\mathbf{u}$, $\mathbf{v}_{j+1} = \mathbf{v}_j + \Delta\mathbf{x}$, and $\lambda_j = \lambda_j + \Delta\lambda$ using the algorithm above.
   (c)
   $$M_{j+1} := \alpha M_j;\ W_{j+1,i} = \begin{cases} \varepsilon,\ |x_i| > M_{j+1}W_{j,i}, \\ 1,\ |x_i| \le M_{j+1}W_{j,i}; \end{cases}$$
   (d) Set $A_{j+1,m,i} := A_{j,m,i}/W_{j+1,i}$, $i = 1,\dots,N$, and $m = 1,\dots,n$.
   (e) Let $w_i := W_{j,i}/W_{j+1,i}$. Set $x_{j+1,i} := x_{j,i}/w_i$, $u_{j+1,i} := u_{j,i}/w_i$, $\lambda_{j+1,i} := \lambda_{j,i}w_i$, $\lambda_{j+1,i+N} := \lambda_{j,i+N}w_i$, $i = 1,\dots,N$.
   (f) $j := j+1$.
   (g) if $j < N_{\max}$, go to 1).

2. Standard weighted ℓ^1-minimization

   (a) Get $\mathbf{x}$ using LPA algorithm with the input $\mathbf{b}$, $A_{N_{max}}$, $\mathbf{x}_{N_{\max}}$.
   (b) Compute the final result $\mathbf{x}$ as $x_i := x_i / W_{N_{\max},i}$, $i = 1,\ldots,N$.

4 Numerical Experiments

In our experiments we used the modified "ℓ^1-MAGIC" MATLAB package. Our settings coincide with those in [3] and [9]. Namely, for each try, we take a random matrix $A \in \mathbb{R}^{n\times N}$ whose columns are uniformly distributed over the unit sphere in $\mathbb{R}^n$. The vector $\mathbf{x} \in \mathbb{R}^N$ has k nonzero components with the standard Gaussian distribution.

We compare five algorithms for finding sparse solutions to (1): the standard ℓ^1-minimization, its reweighted version as described in [3], the fast version of (St)OGA as described in [12], LGA from [9], and the fast LGA described in Sect. 3.

The curves on Figs. 1 and 2 give the dependence of the relative frequency of the success of the recovery on the value k characterizing the sparsity of the vector $\mathbf{x}$. The value n for both cases is 128, whereas $N = 256$ for Fig. 1 and $N = 512$ for Fig. 2.

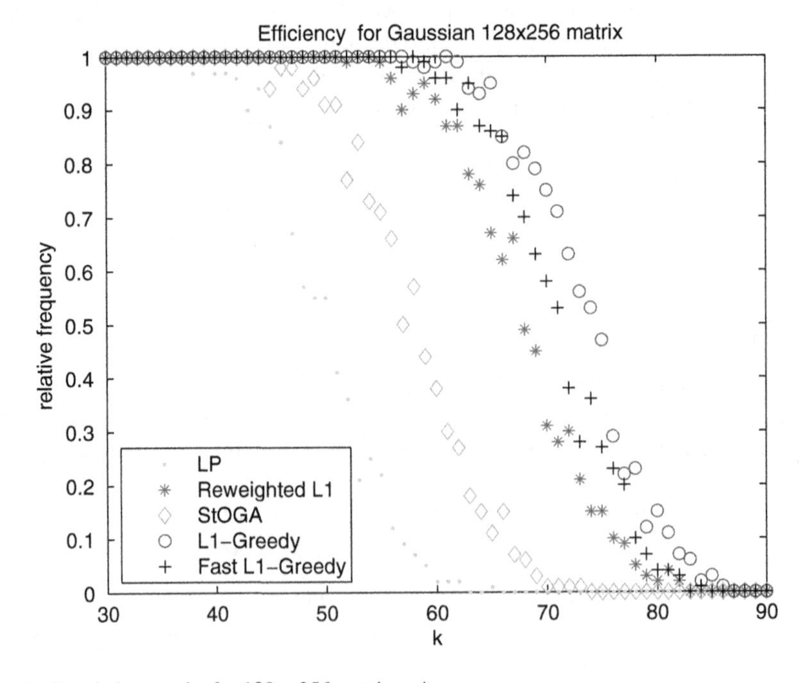

Fig. 1 Simulation results for 128×256 matrices A

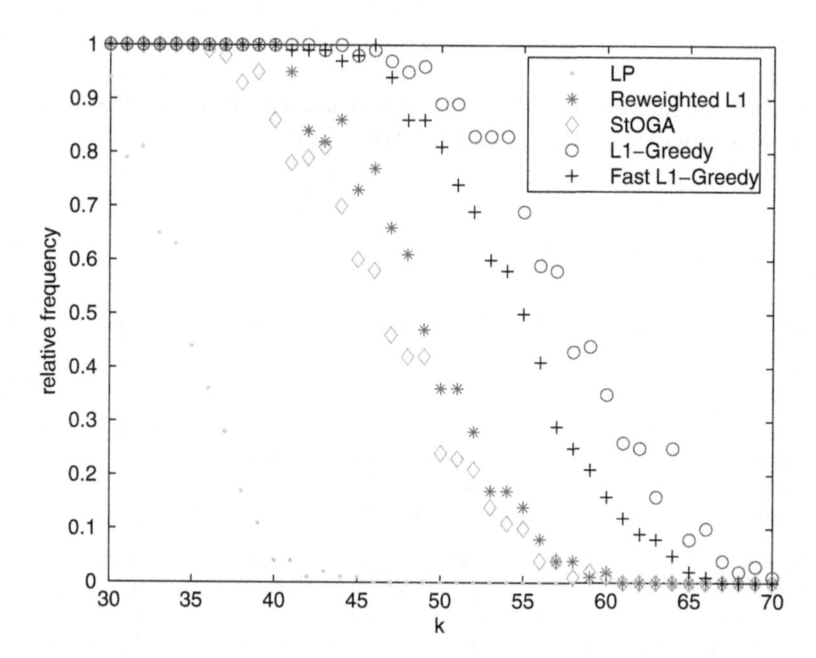

Fig. 2 Simulation results for 128×512 matrices A

Parameters used for the fas LGA are as follows: $\varepsilon = 0.02$, and $\alpha = 0.87$. The number of one-step LPA iterations was $N_{\max} = 30$. We use initial value $\tau = 30$ which is three times larger than τ selected in "ℓ^1-MAGIC."

Thus, the ℓ^1-greedy strategy allows recovering 12% more entries than RLA from [3] if $N = 256$ and 20% more entries for $N = 512$. The fast version considered in this chapter also outperforms the recovery ability of RLA by about 6% and 12%, correspondingly.

Bernoulli distribution (random ± 1) of nonzero input entries of the vector $\mathbf{x}$ was also checked in our numerical experiments. The results do not look optimistic. Both versions of LGAs have insignificant advantage over regular ℓ^1-minimization and have no visible advantage over the reweighted ℓ^1-optimization from [3]. However, it should be noted that the Bernoulli distribution corresponds to the binary data representation. For this reason, it is unnatural to use real-valued arithmetics for encoding such kind of data. At the same time, this is an important test for the algoritms even if it serves as a counterexample showing algorithm restrictions.

What may be even more important, the fast verion of LGA has computational complexity 2.5–3 times lower than RLA and much lower than LGA. The estimate is based on the average number of iterations necessary to reach the result above. MATLAB implementation also confirms this estimate.

We would like to emphasize that the result of the modeling showed high stability to the selection of the algorithm parameters. Say, setting $\varepsilon := 0.1$, the results are practically unchanged.

The bringing τ to the optimal value also has minor influence on the algorithm efficiency.

The higher computational efficiency can be reached by decreasing the number of iterations at cost of some losses in the recover capability.

References

1. Boyd, S., Vandenberghe, L.: Convex Optimization. Cambridge University Press, Cambridge (2004)
2. Candès, E.J., Tao, T.: Decoding by linear programming. IEEE Trans. Inform. Theor. **51**, 4203–4215 (2005)
3. Candes, E.J., Wakin, M.B., Boyd, S.: Enhancing sparsity by reweighted ℓ_1 minimization. J. Fourier Anal. Appl. **14**, 877–905 (2008) (special issue on sparsity)
4. Chartrand, R., Yin, W.: Iteratively reweighted algorithms for compressive sensing. In: Proceedings of International Conference on Acoustics, Speech, Signal Processing (ICASSP), pp. 3869–3872 (2008)
5. Daubechies, I., DeVore, R., Fornasier, M., Güntürk, C.S.: Iteratively re-weighted least squares minimization for sparse recovery. Commun. Pure Appl. Math. **63**, 1–38 (2010)
6. Donoho, D.: Compressed sensing. IEEE Trans. Inform. Theor. **52**, 1289–1306 (2006)
7. Donoho, D., Tsaig, Y., Drori, I., Starck, J.: Sparse solutions of underdetermined linear equations by stagewise orthogonal matching pursuit. Report, Stanford University (2006)
8. Foucart, S., Lai, M.J.: Sparsest solutions of underdetermined linear systems via ℓ_q-minimization for $0 \leq q \leq 1$. Appl. Comp. Harmonic Anal. **26**, 395–407 (2009)
9. Kozlov, I., Petukhov, A.: Sparse solutions for underdetermined systems of linear equations. In: Freeden, W., Nashed, M.Z., Sonar, T. (eds.) Handbook of Geomathematics, pp. 1243–1259. Springer, Berlin (2010)
10. Lai, M.-J., Wang, J.: An unconstrained l_q minimization with $0 < q \leq 1$ for sparse solution of under-determined linear systems. SIAM J. Optim. **21**, 82–101 (2010)
11. Natarajan, B.K.: Sparse approximate solution to linear systems. SIAM J. Comput. **24**, 227–234 (1995)
12. Petukhov, A.: Fast implementation of orthogonal greedy algorithm for tight wavelet frames. Signal Process. **86**, 471–479 (2006)
13. Rudelson, M., Vershynin, R.: Geometric approach to error correcting codes and reconstruction of signals. Int. Math. Res. Not. **64**, 4019–4041 (2005)

Harmonic Analysis and Uniqueness Questions in Convex Geometry

Dmitry Ryabogin, Vlad Yaskin, and Artem Zvavitch

Abstract We discuss some open questions on unique determination of convex bodies.

1 Introduction and Notation

The purpose of this chapter is to give a short overview of known results and open questions on unique determination of convex and star bodies. These questions are usually treated with the aid of techniques of harmonic analysis. We give a typical example of such a problem in the next section. The reader is referred to the books by Groemer [9] and Koldobsky [20] as well as to the articles by Falconer [5] and Schneider [31] for the use of spherical harmonics and the Fourier transform in convex geometry.

First we introduce some notation. For standard notions in convex geometry we refer the reader to the books by Gardner [8] and Schneider [33].

A convex body in $\mathbb{R}^n$ is a compact convex set with nonempty interior. The support function h_K of a convex body K in $\mathbb{R}^n$ is defined by

$$h_K(x) = \max\{\langle x, y\rangle : y \in K\}, \quad x \in \mathbb{R}^n.$$

Clearly, h_K is positively homogeneous of degree 1, and therefore is determined by its values on the unit sphere. If V is a subspace of $\mathbb{R}^n$, then we write $K|V$ for the orthogonal projection of K onto V. It is easy to see that the support function of $K|V$, as a convex body in V, is just the restriction of h_K to V.

D. Ryabogin (✉) • A. Zvavitch
Department of Mathematics, Kent State University, Kent, OH 44242, USA
e-mail: ryabogin@math.kent.edu; zvavitch@math.kent.edu

V. Yaskin
Department of Mathematical and Statistical Sciences, University of Alberta, Edmonton, Alberta T6G 2G1, Canada
e-mail: vladyaskin@math.ualberta.ca

D. Bilyk et al. (eds.), *Recent Advances in Harmonic Analysis and Applications*, Springer Proceedings in Mathematics & Statistics 25, DOI 10.1007/978-1-4614-4565-4_26,

A convex body K is of constant width if $h_K(\xi)+h_K(-\xi)=$ constant for all $\xi\in S^{n-1}$. Let $G(n,k)$ denote the Grassmannian of k-dimensional subspaces of $\mathbb{R}^n$. The kth projection function of a convex body K is the function on $G(n,k)$ that assigns $\mathrm{vol}_k(K|H)$ to every k-dimensional subspace $H\in G(n,k)$. A convex body K has constant k-brightness if $\mathrm{vol}_k(K|H)=c$ for all $H\in G(n,k)$ and some constant c. In the case $k=n-1$ we say that K has constant brightness.

Let K be a compact convex set in $\mathbb{R}^n$. Its intrinsic volumes $V_i(K)$, $1\le i\le n$, can be defined via Steiner's formula:

$$\mathrm{vol}_n(K+\varepsilon B_2^n)=\sum_{i=0}^{n}\kappa_{n-i}V_i(K)\varepsilon^{n-i},$$

where the addition is the Minkowski addition, κ_{n-i} is the volume of the $(n-i)$-dimensional Euclidean ball, and $\varepsilon\ge 0$.

In particular, if K is a convex body in $\mathbb{R}^n$, then $V_n(K)$ is its volume and $V_{n-1}(K)$ is half the surface area. For these and other facts about intrinsic volumes we refer the reader to the book [33].

We say that K is a star body if it is compact, star-shaped at the origin, and its radial function ρ_K defined by

$$\rho_K(x)=\max\{\lambda>0:\lambda x\in K\},\qquad x\in S^{n-1},$$

is positive and continuous.

2 Typical Result

Harmonic analysis is an indispensable tool in convex geometry. Let us demonstrate this with the following well-known result, showing that origin-symmetric star bodies are uniquely determined by the size of their central sections.

Theorem 1. *Let K and L be origin-symmetric star bodies in $\mathbb{R}^n$ such that*

$$\mathrm{vol}_{n-1}(K\cap H)=\mathrm{vol}_{n-1}(L\cap H)$$

for every central hyperplane H. Then $K=L$.

We will give two very similar harmonic analysis proofs of this theorem. The first one uses the spherical Radon transform, which is a linear operator $\mathscr{R}:C(S^{n-1})\to C(S^{n-1})$ defined by

$$\mathscr{R}f(\xi)=\int_{S^{n-1}\cap\xi^{\perp}}f(x)\mathrm{d}x,\qquad \xi\in S^{n-1},$$

where $\xi^{\perp}$ is the hyperplane passing through the origin, and orthogonal to a given direction $\xi \in S^{n-1}$,

$$\xi^{\perp} = \{x \in \mathbb{R}^n : \langle x, \xi \rangle = 0\}.$$

Proof. The spherical Radon transform arises naturally in problems about volumes of central sections of star bodies. If H has unit normal vector ξ, then passing to polar coordinates in H, we obtain

$$\mathrm{vol}_{n-1}(K \cap H) = \frac{1}{n-1} \int_{S^{n-1} \cap \xi^{\perp}} \rho_K^{n-1}(x) \mathrm{d}x = \frac{1}{n-1} \mathscr{R} \rho_K^{n-1}(\xi).$$

Thus, we can reduce the geometric question about sections of star bodies to the question about the injectivity properties of the spherical Radon transform. The latter is known to be injective on even functions; see [9, Sect. 3.4] for details. In our case, both ρ_K^{n-1} and ρ_L^{n-1} are even functions on the unit sphere, since K and L are origin-symmetric. Hence, the condition of the theorem,

$$\frac{1}{n-1} \mathscr{R} \rho_K^{n-1}(\xi) = \frac{1}{n-1} \mathscr{R} \rho_L^{n-1}(\xi) \qquad \forall \xi \in S^{n-1},$$

yields $\rho_K^{n-1}(x) = \rho_L^{n-1}(x)$, and $\rho_K(x) = \rho_L(x)$ for all $x \in S^{n-1}$. This gives the desired result. □

Another known proof of Theorem 1 is based on the Fourier transform of distributions; see [20].

Proof. The main idea is that for an even function f, homogeneous of degree $-n+1$, and continuous on $\mathbb{R}^n \setminus \{0\}$, we have

$$\mathscr{R} f(\xi) = \frac{1}{\pi} \widehat{f}(\xi), \qquad \forall \xi \in S^{n-1}.$$

Thus, the assumption of Theorem 1 can be written as

$$\widehat{\rho_K^{n-1}}(\xi) = \widehat{\rho_L^{n-1}}(\xi), \qquad \forall \xi \in S^{n-1}.$$

By homogeneity, the latter equality is true on $\mathbb{R}^n \setminus \{0\}$. Inverting the Fourier transforms, we get $\rho_K = \rho_L$. □

For bodies that are not necessarily symmetric, Theorem 1 is not true; see [8, Theorem 6.2.18, Theorem 6.2.19].

We would like to mention that we are not aware of any other proof of Theorem 1 that uses ideas, different from the ones we just discussed.

3 Central Sections

What is the answer in Theorem 1 if we replace the $(n-1)$-volume by the surface area or other intrinsic volumes?

Question 1. Let i and k be integers with $1 \le i \le k \le n-1$, and let K and L be origin-symmetric convex bodies in $\mathbb{R}^n$ such that

$$V_i(K \cap H) = V_i(L \cap H), \qquad \forall H \in G(n,k).$$

Is it true that $K = L$?

In the simplest form when $n = 3$, $k = 2$, and $i = 1$, this question was asked by Gardner in his book [8]. Namely, are two origin-symmetric convex bodies K and L in $\mathbb{R}^3$ equal if the sections $K \cap \xi^{\perp}$ and $L \cap \xi^{\perp}$ have equal perimeters for all $\xi \in S^2$? The problem is open, except for some particular cases.

Howard, Nazarov, Ryabogin, and Zvavitch [16] solved the problem in the class of C^1 star bodies of revolution. Rusu [28] settled an infinitesimal version of the problem, when one of the bodies is the Euclidean ball and the other is its one-parameter analytic deformation. Yaskin [36] showed that the answer is affirmative in the class of origin-symmetric convex polytopes in $\mathbb{R}^n$, where in dimensions $n \ge 4$, the perimeter is replaced by the surface area of the sections. On the other hand, Question 1 has a negative answer in the class of general (not necessarily symmetric) convex bodies containing the origin in their interiors; see [29].

There are many interesting questions about the so-called *intersection bodies*, related to the volumes of sections of origin-symmetric star bodies. We refer the reader to the books [20, 22] for these problems.

We finish this section with several uniqueness results and questions about *congruent* sections of convex bodies. We start with the following result of Schneider [32].

Theorem 2. *Let $K \subset \mathbb{R}^n$, $n \ge 3$, be a convex body containing the origin. If for all $\xi \in S^{n-1}$ all the intersections $K \cap \xi^{\perp}$ are congruent, then K is a Euclidean ball.*

A similar problem about two bodies is still open even in the three-dimensional case (cf. [Ga], page 289, Problem 7.3).

Question 2. Let K and L be two convex bodies in $\mathbb{R}^n$, $n \ge 3$, containing zero in their interiors. Assume that the $(n-1)$-dimensional sections of these bodies by the hyperplanes passing through the origin are congruent. Does it follow that K is a translate of $\pm L$?

What happens if we drop the convexity assumption, but require only the "parallel translation congruency" of sections? Gardner notes that the following question has an affirmative answer, [8, p. 290].

Question 3. Let K and L be two star-shaped bodies in $\mathbb{R}^n$ containing the origin in their interiors. Assume that the $(n-1)$-dimensional sections of these bodies by the hyperplanes passing through the origin are translates of each other. Does it follow that K is a translate of L?

We will return to analogous questions about projections in a subsequent section.

4 Maximal Sections

Let K be a convex body in $\mathbb{R}^n$. The *inner section function* m_K is defined by

$$m_K(\xi) = \max_{t \in \mathbb{R}} \mathrm{vol}_{n-1}(K \cap (\xi^\perp + t\xi)),$$

for $\xi \in S^{n-1}$.

In 1969, Klee [18, 19] asked whether a convex body is uniquely determined (up to translation and reflection in the origin) by its inner section function. In [10] Gardner, Ryabogin, Yaskin, and Zvavitch answered Klee's question in the negative by constructing two convex bodies K and L, one of them origin-symmetric and the other is not centrally symmetric, such that $m_K(\xi) = m_L(\xi)$ for all $\xi \in S^{n-1}$. Klee also asked whether a convex body in $\mathbb{R}^n$, $n \geq 3$, whose inner section function is constant, must be a ball. This question was recently answered in the negative in [27].

Since the knowledge of the inner section function is not sufficient for determining a convex body, one can try to put additional assumptions. For example, it is natural to ask the following.

Question 4. Are convex bodies uniquely determined by their inner section functions and the function $t_K(\xi)$ that gives the distance from the origin to the affine hyperplane that contains the maximal section in the direction of ξ?

Motivated by Theorem 2 one can also ask the following question about maximal sections.

Question 5. Let K be a convex body and $t_K(\xi)$ be defined as above. If for all $\xi \in S^{n-1}$ all the intersections $K \cap \{\xi^\perp + t_K(\xi)\xi\}$ are congruent, is then K a Euclidean ball?

We finish this section with the question about maximal sections which is a version of a result of Montejano (see [8] for references and related results).

Question 6. Let K be a convex body in $\mathbb{R}^3$ such that its maximal sections are of constant width. Does it follow that K is a Euclidean ball?

5 *t*-Sections

In [2] Barker and Larman ask the following.

Question 7. Let K and L be convex bodies in $\mathbb{R}^n$ containing a sphere of radius t in their interiors. Suppose that for every hyperplane H tangent to the sphere we have $\mathrm{vol}_{n-1}(K \cap H) = \mathrm{vol}_{n-1}(L \cap H)$. Does this imply that $K = L$?

In [2] the authors obtained several partial results. They showed that in $\mathbb{R}^2$ the uniqueness holds if one of the bodies is a Euclidean disk. (The authors of [2] were apparently unaware of the paper [30] by Santaló, where he obtained an analogous result on the sphere. He then remarks that the limiting case, when the radius of the sphere tends to infinity, gives the result in the Euclidean plane.) In $\mathbb{R}^n$ Barker and Larman proved that the answer to this conjecture is affirmative if hyperplanes are replaced by planes of a larger codimension. However, the answer to the original question is still unknown, even in dimension 2.

Yaskin [35] showed that the answer to the problem is affirmative if both K and L are convex polytopes in $\mathbb{R}^n$. The case $n = 2$ of this result when K and L are polygons in $\mathbb{R}^2$ was earlier settled by Xiong, Ma, and Cheung [34].

Barker and Larman also suggest another generalization of the problem.

Question 8. Let K and L be convex bodies in $\mathbb{R}^n$ containing a convex body M in their interiors. Suppose that for every hyperplane H that supports M we have $\mathrm{vol}_{n-1}(K \cap H) = \mathrm{vol}_{n-1}(L \cap H)$. Does this imply that $K = L$?

Let us mention that in the case when M is just a straight line segment, the answer to the latter problem is affirmative. This is just a reformulation of the result proved independently by Falconer [6] and Gardner [8] that any convex body is uniquely determined by the volumes of hyperplane sections through any two points in the interior of the body. See also [21].

6 Slabs

Let $t > 0$ and $\xi \in S^{n-1}$. The slab of width $2t$ in the direction of ξ is defined by

$$S_t(\xi) = \{x \in \mathbb{R}^n : |\langle x, \xi \rangle| \le t\}.$$

Slabs can be thought of as "thick" sections.

In [29] the following problem was suggested.

Question 9. Let K and L be origin-symmetric convex bodies in $\mathbb{R}^n$ that contain the Euclidean ball of radius t in their interiors. Suppose that for some i $(1 \le i \le n)$

$$V_i(K \cap S_t(\xi)) = V_i(L \cap S_t(\xi)), \qquad \forall \xi \in S^{n-1}.$$

Is it true that $K = L$?

Note that without the symmetry assumption this question has a negative answer; see [29].

7 Projections

The well-known Aleksandrov's projection theorem states that origin-symmetric convex bodies are uniquely determined by the sizes of their projections.

Theorem 3. *Let $1 \le i \le k \le n$ and let K and L be origin-symmetric convex bodies in $\mathbb{R}^n$. If*

$$V_i(K|H) = V_i(L|H), \qquad \forall H \in Gr(n,k),$$

then $K = L$.

This result fails in the absence of symmetry; see [8, Theorem 3.3.17, Theorem 3.3.18] or [12]. In fact, they prove more.

Theorem 4. *There are noncongruent convex bodies K and L in $\mathbb{R}^n$ such that for all i and k with $1 \le i \le k \le n$, we have*

$$V_i(K|H) = V_i(L|H), \qquad \forall H \in Gr(n,k).$$

Moreover, the pair of bodies can be chosen to be C_+^∞ bodies of revolution, or polytopes.

As we can see, even the knowledge of all projection functions does not allow to determine a convex body uniquely. Suppose that some projection function of convex body is known to be constant, is then the body a ball? No, since there are nonspherical bodies of constant k-brightness; see [7].

The following is an old question of Bonnesen [4].

Question 10. Let $K \subset \mathbb{R}^n$, $n \ge 3$, be a convex body whose inner section function and brightness function are constant. Does it follow that K is a Euclidean ball?

If the inner section function and the brightness function of a body K are both equal to the same constant, then the answer is known to be affirmative. A simple proof of this result was communicated to us by Nazarov.

Assume now that *two* projection functions are constant. Is then the body a ball? The question of whether a convex body in $\mathbb{R}^3$ of constant width and constant brightness must be a ball is known as the Nakajima problem. Back in 1926, Nakajima [26] gave an affirmative answer to the problem under the additional assumption that the boundary of the body is of class C_+^2. The general case was resolved by Howard [13].

Theorem 5. *Let K be a convex body in $\mathbb{R}^3$ of constant width and constant brightness. Then K is a Euclidean ball.*

Generalizations were obtained by Howard and Hug [14, 15] and by Hug [17].

Theorem 6. *Let K be a convex body in $\mathbb{R}^n$. Let $1 \le i < j \le n-2$ and $(i,j) \ne (1,n-2)$. Assume that K has constant i-brightness and constant j-brightness. Then K is a Euclidean ball.*

However, the following, for example, is still unknown, even in the smooth case.

Question 11. Let K be a convex body in $\mathbb{R}^n$, $n \geq 4$, with constant $(n-1)$-brightness and constant $(n-2)$-brightness. Is K a Euclidean ball? What about bodies of constant width and $(n-1)$-brightness?

There are questions where one is interested in the shape of projections, rather than their size. The following problem is a "dual" version of Question 2.

Question 12. Let K and L be two convex bodies in $\mathbb{R}^n$, $n \geq 3$. Assume that the projections of these bodies (on corresponding hyperplanes) are congruent. Does it follow that K is a translate of $\pm L$?

A beautiful Fourier analytic lemma was obtained by Golubyatnikov [11], who showed that in the three-dimensional case, the corresponding projections could only be of three types: translations, reflections, and two-dimensional bodies of constant width. Using this lemma, Golubyatnikov, in particular, proved the following.

Theorem 7. *Let K and L be convex bodies in $\mathbb{R}^3$ with twice continuously differentiable support functions and such that for all $\xi \in S^2$, the projections $K|\xi^\perp$ and $L|\xi^\perp$ are $SO(2)$-congruent and one of the following is true:*

(i) The projections are discs.
(ii) The projections are not of constant width.
(iii) The projections have no $SO(2)$ symmetries.

Then K is parallel to $\pm L$.

This result is a generalization of the so-called Süss' Lemma (an analogue of Theorem 7 when the projections $K|\xi^\perp$ and $L|\xi^\perp$ are translates of each other). We would also like to mention that a beautiful and elementary proof of Süss' Lemma was obtained by Lieberman [1].

There is no doubt that the reader is now able to come up with other questions of mixed nature, involving sections and projections. We would like to add one more suggested to us by Gardner.

Question 13. Let K in $\mathbb{R}^3$ be a convex body containing the origin in its interior. If all one-dimensional central sections of K have constant length and K is of constant brightness, does it follow that K is a ball?

We finish our chapter with several questions about symmetry of convex bodies.

8 Symmetry

Let K be a convex body in $\mathbb{R}^n$. The parallel section function of K in the direction of $\xi \in S^{n-1}$ is defined by

$$A_{K,\xi}(t) = \mathrm{vol}_{n-1}(K \cap (\xi^\perp + t\xi)), \qquad t \in \mathbb{R}.$$

It is a consequence of the Brunn–Minkowski inequality that the maximal sections of an origin-symmetric convex body pass through the origin. Makai, Martini, and Ódor [25] have shown that the converse statement is also true.

Theorem 8. *Let K be a convex body in $\mathbb{R}^n$ such that $A_{K,\xi}(0) \geq A_{K,\xi}(t)$ for all $\xi \in S^{n-1}$ and all $t \in \mathbb{R}$, then K is symmetric with respect to the origin.*

They also posed a similar question for lower intrinsic volumes of sections (see also [23] for other results in this direction).

Question 14. Let K be a convex body in $\mathbb{R}^n$ such that $V_i(K \cap \xi^{\perp}) \geq V_i(K \cap \{\xi^{\perp} + t\xi\})$ for all $\xi \in S^{n-1}$ and all $t \in \mathbb{R}$. Is K an origin-symmetric body?

The problem is open unless K is a smooth perturbation of the Euclidean ball. In the latter case, Makai and Martini [24] have shown that the answer is affirmative.

We suggest a question in the same spirit.

Question 15. Let $t_K(\xi)$ be such a function on the sphere that $A_{K,\xi}(t_K(\xi)) = \max_{t \in \mathbb{R}} A_{K,\xi}(t)$. Assume that for every $\xi \in S^{n-1}$ the hyperplane $\{\xi^{\perp} + t_K(\xi)\xi\}$ divides the surface of K into two parts of equal area. Is K centrally symmetric?

What are other criteria that allow to determine the symmetry of a given body? In the spirit of a conjecture of Bianchi and Gruber [3], one can ask the following.

Question 16. Let K be a convex body and t a continuous function on the sphere S^{n-1}. Assume that for every $\xi \in S^{n-1}$ the $(n-1)$-dimensional body $K \cap \{x \in \mathbb{R}^n : \langle x, \xi \rangle = t(\xi)\}$ has a center of symmetry. Is then K centrally symmetric?

In particular, is the following true?

Question 17. Let $t_K(\xi)$ be such that $A_{K,\xi}(t_K(\xi)) = \max_{t \in \mathbb{R}} A_{K,\xi}(t)$. If for every $\xi \in S^{n-1}$ the $(n-1)$-dimensional body $K \cap \{x \in \mathbb{R}^n : \langle x, \xi \rangle = t_K(\xi)\}$ is centrally symmetric, does it follow that K is centrally symmetric?

Such questions are directly related to questions about t-sections, mentioned before.

Question 18. Let K be convex body in $\mathbb{R}^n$ containing the Euclidean ball of radius $t > 0$ in its interior. Suppose that the sections $K \cap \{\xi^{\perp} + t\xi\}$ are $(n-1)$-dimensional centrally symmetric bodies for all $\xi \in S^{n-1}$. Is it true that K is origin symmetric? Is it an ellipsoid?

Question 19. Let K be convex body in $\mathbb{R}^n$ containing the Euclidean ball of radius $t > 0$ in its interior. Suppose that $A_{K,\xi}(t) = A_{K,\xi}(-t)$ for every $\xi \in S^{n-1}$. Is K origin symmetric?

As one can check, the latter question is equivalent to the following question about slabs.

Question 20. Let K be a convex body in $\mathbb{R}^n$ containing the Euclidean ball of radius $t > 0$ in its interior. Suppose that

$$\mathrm{vol}_n(K \cap \{x : |\langle x, \xi \rangle| \leq t\}) = \max_{a \in \mathbb{R}} \mathrm{vol}_n(K \cap \{x : -t + a \leq \langle x, \xi \rangle \leq t + a\})$$

for all $\xi \in S^{n-1}$. Does this imply that K is origin symmetric?

Acknowledgments First author supported in part by U.S. National Science Foundation Grants DMS-0652684 and DMS-1101636. Second author supported in part by NSERC. Third author supported in part by U.S. National Science Foundation Grant DMS-1101636.

References

1. Aleksandrov, A.D.: On the theory of mixed volumes of convex bodies, Part II, New inequalities between mixed volumes and their applications. Mat. Sbornik (N. S.) **2**(44), 1205–1238 (1937)
2. Barker, J.A., Larman, D.G.: Determination of convex bodies by certain sets of sectional volumes. Disc. Math. **241**, 79–96 (2001)
3. Bianchi, G., Gruber, P.M.: Characterization of ellipsoids. Arch. Math. (Basel) **49**, 344–350 (1987)
4. Bonnesen, T.: Om Minkowskis uligheder for konvekse legemer. Mat. Tidsskr. B 74–80 (1926)
5. Falconer, K.J.: Applications of a result of spherical integration to the theory of convex sets. Am. Math. Mon. **90**(10), 690–695 (1983)
6. Falconer, K.J.: X-ray problems for point sources. Proc. London Math. Soc. **46**, 241–262 (1983)
7. Firey, W.J.: Convex bodies of constant outer p-measure. Mathematika **17**, 21-27 (1970)
8. Gardner, R.J.: Geometric Tomography, 2nd edn. Encyclopedia of Mathematics and its Applications, vol. 58. Cambridge University Press, New York (2006)
9. Groemer, H.: Geometric Applications of Fourier Series and Spherical Harmonics. Cambridge University Press, New York (1996)
10. Gardner, R.J., Ryabogin, D., Yaskin, V., Zvavitch, A.: A problem of Klee on inner section functions of convex bodies. J. Diff. Geom., to appear
11. Golubyatnikov, V.P.: Uniqueness questions in reconstruction of multidimensional objects from tomography type projection data. In: Inverse and Ill-Posed Problems Series. Utrecht-Boston-Köln-Tokyo (2000)
12. Goodey, P., Schneider, R., Weil, W.: On the determination of convex bodies by projection functions. Bull. London Math. Soc. **29**, 82–88 (1997)
13. Howard, R.: Convex bodies of constant width and constant brightness. Adv. Math. **204**, 241261 (2006)
14. Howard, R., Hug, D.: Smooth convex bodies with proportional projection functions. Israel J. Math. **159**, 317–341 (2007)
15. Howard, R., Hug, D.: Nakajima's problem: convex bodies of constant width and constant brightness. Mathematika **54**, 15–24 (2007)
16. Howard, R., Nazarov, F., Ryabogin, D., Zvavitch, A.: Determining starlike bodies by the perimeters of their central sections. Preprint
17. Hug, D.: Nakajima's problem for general convex bodies. Proc. Amer. Math. Soc. **137**, 255–263 (2009)
18. Klee, V.: Is a body spherical if its HA-measurements are constant? Am. Math. Mon. **76**, 539-542 (1969)
19. Klee, V.: Shapes of the future. Am. Sci. **59**, 84-91 (1971)

20. Koldobsky, A.: Fourier Analysis in Convex Geometry. American Mathematical Society, Providence, RI (2005)
21. Koldobsky, A., Shane, C.: The determination of convex bodies from derivatives of section functions. Arch. Math. **88**, 279–288 (2007)
22. Koldobsky, A., Yaskin, V.: The interface between convex geometry and harmonic analysis. CBMS Regional Conference Series, vol. 108. American Mathematical Society, Providence, RI (2008)
23. Makai, E., Martini, H.: On maximal k-sections and related common transversals of convex bodies. Canad. Math. Bull. **47**, 246–256 (2004)
24. Makai, E., Martini, H.: Centrally symmetric convex bodies and sections having maximal quermassintegrals. Studia Sci. Math. Hungar. **49**, 189–199 (2012)
25. Makai, E., Martini, H., Ódor, T.: Maximal sections and centrally symmetric bodies. Mathematika **47**, 19–30 (2000)
26. Nakajima, S.: Eine charakteristische Eigenschaft der Kugel. Jber. Deutsche Math.-Verein **35**, 298-300 (1926)
27. Nazarov, F., Ryabogin, D., Zvavitch, A.: Non-uniqueness of convex bodies with prescribed volumes of sections and projections, accepted to Mathematika
28. Rusu, A.: Determining starlike bodies by their curvature integrals. Ph.D. Thesis, University of South Carolina (2008)
29. Ryabogin, D., Yaskin, V.: On counterexamples in questions of unique determination of convex bodies. Proc. Amer. Math. Soc., to appear
30. Santaló, L.A.: Two characteristic properties of circles on a spherical surface (Spanish). Math. Notae **11**, 73-78 (1951)
31. Schneider, R.: Functional equations connected with rotations and their geometric applications. Enseign. Math. **16**, 297–305 (1970)
32. Schneider, R.: Convex bodies with congruent sections. Bull. London Math. Soc. **12**, 52-54 (1980)
33. Schneider, R.: Convex Bodies: The Brunn-Minkowski Theory. Cambridge University Press, Cambridge (1993)
34. Xiong, G., Ma, Y.-W., Cheung, W.-S.: Determination of convex bodies from Γ-section functions. J. Shanghai Univ. **12**(3), 200–203 (2008)
35. Yaskin, V.: Unique determination of convex polytopes by non-central sections. Math. Ann. **349**, 647–655 (2011)
36. Yaskin, V.: On perimeters of sections of convex polytopes. J. Math. Anal. Appl. **371**, 447–453 (2010)

Moduli of Smoothness and Rate of a.e. Convergence for Some Convolution Operators

Alexander M. Stokolos and Walter Trebels

Dedicated to Konstantin Oskolkov

Abstract One purpose of this chapter is to establish results on the rate of almost everywhere convergence of approximation processes of convolution type in $L^p(\mathbb{R}^n)$, where instead of a particular rate (like t^μ, $\mu > 0$, $t \to 0+$), fractional moduli of smoothness are employed. An essential tool is a modified K-functional. Away from saturation orders these results are nearly optimal. A second purpose is to illustrate that the methods applied also work in other settings which feature a convolution/multiplier structure.

2000 Mathematics Subject Classification: Primary 41A25; Secondary 41A40, 42B15, 42C10.

1 Introduction

Modify to "Starting point is a paper by K.I. Oskolkov [15] and two papers by the authors [13, 21] on the rate of a.e. convergence of certain integral means of convolution type on $L^p(\mathbb{R})$, $p \geq 1$. One application for the Weierstrass means of the general results there reads as follows.

A.M. Stokolos (✉)
Department of Mathematical Sciences, Georgia Southern University, PO Box 8093, Statesboro, GA 30460, USA
e-mail: astokolos@georgiasouthern.edu

W. Trebels
AG AGF, Fb. Mathematik, TU Darmstadt, Schlossgartenstr. 7, 64289 Darmstadt, Germany
e-mail: trebels@mathematik.tu-darmstadt.de

D. Bilyk et al. (eds.), *Recent Advances in Harmonic Analysis and Applications*, Springer Proceedings in Mathematics & Statistics 25, DOI 10.1007/978-1-4614-4565-4_27,

Let $\varepsilon > 0$ *and* $f \in L^p(\mathbb{R})$, $1 < p < \infty$, *satisfy* $\sup_{|h|<t} \|f(\cdot+h) - f(\cdot)\|_p = O(t^\mu), 0 < \mu < 1$, *then*

$$\frac{1}{2\pi t}\int_{\mathbb{R}} f(x-y)\,\mathrm{e}^{-(|y|/t)^2}\,\mathrm{d}y - f(x) = o_x(t^\mu|\log t|^{1/p+\varepsilon}) \quad \textit{a.e. as } t \to 0+.$$

It is impossible to choose $\varepsilon = 0$.

Here we considerably generalize the results of [21] in the multivariate framework $L^p(\mathbb{R}^n)$, $1 < p < \infty$, by relating the intrinsic smoothness of f, given by its norm modulus of smoothness, with the rate of a.e. convergence directly; in particular, moduli of smoothness of fractional order are needed to describe the rate of a.e. convergence for saturated approximation processes like the Riesz or the general Weierstrass/Abel–Cartwright means adequately. Further, we indicate that the method employed also works for summability methods of certain orthogonal expansions.The methods of proof themselves were motivated by a programmatic remark due to Shapiro [17, p. 120]: "Another interesting area for study is how far *pointwise* (rather than norm) approximation theorems can be inferred from the Fourier transform of the kernel." This means to look at this area from a multiplier point of view.

Here and in the next two sections we deal with the $L^p(\mathbb{R}^n)$-setting. First define the modulus of smoothness $\omega_\lambda(\cdot,f)_p$ of order $\lambda > 0$ of $f \in L^p$, $1 < p < \infty$, by

$$\omega_\lambda(t,f)_p := \sup_{|h|<t} \|\Delta_h^\lambda f\|_p, \qquad \Delta_h^\lambda f(x) = \sum_{j=0}^{\infty}(-1)^j\binom{\lambda}{j} f(x+jh), \quad \lambda > 0.$$

Approximation processes $(T_t f)_{t>0}$ of convolution type on $L^p(\mathbb{R}^n), 1 < p < \infty$, are considered, where

$$T_t f := t^{-n}\int_{\mathbb{R}^n} K(y/t)f(x-y)\,\mathrm{d}y, \qquad K \in L^1(\mathbb{R}^n), \quad \int_{\mathbb{R}^n} K(y)\,\mathrm{d}y = 1. \tag{1}$$

It is well known (see, e.g., [18, p.62]) that $\sup_{|y|\geq|x|}|K(y)| \in L^1$ implies

$$\lim_{t\to 0+} T_t f(x) - f(x) = 0 \quad \text{a. e.}$$

Here we are looking for a refinement of this result in the sense of

$$T_t f(x) - f(x) = o(1)\,\omega_\lambda(t,f)_p F(\omega_\lambda(t,f)_p) \qquad \text{a. e.} \tag{2}$$

Also interesting is the question how the rate of norm convergence directly entails a certain rate of a.e. convergence. The central problem is to find a "good" (decreasing) adjusting function $F : (0,1) \to (0,\infty)$ ensuring (2). It turns out advantageously to replace $\omega_\lambda(t,f)$ by Peetre's (modified) K-functional. This we define as follows.

Let $\varphi \in \mathscr{S}(\mathbb{R}^n)$ be such that $\operatorname{supp}\varphi = \{\xi \in \mathbb{R}^n : 1/2 \le |\xi| \le 2\}$, $\varphi(\xi) > 0$ for $1/2 < |\xi| < 2$, and $\sum_{k=-\infty}^{\infty} \varphi_k(\xi) = 1$, $\xi \neq 0$, where $\varphi_k(\xi) = \varphi(2^{-k}\xi)$. Then, for $1 \le p \le \infty$ and $\lambda > 0$, define the Riesz space $H_\lambda^p(\mathbb{R}^n)$ by (cf. [2, p. 147])

$$H_\lambda^p(\mathbb{R}^n) := \{f \in L^p : \|f\|_p + |f|_{H_\lambda^p} < \infty\}, \quad |f|_{H_\lambda^p} := \left\| \sum_{k=-\infty}^{\infty} \mathscr{F}^{-1}[|\xi|^\lambda \varphi_k] * f \right\|_p .$$

Here $\mathscr{F}$ denotes the Fourier transform ($\mathscr{F}^{-1}$ its inverse) defined on the Schwarz test function space $\mathscr{S}(\mathbb{R}^n)$ by

$$\mathscr{F}[f](\xi) \equiv \widehat{f}(\xi) := \int_{\mathbb{R}^n} f(x) \mathrm{e}^{-i\xi x}\,\mathrm{d}x, \qquad f \in \mathscr{S}.$$

If one introduces an associated K-functional by

$$K(t,f) \equiv K(t,f;L^p,H_\lambda^p) := \inf_{g \in H_\lambda^p} (\|f-g\|_p + t|g|_{H_\lambda^p}), \quad t > 0 \tag{3}$$

(note that it is increasing), then Wilmes [29] has shown that

$$K(t^\lambda, f; L^p(\mathbb{R}^n), H_\lambda^p) \approx \omega_\lambda(t,f)_p, \quad 1 < p < \infty,\ \lambda > 0. \tag{4}$$

Observing that for $\lambda = N \in \mathbb{N}$ and $1 < p < \infty$ the Riesz space H_N^p can be identified with the Sobolev space $W_N^p : H_N^p = W_N^p$, Wilmes' result is not surprising in view of the well-known characterization [1, p. 341].

In the following we will only consider approximation processes of type (1) with radial kernels, i.e., $K(y) := k(|y|)$. Since $\widehat{K}$ is also radial, we write $\widehat{K}(\xi) := m(|\xi|)$. Thus, the approximation processes to be discussed may be written in the form

$$T_{m_t} f = \mathscr{F}^{-1}[m(t|\xi|)] * f, \quad \mathscr{F}^{-1}[m(|\xi|)] \in L^1(\mathbb{R}^n), \quad m(0) = 1. \tag{5}$$

Also we restrict ourselves to two cases: the saturated means which satisfy

$$\lim_{s \to 0+} \frac{1 - m(s)}{s^\gamma} = c, \qquad c \neq 0,\ c \in \mathbb{C},\ \text{some } \gamma > 0, \tag{6}$$

and the non-saturated ones for which $c = 0$ for all $\gamma > 0$. This automatically leads to the Riesz spaces H_γ^p, $\gamma > 0$ as saturation classes. Typical examples of saturated processes with saturation class H_γ^p are the general Riesz means $(R_t^{\delta,\gamma})_{t>o}$ and the general Weierstrass means $(W_t^\gamma)_{t>o}$. In the first case there is $m_{\delta,\gamma}(u) = (1-u^\gamma)_+^\delta$, where $\gamma > 0$ and $\delta > (n-1)/2$, in the second $m_\gamma(u) = \mathrm{e}^{-u^\gamma}$, $\gamma > 0$. Both means can be used [25] to characterize the K-functional

$$K(t^\gamma, f; L^p(\mathbb{R}^n), H_\gamma^p) \approx \|R_t^{\delta,\gamma} f - f\|_p \approx \|W_t^\gamma f - f\|_p, \qquad 1 \le p < \infty. \tag{7}$$

Before formulating a first result we quote two lemmas concerning multipliers useful when estimating norms of convolution operators and maximal functions. By purpose, we restrict ourselves to derivatives of integer order—for fractional derivatives see the corresponding references. We employ the following notation $A \approx B$ if $A \lesssim B$ and $A \gtrsim B$, where the former (the latter) is the shortcut of $A \leq cB$ ($A \geq cB$) for some constant $c > 0$ independent of essential quantities.

Lemma 1.1 ([24]). *Let m be a measurable, bounded, sufficiently smooth function on $(0,+\infty)$ which vanishes at infinity and which satisfies for some integer $N > (n-1)/2$*

$$A_N(m) \equiv \int_0^\infty u^N \,|m^{(N+1)}(u)|\, du < \infty.$$

Then $\mathscr{F}^{-1}[m(|\xi|)] \in L^1(\mathbb{R}^n)$ and $\|\mathscr{F}^{-1}[m]\|_1 \lesssim A_N(m)$.

Lemma 1.2 ([5,6]). *Let m be a measurable, bounded, sufficiently smooth function on $(0,+\infty)$ which vanishes at infinity and which satisfies for some $N \in \mathbb{N}, N > n|1/p-1/2|+1/2$,*

$$B_N(m) \equiv \left(\int_0^\infty |u^N m^{(N)}(u)|^2 \frac{du}{u}\right)^{1/2} + \int_0^\infty u^{N-1}|m^{(N)}(u)|\, du < \infty. \tag{8}$$

Define the maximal operator T_m^* on $L^2(\mathbb{R}^n)$ by

$$T_m^* f(x) = \sup_{t>0} |T_{m_t} f(x)|, \quad T_{m_t} f(x) = \mathscr{F}^{-1}[m(t|\xi|)\widehat{f}](x).$$

Then T_m^* is of strong type $(p,p), 1 < p < \infty$, with the operator norm estimate $\|T_m^*\|_{p\to p} \lesssim B_N(m)$; also T_m^* is of weak type $(1,1)$.

Our main result concerning the problem (2) reads as follows.

Theorem 1.3. *Let $f \in L^p(\mathbb{R}^n)$, $1 < p < \infty$. For $\lambda > 0$, choose a sequence $(\theta_\ell)_{\ell \in \mathbb{N}_0}$ via $K(\theta_\ell^\lambda, f) \equiv K(\theta_\ell^\lambda, f; L^p(\mathbb{R}^n), H_\lambda^p(\mathbb{R}^n)) = 2^{-\ell}$, and a decreasing function $F : (0,1) \to (0,\infty)$ with $\sum_{\ell=0}^\infty 1/F^p(2^{-\ell}) < \infty$. Further, assume that the function $tF(t)$ is increasing. Let $\psi \geq 0$ be a smooth cutoff function with $\psi(t) = 1$ if $0 \leq t \leq 1/2$, and $= 0$ if $t \geq 3/4$. Let $h \geq 0$ be a bump function with $\operatorname{supp} h \subset [1/2, 2]$ such that $\sum_{j=0}^\infty h(2^j s) = 1, 0 < s < 1$, and let m be sufficiently smooth with finite*

$$B_N(m(1-\psi)), \quad B_N((1-m(2^{-j}\cdot))h) =: B_j, \; j \in \mathbb{N}_0, \; N \in \mathbb{N}, \; N \geq (n+1)/2.$$

(a) If

$$\sum_{j=0}^\infty B_j 2^{j\lambda} \left(\sum_{\ell:\, \theta_\ell < 2^{-j}}^\infty \left(\frac{1}{F(2^{-\ell-1})} \right)^p \right)^{1/p} < \infty, \tag{9}$$

then we have

$$\mathscr{F}^{-1}[m(t|\xi|)] * f(x) - f(x) = o_x(1) K(t^\lambda, f) F(K(t^\lambda, f)) \text{ a.e., } t \to 0+,$$

where $o_x(1) \in L^p(\mathbb{R}^n)$.

(b) If f *additionally satisfies* $K(t^\lambda, f) \lesssim t^\eta$ *for some fixed* $\eta, 0 < \eta < \lambda$, *and* $F(t) := 1 + |\log t|^{1+1/p+\varepsilon}, \varepsilon > 0$, *with* $\sum_{j=0}^{\infty}(j+1)^{-1-\varepsilon} B_j 2^{j\lambda} < \infty, \ \varepsilon > 0$, *then, for* $t \to 0+$,

$$\mathscr{F}^{-1}[m(t|\xi|)] * f(x) - f(x) = o_x(1) K(t^\lambda, f) |\log K(t^\lambda, f)|^{1+1/p+\varepsilon} \text{ a.e.,}$$

where again $o_x(1) \in L^p(\mathbb{R}^n)$.

In view of (4), the assumption $K(t^\lambda, f) \lesssim t^\eta$ means that f should have some Hölder smoothness.

Example 1.4. (a) Standard choices for the adjusting function F are $F(t) = 1 + |\log t|^{1/p+\varepsilon}, 0 < t < 1$, or $F(t) = 1 + |\log t|^{1/p} |\log|\log t||^{1/p+\varepsilon}$, where $0 < t < 1/2$, $\varepsilon > 0$, etc.; in these cases, certainly $tF(t)$ is increasing.

(b) We start by giving two examples of non-saturated approximation processes.

(i) *De La Vallée Poussin Means.* Consider a C^∞-cutoff-function $m : [0,\infty) \to [0,1]$, $m(u) = 1$ for $0 \le u \le 1$ and $m(u) = 0$ for $2 \le u < \infty$. By Lemma 1.1 it is clear that $m(|\xi|)$ is the Fourier transform of an admissible approximation kernel. Further, condition (6) is satisfied. Now define $V_t f := \mathscr{F}^{-1}[m(t|\xi|)] * f$ as de La Vallée Poussin means and observe that, by Lemma 2.2, $B_j = 0$ for $j \ge 2$, the remaining B's are finite. Thus, by the characterization of the K-functional in [1, p. 341], we obtain for any fixed $k \in \mathbb{N}$, $\varepsilon > 0$,

$$V_t f(x) - f(x) = o_{x,k}(1)\, \omega_k(t,f)_p \,|\log \omega_k(t,f)_p|^{1/p+\varepsilon} \quad \text{a.e.,} \quad t \to 0+.$$

For comparison reasons we note the associated norm convergence result

$$\|V_t f - f\|_p \le C_k\, \omega_k(t,f)_p, \qquad f \in L^p, \ k \in \mathbb{N}.$$

(ii) Choose $m : (0,\infty) \to [0,1]$, $m(u) = 1 - \mathrm{e}^{-1/u}$, $m(0) = 1$. Since $m \in C^\infty$ and $\lim_{u\to\infty} m(u) = 0$, by Lemma 1 we have $\mathscr{F}^{-1}[m(|\xi|)] \in L^1(\mathbb{R}^n)$. Further, $B_N((1-\mathrm{e}^{-1/t})(1-\psi(t)) < \infty$ and $B_j = O(2^j \mathrm{e}^{-2^{j-1}})$, $j \in \mathbb{N}_0$, so that by Lemma 2 we see that, as in the previous example, we have for any fixed $k \in \mathbb{N}$ and $t \to 0+$

$$\mathscr{F}^{-1}[1 - \mathrm{e}^{-1/(t|\xi|)}] * f(x) - f(x) = o_{x,k}(1)\, \omega_k(t,f)_p \,|\log \omega_k(t,f)_p|^{1/p+\varepsilon} \quad \text{a.e.}$$

As already mentioned, typical examples of saturated approximation processes are the general Riesz means, the general Weierstrass means, but also the integral means occuring in Lebesgue's differentiation theorem. Theorem 1.3 now yields the following results.

(iii) *General Riesz Means.* Consider the general Riesz means from (7), i.e., $m(u) \equiv m_{\delta,\gamma}(u) = (1-u^\gamma)_+^\delta$. Then the assumptions of Theorem 1.3(a) are satisfied, since $B_j = O(2^{-j\gamma})$ and $B_N(m(u)(1-\psi)) < \infty$ when $0 < \lambda < \gamma$. Thus, we get with $F(t) = |\log t|^{1/p+\varepsilon}$ that

$$R_t^{\delta,\gamma} f(x) - f(x) = o_{x,\lambda,\gamma}(1)\, \omega_\lambda(t,f)_p |\log \omega_\lambda(t,f)_p|^{1/p+\varepsilon} \quad \text{a.e.}, \quad t \to 0+.$$

In view of (4) and (7) the associated norm convergence result reads

$$\|R_t^{\delta,\gamma} f - f\|_p \approx \omega_\gamma(t,f)_p\,, \qquad f \in L^p. \tag{10}$$

When we choose $F(t) = |\log t|^{1+1/p+\varepsilon}$, then the hypotheses of Theorem 1.3(b) are satisfied even when $\lambda = \gamma$ provided the additional smoothness assumption $\omega_\gamma(t,f)_p \lesssim t^\eta$ for some $\eta, 0 < \eta < \gamma$ holds. Therefore, we have for almost all $x \in \mathbb{R}^n$ that

$$R_t^{\delta,\gamma} f(x) - f(x) = o_x(1)\, \omega_\gamma(t,f)_p |\log \omega_\gamma(t,f)_p|^{1+1/p+\varepsilon} \quad \text{a.e.}, \quad t \to 0+,$$

or, when inferring pointwise convergence from strong convergence,

$$R_t^{\delta,\gamma} f(x) - f(x) = o_x(1) \|R_t^{\delta,\gamma} f - f\|_p |\log \|R_t^{\delta,\gamma} f - f\|_p|^{1+1/p+\varepsilon} \text{ a.e.}, t \to 0+.$$

We point out that, probably caused by the method of proof, the above result involving $\omega_\gamma(t,f)_p$ does not look natural "near" the saturation order t^γ (like $t^\gamma|\log t|$), since in the saturation case $R_t^{\delta,\gamma} f(x) - f(x) = O_x(t^\gamma)$—see [21].

(iv) *General Weierstrass Means.* In this case the choice $m_\gamma(u) = \mathrm{e}^{-u^\gamma}$ yields the same approximation results as in (iii) for the the general Riesz means.

(v) *Lebesgue's Differentiation Theorem.* This can also be looked at under the aspect of the rate of a.e.-convergence. Observe that (cf. [19, p. 171]

$$\frac{1}{|B_t(0)|} \int_{B_t(0)} f(x+y)\,\mathrm{d}y = \mathscr{F}^{-1}[m(t|\xi|)] * f(x), \qquad m(u) = c_n\, u^{-n/2} J_{n/2}(u),$$

where J_α denotes the Bessel function of order α and c_n is given by $m(0) = 1$. Clearly, $T^*_{m(1-\psi)}$ essentially is the Hardy–Littlewood maximal function, thus $\|T^*_{m(1-\psi)}\|_{p\to p} < \infty$, which replaces the condition $B_N(m(1-\psi)) < \infty$ in Theorem 1.3. Using the power series representation of the Bessel function it immediately follows that $B_j = O(2^{-2j})$. Hence, for $0 < \lambda < 2$, we obtain almost everywhere

$$\frac{1}{|B_t(0)|}\int_{B_t(0)} f(x+y)\,dy = f(x) + o_x(1)\,\omega_\lambda(t,f)_p|\log \omega_\lambda(t,f)_p|^{1/p+\varepsilon},\ t\to 0+,$$

or, when f satisfies a Hölder condition in L^p-norm, we get for $t\to 0+$

$$\frac{1}{|B_t(0)|}\int_{B_t(0)} f(x+y)\,dy = f(x) + o_x(1)\,\omega_2(t,f)_p|\log \omega_2(t,f)_p|^{1+1/p+\varepsilon} \quad \text{a.e.}$$

2 Proof of Theorem 1.3

Consider a smooth cutoff function ψ with $\psi(u)=1$ for $u\le 1/2$ and $=0$ for $u\ge 3/4$. Then

$$T_{m_t}f - f = T_{(m_t-1)\psi_t}f + T_{(m_t-1)(1-\psi_t)}f. \tag{11}$$

First consider the contribution given by the multiplier $(m-1)(1-\psi)$. By the choice of ψ we have

$$\operatorname{supp}(m-1)(1-\psi)\subset\{\xi : |\xi|\ge 1/2\}.$$

Introduce $\chi\in C_0^\infty$ such that $\chi(\xi)=1$ for $|\xi|\le 1/2$ and $\chi(\xi)=0$ for $|\xi|\ge 1$, then

$$||T_{1-\chi_t}f||_p \le ||T_{\chi_t}f-f||_p \lesssim K(t^\lambda,f) \quad 0<t<1. \tag{12}$$

This follows from the standard estimate: For all $g\in S$ there holds

$$||T_{1-\chi_t}f||_p \le ||T_{1-\chi_t}(f-g)||_p + ||T_{\chi_t}g-g||_p \lesssim ||f-g||_p + t^\lambda||\mathscr{F}^{-1}[|\xi|^\lambda\,\widehat{g}]\,||_p$$

and hence also for the infimum over all $g\in S$. Set

$$w(t) := K(t^\lambda,f)F(K(t^\lambda,f))$$

and assume $\theta_{\ell+1}<t\le\theta_\ell$. Since

$$\operatorname{supp}(m_t-1)(1-\psi_t)\subset\left\{\xi : |\xi|\ge\frac{1}{2\theta_\ell}\right\},\quad 1-\chi_{2\theta_\ell}(\xi)=1\ \text{ for } |\xi|\ge\frac{1}{2\theta_\ell},$$

we have

$$T_{(m_t-1)(1-\psi_t)}f = T_{(m_t-1)(1-\psi_t)(1-\chi_{2\theta_\ell})}f = T_{(m_t-1)(1-\psi_t)}(T_{1-\chi_{2\theta_\ell}}f).$$

This implies the key relation

$$|T_{(m_t-1)(1-\psi_t)}f(x)| \le C\sup_{\ell\ge 0}\frac{T^*_{(m-1)(1-\psi)}(T_{1-\chi_{2\theta_\ell}}f)(x)}{w(\theta_{\ell+1})}w(t),\quad t>0, \tag{13}$$

for almost all $x \in \mathbb{R}^n$. But

$$T_{(m_t-1)(1-\psi_t)}f = T_{m_t(1-\psi_t)}f + T_{\psi_t}f - f,$$

so

$$T^*_{(m-1)(1-\psi)}f \le T^*_{m(1-\psi)}f + T^*_{\psi}f + |f|.$$

Now $T^*_{\psi}, T^*_{m(1-\psi)} : L^p \to L^p$ are bounded maximal operators by Lemma 1.2 and the hypothesis. So we may conclude that $||T^*_{(m-1)(1-\psi)}||_{p\to p} < \infty$ and, therefore,

$$\left\| \sup_{\ell\ge 0} \frac{T^*_{(m-1)(1-\psi)}(T_{1-\chi_{2\theta_\ell}}f)}{w(\theta_{\ell+1})} \right\|_p^p \lesssim \sum_{\ell\ge 0} \frac{||T^*_{(m-1)(1-\psi)}(T_{1-\chi_{2\theta_\ell}}f)||_p^p}{w(\theta_{\ell+1})^p}$$
$$\lesssim ||T^*_{(m-1)(1-\psi)}||^p_{p\to p} \sum_{\ell\ge 0} \frac{||T_{1-\chi_{2\theta_\ell}}f||_p^p}{w(\theta_{\ell+1})^p}.$$

The estimate (12) and the definition of $w(t)$ give

$$\left\| \sup_{\ell\ge 0} \frac{T^*_{(m-1)(1-\psi)}(T_{1-\chi_{2\theta_\ell}}f)}{w(\theta_{\ell+1})} \right\|_p^p \le C_p(m,\psi) \sum_{\ell=1}^{\infty} \left(\frac{1}{F(K(\theta_\ell^\lambda, f))} \right)^p \tag{14}$$

which is finite by hypothesis (9) (only the contribution $j=0$ is needed here).

Now we estimate the contribution caused by $T_{(m_t-1)\psi_t}f(x)$.

Set $\mu \equiv (m-1)\psi$ and remember that $\mu = 0$ for $|\xi| \ge 3/4$ and $\mu(\xi) = m(\xi) - 1$ for $|\xi| \le 1/2$. Recall the partition of unity from the assumptions of Theorem 1.3:

$$\sum_{j=0}^{\infty} h(2^j t|\xi|) = 1 \quad \text{for} \quad |t\xi| < 1,\ 0 < t < 1.$$

Without loss of generality we may assume $\theta_0 = 1$ and choose $\ell_0 = \ell_0(j) \in \mathbb{N}$ such that $\theta_{\ell_0} \ge 2^{-j}$ but $\theta_{\ell_0+1} < 2^{-j}$. Then

$$\sup_{0<t<1} \frac{|T_{\mu_t}f|}{w(t)} \le \sum_{j=0}^{\infty} \sup_{0<t<1} \frac{|T_{\mu_t h_{t2^j}}f|}{w(t)}$$
$$\le \sum_{j=0}^{\infty} \sup_{\ell\ge\ell_0} \sup_{\theta_{\ell+1}<t\le\theta_\ell} \frac{|T_{\mu_t h_{t2^j}}f|}{w(t)} + \sum_{j=0}^{\infty} \sup_{2^{-j}<t\le 1} \frac{|T_{\mu_t h_{t2^j}}f|}{w(2^{-j})}. \tag{15}$$

In the last term we used that w is increasing. We begin to discuss the critical first sum on the right side of (15). It is clear that $\theta_{\ell+1} < t \le \theta_\ell$ implies $\operatorname{supp}(h_{t2^j}) \subset [(2^{j+1}\theta_\ell)^{-1}, (2^{j-1}\theta_{\ell+1})^{-1}]$. Now choose a nondecreasing function $\Phi \in C^\infty[0,\infty)$ such that $\Phi(|\xi|) = 1$ for $|\xi| \ge 1/2$ and $\Phi(|\xi|) = 0$ for $|\xi| \le 1/4$ and set

$\Phi_{\ell,j}(|\xi|) = \Phi(2^j\theta_\ell|\xi|)$. Hence $\Phi_{\ell,j}(|\xi|) = 1$ for $|\xi| \geq (2^{j+1}\theta_\ell)^{-1}$, and, therefore, we have $\Phi_{\ell,j}h_{t2^j} = h_{t2^j}$ provided $\theta_{\ell+1} < t \leq \theta_\ell$. Thus,

$$\left\| \sum_{j=0}^{\infty} \sup_{\ell\geq\ell_0} \frac{\sup_{\theta_{\ell+1}<t\leq\theta_\ell} |T_{\mu_t h_{t2^j}} f|}{w(\theta_{\ell+1})} \right\|_p \leq \sum_{j=0}^{\infty} \left\| \sup_{\ell\geq\ell_0} \frac{T^*_{\mu h_{2^j}}(T_{\Phi_{\ell,j}} f)}{w(\theta_{\ell+1})} \right\|_p$$

$$\leq \sum_{j=0}^{\infty} \left(\sum_{\ell\geq\ell_0} \left(\frac{||T^*_{\mu h_{2^j}}(T_{\Phi_{\ell,j}} f)||_p}{w(\theta_{\ell+1})} \right)^p \right)^{1/p}$$

$$\leq \sum_{j=0}^{\infty} ||T^*_{\mu h_{2^j}}||_{p\to p} \left(\sum_{\ell\geq\ell_0} \left(\frac{||T_{\Phi_{\ell,j}} f||_p}{w(\theta_{\ell+1})} \right)^p \right)^{1/p}.$$

Now observe that $||T_{\Phi_{\ell,j}} f||_p = ||T_{1-\Phi_{\ell,j}} f - f||_p \lesssim K((2^j\theta_\ell)^\lambda, f)$ analogous to (12). Therefore, we can continue the preceding estimate by

$$\lesssim \sum_{j=0}^{\infty} ||T^*_{\mu h_{2^j}}||_{p\to p} \left(\sum_{\ell\geq\ell_0} \left(\frac{K((2^j\theta_\ell)^\lambda, f)}{w(\theta_{\ell+1})} \right)^p \right)^{1/p}$$

$$\lesssim \sum_{j=0}^{\infty} B_j 2^{j\lambda} \left(\sum_{\ell\geq\ell_0} \left(\frac{K(\theta_\ell^\lambda, f)}{w(\theta_{\ell+1})} \right)^p \right)^{1/p}$$

$$\lesssim \sum_{j=0}^{\infty} B_j 2^{j\lambda} \left(\sum_{\ell\geq\ell_0} \left(\frac{1}{F(2^{-\ell})} \right)^p \right)^{1/p}. \tag{16}$$

For the last estimates we used the dilation invariance of the integral conditions in (8), namely $B_N(m_t) = B_N(m), t > 0$, and $w(\theta_{\ell+1}) \approx 2^{-\ell\lambda}F(2^{-\ell})$.

The second series on the right side of (15) is even easier to estimate, since

$$\left\| \sum_{j=0}^{\infty} \sup_{2^{-j}<t<1} \frac{|T_{\mu_t h_{t2^j}} f|}{w(2^{-j})} \right\|_p \lesssim \sum_{j=0}^{\infty} ||T^*_{\mu h_{2^j}}||_{p\to p} ||f||_p \frac{1}{w(2^{-j})}$$

$$\lesssim \sum_{j=0}^{\infty} B_j 2^{j\lambda} \frac{||f||_p}{F(2^{-j\lambda})}, \tag{17}$$

which is finite by (9) (for the simplification of the denominator we used that $K(2^{-j\lambda}, f) \geq 2^{-j\lambda}K(1, f)$).

Summarizing, in view of (11), we see that by the hypothesis (9) the estimates (15), (16) (clearly $\ell_0 \geq 0$) in combination with (12), (14) show that

$$\sup_{0<t<1} \frac{T_{m_t} f(x) - f(x)}{K(t^\lambda, f)F(K(t^\lambda, f))} < \infty \qquad \text{a.e.}, \ t \to 0+,$$

which proves Theorem 1.3(a).

Concerning Part (b) observe that, in view of the above proof, we have only to estimate the last double sum in (16) conveniently. By the the smoothness of f the choice of $\theta_{\ell_0(j)}$ yields $\ell_0(j) \gtrsim j+1$ since

$$2^{-\ell_0(j)} = K(\theta^\lambda_{\ell_0(j)}, f) = 2K(\theta^\lambda_{\ell_0(j)+1}, f) \lesssim \theta^\eta_{\ell_0(j)+1} < 2^{-j\eta}.$$

Therefore, with $F(t) = |\log t|^{1+1/p+\varepsilon}$,

$$\Big(\sum_{\ell\ge\ell_0}\Big(\frac{1}{F(2^{-\ell})}\Big)^p\Big)^{1/p} \lesssim \Big(\sum_{\ell\ge\ell_0}\ell^{-p-1-\varepsilon p}\Big)^{1/p} \lesssim \ell_0^{-1-\varepsilon} \lesssim (j+1)^{-1-\varepsilon},$$

which gives the assertion (b). □

Remark 2.1. (a) On account of [8, Theorem 3.14] the results in Sect. 1 also hold in the framework of Lorentz–Zygmund spaces discussed there.

(b) Further, these results easily carry over to the $L^p(\mathbb{T}^n)$-setting, $1 < p < \infty$, in view of the results in [19, Chap. VII, Sect. 3], [14, 30].

3 Generalizations

We want to show that the above method to obtain rates for a.e. convergence also works in other settings. For this purpose note that recalling the methods of Sect. 2 one realizes that we have only needed control over norms of operators given via multipliers, some norm estimates for the rate of convergence of de La Vallée Poussin type means via the K-functional and characterizations of the K-functional. Concerning a characterization via moduli of smoothness based on (generalized) translations we have needed the special structure of $\mathbb{R}^n$. Apart from that $\mathbb{R}^n$ did not appear, once one has the required multiplier criteria.

We start by briefly describing the abstract setting concerning eigenfunction expansions given in [4]. For a Lebesgue measurable set, $E \subset \mathbb{R}^n$ and a nonnegative measure μ on E define the space $L^p(E, \mathrm{d}\mu)$, $1 < p < \infty$, as the Banach space of all (Lebesgue) measurable functions on E with finite norm

$$\|f\|_p := \Big(\int_E |f(x)|^p \,\mathrm{d}\mu(x)\Big)^{1/p}.$$

Assume that there exists a sequence of projections $(P_k)_{k\in\mathbb{N}_0}$, $P_k : L^p \to L^p$, being linear, bounded, and satisfying

(i) $P_j P_k = \delta_{j,k} P_k$, $\delta_{j,k}$ being Kronecker's symbol (mutual orthogonality).
(ii) The sequence (P_k) is total, i.e., $P_k f = 0$ for all $k \in \mathbb{N}_0$ implies $f = 0$.
(iii) The sequence (P_k) is fundamental, i.e., the linear span of the ranges $P_k(L^p)$ is dense: $\overline{\bigcup_{k\in\mathbb{N}_0} P_k(L^p)} = L^p$.

Then, with each $f \in L^p$, one may associate its (formal) Fourier series expansion $f \sim \sum_{k=0}^{\infty} P_k f$. Denote by $\mathfrak{P} \subset L^p$ the set of all "polynomials," i.e., those $f \in X$, for which only finitely many $P_k f \neq 0$. Given a scalar-valued sequence $m = (m_k)_{k \in \mathbb{N}_0}$, we define the linear operator $T_m : \mathfrak{P} \to \mathfrak{P}$ by $T_m f = \sum_{k=0}^{\infty} m_k P_k f$. In particular, employing the notation $\phi_c^\gamma : [0,\infty) \to \mathbb{R}$, $\phi_c^\gamma(u) = (u(u+c))^{\gamma/2}$, we define an abstract derivative $D_c^\gamma : D_p(D_c^\gamma) \to L^p$, first on $\mathfrak{P}$ by

$$D_c^\gamma f := \sum_{k=0}^{\infty} \phi_c^\gamma(k) P_k f, \qquad \gamma > 0,\ c \geq 0,\ f \in \mathfrak{P},$$

and then take its closure in L^p, i.e., to each $f \in D_p(D_c^\gamma)$ there exist $f_j \in \mathfrak{P}$ with $\|f - f_j\|_p \to 0$ and $(D_c^\gamma f_j)_j$ is a Cauchy sequence in L^p. We call m a multiplier on L^p, notation $m \in M^p$, if

$$\|m\|_M := \sup\{\|T_m f\|_p : \|f\|_p \leq 1, f \in L^p\}$$

is finite (on account of (iii) there is a unique extension of T_m from $\mathfrak{P}$ to L^p).

To ensure control over maximal functions, which were essentially used in the $L^p(\mathbb{R}^n)$-case, we assume for some $N \in \mathbb{N}_0$ the following property:

$$\|\sup_k (C,N)_k f\|_p \leq C_N \|f\|_p \quad \text{for all } f \in L^p,\ 1 < p < \infty, \tag{18}$$

where the $(C,N)_k f$ denote the Cesàro means of order N of $f \in L^p$. These are given by

$$(C,N)_k f := (A_k^N)^{-1} \sum_{i=0}^{k} A_{k-i}^N P_i f, \quad A_k^N = \binom{k+N}{k}. \tag{19}$$

Clearly, (18) implies the boundedness of the Cesàro means:

$$\|(C,N)_k f\|_p \leq C_N \|f\|_p \quad \text{for all } f \in L^p,\ 1 < p < \infty, \tag{20}$$

which immediately leads to a substitute of Lemma 1.1—for the idea of the proof see the proof of Lemma 3.1 (cf. [4, II]).

Lemma 3.1. *(a) Let the projections $(P_k)_{k \in \mathbb{N}_0} \subset \mathscr{L}(L^p)$ be as above and (20) hold. Then*

$$\|T_m f\| \leq C_N \sum_{k=0}^{\infty} \binom{k+N}{k} |\Delta^{N+1} m_k| \, \|f\|_p \quad \textit{for all } f \in L^p.$$

(b) If one defines for continuous $m : [0,\infty) \to \mathbb{C}$ with $\lim_{t \to \infty} m(t) = 0$, an operator $T_{m_{\phi;t}}$ via the family of sequences $m_{\phi;t} := \left(m(\rho(t)\phi_c^\gamma(k))\right)_{k \in \mathbb{N}_0}$, $\gamma > 0$, with $\rho : [0,\infty) \to [0,\infty)$, $\rho(0) = 0$, strictly increasing then, for sufficiently smooth m,

$$\|T_{m_{\phi;t}} f\|_p \lesssim \int_0^\infty s^N \, |m^{(N+1)}(s)| \, \mathrm{d}s \, \|f\|_p$$

uniformly in t.

The latter estimate follows by a variant of the second theorem of consistency—see [23, Lemma 3.8]. Recall $\phi_c^\gamma(u) = (u(u+c))^{\gamma/2}$ and note that $\rho(t)$ will, e.g., be specialized to $2^j \phi_c^\gamma(t)$.

One substitute of Lemma 1.2 reads as follows.

Lemma 3.2. *Let the projections* $(P_k)_{k\in\mathbb{N}_0} \subset \mathcal{L}(L^p)$ *be as above, let* (18) *hold, and let* m *be as in* Lemma 3.1(b). *Define the maximal operator* $T_{m_\phi}^*$ *by*

$$T_{m_\phi}^* f := \sup_{t>0} |T_{m_{\phi;t}} f|.$$

Then, for sufficiently smooth m *and all* $f \in L^p$,

$$\|T_{m_\phi}^* f\|_p \lesssim B_N^*(m) \|f\|_p, \qquad B_N^*(m) := \int_0^\infty s^N \, |m^{(N+1)}(s)| \, \mathrm{d}s.$$

In the special case $\rho(t) = \phi_c^\gamma(t)$, $c = 0$, $\gamma = 1$, the maximal function is described on $\mathfrak{P}$ by $\sup_{t>0} |\sum_{k=0}^\infty m(tk) P_k f|$.

Proof. As is well known (cf. Part II of [4])

$$\begin{aligned}
|T_{m_{\phi;t}} f| &= \Big| \sum_{k=0}^\infty \binom{k+N}{k} \Delta^{N+1} m(\rho(t)\phi_c^\gamma(k)) \, (C,N)_k f \Big| \\
&\le |\sup_k (C,N)_k f| \sum_{k=0}^\infty \binom{k+N}{k} |\Delta^{N+1} m(\rho(t)\phi_c^\gamma(k))| \\
&\lesssim |\sup_k (C,N)_k f| \int_0^\infty s^N \, |(\mathrm{d}/\mathrm{d}s)^{(N+1)} m(\rho(t)\phi_c^\gamma(s))| \, \mathrm{d}s \\
&\lesssim |\sup_k (C,N)_k f| \int_0^\infty s^N \, |m^{(N+1)}(s)| \, \mathrm{d}s.
\end{aligned}$$

Now take the supremum over $t > 0$ on the left-hand side, then the L^p-norm of this inequality to get the assertion by the hypothesis (18). □

We point out that this simple derivation costs a certain trade-off concerning the size of the parameter N. But here we do not care for this aspect.

Similarly to [27] we define the (modified) K-functional in the present setting by

$$K_\gamma(t, f; L^p, D_p(D_c^\gamma)) := \inf_{g \in D_p(D_c^\gamma)} \{\|f-g\|_p + (t(t+c))^{\gamma/2} \|D_c^\gamma g\|_p\},$$

where $\gamma > 0$ and $c \geq 0$. In essential it is shown in [26] that

$$K_\gamma(t,f;L^p,D_p(D_c^\gamma)) \approx \left\| f - \sum_{k=0}^{\infty} \mathrm{e}^{-\phi_c^\gamma(t)\phi_c^\gamma(k)} P_k f \right\|_p \tag{21}$$

$$\approx \left\| f - \sum_{k=0}^{\infty} \left(1 - \phi_c^\gamma(t)\phi_c^\gamma(k)\right)_+^\kappa P_k f \right\|_p , \tag{22}$$

where $\kappa > N$. Also norm estimates of the de La Vallée Poussin means by the K-functional easily follow. Let m be the smooth cutoff function from Example 1.4(b)(i). Set

$$V_{\phi;t} f := \sum_{k=0}^{\infty} m(\phi_c^\gamma(t)\phi_c^\gamma(k)) P_k f , \qquad f \in L^p .$$

Then $\|V_{\phi;t}f - f\|_p \lesssim \|f\|_p$ and $\|V_{\phi;t}g - g\|_p \lesssim \phi_c^\gamma(t)\|D_c^\gamma g\|_p$ by Lemma 3.1(b), just as in the $L^p(\mathbb{R}^n)$-case. Therefore,

$$\|V_{\phi;t}f - f\|_p \lesssim K_\gamma(t,f;L^p,D_p(D_c^\gamma)), \qquad f \in L^p . \tag{23}$$

Thus, all ingredients of Sects. 1 and 2 are available up to a characterization of the K-functional involving moduli of smoothness with respect to generalized translations.

Theorem 3.3. *Let $f \in L^p$, $1 < p < \infty$. Choose a positive sequence $(\theta_\ell)_{\ell \in \mathbb{N}_0}$ via*

$$K_\gamma(\theta_\ell, f; L^p, D_p(D_c^\gamma)) \equiv K(\phi_c^\gamma(\theta_\ell), f; L^p, D_p(D_c^\gamma)) = 2^{-\ell}, \quad 1 < p < \infty,$$

and a decreasing function $F : (0,1) \to (0,\infty)$ with $\sum_{\ell=0}^{\infty} 1/F^p(2^{-\ell}) < \infty$. Let $\psi \geq 0$ be a smooth cutoff function with $\psi(t) = 1, 0 \leq t \leq 1/2$, and $= 0, t \geq 3/4$. Let $h \geq 0$ be a bump function with $\operatorname{supp} h \subset [1/2,2]$ such that

$$\sum_{j=0}^{\infty} h(2^j \phi_c^\gamma(t)\phi_c^\gamma(k)) = 1 \quad \text{if} \quad 0 < \phi_c^\gamma(t)\phi_c^\gamma(k) < 1.$$

Let m be sufficiently smooth and decaying with finite

$$B_N^*(m(1-\psi)), \quad B_N^*((1-m)h(2^j \cdot)) =: B_j^*, \ j \in \mathbb{N}_0, \quad N \text{ as in } (18).$$

(a) If $tF(t)$ is increasing and

$$\sum_{j=0}^{\infty} B_j^* 2^j \left(\sum_{\ell:\theta_\ell < 2^{-j}} \left(\frac{1}{F(2^{-\ell-1})} \right)^p \right)^{1/p} < \infty, \tag{24}$$

then, for every $f \in L^p$, $1 < p < \infty$,

$$T_{m_{\phi,t}} f(x) - f(x) = o_x(1) K_\gamma(t,f) F(K_\gamma(t,f)) \quad \text{a.e.}, \quad t \to 0+,$$

where $o_x(1) \in L^p$ *(recall* $T_{m_{\phi,t}} f = \sum_{k=0}^\infty m(\phi_c^\gamma(t)\phi_c^\gamma(k) P_k f$ *on* $\mathfrak{P}$*).*

(b) *Let* f *satisfy* $K_\gamma(t,f) \lesssim t^\eta$ *for some fixed* $\eta, 0 < \eta < 1$. *If* $F(t) = 1 + |\log t|^{1+1/p+\varepsilon}$, $\varepsilon > 0$, *and* $\sum_{j=1}^\infty (j+1)^{-1-\varepsilon} B_j^* 2^j < \infty$, *then, for every* $f \in L^p$, $1 < p < \infty$, *we have*

$$T_{m_{\phi,t}} f(x) - f(x) = o_x(1) K_\gamma(t,f) \,|\log K_\gamma(t,f)\,|^{1+1/p+\varepsilon} \quad \text{a.e.}, \quad t \to 0+,$$

where again $o_x(1) \in L^p$.

As already suggested above, the proof of Theorem 1.3 carries over without difficulties. Replace $t|\phi|$ there by $\phi_c^\gamma(t)\phi_c^\gamma(k)$ here; then the analogs of (14) and (17) immediately follow. Concerning the critical contribution, analogous to the left term in (15), one gets the estimate

$$\lesssim \sum_{j=0}^\infty \|T^*_{((m-1)\psi)_\phi h_{2^j\phi}}\|_{p\to p} \left(\sum_{\ell \geq \ell_0} \left(\frac{K(2^j \phi_c^\gamma(\theta_\ell), f)}{K(\phi_c^\gamma(\theta_{\ell+1}), f) F(K(\phi_c^\gamma(\theta_{\ell+1}), f)} \right)^p \right)^{1/p}$$

and can reason as in Sect. 2 to arrive at the assertions of Theorem 3.3.

Example 3.5. (a) *Expansions into spherical harmonics.* Let $\mathbb{S}^n$ be the unit sphere in $\mathbb{R}^{n+1}$, $n \in \mathbb{N}$, with the origin as center and elements $\xi, \theta, \ldots$, where $\xi = (\xi_1, \ldots, \xi_{n+1})$, $|\xi| = 1$. We denote by $L^p(\mathbb{S}^n)$, $1 \leq p < \infty$, the Banach space of all functions $f(\xi)$ which are pth power integrable on $\mathbb{S}^n$, i.e., for which

$$\|f\|_p = \left(\frac{1}{|\mathbb{S}^n|} \int_{\mathbb{S}^n} |f(\xi)|^p \, \mathrm{d}\sigma(\xi) \right)^{1/p} < \infty, \qquad |\mathbb{S}^n| = \frac{2\pi^{(n+1)/2}}{\Gamma((n+1)/2)},$$

where $\mathrm{d}\sigma$ is the surface element of $\mathbb{S}^n$. Let

$$f(\xi) \sim \sum_{k=0}^\infty Y_k(f;\xi), \qquad Y_k(f;\xi) = \sum_{m=1}^{H(k,n)} a_{k,m} Y_{k,m}(\xi), \quad f \in L^p(\mathbb{S}^n),$$

be the expansion in a series of spherical harmonics. Here $H(k,n)$ is the number of linear independent spherical harmonics of degree k on $\mathbb{S}^n$, $H(k,n) = (2k + n - 1)(k+n-2)!/(k!\,(n-1)!)$. For the Beltrami–Laplace operator δ, one has $\delta Y_{k,m} = -k(k+n-1) Y_{k,m}$, thus set $D_{n-1}^\gamma := (-\delta)^\gamma$—for this setting see, e.g., [28]. The basic property (18) with $N > (n-1)/2$ for spherical harmonics expansions can be found in Bonami and Clerc [3, Sect. 3].

Concerning the characterization of the K-functional in the present situation, introduce the shift operator S_t,

$$(S_t f)(\xi) = \frac{1}{|\mathbb{S}^{n-1}|\,(\sin t)^{n-1}} \int_{\xi\cdot\theta=\cos t} f(\theta)\,\mathrm{d}t(\theta), \quad 0<t<\pi,$$

where $\mathrm{d}t(\theta)$ is the element of the section $\{\theta : \xi\cdot\theta = \cos t\}$. The shift operator may equivalently be defined by the generalized translation operator for ultraspherical expansions, since $Y_k((S_t f);\xi) = (P_k^{(n-1)/2}(\cos t)/P_k^{(n-1)/2}(1))Y_k(f;\xi)$. If I is the identity operator, introduce a difference operator Δ_t^γ by $\Delta_t^\gamma = (I-S_t)^{\gamma/2}$ and an γth order modulus of smoothness by [28, Sect. 4.5]

$$\omega_{\gamma,\mathbb{S}^n}(t,f)_p = \sup_{0<s\leq t} \|\Delta_t^\gamma f\|_{L^p(\mathbb{S}^n)}, \quad \Delta_t^\gamma = \sum_{k=0}^{\infty}(-1)^k\binom{\gamma/2}{k}(S_t)^k, \ \gamma>0.$$

Then, by Wang and Li [28, Theorem 5.1.2] (see Kaljabin [12] for $\gamma\in\mathbb{N}$),

$$K_\gamma(t,f;L^p(\mathbb{S}^n,D(D_{n-1}^\gamma)) \approx \omega_{\gamma,\mathbb{S}^n}(t,f)_p, \quad 1<p<\infty, \ \gamma>0. \qquad (25)$$

Hence we can formulate results analogous to Example 1.4(b). For example, in the case of the de La Vallée Poussin means, one has for each fixed $L\in\mathbb{N}$

$$V_{\phi;t}f(x) - f(x) = o_x(1)\,\omega_{L,\mathbb{S}^n}(t,f)_p\,|\log \omega_{L,\mathbb{S}^n}(t,f)_p|^{1/p+\varepsilon} \quad \text{a.e.}, \quad t\to 0+,$$

or for the Riesz means $R_t^{\kappa,\lambda}f(\xi) = \Sigma_{k=0}^{\infty}(1-\phi_c^\lambda(t)\phi_c^\lambda(k))_+^\kappa Y_k(f;\xi)$, $\kappa>N$, when $0<\gamma<\lambda$

$$R_t^{\kappa,\lambda}f(\xi) - f(\xi) = o_\xi(1)\,\omega_{\gamma,\mathbb{S}^n}(t,f)_p|\log \omega_{\gamma,\mathbb{S}^n}(t,f)_p|^{1/p+\varepsilon} \quad \text{a.e.}, \quad t\to 0+.$$

(Observe that we replace $2^j t|\xi|$ of Sect. 2 here by $2^j\phi_c^\gamma(t)\phi_c^\gamma(k)$ which leads to $B_j^* \lesssim 2^{-j\lambda/\gamma}$ in the application of Theorem 3.1 (a).)

For every $f\in L^p(\mathbb{S}^n)$, $1<p<\infty$, with $\omega_{\gamma,\mathbb{S}^n}(t,f)_p \lesssim \phi_c^\eta(t)$ for some fixed $\eta, 0<\eta<\gamma$, we obtain

$$R_t^{\kappa,\gamma}f(\xi) - f(\xi) = o_\xi(1)\,\omega_{\gamma,\mathbb{S}^n}(t,f)_p|\log \omega_{\gamma,\mathbb{S}^n}(t,f)_p|^{1+1/p+\varepsilon} \quad \text{a.e.}, \quad t\to 0+.$$

In view of (25) and (22) a reformulation of this gives a direct estimate of the rate of a.e. pointwise convergence by the rate of strong convergence, i.e., for $t\to 0+$

$$R_t^{\kappa,\gamma}f(\xi) - f(\xi) = o(1)\|R_t^{\kappa,\gamma}f - f\|_{L^p(\mathbb{S}^n)}\,|\log\|R_t^{\kappa,\gamma}f - f\|_{L^p(\mathbb{S}^n)}|^{1+1/p+\varepsilon} \ \text{a.e.}.$$

This estimate has the advantage that it does not need a characterization of the K-functional via a modulus of smoothness based on generalized translations.

We mention that for even integers $2r$, $r \in \mathbb{N}$, one can find in [7] the following estimate for the K-functional by a different type of modulus of smoothness:

$$K(t^{2r}, f) \lesssim \sup_{\rho \in O_t} \|\Delta_\rho^{2r} f\|_{L^p(\mathbb{S}^n)}, \quad \Delta_\rho f(\xi) = f(\rho\xi) - f(\xi), \ \Delta_\rho^r = \Delta_\rho \Delta_\rho^{r-1}.$$

Here $\rho \in O_t$ are described by those orthogonal matrices with $n \times n$ real entries and determinant 1, for which $(\rho\theta \cdot \theta) \geq \cos t$ for all $\theta \in \mathbb{S}^n$.

(b) *Gegenbauer Expansions.* For the definitions see, e.g., [3, Sect. 6] where in particular the crucial property concerning the maximal function of the Cesàro means (18) with $N > \lambda + 1/2$ is proved. For the characterizations (21), (22), and the estimate (23) see [26] ($\alpha = \beta = \lambda - 1/2$); here, the differential operator is

$$D = -(1 - x^2)\frac{\mathrm{d}^2}{\mathrm{d}x^2} + (2\lambda + 1)x\frac{\mathrm{d}}{\mathrm{d}x}.$$

For the characterization of the K-functional based on generalized translations we refer to [16].

(c) *Laguerre and Hermite expansions on $\mathbb{R}_+$ and $\mathbb{R}$*, Respectively

Concerning the basic property (18) for Laguerre expansions, see [22, p. 126], and for Hermite expansions, see [22, p. 113].

Remark 3.5. (a) Since our intention in Sect. 3 is only to indicate the method, we do not care for "smallest" admissible Ns (which would have to be substituted by some critical $\kappa_{crit} \in \mathbb{R}_+$ and would require a fractional calculus—cf. [23]).

(b) Estimates for the maximal function of the Cesàro or Riesz means have been established in many settings, in particular, in connection with a.e. summability: For example, for the Fourier–Bessel transform in [20], for compact symmetric spaces of rank 1 in [3, Sect. 7], for compact Riemannian manifolds in [9, 10], and for some Lie groups in [11].

References

1. Bennett, C., Sharpley, R.: Interpolation of Operators. Academic Press, Boston (1988)
2. Bergh, J., Löfström, J.: Interpolation Spaces. An Introduction. Springer, Berlin (1976)
3. Bonami, A., Clerc, J.-L.: Sommes de Cesàro et multiplicateurs de developpements en harmoniques sphériques. Trans. Amer. Math. Soc. **183**, 223–263 (1973)
4. Butzer, P.L., Nessel, R.J., Trebels, W.: On summation processes of Fourier expansions in Banach spaces. Tôhoku Math. J. I. Comparison theorems. **24**, 127–140 (1972); Tôhoku Math. J. II. Saturation theorems. Tohoku Math. J. **24**, 551–569 (1972)
5. Carbery, A.: Radial Fourier multipliers and associated maximal functions. In: Peral, I., Rubio-de-Francia, J.-L. (eds.) Recent Progress in Fourier Analysis, pp. 49–56. North-Holland, Amsterdam (1985)
6. Dappa, H., Trebels, W.: On maximal functions generated by Fourier multipliers. Ark. Mat. **23**, 241–259 (1985)

7. Ditzian, Z.: Jackson-type inequality on the sphere. Acta Math. Hungar. **102**, 1–35 (2004)
8. Edmunds, D.E., Gurka, P., Opic, B.: On embeddings of logarithmic Bessel potential spaces. J. Functional Anal. **146**, 116–150 (1997)
9. Hörmander, L.: The spectral function of an elliptic operator. Acta Math. **121**, 193–218 (1968)
10. Hörmander, L.: On the Riesz means of spectral functions and eigenfunction expansions for elliptic operators. 1969 Some Recent Advances in the Basic Sciences, vol. 2. Proceedings of Annual Scientific Conference. Belfer Graduate School Science, Yeshiva University, New York, pp. 155–202 (1965–1966)
11. Hulanicki, A., Jenkins, J.W.: Almost everywhere summability on nilmanifolds. Trans. Amer. Math. Soc. **278**, 703–715 (1983)
12. Kalyabin, G.A.: On moduli of smoothness of functions given on the sphere. Soviet Math. Dokl. **35**, 619–622 (1987)
13. Kamaly, A., Stokolos, A., Trebels, W.: On the rate of almost everywhere convergence of certain integral means. II. J. Approx. Theory **101**, 240–264 (1999)
14. Kenig, C.E., Tomas, P.A.: Maximal operators defined by Fourier multipliers. Studia Math. **68**, 79–83 (1980)
15. Oskolkov, K.I.: Approximative properties of summable function on sets of full measure. Mat. Sb. **103(145)**(4), 563–589 (1977) English translation in Math. USSR-Sb. **32**, 489–514 (1977)
16. Rustamov, Kh.P.: Moduli of smoothness of higher orders related to the Fourier–Jacobi expansion and the approximation of functions by algebraic polynomials. Dokl. Math. **52**, 244–247 (1995)
17. Shapiro, H.S.: Smoothing and Approximation of Functions. Van Nostrand Reinhold Co., New York (1969)
18. Stein, E.M.: Singular integrals and differentiability properties of functions. Princeton University Press, New Jersey (1970)
19. Stein, E.M., Weiss, G.: Introduction to Fourier Analysis on Euclidean Spaces. Princeton University Press, New Jersey (1971)
20. Stempak, K.: La théorie de Littlewood-Paley pour la transformation de Fourier-Bessel. C.R. Acad. Sci. Paris Sér I Math. **303**, 15–18 (1986)
21. Stokolos, A., Trebels, W.: On the rate of almost everywhere convergence of certain integral means. J. Approx. Theory **98**, 203–222 (1999)
22. Thangavelu, S.: Lectures on Hermite and Laguerre expansions. Mathematical Notes, vol. 42. Princeton University Press, New Jersey (1993)
23. Trebels, W.: Multipliers for (C,α)-bounded Fourier expansions in Banach spaces and approximation theory. Lecture Notes in Mathematics, vol. 329. Springer, Berlin (1973)
24. Trebels, W.: Some Fourier multiplier criteria and the spherical Bochner–Riesz kernel. Rev. Roumaine Math. Pures Appl. **20** 1173–1185 (1975)
25. Trebels, W.: On the approximation behavior of the Riesz means in $L^p(\mathbb{R}^n)$. In: Lecture Notes in Mathematics, vol. 556, pp. 428–438. Springer, Berlin (1976)
26. Trebels, W.: Equivalence of a K-functional with the approximation behavior of some linear means for abstract Fourier series. Proc. Amer. Math. Soc. **127**, 2883–2887 (1999)
27. Trebels, W., Westphal, U.: On Ulyanov inequalities in Banach spaces and semigroups of linear operators. J. Approx. Theory **160**, 154–170 (2009)
28. Wang, K., Li, L.: Harmonic Analysis and Approximation on the Unit Sphere. Science Press, Beijing (2000)
29. Wilmes, G.: On Riesz-type inequalities and K-functionals related to Riesz potentials in $\mathbb{R}^N$. Numer. Funct. Anal. Optim. **1**, 57–77 (1979)
30. Wilmes, G.: Some inequalities for Riesz potentials of trigonometric polynomials of several variables. In: Proceedings of Symposia in Pure Mathematics, vol. 35, Part 1, pp. 175–182. American Mathematical Society, Providence, RI (1979)

On the Littlewood–Paley Inequalities for Subharmonic Functions on Domains in $\mathbb{R}^n$

Manfred Stoll

Abstract For the unit disk $\mathbb{D}$ in $\mathbb{C}$, the classical Littlewood–Paley inequalities (1936) are as follows: Let h be harmonic on $\mathbb{D}$. There exists a positive constant C, independent of h, such that for all p, $2 \le p < \infty$,

$$\int_{\mathbb{D}} (1-|z|)^{p-1} |\nabla h(z)|^p dx dy \le C \left[|h(0)|^p + \sup_{0<r<1} \int_0^{2\pi} |h(re^{i\theta})|^p d\theta \right],$$

with the reverse inequality valid for all p, $1 < p \le 2$.

In this chapter we will consider various extensions of the Littlewood–Paley inequalities to subharmonic functions on domains in $\mathbb{R}^n$, $n \ge 2$. Specifically we will prove that if f is a nonnegative C^2 subharmonic function on a bounded domain Ω in $\mathbb{R}^n$ with $C^{1,1}$ boundary for which Δf is subharmonic or has subharmonic behavior, then for $1 \le p < \infty$, there exists a constant C independent of f, such that

$$\int_{\Omega} \delta(x)^{2p-1} (\Delta f(x))^p dx \le C \sup_{0<r<r_o} \int_{\partial\Omega} f^p(t - r\mathbf{n}_t) ds(t), \tag{1}$$

where $\delta(x)$ is the distance from x to $\partial\Omega$ and $\mathbf{n}_t$ is the unit outward normal at $t \in \partial\Omega$. We will also present the analogue of Eq. (1) for the case $0 < p < 1$. Taking $f = h^2$, where h is harmonic on Ω, gives the usual Littlewood–Paley inequalities for harmonic functions. We will also consider analogues of the Littlewood–Paley inequalities for nonnegative subharmonic functions f for which $|\nabla f|$ is subharmonic or has subharmonic behavior.

M. Stoll (✉)
Department of Mathematics, University of South Carolina, Columbia, SC 29208, USA
e-mail: stoll@math.sc.edu

D. Bilyk et al. (eds.), *Recent Advances in Harmonic Analysis and Applications*, Springer Proceedings in Mathematics & Statistics 25, DOI 10.1007/978-1-4614-4565-4_28,

1 Introduction

For the unit disk $\mathbb{D}$ in $\mathbb{C}$, the harmonic Hardy spaces $\mathscr{H}^p$, $1 \le p < \infty$, are defined as the set of harmonic functions h on $\mathbb{D}$ satisfying

$$||h||_p^p = \sup_{0<r<1} \frac{1}{2\pi} \int_0^{2\pi} |h(re^{i\theta})|^p d\theta < \infty.$$

The classical Littlewood–Paley inequalities for harmonic functions [9] in $\mathbb{D}$ are as follows: Let h be harmonic on $\mathbb{D}$. Then there exist positive constants C_1, C_2, independent of h, such that:

(a) For $1 < p \le 2$,

$$||h||_p^p \le C_1 \left[|h(0)|^p + \int_{\mathbb{D}} (1-|z|)^{p-1} |\nabla h(z)|^p dx\,dy \right]. \tag{2}$$

(b) For $p \ge 2$, if $h \in \mathscr{H}^p$, then

$$\int_{\mathbb{D}} (1-|z|)^{p-1} |\nabla h(z)|^p dx\,dy \le C_2 ||h||_p^p. \tag{3}$$

In 1956 Flett [5] proved that for analytic functions inequality (2) is valid for all $p, 0 < p \le 2$. Hence if $u = \operatorname{Re} h$, h analytic, then since $|\nabla u| = |h'|$ it immediately follows that inequality (2) also holds for harmonic functions in $\mathbb{D}$ for all $p, 0 < p \le 2$. A new proof of the Littlewood–Paley inequalities for analytic functions was given by Luecking in [10]. Also, a short proof of the inequalities for harmonic functions in $\mathbb{D}$ valid for all $p, 0 < p < \infty$ has been given recently by Pavlović in [15].

The Littlewood–Paley inequalities are also known to be valid for harmonic functions in the unit ball in $\mathbb{R}^n$ when $p > 1$. In [17] Stević proved that for $n \ge 3$, inequality (2) is valid for all $p \in [\frac{n-2}{n-1}, 1]$. In [20] the author proved that the analogue of Eq. (2) was valid for harmonic functions on bounded domains with $C^{1,1}$ boundary for all $p, 0 < p \le 2$. In the case of the unit ball, this result was subsequently improved upon by Djordjević and Pavlović. In [3] they proved that if u is a harmonic function on the unit ball B in $\mathbb{R}^N$ and if $0 < p \le 1$, then

$$\int_{\partial B} u^*(y)^p d\sigma(y) \le C \left(|u(0)|^p + \int_B (1-|x|)^{p-1} |\nabla u(x)|^p dx \right), \tag{4}$$

where u^* is the non-tangential maximal function of u. The Littlewood–Paley inequalities have also been extended by the author to Hardy–Orlicz spaces of harmonic functions on general domains in $\mathbb{R}^n$, including Lipschitz domains [21].

The results of this chapter were motivated by a short paper of Miroslav Pavlović [14] in which he proved variations of the Littlewood–Paley inequalities for nonnegative subharmonic functions on $\mathbb{D}$ for which the Laplacian Δf is also subharmonic. This of course is the case when $f = h^2$ where h is harmonic, in which case $\Delta f = 2|\nabla h|^2$, which is subharmonic on $\mathbb{D}$. The results of [14], as well as those of this chapter, are also applicable to subharmonic functions f of the form

$$f(x) = \sum_{n=1}^{\infty} |h_n(x)|^2,$$

where $h = \{h_n\}$ is a harmonic function in ℓ^2. In this case, $\Delta f(x) = 2\sum |\nabla h_n(x)|^2$ which again is subharmonic. The results of [14] have also been extended by Jevtić [7] to functions that are subharmonic with respect to the Laplace–Beltrami operator $\widetilde{\Delta}$ on the unit ball in $\mathbb{C}^N$.

In this chapter we consider two generalizations of the Littlewood–Paley inequalities to domains in $\mathbb{R}^n$, namely to:

1. Nonnegative subharmonic functions f for which Δf is subharmonic or has subharmonic behavior
2. Nonnegative subharmonic functions f for which $|\nabla f|$ is subharmonic or has subharmonic behavior

We begin with the following definition.

Definition 1. An upper semicontinuous function g on a domain $\Omega \subsetneq \mathbb{R}^n$ is said to have subharmonic behavior, if there exists a constant C, depending only on n and Ω, such that

$$g(x) \le \frac{C}{r^n} \int_{B(x,r)} g(y)\,dy \tag{5}$$

for all $x \in \Omega$ and $r > 0$ such that $B(x,r) \subset \Omega$.

Clearly every subharmonic function g has subharmonic behavior. Furthermore, by Jensen's inequality, if $g \ge 0$ has subharmonic behavior, then g^p has subharmonic behavior for all p, $1 \le p < \infty$. In fact however, as we will see in Lemma 1, if $g \ge 0$ has subharmonic behavior, then g^p also has subharmonic behavior for all $p, 0 < p < \infty$.

Throughout this chapter we will assume that $\Omega \subsetneq \mathbb{R}^n$, $n \ge 2$, is a domain with Green function G satisfying the following conditions: For fixed $t_o \in \Omega$, there exist constants C_1 and C_2, depending only on t_o, such that

$$C_1\delta(x) \le G(t_o,x) \quad \text{for all} \quad x \in \Omega, \tag{6}$$

$$G(t_o,x) \le C_2\delta(x) \quad \text{for all} \quad x \in \Omega \setminus B(t_o, \tfrac{1}{2}\delta(t_o)), \tag{7}$$

where $\delta(x)$ denotes the distance from x to the boundary of Ω,[1] and $B(t_o,r)$ denotes the ball of radius r with center at t_o.

[1]If the boundary of Ω is C^2 or $C^{1,1}$, then the inequalities can be established by comparing the Green function G to the Green function of balls that are internally and externally tangent to the boundary of Ω. By the results of Widman [23], the inequalities are also valid for domains with $C^{1,\alpha}$ or Liapunov–Dini boundaries.

2 Notation and Statement of Results

Our setting throughout this chapter is $\mathbb{R}^n$, $n \geq 2$, the points of which are denoted by $x = (x_1, \ldots, x_n)$ with euclidean norm $|x| = \sqrt{x_1^2 + \cdots + x_n^2}$. For $r > 0$ and $x \in \mathbb{R}^n$, set $B_r(x) = B(x,r) = \{y \in \mathbb{R}^n : |x-y| < r\}$ and $S_r(x) = S(x,r) = \{y \in \mathbb{R}^n : |x-y| = r\}$. For convenience we denote the ball $B(0,\rho)$ by B_ρ, and the unit sphere $S_1(0)$ by S. Lebesgue measure in $\mathbb{R}^n$ will be denoted by $d\lambda$ or simply dx, and the normalized surface measure on S by $d\sigma$. The volume of the unit ball B_1 in $\mathbb{R}^n$ will be denoted by ω_n. Finally, for a real (or complex) valued C^1 function f, the gradient of f is denoted by ∇f, and if f is C^2, the Laplacian Δf of f is given by

$$\Delta f = \sum_{j=1}^{N} \frac{\partial^2 f}{\partial x_j^2}.$$

Let Ω be an arbitrary domain in $\mathbb{R}^n$, $n \geq 2$. For $0 < p < \infty$, we denote by $\mathcal{S}^p(\Omega)$ the set of non-negative subharmonic functions f on Ω for which f^p is subharmonic[2] and f^p has a harmonic majorant. For $f \in \mathcal{S}^p(\Omega)$ we denote the least harmonic majorant of f^p by H_{f^p}, and for fixed $t_o \in \Omega$ we set

$$N_p(f) = (H_{f^p}(t_o))^{1/p}. \tag{8}$$

It is easily shown that

$$N_p(f) = \lim_{n\to\infty} \left(\int_{\partial\Omega_n} f^p(t)\, d\omega_n^{t_o}(t) \right)^{1/p}, \tag{9}$$

where $\{\Omega_n\}$ is a regular exhaustion of Ω and $\omega_n^{t_o}$ is the harmonic measure on $\partial\Omega_n$ with respect to the point t_o. Here we assume that $t_o \in \Omega_n$ for all n.

Although $\mathcal{S}^p(\Omega)$ is not a linear space, it is a cone; namely, if $f, g \in \mathcal{S}^p(\Omega)$ and $\alpha, \beta \in (0,\infty)$, then $\alpha f + \beta g \in \mathcal{S}^p(\Omega)$ with

$$\begin{aligned} N_p(\alpha f + \beta g) &\leq \alpha N_p(f) + \beta N_p(g), && 1 \leq p < \infty, \\ N_p^p(\alpha f + \beta g) &\leq \alpha^p N_p^p(f) + \beta^p N_p^p(g), && 0 < p < 1. \end{aligned}$$

Also, for $1 \leq p < \infty$, we denote by $\mathcal{H}^p(\Omega)$ the set of harmonic functions h on Ω for which $|h| \in \mathcal{S}^p(\Omega)$. In this case, N_p is a norm on $\mathcal{H}^p(\Omega)$ with respect to which it becomes a Banach space. If Ω is a bounded domain with $C^{1,1}$ boundary, then for $f \in \mathcal{S}^p(\Omega)$, the "norm" $N_p(f)$ can be replaced by $\|f\|_p$, where for $0 < p < \infty$,

$$\|f\|_p^p = \sup_{0<r<r_o} \int_{\partial\Omega} f^p(t - r\mathbf{n}_t)\, ds(t). \tag{10}$$

---

[2] This hypothesis is only required when $0 < p < 1$.

In the above, for each $t \in \partial\Omega$, $\mathbf{n}_t$ denotes the unit outward normal at t, and ds denotes surface measure on $\partial\Omega$ (see [19] for details). Using the Poisson integral formula for Ω it is easily shown that for fixed $t_o \in \Omega$ there exist constants c_1 and c_2, depending only on t_o and n, such that

$$c_1 N_p^p(f) \le \|f\|_p^p \le c_2 N_p^p(f)$$

whenever $f \in \mathscr{S}^p$.

In this chapter we prove the following theorems.

Theorem 1. *Let Ω be a domain in $\mathbb{R}^n, n \ge 2$, with $\Omega \subsetneq \mathbb{R}^n$, and let f be a nonnegative subharmonic function on Ω:*

(a) Suppose Δf has subharmonic behavior on Ω. Then for $1 \le q \le p < \infty$ and $\gamma \in \mathbb{R}$, there exists a constant C, independent of f, such that

$$\int_\Omega \delta(x)^\gamma (\Delta f(x))^p dx \le C \int_\Omega \delta(x)^{\gamma-2p+2\frac{p}{q}} (\Delta f^q(x))^{p/q} dx.$$

(b) Suppose $0 < p \le q < \infty$ are such that f^p is C^2 and subharmonic and Δf^q has subharmonic behavior, then for $\gamma \in \mathbb{R}$ there exists a constant C, independent of f, such that

$$\int_\Omega \delta(x)^\gamma \Delta f^p(x) dx \le C \int_\Omega \delta(x)^{\gamma-2+2\frac{p}{q}} (\Delta f^q(x))^{p/q} dx.$$

As a consequence of Theorem 1 we obtain the following generalization of the Littlewood–Paley inequalities for subharmonic functions on Ω.

Theorem 2. *Let $\Omega \subsetneq \mathbb{R}^n$ be a domain with Green function G satisfying Eqs. (6) and (7), and let $t_o \in \Omega$ be fixed. Let f be a nonnegative C^2 subharmonic function on Ω:*

(a) Suppose $f \in S^p(\Omega)$, $1 \le p < \infty$, is such that Δf has subharmonic behavior. Then there exists a constant C, independent of f, such that

$$\int_\Omega \delta(x)^{2p-1} (\Delta f(x))^p dx \le C N_p^p(f).$$

(b) Suppose $f \in S^p(\Omega)$, $0 < p \le 1$, is such that f^p is C^2 and Δf has subharmonic behavior. Then there exists a constant C, independent of f, such that

$$N_p^p(f) \le C \left[f^p(t_o) + \int_\Omega \delta(x)^{2p-1} (\Delta f(x))^p dx \right].$$

As a consequence of Theorem 2 we obtain the following corollary for harmonic functions.

Corollary 1. *Let $\Omega \subsetneq \mathbb{R}^n$ be a domain with Green function G satisfying Eqs. (6) and (7), and let $t_o \in \Omega$ be fixed. Then for h a real-valued harmonic function on Ω, there exist constants C_1 and C_2, independent of h, such that:*

(a) For $2 \le q < \infty$,

$$\int_\Omega \delta(x)^{q-1} |\nabla h(x)|^q dx \le C_1 N_q^q(h).$$

(b) For $1 < q \le 2$,

$$N_q^q(h) \le C_2 \left[|h(t_o)|^p + \int_\Omega \delta(x)^{q-1} |\nabla h(x)|^q dx \right].$$

If Ω is a domain in $\mathbb{C}^N$, then for holomorphic functions on Ω, (b) is valid for all $q, 0 < q \le 2$.

Remark 1. Suppose $g \ge 0$ is subharmonic on Ω. Then

$$\Delta g^2 = 2|\nabla g|^2 + 2g\Delta g.$$

Since g and Δg are nonnegative, we have

$$\Delta g^2 \ge 2|\nabla g|^2.$$

Hence if Δg^2 has subharmonic behavior, by Theorem 2(a), for $2 \le q < \infty$, we have

$$\int_\Omega \delta(x)^{q-1} |\nabla g(x)|^q dx \le C N_q^q(g). \tag{11}$$

However, as we will prove in the following theorem, this result is still valid under the weaker assumption that only $|\nabla g|$ has subharmonic behavior.

Theorem 3. *Let g be a nonnegative C^2 subharmonic function on Ω for which $|\nabla g|$ has subharmonic behavior. Then:*

(a) For $2 \le q < \infty$,

$$\int_\Omega \delta(x)^{q-1} |\nabla g(x)|^q dx \le C_1 N_q^q(g),$$

where C_1 is a constant independent of g.

(b) If in addition $\Delta g^2 \le c|\nabla g|^2$ for some $c \ge 2$, then for $1 < q \le 2$,

$$N_q^q(g) \le C_2 \left[g^p(t_o) + \int_\Omega \delta(x)^{q-1} |\nabla g(x)|^q dx \right],$$

where $t_o \in \Omega$ is fixed and C_2 is a constant independent of g.[3]

[3]If $g \ge 0$ is subharmonic with $\Delta g^2 \le c|\nabla g|^2$ for some $c < 2$, then it is easily seen that g is constant. The case $c = 2$ corresponds to g harmonic. An example of a subharmonic function on B_1 satisfying $\Delta g^2 = c|\nabla g|^2$ is given by $g(x) = |y - x|^{-\beta}$ with $y \in \mathbb{R}^n \setminus B_1$ and $\beta \ge n - 2$.

In Sect. 5 we will briefly indicate the appropriate analogue of Theorem 2 for bounded k-Lipschitz domains in $\mathbb{R}^n$. Although it is not known to the author whether the analogue of Corollary 1(b) is valid for nonnegative subharmonic functions f for which $|\nabla f|$ has subharmonic behavior, in Sect. 6 we will prove the following generalization of the Littlewood–Paley inequality valid for all p, $0 < p \leq 1$. In view of this result it is conjectured that part (b) of Theorem 3 is still valid without the assumption that $\Delta g^2 \leq c|\nabla g|^2$.

Theorem 4. *Let $\Omega \subset \mathbb{R}^n$ be a bounded domain with $C^{1,1}$ boundary, and let f be a C^1 function on Ω for which $|f|$ and $|\nabla f|$ have subharmonic behavior on Ω. Then for $0 < p \leq 1$, $\alpha > 1$, and $t_o \in \Omega$ fixed, there exists a constant $C_{p,\alpha}$, depending only on p and α, such that*

$$\int_{\partial\Omega} (M_\alpha f)^p(\zeta)\,ds(\zeta) \leq C\left[|f(t_o)|^p + \int_\Omega \delta(y)^{p-1}|\nabla f(y)|^p dy\right],$$

where $M_\alpha f$ is the non-tangential maximal function of f.

3 Preliminary Results and Key Lemmas

Let Ω be an open subset of $\mathbb{R}^n$, $n \geq 2$, with $\Omega \subsetneq \mathbb{R}^n$. For $x \in \Omega$ set

$$B(x) = B(x, \tfrac{1}{2}\delta(x)) = \left\{ y \in \Omega : |y - x| < \frac{1}{2}\delta(x) \right\}. \tag{12}$$

Then for all $y \in B(x)$ we have

$$\frac{1}{2}\delta(x) \leq \delta(y) \leq \frac{3}{2}\delta(x). \tag{13}$$

For completeness we state the following result of Pavlović [13] as a lemma.

Lemma 1. *If $f \geq 0$ has subharmonic behavior in Ω, then f^p has subharmonic behavior for all p, $0 < p < \infty$.*

Remark 2. For $p \geq 1$, the inequality follows immediately from Hölder's inequality. For $|h|$ where h is harmonic in Ω, the result was originally proved by Fefferman and Stein in [4] and independently by Kuran in [8]. For a nonnegative subharmonic function f, Lemma 1 has previously been stated by Riihentaus in [16] and by Suzuki in [22]. The inequality was also proved by Pavlović [12] for nonnegative continuous functions on the unit ball in $\mathbb{C}^n$.

For the proofs of the main results we require several preliminary lemmas.

Lemma 2. *For $f \in L^1(\Omega)$ and $\gamma \in \mathbb{R}$,*

$$\int_\Omega \delta(x)^\gamma |f(x)|\,dx \approx^4 \int_\Omega \delta(w)^{\gamma-n}\left[\int_{B(w)} |f(x)|\,dx\right] dw.$$

Proof. For $E \subset \Omega$, let χ_E denote the characteristic function of E. Thus

$$\int_\Omega \delta(y)^{\gamma-n}\left[\int_{B(y)} |f(x)|\,dx\right] dy = \int_\Omega\int_\Omega \delta(y)^{\gamma-n}\chi_{B(y)}(x)|f(x)|\,dxdy.$$

But by Eq. (13), $\chi_{B(y)}(x) \le \chi_{2B(x)}(y)$. Thus

$$\begin{aligned}\int_\Omega\int_\Omega &\delta(x)^{\gamma-n}\chi_{B(y)}(x)|f(x)|dxdy\\ &\le C\int_\Omega\int_\Omega \delta(x)^{\gamma-n}\chi_{2B(x)}(y)|f(x)|\,dydx\\ &\le C\int_\Omega \delta(x)^\gamma|f(x)|\,dx.\end{aligned}$$

The reverse inequality follows likewise. □

Lemma 3. *For $u \in C^2(\overline{B_\rho})$, $\rho > 0$,*

$$\int_S u(\rho\zeta)\,d\sigma(\zeta) = u(0) + \int_{B_\rho} \Delta u(x)G_\rho(x)\,dx,$$

where for $0 < |x| \le \rho$,

$$G_\rho(x) = \begin{cases} \dfrac{1}{n(n-2)\omega_n}\left[\dfrac{1}{|x|^{n-2}} - \dfrac{1}{\rho^{n-2}}\right], & \text{if } n \ge 3,\\[2ex] \dfrac{1}{2\pi}\log\dfrac{\rho}{|x|}, & \text{if } n = 2,\end{cases} \tag{14}$$

is the Green function of $B(0,\rho)$ with singularity at 0.

Proof. The proof is an immediate consequence of Green's formula and hence is omitted. □

---

[4]The notation $A \approx B$ means that there exist constants c_1 and c_2 such that $c_1A \le B \le c_2A$.

Lemma 4. *Let f be a positive C^2 subharmonic function on $\overline{B(0,\rho)}$, $\rho > 0$.*

(a) Suppose Δf has subharmonic behavior on $B(0,\rho)$. Then for $1 \le q \le p < \infty$

$$\int_{B_{\rho/2}} (\Delta f(y))^p dy \le C_n \rho^{2\frac{p}{q}-2p} \int_{B_\rho} (\Delta f^q(y))^{p/q} dy,$$

where C_n is a constant depending only on n.

(b) Suppose $0 < p \le q < \infty$ are such that f^p is C^2 and subharmonic and Δf^q has subharmonic behavior, then

$$\int_{B_{\rho/4}} \Delta f^p(y) dy \le C_n \rho^{2\frac{p}{q}-2} \int_{B_\rho} (\Delta f^q(y))^{p/q} dy,$$

where C_n is a constant depending only on n.

Remark 3. In part (b), if we assume Δf^p has subharmonic behavior, then by part (a), for $1 \le r \le s < \infty$, we obtain

$$\int_{B_{\rho/2}} (\Delta f^p(y))^r dy \le C\rho^{2\frac{r}{s}-2r} \int_{B_\rho} (\Delta f^{ps}(y))^{r/s} dy. \tag{15}$$

Then with $r = 1$ and $s = q/p$, we obtain the conclusion of (b). However, our primary interest in (b) will be the case where $q = 1$; namely, $0 < p \le 1$, f^p is subharmonic, and Δf has subharmonic behavior.

Proof. We only prove the Lemma for $n \ge 3$, the special case $n = 2$ being similar.

(a) Since $f(x) > 0$ for all $x \in \overline{B(0,\rho)}$, f^q is C^2 for all $q \ge 1$. For $|x| \le \frac{\rho}{4}$,

$$\begin{aligned} G_{\frac{\rho}{2}}(x) &= \frac{1}{n(n-2)\omega_n}\left[\frac{1}{|x|^{n-2}} - \frac{2^{n-2}}{\rho^{n-2}}\right] \\ &\ge \frac{1}{n(n-2)\omega_n}\left[\frac{4^{n-2}}{\rho^{n-2}} - \frac{2^{n-2}}{\rho^{n-2}}\right] = c_n \rho^{2-n}. \end{aligned}$$

Since Δf has subharmonic behavior,

$$\begin{aligned} (\Delta f(0))^p &\le C_n \rho^{-np} \left(\int_{B_{\rho/4}} \Delta f(x) dx\right)^p \\ &\le C_n \rho^{-np} \left(\rho^{n-2} \int_{B_{\rho/2}} \Delta f(x) G_{\frac{\rho}{2}}(x)\, dx\right)^p, \end{aligned}$$

which by Lemma 3

$$= C_n\rho^{-2p}\left(\int_S f\left(\frac{1}{2}\rho\zeta\right)d\sigma(\zeta) - f(0)\right)^p$$

$$= C_n\rho^{-2p}\left[\left(\int_S f\left(\frac{1}{2}\rho\zeta\right)d\sigma(\zeta) - f(0)\right)^q\right]^{p/q}.$$

Since $(a-b)^q \leq a^q - b^q$ whenever $q \geq 1$ and $0 \leq b \leq a$, we have

$$\left(\int_S f\left(\frac{1}{2}\rho\zeta\right)d\sigma(\zeta) - f(0)\right)^q \leq \left(\int_S f\left(\frac{1}{2}\rho\zeta\right)d\sigma(\zeta)\right)^q - f^q(0)$$

$$\leq \int_S f^q\left(\frac{1}{2}\rho\zeta\right)d\sigma(\zeta) - f^q(0).$$

The last inequality follows by Hölder's inequality. Thus by Lemma 3 and the above

$$(\Delta f(0))^p \leq C_n\rho^{-2p}\left(\int_{B_{\rho/2}} \Delta f^q(x)G_{\frac{\rho}{2}}(x)\,dx\right)^{p/q}. \tag{16}$$

For $w \in B_{\rho/2}$ set $v(x) = f(w+x)$. Then by Eq. (16)

$$(\Delta v(0))^p = (\Delta f(w))^p \leq C_n\rho^{-2p}\left(\int_{B_{\rho/2}} \Delta_x f^q(w+x)G_{\frac{\rho}{2}}(x)\,dx\right)^{p/q}$$

$$\leq C_n\rho^{-2p}\left(\int_{B_{\rho/2}} \Delta_x f^q(w+x)\,|x|^{-(n-2)}\,dx\right)^{p/q}.$$

Finally, by the change of variable $y = w+x$, we have

$$(\Delta f(w))^p \leq C_n\rho^{-2p}\left(\int_{B(w,\frac{1}{2}\rho)} \frac{\Delta f^q(y)}{|y-w|^{(n-2)}}dy\right)^{p/q}. \tag{17}$$

Suppose $p > q$. Set $r = p/q$ and $r' = p/(p-q)$ the conjugate exponent of r. If $r \geq n/(n-2)$, choose α, $0 < \alpha < 1$, such that

$$\alpha r < \frac{n}{n-2} \quad \text{and} \quad (1-\alpha)r' < \frac{n}{n-2}.$$

Such an α exists, since if we let $\beta = n/(n-2)$, then the above is equivalent to choosing an α, $0 < \alpha < 1$, such that

$$\frac{\beta}{r} - (\beta - 1) < \alpha < \frac{\beta}{r}.$$

If $r < n/(n-2)$ then we simply take $\alpha = 1$. By Hölder's inequality we now have

$$\int\limits_{B(w,\frac{1}{2}\rho)} \frac{\Delta f^q(y)\,dy}{|y-w|^{n-2}} \leq \left[\int\limits_{B(w,\frac{1}{2}\rho)} \frac{(\Delta f^q(y))^r dy}{|y-w|^{r\alpha(n-2)}}dy\right]^{\frac{1}{r}} \left[\int\limits_{B(w,\frac{1}{2}\rho)} \frac{dy}{|y-w|^{r'(1-\alpha)(n-2)}}\right]^{\frac{1}{r'}}.$$

But by the change of variable $x = y - w$,

$$\begin{aligned}\int\limits_{B(w,\frac{1}{2}\rho)} |y-w|^{-r'(1-\alpha)(n-2)}dy &= \int_{B_{\rho/2}} |x|^{-r'(1-\alpha)(n-2)}dx \\ &= n\int_0^{\rho/2} t^{n-1-r'(1-\alpha)(n-2)}dt \\ &= C_n\rho^{n-r'(1-\alpha)(n-2)},\end{aligned}$$

provided of course that $n - r'(1-\alpha)(n-2) > 0$. Therefore, by inequality (17),

$$(\Delta f(w))^p \leq C_n\rho^{-2p+(\frac{p-q}{q})n-\frac{p}{q}(1-\alpha)(n-2)} \int\limits_{B(w,\frac{1}{2}\rho)} \frac{(\Delta f^q(y))^{\frac{p}{q}}dy}{|y-w|^{\frac{p}{q}\alpha(n-2)}}dy.$$

Integrating the above over $B_{\rho/2}$ gives

$$\int\limits_{B_{\rho/2}} (\Delta f(w))^p dw \leq C_n\rho^{-2p+(\frac{p-q}{q})n-\frac{p}{q}(1-\alpha)(n-2)} \int\limits_{B_{\rho/2}}\int\limits_{B(w,\frac{1}{2}\rho)} \frac{(\Delta f^q(y))^{\frac{p}{q}}}{|y-w|^{\frac{p}{q}\alpha(n-2)}}dydw. \tag{18}$$

Since $B(w,\frac{1}{2}\rho) \subset B_\rho$ for all $w \in B_{\rho/2}$, and $y \in B(w,\frac{1}{2}\rho)$ if and only if $w \in B(y,\frac{1}{2}\rho)$, by Fubini's theorem, we have

$$\int\limits_{B_{\rho/2}}\int\limits_{B(w,\frac{1}{2}\rho)} \frac{(\Delta f^q(y))^{\frac{p}{q}}}{|y-w|^{\frac{p}{q}\alpha(n-2)}}dydw = \int\limits_{B_\rho}(\Delta f^q(y))^{\frac{p}{q}} \int\limits_{B(y,\frac{1}{2}\rho)} \frac{1}{|y-w|^{\frac{p}{q}\alpha(n-2)}}dwdy.$$

As above, by the change of variable $x = w - y$,

$$\int\limits_{B(y,\frac{1}{2}\rho)} |y-w|^{-\frac{p}{q}\alpha(n-2)}\,dw = \int\limits_{B_{\rho/2}} |x|^{-\frac{p}{q}\alpha(n-2)}$$

$$= C_n \rho^{n-\alpha\frac{p}{q}(n-2)}.$$

Combining this with Eq. (18) gives

$$\int\limits_{B_{\rho/2}} (\Delta f(w))^p\,dw \le C_n \rho^{-2p+(\frac{p-q}{q})n-\frac{p}{q}(1-\alpha)(n-2)+n-\alpha\frac{p}{q}(n-2)} \int\limits_{B_\rho} (\Delta f^q(y))^{\frac{p}{q}}\,dy$$

$$= C_n \rho^{-2p+2\frac{p}{q}} \int\limits_{B_\rho} (\Delta f^q(y))^{\frac{p}{q}}\,dy,$$

which proves the result when $p > q$. If $p = q$, then the result follows by integrating inequality (17) directly.

(b) As in part (a),

$$\int_{B_{\rho/4}} \Delta f^p(x)\,dx \le C_n \rho^{n-2} \int_{B_{\rho/2}} \Delta f^p(x)\,G_{\frac{\rho}{2}}(x)\,dx,$$

which by Lemma 3

$$= C_n \rho^{n-2} \left[\int_S f^p\left(\frac{1}{2}\rho\zeta\right) d\sigma(\zeta) - f^p(0)\right]$$

$$= C_n \rho^{n-2} \left[\int_S \left(f^q(\frac{1}{2}\rho\zeta)\right)^{p/q} d\sigma(\zeta) - (f^q(0))^{p/q}\right].$$

For $0 < \alpha \le 1$ and $0 \le b \le a$ we have $a^\alpha - b^\alpha \le (a-b)^\alpha$. Thus since $q \ge p$, f^q is also subharmonic, and since $p/q \le 1$ we have

$$\int_{B_{\rho/4}} \Delta f^p(x)\,dx \le C_n \rho^{n-2} \left(\int_S f^q\left(\frac{1}{2}\rho\zeta\right) d\sigma(\zeta)\right)^{p/q} - (f^q(0))^{p/q}$$

$$\le C_n \rho^{n-2} \left[\int_S f^q\left(\frac{1}{2}\rho\zeta\right) d\sigma(\zeta) - f^q(0)\right]^{p/q},$$

which by Lemma 3

$$= C_n \rho^{n-2} \left[\int_{B_{\rho/2}} \Delta f^q(x) G_{\frac{\rho}{2}}(x)\,dx\right]^{p/q}.$$

Therefore,

$$\int_{B_{\rho/4}} \Delta f^p(x)\,dx \le C_n\rho^{n-2} \sup_{x\in B_{\rho/2}} (\Delta f^q(x))^{p/q} \left[\int_{B_{\rho/2}} G_{\frac{\rho}{2}}(x)\,dx\right]^{p/q}$$

$$= C_n\rho^{n-2+2\frac{p}{q}} \sup_{x\in B_{\rho/2}} (\Delta f^q(x))^{p/q}.$$

By hypothesis Δf^q has subharmonic behavior and hence so does $(\Delta f^q)^{p/q}$. Consequently, since $B(x,\frac{1}{2}\rho)\subset B_\rho$,

$$\sup_{x\in B_{\rho/2}} (\Delta f^q(x))^{p/q} \le \frac{C}{\rho^n}\int_{B\rho} (\Delta f^q(y))^{p/q}dy.$$

Therefore,

$$\int_{B_{\rho/4}} \Delta f^p(x)\,dx \le C_n\rho^{2\frac{p}{q}-2}\int_{B_\rho} (\Delta f^q(y))^{p/q}dy. \qquad \square$$

Remark 4. Suppose $g\ge 0$ is subharmonic on Ω. Then

$$\Delta g^2 = 2|\nabla g|^2 + 2g\,\Delta g.$$

Since g and Δg are nonnegative we have

$$\Delta g^2 \ge 2|\nabla g|^2.$$

Hence if Δg^2 has subharmonic behavior, by Lemma 4(a), for $1\le r\le s<\infty$,

$$\int_{B_{\rho/2}} |\nabla g(x)|^{2r} \le C\rho^{2\frac{r}{s}-2r}\int_{B_\rho} (\Delta g^{2s}(x))^{r/s}dx.$$

In particular, if $2\le p<\infty$, taking $r=s=p/2$ gives

$$\int_{B_{\rho/2}} |\nabla g(x)|^p dx \le C\rho^{2-p}\int_{B_\rho} \Delta g^p(x)\,dx. \tag{19}$$

Inequality (19), however, is still true if we only assume that $|\nabla g|$ has subharmonic behavior.

Lemma 5. *Let g be a nonnegative C^2 subharmonic function on $\overline{B(0,\rho)}$ for which $|\nabla g|$ has subharmonic behavior. Then*

(a) For $2\le p<\infty$, there exists a constant C_1, independent of g, such that

$$\int_{B_{\rho/2}} |\nabla g(x)|^p dx \le C_1\rho^{2-p}\int_{B_\rho} \Delta g^p(x)\,dx.$$

(b) *If in addition $g(x) > 0$ and $\Delta g^2 \leq c|\nabla g|^2$ for some $c \geq 2$, then for $1 < p \leq 2$,*

$$\int_{B_{\rho/4}} \Delta g^p(x) dx \leq C_2 \rho^{p-2} \int_{B_\rho} |\nabla g(x)|^p dx,$$

where C_2 is a constant independent of g.

Proof. (a) Since $|\nabla g|$ has subharmonic behavior, so does $|\nabla g|^2$. Therefore,

$$|\nabla g(0)|^p \leq C\rho^{-pn/2} \left(\int_{B_{\rho/4}} |\nabla g(x)|^2 dx \right)^{p/2},$$

which since $\Delta g^2 \geq 2|\nabla g|^2$,

$$\leq C\rho^{-pn/2} \left(\int_{B_{\rho/4}} \Delta g^2(x)\, dx \right)^{p/2}.$$

Since $G_{\frac{\rho}{2}}(x) \geq c_n \rho^{2-n}$ for all $x \in B_{\rho/4}$,

$$|\nabla g(0)|^p \leq C\rho^{-pn/2} \left(\rho^{n-2} \int_{B_{\rho/2}} \Delta g^2(x) G_\delta(x)\, dx \right)^{p/2}.$$

Thus by Lemma 3

$$|\nabla g(0)|^p \leq C_n \rho^{-p} \left(\int_S g^2 \left(\frac{1}{2}\rho\zeta \right) d\sigma(\zeta) - g^2(0) \right)^{p/2}.$$

Since $p/2 \geq 1$, we obtain as in Lemma 4 that

$$|\nabla g(0)|^p \leq C_n \rho^{-p} \left(\int_S g^p \left(\frac{1}{2}\rho\zeta \right) d\sigma(\zeta) - g^p(0) \right),$$

which by Lemma 3

$$= C_n \rho^{-p} \int_{B_{\rho/2}} \Delta g^p(x) G_{\frac{\rho}{2}}(x)\, dx.$$

Applying the above to $v(x) = g(w+x)$, $w \in B_{\rho/2}$, $x \in B_{\rho/4}$, gives

$$|\nabla g(w)|^p \leq C_n \rho^{-p} \int_{B_{\rho/2}} \Delta_x g^p(w+x) G_{\frac{\rho}{2}}(x)\, dx$$

$$\leq C_n \rho^{-p} \int_{B_{\rho/2}} \Delta_x g^p(w+x) |x|^{-(n-2)} dx.$$

Thus by the change of variable $y = w + x$,

$$|\nabla g(w)|^p \le C_n \int_{B(w,\frac{1}{2}\rho)} \Delta g^p(y)|w-y|^{-(n-2)}dy.$$

The result now follows as in Lemma 4 by integrating the above over $B_{\rho/2}$.

(b) Since $g(x)>0$, g^p is C^2 for all $p>1$. As in Lemma 4 we have $G_{\frac{\rho}{2}}(x) \ge c_n\rho^{2-n}$ for all x, $|x| \le \rho/4$. Therefore,

$$\begin{aligned}
\int_{B_{\rho/4}} \Delta g^p(x)dx &\le C_n\rho^{n-2}\int_{B_{\rho/2}} \Delta g^p(x)G_\delta(x)dx, \quad \text{which by Lemma 3,}\\
&= C_n\rho^{n-2}\left[\int_S g^p\left(\frac{1}{2}\rho\zeta\right)d\sigma(\zeta) - g^p(0)\right], \quad \text{which since } p \le 2,\\
&\le C_n\rho^{n-2}\left[\left(\int_S g^2\left(\frac{1}{2}\rho\zeta\right)d\sigma(\zeta)\right)^{p/2} - (g^2(0))^{p/2}\right].
\end{aligned}$$

For $0 < \alpha \le 1$ and $0 \le b \le a$ we have $a^\alpha - b^\alpha \le (a-b)^\alpha$. Thus since g^2 is subharmonic, we have

$$\int_{B_{\rho/4}} \Delta g^p(x)dx \le C_n\rho^{n-2}\left[\int_S g^2\left(\frac{1}{2}\rho\zeta\right)d\sigma(\zeta) - g^2(0)\right]^{p/2},$$

which again by Lemma 3

$$= C_n\rho^{n-2}\left[\int_{B_{\rho/2}} \Delta g^2(x)G_\delta(x)dx\right]^{p/2}.$$

Thus, by our hypothesis on g,

$$\begin{aligned}
\int_{B_{\rho/4}} \Delta g^p(x)dx &\le C_n\rho^{n-2}\left[\int_{B_{\rho/2}} |\nabla g(x)|^2 G_{\frac{\rho}{2}}(x)dx\right]^{p/2}\\
&\le C_n\rho^{n-2}\sup_{x\in B_{\rho/2}}|\nabla g(x)|^p\left[\int_{B_{\rho/2}} G_{\frac{\rho}{2}}(x)dx\right]^{p/2}\\
&\le C_n\rho^{n+p-2}\sup_{x\in B_{\rho/2}}|\nabla g(x)|^p.
\end{aligned}$$

Since $|\nabla g(x)|$ has subharmonic behavior, the conclusion now follows as in the proof of Lemma 4. □

4 Proofs of the Main Results

Proof of Theorem 1. (a) By Lemma 2

$$\int_{\Omega} \delta(x)^{\gamma} (\Delta f(x))^p dx \leq C \int_{\Omega} \delta(w)^{\gamma - n} \left[\int_{B(w, \frac{1}{4}\delta(w))} (\Delta f(y))^p dy \right] dw.$$

Set $\rho = \frac{1}{2}\delta(w)$ and $h(x) = f(w+x)$. Then by the change of variable $y = w + x$,

$$\int_{B(w, \frac{1}{4}\delta(w))} (\Delta f(y))^p dy = \int_{B_{\rho/2}} (\Delta h(x))^p dx,$$

which by Lemma 4(a)

$$\leq C \rho^{2\frac{p}{q} - 2p} \int_{B_\rho} (\Delta h^q(x))^{p/q} dx.$$

Therefore,

$$\int_{B(w, \frac{1}{4}\delta(w))} (\Delta f(y))^p dy \leq C \delta(w)^{2\frac{p}{q} - 2p} \int_{B(w)} (\Delta f^q(y))^{p/q} dy,$$

where $B(w) = B\left(w, \frac{1}{2}\delta(w)\right)$. But by Lemma 2

$$\int_{\Omega} \delta(w)^{\gamma - n + 2\frac{p}{q} - 2p} \left[\int_{B(w)} (\Delta f^q(y))^{p/q} dy \right] dw \leq C \int_{\Omega} \delta(x)^{\gamma + 2\frac{p}{q} - 2p} (\Delta f^q(x))^{p/q} dx,$$

which proves part (a). The proof of (b) is similar. □

Before proving Theorem 2 we require some preliminary results about subharmonic functions. As in Sect. 2 we denote by $\mathscr{S}^1(\Omega)$ the set of nonnegative subharmonic functions on Ω that have a harmonic majorant on Ω. For $f \in \mathscr{S}^1(\Omega)$, we let H_f denote the least harmonic majorant of f on Ω. For convenience we will assume that $f \in C^2(\Omega)$. As in [18, 19], we have the following.

Lemma 6. *Let Ω be a domain in $\mathbb{R}^n$, $n \geq 2$, with Green function G, and let $f \in C^2(\Omega)$. Then $f \in \mathscr{S}^1(\Omega)$ if and only if there exists $t_o \in \Omega$ such that*

$$\int_{\Omega} G(t_o, x) \Delta f(x) dx < \infty. \tag{20}$$

If this is the case, then by the Riesz decomposition theorem,

$$H_f(x) = f(x) + \int_\Omega G(x,y)\Delta f(y)\,dy. \tag{21}$$

If the Green function G satisfies Eqs. (6) and (7), then it is easily seen that Eq. (20) is equivalent to

$$\int_\Omega \delta(x)\Delta f(x)\,dx < \infty. \tag{22}$$

If the subharmonic function f is not C^2, then the quantity $\Delta f(x)\,dx$ may be replaced by $d\mu_f$, where μ_f is the Riesz measure of the subharmonic function f.

Lemma 7. *Let Ω be a domain in $\mathbb{R}^n, n \geq 2$, with Green function G satisfying Eqs. (6) and (7), and let $t_o \in \Omega$ be arbitrary. Then there exist constants C_1 and C_2, depending only on t_o and Ω, such that for all $f \in \mathscr{S}^1(\Omega) \cap C^2(\Omega)$,*

$$C_1 \int_\Omega \delta(x)\Delta f(x)dx \leq H_f(t_o) \leq C_2 \left[\int_{B(t_o)} f(x)dx + \int_\Omega \delta(x)\Delta f(x)dx \right]. \tag{23}$$

Proof. The left side of inequality (23) is an immediate consequence of Eq. (21) and inequality (6). The right side is obtained by integrating equation (21) over $B(t_o)$ and then proving that

$$\int_{B(t_o)} G(x,y)\,dx \leq C\delta(y) \tag{24}$$

for all $y \in \Omega$, where C is a constant depending only on $\delta(t_o)$ and n. Inequality (24) follows from the fact that for $y \notin B(t_o)$, $x \to G(x,y)$ is harmonic on $B(t_o)$. For $y \in B(t_o)$, we use the fact that $G(x,y) \leq c_n|x-y|^{2-n}$. For further details, the reader is referred to the proof of Proposition 4.3 of [19], where the result was proved for bounded domains in $\mathbb{R}^n$. The same proof works in general. □

Proof of Theorem 2. (a) Let f be a nonnegative subharmonic function on Ω for which Δf has subharmonic behavior and $1 \leq p < \infty$. Assume first that $f(x) > 0$ for all x.[5] Then by Theorem 1(a) with $q = p$ and $\gamma = 2p - 1$ we have

$$\int_\Omega \delta(x)^{2p-1}(\Delta f(x))^p dx \leq C \int_\Omega \delta(x)\Delta f^p(x)\,dx,$$

which by Lemma 7

$$\leq CH_{f^p}(t_o) = CN_p^p(f).$$

[5] This assumption is only required for the case $1 < p < 2$.

In the case $1 < p < 2$, set $f_\varepsilon = f + \varepsilon$. Then f_ε is subharmonic with $f_\varepsilon(x) > 0$ for all x and $\Delta f_\varepsilon = \Delta f$. Hence by the above we have

$$\int_\Omega \delta(x)^{2p-1}(\Delta f(x))^p dx \leq C \int_\Omega \delta(x)\Delta f_\varepsilon^p(x)\, dx \leq C N_p^p(f_\varepsilon).$$

But $N_p^p(f_\varepsilon) \leq N_p^p(f) + \varepsilon$. Letting $\varepsilon \to 0$ proves the result.

(b) Assume that $f \in \mathscr{S}^p$ is such that f^p is C^2 and Δf has subharmonic behavior. By translation we can assume without loss of generality that $t_o = 0$. Let $\rho_o = \frac{1}{2}\delta(0)$. Then by Lemma 7

$$H_{f^p}(0) \leq C\left[\int_{B_{\rho_o/2}} f^p(x)\, dx + \int_\Omega \delta(x)\Delta f^p(x)\, dx\right]. \tag{25}$$

By Theorem 1(b) with $q = 1$ and $\gamma = 1$,

$$\int_\Omega \delta(x)\Delta f^p(x)\, dx \leq C \int_\Omega \delta(x)^{2p-1}(\Delta f(x))^p dx. \tag{26}$$

For $0 < r < \frac{1}{2}\rho_o$, since $0 < p \leq 1$,

$$\int_S f^p(r\zeta) d\sigma(\zeta) \leq \left[\int_S f(r\zeta)\, d\sigma(\zeta)\right]^p,$$

which by Lemma 3

$$\begin{aligned} &\leq f^p(0) + \left[\int_{B_r} \Delta f(x) G_r(x)\, dx\right]^p \\ &\leq f^p(0) + C\rho_o^{2p} \sup_{x\in B_r} (\Delta f(x))^p. \end{aligned}$$

Since $(\Delta f(x))^p$ has subharmonic behavior we obtain

$$\int_S f^p(r\zeta)\, d\sigma(\zeta) \leq f^p(0) + C\rho_o^{2p-n} \int_{B_{\rho_o}} (\Delta f(x))^p\, dx.$$

Finally since $\delta(x) \approx \rho_o$ for all $x \in B_{\rho_o/2}$,

$$\int_{B_{\rho_o/2}} f^p(x) dx \leq C(\rho_o)\left[f^p(0) + \int_{B_{\rho_o}} \delta(x)^{2p-1}(\Delta f(x))^p dx\right].$$

Combining the above with Eq. (26) gives

$$H_{f^p}(0) \le C\left[f^p(0) + \int_\Omega \delta(x)^{2p-1}(\Delta f(x))^p\,dx\right],$$

which proves the result. □

Proof of Corollary 1. (a) For $q \ge 2$, the result follows directly from Theorem 2(a) with $p = q/2$ by taking $f = |h|^2$ in which case $\Delta f = 2|\nabla h|^2$.

(b) For $1 < q < 2$ we consider $f_\varepsilon = |h + i\varepsilon|^2$. Then f_ε is a positive C^2 subharmonic function for which $\Delta f_\varepsilon = 2|\nabla h|^2$. Thus by Theorem 2(b) with $p = q/2$,

$$N_p^p(f_\varepsilon) \le C\left[|h(t_o + i\varepsilon)|^q + \int_\Omega \delta(x)^{q-1}|\nabla h(x)|^q dx\right].$$

The result now follows by letting $\varepsilon \to 0$.

Suppose now that Ω is a domain in $\mathbb{C}^N = \mathbb{R}^{2n}$ and h is a holomorphic function on Ω. If we set $f_\varepsilon = |h|^2 + \varepsilon^2$, then f_ε is subharmonic on Ω. Furthermore, for $0 < q \le 2$,

$$\begin{aligned}\Delta f_\varepsilon^{q/2} &= 4\sum_{j=1}^N \frac{\partial^2 f_\varepsilon^{q/2}}{\partial z_j \partial \bar{z}_j}\\ &= q(q-2)(|h|^2+\varepsilon^2)^{\frac{1}{2}q-2}|h|^2|\partial h|^2 + 2q(|h|^2+\varepsilon^2)^{\frac{1}{2}q-1}|\partial h|^2,\end{aligned}$$

where $\partial h = (\frac{\partial h}{\partial z_1}, \cdots, \frac{\partial h}{\partial z_N})$. Thus $f_\varepsilon^{q/2}$ is subharmonic for all $q, 0 < q \le 2$, with $\Delta f_\varepsilon = 4|\partial h|^2$. Thus Δf_ε has subharmonic behavior. The result now follows as above. □

Proof of Theorem 3. (a) As in the proof of Theorem 1, by Lemma 5,

$$\int_\Omega \delta(x)^\gamma |\nabla g(x)|^q dx \le C\int_\Omega \delta(x)^{\gamma+2-q}\Delta g^q(x)\,dx.$$

Thus with $\gamma = q-1$, by Lemma 7 we have

$$\int_\Omega \delta(x)^{q-1}|\nabla g(x)|^q dx \le C\int_\Omega \delta(x)\Delta g^q(x)\,dx \le CN_q^q(g),$$

which proves the result. The proof of (b) follows similarly. □

5 The Littlewood–Paley Inequalities for Lipschitz Domains

In this section we extend the Littlewood–Paley inequalities to bounded Lipschitz domains in $\mathbb{R}^n, n \geq 2$. As in [1, 11] a bounded domain Ω is a k-Lipschitz domain $(k > 0)$ if Ω and its boundary $\partial\Omega$ are given locally by a Lipschitz function whose Lipschitz constant is less than or equal to k. For such domains one can find an exterior and interior cone at each boundary point with fixed aperture ψ. This aperture ψ is related to the Lipschitz constant k by $\psi = \arctan(1/k) \in (0, \frac{1}{2}\pi)$. Maeda and Suzuki proved [11] that for such domains Ω there exist constants α and β, $0 < \beta \leq 1 \leq \alpha$, such that the Green function G of Ω satisfies the following: for fixed $t_o \in \Omega$, there exist positive constants C_1, C_2, such that

$$C_1\delta(x)^\alpha \leq G(t_o,x) \quad \text{for all } x \in \Omega, \text{ and} \tag{27}$$

$$G(t_0,x) \leq C_2\delta(x)^\beta \quad \text{for all } x \in \Omega \setminus B\left(t_o, \frac{1}{2}\delta(t_o)\right). \tag{28}$$

The constants α and β are given by

$$\alpha = \alpha_n(\psi) \quad \text{and} \quad \beta = \alpha_n(\pi - \psi),$$

where α_n is a strictly decreasing continuous function on $(0, \pi)$ with $\alpha_n(\frac{1}{2}\pi) = 1$, $\alpha_n(\theta) \to \infty$ as $\theta \to 0$, and $\alpha_n(\theta) \to 0$ as $\theta \to \pi$ $(n \geq 3)$. If $\partial\Omega$ is $C^{1,1}$, then $\alpha = \beta = 1$.

Using the same techniques as in the proof of Theorem 2 we obtain the following version of the Littlewood–Paley inequalities for subharmonic functions on bounded Lipschitz domains.

Theorem 5. *Let Ω be a Lipschitz domain in $\mathbb{R}^n$ with Lipschitz constant $k > 0$, and let G be the Green function of Ω satisfying Eqs. (27) and (28) for fixed $t_o \in \Omega$.*

(a) Suppose $f \in \mathscr{S}^p(\Omega)$, $1 \leq p < \infty$, is such that Δf has subharmonic behavior. Then there exists a constant C, independent of f, such that

$$\int_\Omega \delta(x)^{\alpha+2p-2}(\Delta f(x))^p dx \leq C N_p^p(f).$$

(b) Suppose $f \in \mathscr{S}^p(\Omega)$, $0 < p \leq 1$, is such that f^p is C^2 and Δf has subharmonic behavior. Then there exists a constant C, independent of f, such that

$$N_p^p(f) \leq C\left[f^p(t_o) + \int_\Omega \delta(x)^{\beta+2p-2}(\Delta f(x))^p dx\right].$$

6 The Case $0 < p \le 1$

Suppose now that $\Omega \subset \mathbb{R}^n$ is a bounded domain with $C^{1,1}$ boundary. For each $t \in \partial\Omega$, let $\mathbf{n}_t$ denote the unit exterior normal at t, and for $r > 0$, set

$$\Omega_r = \{x \in \Omega : \delta(x) > r\}.$$

Since $\partial\Omega$ is $C^{1,1}$, there exists a $\rho_o > 0$ such that for each r, $0 < r < \rho_o$, $\partial\Omega_r$ is of class $C^{1,1}$ and the mapping $t \to t - r\mathbf{n}_t$ is a 1–1 transformation of $\partial\Omega$ onto $\partial\Omega_r$. (See [2, 23] for details.) We denote the element of surface measure on $\partial\Omega$ by ds.

Let ρ_o be as above. For $\alpha > 1$ and $\zeta \in \partial\Omega$, we denote by $\Gamma_{\rho_o,\alpha}(\zeta)$ the *non-tangential approach region* defined by

$$\Gamma_{\rho_o,\alpha}(\zeta) = \{y \in \Omega \setminus \Omega_{\rho_o} : |y - \zeta| < \alpha\delta(y)\}.$$

For f continuous on Ω and $\alpha > 1$, the *non-tangential maximal function* $M_\alpha f$ on $\partial\Omega$ is defined by

$$M_\alpha f(\zeta) = \sup\{|f(y)| : y \in \Gamma_{\rho_o,\alpha}(\zeta)\}.$$

The *radial maximal function* of f is given by

$$M_{rad} f(\zeta) = \sup\{|f(\zeta - r\mathbf{n}_\zeta)| : 0 < r < \rho_o\}.$$

Also, for a locally integrable function f on $\partial\Omega$, the *Hardy–Littlewood maximal function* $\widetilde{M}f$ of f is defined by

$$\widetilde{M}f(\zeta) = \sup_{r>0} \frac{1}{r^{n-1}} \int_{S(\zeta,r)} |f(t)|\, ds(t),$$

where $S(\zeta,r) = B(\zeta,r) \cap \partial\Omega$.

For the proof of the theorem we require the following lemma.

Lemma 8. *Let $\Omega \subset \mathbb{R}^n$ be a bounded domain with $C^{1,1}$ boundary, and let $f \ge 0$ be a continuous function that has subharmonic behavior on Ω. Then for $0 < p < \infty$ and $\alpha > 1$, we have*

$$\int_{\partial\Omega} (M_\alpha f)^p(\zeta)\, ds(\zeta) \le C_{p,\alpha} \int_{\partial\Omega} (M_{rad} f)^p(\zeta)\, ds(\zeta),$$

where $C_{p,\alpha}$ is a positive constant independent of f.

The proof provided is a variation of a proof given in Theorem 3.6 of [6] for harmonic functions in the unit disk in $\mathbb{C}$.

Proof. Let $x \in \Gamma_{\rho_o,\alpha}(\zeta)$ with $\delta(x) < \rho_o$. Write x as $x = \eta - r\mathbf{n}_\eta$ with $r = \delta(x)$. Let $0 < q < p$. Since f has subharmonic behavior,

$$f^q(x) \le \frac{C}{r^n} \int_{B(x,\frac{1}{2}r)} f^q(y)\,dy.$$

Very crude estimates show that $B(x,\frac{1}{2}r) \subset S(\eta,4\alpha r) \times [0,\frac{3}{2}r]$. Thus by Fubini's theorem

$$\begin{aligned} f^q(x) &\le \frac{C}{r^n} \int_{S(\eta,4\alpha r)} \left[\int_0^{\frac{3}{2}r} f^q(y - t\mathbf{n}_y)\,dt \right] ds(y) \\ &\le \frac{C}{r^{n-1}} \int_{S(\eta,4\alpha r)} (M_{rad} f)^q(y)\,ds(y) \\ &\le \frac{C}{r^{n-1}} \int_{S(\zeta,8\alpha r)} (M_{rad} f)^q(y)\,ds(y) \le C_\alpha \widetilde{M} g(\zeta), \end{aligned}$$

where $g(\zeta) = (M_{rad} f)^q(\zeta)$. Therefore, $(M_\alpha f)^p(\zeta) \le C_\alpha (\widetilde{M} g)^{p/q}(\zeta)$. Since $\partial\Omega$ is $C^{1,1}$ and $p/q > 1$, by the Hardy–Littlewood maximal theorem,

$$\int_{\partial\Omega} (\widetilde{M} g)^{p/q}(\zeta)\,ds(\zeta) \le C_p \int_{\partial\Omega} g^{p/q}(\zeta)\,ds(\zeta) = C_p \int_{\partial\Omega} (M_{rad} f)^p(\zeta)\,ds(\zeta),$$

from which the result follows. □

Proof of Theorem 4. Let $0 < r < \rho_o$, and let f be as in the hypothesis of the theorem. By the fundamental theorem of calculus, for fixed $\zeta \in \partial\Omega$,

$$f(\zeta - r\mathbf{n}_\zeta) - f(\zeta - \rho_o \mathbf{n}_\zeta) = \int_r^{\rho_o} \langle \nabla f(\zeta - t\mathbf{n}_\zeta), \mathbf{n}_\zeta \rangle\,dt.$$

Therefore,

$$|f(\zeta - r\mathbf{n}_\zeta)| \le |f(\zeta - \rho_o \mathbf{n}_\zeta)| + \int_r^{\rho_o} |\nabla f(\zeta - t\mathbf{n}_\zeta)|\,dt. \tag{29}$$

For $k = 0, \ldots, N$, set $t_k = \rho_o 2^{-k}$, where N is the smallest integer such that $t_N < r$. Then

$$\begin{aligned} \int_r^{\rho_o} |\nabla f(\zeta - t\mathbf{n}_\zeta)|\,dt &\le \sum_{k=1}^{N} \int_{t_k}^{t_{k-1}} |\nabla f(\zeta - t\mathbf{n}_\zeta)|\,dt \\ &\le \sum_{k=1}^{N} \sup_{t \in (t_k, t_{k-1}]} |\nabla f(\zeta - t\mathbf{n}_\zeta)|(t_{k-1} - t_k) \\ &= \sum_{k=1}^{N} t_k \sup_{t \in (t_k, t_{k-1}]} |\nabla f(\zeta - t\mathbf{n}_\zeta)|. \end{aligned}$$

Hence, for $0 < p \le 1$,

$$|f(\zeta - r\mathbf{n}_\zeta)|^p \le |f(\zeta - \rho_o\mathbf{n}_\zeta)|^p + \sum_{k=1}^{N} t_k^p \sup_{t\in(t_k,t_{k-1}]} |\nabla f(\zeta - t\mathbf{n}_\zeta)|^p. \tag{30}$$

For $N = 1,2,\ldots$, and $\zeta \in \partial\Omega$, set

$$S_N(\zeta) = |f(\zeta - \rho_o\mathbf{n}_\zeta)|^p + \sum_{k=1}^{N} t_k^p \sup_{t\in(t_k,t_{k-1}]} |\nabla f(\zeta - t\mathbf{n}_\zeta)|^p, \tag{31}$$

and

$$S(\zeta) = |f(\zeta - \rho_o\mathbf{n}_\zeta)|^p + \sum_{k=1}^{\infty} t_k^p \sup_{t\in(t_k,t_{k-1}]} |\nabla f(\zeta - t\mathbf{n}_\zeta)|^p. \tag{32}$$

Then $S_N(\zeta) \uparrow S(\zeta)$ as $N \to \infty$, and by Eq. (30) we have $(M_{rad}f)^p(\zeta) \le S(\zeta)$ for all $\zeta \in \partial\Omega$ for which $S(\zeta) < \infty$.

Since $|\nabla f|$ has subharmonic behavior, by Lemma 1, for all $t \in (r,\rho_o]$,

$$|\nabla f(\zeta - t\mathbf{n}_\zeta)|^p \le \frac{C}{t^n}\int_{B(\zeta - t\mathbf{n}_\zeta)} |\nabla f(y)|^p\,dy. \tag{33}$$

For each $k = 1,\ldots,N$ and $\zeta \in S$, set

$$B_k(\zeta) = B\left(\zeta - t_{k-1}\mathbf{n}_\zeta, \frac{7}{4}t_k\right).$$

We claim that $B(\zeta - t\mathbf{n}_\zeta) \subset B_k(\zeta)$ for all $t \in (t_k,t_{k-1}]$. Suppose $y \in B(\zeta - t\mathbf{n}_\zeta)$. Then

$$|y - (\zeta - t_{k-1}\mathbf{n}_\zeta)| \le |y - (\zeta - t\mathbf{n}_\zeta)| + |t_{k-1} - t| < \frac{1}{2}t + (t_{k-1} - t).$$

If $t \in (t_k, \frac{3}{2}t_k]$, then

$$|y - (\zeta - t_{k-1}\mathbf{n}_\zeta)| < \frac{3}{4}t_k + (t_{k-1} - t_k) = \frac{7}{4}t_k.$$

On the other hand, if $t \in (\frac{3}{2}t_k, t_{k-1}]$, then

$$|y - (\zeta - t_{k-1}\mathbf{n}_\zeta)| < \frac{1}{2}t_{k-1} + \left(t_{k-1} - \frac{3}{2}t_k\right) = \frac{3}{2}t_k,$$

which proves the claim. Also, for each $k = 1,\ldots,N$, we have

$$B_k(\zeta) \subset \Omega_{t_{k+2}} \setminus \Omega_{t_{k-2}}.$$

To see this, for $y \in B_k(\zeta)$ and $x \in \partial\Omega$,

$$\begin{aligned}|y-x| &\geq |(\zeta - t_{k-1}\mathbf{n}_\zeta) - x| - |y - (\zeta - t_{k-1}\mathbf{n}_\zeta)| \\ &> \delta(\zeta - t_{k-1}\mathbf{n}_\zeta) - \frac{7}{4}t_k = t_{k-1} - \frac{7}{4}t_k = t_{k+2}.\end{aligned}$$

Hence $\delta(y) > t_{k+2}$, i.e., $B_k(\zeta) \subset \Omega_{t_{k+2}}$. Also, for $y \in B_k(\zeta)$,

$$\delta(y) \leq |y-\zeta| \leq |y - (\zeta - t_{k-1}\mathbf{n}_\zeta)| + t_{k-1} < \frac{7}{4}t_k + 2t_k < t_{k-2}.$$

Hence $y \notin \Omega_{t_{k-2}}$. Thus $B_k(\zeta) \subset \Omega_{t_{k+2}} \setminus \Omega_{t_{k-2}}$. For convenience we set

$$A_k = \Omega_{t_{k+2}} \setminus \Omega_{t_{k-2}}.$$

From inequality (33) and the above we now obtain

$$\sup_{t \in (t_k, t_{k-1}]} |\nabla f(\zeta - t\mathbf{n}_\zeta)|^p \leq \frac{C}{t_k^n} \int_{B_k(\zeta)} |\nabla f(y)|^p dy.$$

Thus by Eq. (30)

$$S_N(\zeta) \leq |f(\zeta - \rho_o \mathbf{n}_\zeta)|^p + C \sum_{k=1}^{N} t_k^{p-n} \int_{B_k(\zeta)} |\nabla f(y)|^p dy.$$

Integrating the above inequality over $\partial\Omega$ gives

$$\begin{aligned}\int_{\partial\Omega} S_N(\zeta)\, ds(\zeta) &\leq \int_{\partial\Omega} |f(\zeta - \rho_o \mathbf{n}_\zeta)|^p ds(\zeta) \\ &\quad + C \sum_{k=1}^{N} t_k^{p-n} \int_{\partial\Omega} \int_{B_k(\zeta)} |\nabla f(y)|^p dy\, ds(\zeta).\end{aligned}$$

For $y \in A_k$, let

$$\widetilde{B}_k(y) = \{\zeta \in \partial\Omega : y \in B_k(\zeta)\}.$$

Then $\chi_{B_k(\zeta)}(y) \leq \chi_{\widetilde{B}_k(y)}(\zeta)$. Thus

$$\begin{aligned}\int_{\partial\Omega} \int_{B_k(\zeta)} |\nabla f(y)|^p dy\, ds(\zeta) &= \int_{\partial\Omega} \int_{A_k} \chi_{B_k(\zeta)}(y) |\nabla f(y)|^p dy\, ds(\zeta) \\ &\leq \int_{\partial\Omega} \int_{A_k} \chi_{\widetilde{B}_k(y)}(\zeta) |\nabla f(y)|^p dy\, ds(\zeta),\end{aligned}$$

which by Fubini's theorem

$$\leq \int_{A_k} |\widetilde{B}_k(y)||\nabla f(y)|^p dy,$$

where for $E \subset \partial\Omega$, $|E|$ denotes the surface measure of E.

Fix $y \in A_k$, and suppose $\zeta \in \widetilde{B}_k(y)$. Write $y = x - \delta(y)\mathbf{n}_x$, $x \in \partial\Omega$. Since $y \in B_k(\zeta)$,

$$\begin{aligned}|\zeta - x| &\leq |(\zeta - t_{k-1}\mathbf{n}_\zeta) - y| + |t_{k-1}\mathbf{n}_\zeta - \delta(y)\mathbf{n}_x| \\ &< \tfrac{7}{4}t_k + |t_{k-1}\mathbf{n}_\zeta - \delta(y)\mathbf{n}_x|. \end{aligned} \tag{34}$$

Let $\theta(\zeta,x)$ denote the angle between the vectors $\mathbf{n}_\zeta$ and $\mathbf{n}_x$. Then

$$|t_{k-1}\mathbf{n}_\zeta - \delta(y)\mathbf{n}_x|^2 = (t_{k-1} - \delta(y))^2 + 4t_{k-1}\delta(y)\sin^2\frac{1}{2}\theta(\zeta,x) \leq Ct_k^2$$

for all $y \in A_k$. Therefore, $\widetilde{B}_k(y) \subset B(x,Ct_k) \cap \partial\Omega$, and hence $|\widetilde{B}_k(y)| \leq Ct_k^{n-1}$. Thus

$$\begin{aligned} t_k^{p-n}\int_{\partial\Omega}\int_{B_k(\zeta)} |\nabla f(y)|^p dy\,ds &\leq Ct_k^{p-1}\int_{A_k}|\nabla f(y)|^p dy \\ &\leq C\int_{A_k}\delta(y)^{p-1}|\nabla f(y)|^p dy. \end{aligned}$$

Finally, since

$$\sum_{k=1}^{N}\int_{A_k}\delta(y)^{p-1}|\nabla f(y)|^p dy \leq C\int_\Omega \delta(y)^{p-1}|\nabla f(y)|^p dy,$$

we obtain

$$\int_{\partial\Omega} S_N(\zeta)\,ds(\zeta) \leq \int_{\partial\Omega}|f(\zeta - \rho_o\mathbf{n}_\zeta)|^p\,ds(\zeta) + C\int_\Omega \delta(y)^{p-1}|\nabla f(y)|^p dy$$

for all N. Hence by the monotone convergence theorem

$$\int_{\partial\Omega} S(\zeta)\,ds(\zeta) \leq \int_{\partial\Omega}|f(\zeta - \rho_o\mathbf{n}_\zeta)|^p\,ds(\zeta) + C\int_\Omega \delta(y)^{p-1}|\nabla f(y)|^p dy.$$

Hence $S \in L^1(\partial\Omega)$ and thus is finite a.e. Thus we have $M_{rad}f(\zeta) \leq S(\zeta)$ a.e., and

$$\int_{\partial\Omega}(M_{rad}f)^p(\zeta)\,ds(\zeta) \leq \int_{\partial\Omega}|f(\zeta - \rho_o\mathbf{n}_\zeta)|^p ds(\zeta) + C\int_\Omega \delta(y)^{p-1}|\nabla f(y)|^p dy.$$

To conclude the proof it remains to be shown that for $t_o \in \Omega_{\rho_o}$ fixed,

$$\int_{\partial\Omega} |f(\zeta - \rho_o \mathbf{n}_\zeta)|^p\, ds(\zeta) \le C\left[|f(t_o)|^p + \int_\Omega \delta(y)^{p-1}|\nabla f(y)|^p dy\right]. \qquad (35)$$

As in the proof of Theorem 5.1 [19], since $\overline{\Omega}_{\rho_o}$ is connected and $\partial\Omega_{\rho_o}$ is $C^{1,1}$, there exists a constant $C(\rho_o)$ such that for every $y \in \overline{\Omega}_{\rho_o}$ there exists a polygonal path $\gamma(t), 0 \le t \le 1$, in $\overline{\Omega}_{\rho_o}$ with $\gamma(0) = t_o$, $\gamma(1) = y$ and $\int_0^1 |\gamma'(t)|dt \le C(\rho_o)$. Thus

$$\begin{aligned}|f(y)| &\le |f(t_o)| + \int_0^1 |\nabla f(\gamma(t))||\gamma'(t)|\,dt \\ &\le |f(t_o)| + C(\rho_o)\sup\left\{|\nabla f(w)| : w \in \overline{\Omega}_{\rho_o}\right\}.\end{aligned}$$

Inequality (35) now follows by applying Lemma 1 to $|\nabla f(w)|$ on $B(w, \frac{1}{2}\rho_o)$, $w \in \overline{\Omega}_{\rho_o}$, and using the fact that $\frac{1}{2}\rho_o \le \delta(y) \le C$ for all $y \in \Omega_{\rho_o/2}$. □

References

1. Aikawa, H., Essén, M.: Potential Theory - Selected Topics. In: Lecture Notes in Mathematics, vol. 1633. Springer, Berlin (1996)
2. Aronszajn, N., Smith, K.T.: Functional spaces and functional completion. Ann. Inst. Fourier, Grenoble **6**, 125–185 (1955)
3. Djordjević, O., Pavlović, M.: On a Littlewood–Paley type inequality. Proc. Amer. Math. Soc. **135**, 3607–3611 (2007)
4. Fefferman, C., Stein, E.: H^p spaces of several variables. Acta Math **129**, 137–193 (1972)
5. Flett, T.M.: On some theorems of Littlewood and Paley. J. London Math. Soc. **31**, 336–344 (1956)
6. Garnett, J.B.: Bounded analytic functions. In: Pure and Applied Mathematics. Academic, New York (1981)
7. Jevtić, M.: Littlewood–Paley theorems for M–subharmonic functions. J. Math. Anal. Appl. **274**, 685–695 (2002)
8. Kuran, Ü.: Subharmonic behavior of $|h|^p$ $(p > 0, h$ harmonic). J. London Math. Soc. **8**, 529–538 (1974)
9. Littlewood, J.E., Paley, R.E.A.C.: Theorems on Fourier series and power series. Proc. London Math. Soc. **42**, 52–89 (1936)
10. Luecking, D.H.: A new proof of an inequality of Littlewood and Paley. Proc. Amer. Math. Soc. **103**(3), 887–893 (1988)
11. Maeda, F.-Y., Suzuki, N.: The integrability of superharmonic functions on Lipschitz domains. Bull. London Math. Soc. **21**, 270–278 (1989)
12. Pavlović, M.: Inequalities for the gradient of eigenfunctions of the invariant laplacian in the unit ball. Indag. Mathem., N. S. **2**, 89–98 (1991)
13. Pavlovic, M.: On subharmonic behavior and oscillation of functions in balls in $\mathbb{R}^n$. Publ. Inst. Math. (N.S.) **69**, 18–22 (1994)
14. Pavlović, M.: A Littlewood–Paley theorem for subharmonic functions. Publ. Inst. Math. (Beograd) **68**(82), 77–82 (2000)

15. Pavlović, M.: A short proof of an inequality of Littlewood and Paley. Proc. Amer. Math. Soc. **134**, 3625–3627 (2006)
16. Riihentaus, J.: On a theorem of Avanissian-Arsove. Exposition. Math. **7**, 69–72 (1989)
17. Stević, S.: A Littlewood-Paley type inequality. Bull. Braz. Math. Soc. **34**, 1–7 (2003)
18. Stoll, M.: A characterization of Hardy-Orlicz spaces on planar domains. Proc. Amer. Math. Soc. **117**(4), 1031–1038 (1993)
19. Stoll, M.: Harmonic majorants for eigenfunctions of the Laplacian with finite Dirichlet integral. J. Math. Analysis and Appl. **274**, 788–811 (2002)
20. Stoll, M.: On generalizations of the Littlewood-Paley inequalities to domains in $\mathbb{R}^n$ ($n \geq 2$). Unpublished Manuscript (2004) http://www.math.sc.edu/people/faculty/stoll/li-paley.pdf
21. Stoll, M.: The Littlewood–Paley inequalities for Hardy–Orlicz spaces of harmonic function on domains in $\mathbb{R}^n$. Adv. Stud. Pure Math. **44**, 363–376 (2006)
22. Suzuki, N.: Nonintegrability of harmonic functions in a domain. Japan J. Math. **16**, 269–278 (1990)
23. Widman, K.-O.: Inequalities for the Green function and boundary continuity of the gradient of solutions of elliptic differential equations. Math. Scand. **21**, 17–37 (1967)

The Path to Λ-Bounded Variation

Daniel Waterman

Dedicated to the memory of Casper Goffman, my friend, colleague, and coworker

Abstract This is a personal account describing the evolution of the concept of Λ-bounded variation. Earlier generalizations of bounded variation are described. A test for the convergence of Fourier series due to Salem was pivotal, leading to a condition for preservation of convergence of Fourier series under change of variable and, finally, to a new notion of bounded variation. Some applications and generalizations to higher dimensions are indicated.

1 Introduction

At some point in the mid-1960s, Casper Goffman and I became interested in determining the class of functions whose Fourier series remained convergent after any change of variable, i.e., after composition with a homeomorphism of the circle group $\mathbb{T}$ with itself.

Various aspects of the study of homeomorphisms in real analysis, including this, are studied in our book with Togo Nishiura [15]. Here I will tell of how our study of this particular problem led to the formulation of a new generalization of bounded variation, ΛBV.

We knew of several classes that have this convergence-preserving property, e.g., the functions of bounded variation (BV), for the definition of the bounded variation of a function f,

D. Waterman (✉)
Syracuse University and Florida Atlantic University, 7739 Majestic Palm Drive, Boynton Beach, FL 33437, USA,
e-mail: dan.waterman@gmail.com

D. Bilyk et al. (eds.), *Recent Advances in Harmonic Analysis and Applications*, Springer Proceedings in Mathematics & Statistics 25, DOI 10.1007/978-1-4614-4565-4_29,

$$\sup_{\{I_i\}} \sum |f(I_i)| < \infty$$

is clearly independent of a change of variable. Here the $I_i = [a_i, b_i]$ are intervals in $\mathbb{T}$,while $f(I_i) = f(b_i) - f(a_i)$, and $\{I_i\}$ denotes a finite or infinite collection of nonoverlapping intervals. Indeed the study of Fourier series has provided the principal motivation for the development of different notions of bounded variation. The convergence of the Fourier series of f, $S[f]$, for $f \in BV$ is a consequence of the classical Dirichlet–Jordan theorem, which also asserts uniform convergence on compact sets of points of continuity [36].

As early as 1924, Wiener [32] knew that $S[f]$ would converge if, in the definition of BV, $|f(I_i)|$ were replaced by $|f(I_i)|^2$, but for $|f(I_i)|^p, p > 1$, he could not prove this result. We say that this class of functions is of bounded Wiener p-variation, W_p or BV^p. Young proved convergence in 1936 [33]. He and Love [16, 33, 34] were responsible for the next significant generalization, the introduction of ΦBV.

If $\Phi(x)$ is a nonnegative function defined on $[0, \infty)$, continuous, increasing, and such that $\Phi(0) = 0$, the Φ-variation of f is defined to be the

$$\sup_{\{I_i\}} \sum \Phi(|f(I_i)|)$$

if this is finite. The class of such functions is referred to as ΦBV. Clearly the class BV^p is $x^p BV$. For various Φ, Young [34, 35] showed that a Dirichlet–Jordan type theorem held.

Without further restrictions on Φ, this definition suffers from serious defects. The classes thus defined need not be linear spaces. The previous classes, with appropriate norms defined, are Banach spaces and Helly's selection theorem holds in these spaces. The necessary conditions were fully discussed in a seminal paper of Musielak and Orlicz [17]. A critical condition is the Δ_2-condition,

$$\text{for some } x_0 > 0 \text{ and } d \geq 2,\ \Phi(2x) \leq d\Phi(x) \text{ for } 0 \leq x \leq x_0.$$

For some interesting observations on this see [20].

If Φ is as we have described, then Φ and the function

$$\Psi(y) = \max_{x \geq 0}(xy - \Phi(x))$$

are said to be complementary in the sense of W. H. Young [5, 17]. The inequality

$$xy \leq \Phi(x) + \Psi(y)$$

is known as Young's inequality. Note that if $\phi(x)$ is a strictly increasing positive function on $(0, \infty)$, then $\Phi(x) = \int_0^x \phi(t)dt$ is convex, and it and $\Psi(y) = \int_0^y \phi^{-1}(t)dt$ are complementary Young's functions.

Raphael Salem made many important contributions to the theory of Fourier series, but those concerning convergence theory are less well known. The following result [24, Chap.VI] supplied the lead to the discovery of the condition for the preservation of convergence of $S[f]$ under change of variable and to the formulation of the concept of ΛBV.

For odd integers n, let

$$T_n(x) = \frac{f(x) - f(x + \frac{\pi}{n})}{1} + \frac{f(x + 2\frac{\pi}{n}) - f(x + 3\frac{\pi}{n})}{3} + \cdots + \frac{f(x + (n-1)\frac{\pi}{n}) - f(x + \pi)}{n}$$

and let $Q_n(x)$ be obtained from $T_n(x)$ by substituting $-\pi$ for π.

2 The Salem Test

If f is continuous on the circle group $\mathbb{T}$ and $T_n(x)$ and $Q_n(x)$ converge uniformly to 0, then $S[f]$ converges uniformly.

In the same paper he used this to prove an important theorem concerning ΦBV. He showed:

If a continuous $f \in \Phi BV$ and $\sum_1^\infty \Psi(1/n) < \infty$, then $S[f]$ converges uniformly.

In a survey paper, Goffman and I asked if the condition on Ψ is necessary [12]. This was answered in the affirmative by Oskolkov [18] and Baernstein [2] simultaneously. Oskolkov showed that

$$\int_0^1 \log \frac{1}{\Phi(t)} dt < \infty$$

is equivalent to Salem's condition. We note that, according to a result of Chanturiya [6],

$$\sum_1^\infty \frac{1}{n} \Phi^{-1}\left(\frac{1}{n}\right) < \infty$$

is also equivalent. If $N_f(t)$ is the Banach indicatrix of f, Oskolkov also showed that Garsia and Sawyer's result [10], that $\int N_f(t)dt < \infty$ implies uniform convergence of $S[f]$, is a consequence of Salem's result.

It is interesting that Salem's result remained unnoticed for many years. When I called it to Zygmund's attention in the mid-1960s, I found that he had never seen it, although he and Salem were close friends. He immediately said that what Salem had done was to linearize the Dirichlet kernel.

In his introduction to Salem's *Œuvres Mathématiques* [25], published at about the time Goffman and I were working in this area, Kahane remarks of T_n,

qu'on peut considérer comme une analogue discrète de l'intêgrale de Dirichlet con n'inclut dans la somme que des $f(x + k\pi/n)$ avec $k \leq n$ et il l'applique à l'étude des séries de Fourier. Il nous paraît qu'il y a ici une idée digne d'intérèt et qui pourrait donner des résultats intéressants.

I have been able to generalize the Salem test to give a test for convergence of $S[f]$ at a symmetric Lebesgue point and a necessary and sufficient condition for convergence at such a point [30].

The condition that Goffman and I gave for preservation of convergence was stated in terms of systems of intervals [11]. For each n, let I_{nm}, $m = 1, 2, \cdots, k_n$ be disjoint closed intervals such that $I_{n,m-1}$ is to the left of I_{nm} . If for an $x \in T$ and any $\delta > 0$, there is an N such that $I_{nm} \subset (x, x+\delta)$ when $n > N$, then the collection $\mathbb{I} = \{I_{nm}, n = 1, 2, \cdots, m = 1, 2, \cdots, k_n\}$ is called a right system of intervals at x. Left systems are defined similarly. In our result we defined "system" somewhat differently. We required that, as $n \to \infty$, $k_n \to \infty$ and $k_n/n \to 0$. The equivalence of the two definitions was shown in [3]. It is easy to show that "disjoint" may be replaced by "nonoverlapping."

We showed that for a continuous function f, $S[f \circ g]$ converges everywhere for every homeomorphism g if and only if

$$\lim_{n \to \infty} \sum_{i=1}^{k_n} \frac{1}{i} f(I_{ni}) = 0 \qquad \text{(GW condition)}$$

for every system $\mathbb{I}$.

The GW condition was suggested by the observation that changes of variable can transform sums $\sum_{i=1}^{k_n} \frac{1}{i} f(I_{ni})$ into sums resembling the T_n and Q_n in the Salem test.

Baernstein and I found the necessary and sufficient condition, UGW, for preservation of uniform convergence [3]. It is that the GW condition and also

$$\lim_{n \to \infty} \sum_{i=1}^{k_n} \frac{1}{k_n + 1 - i} f(I_{ni}) = 0$$

hold for every system. It is usual to use GW and UGW to denote both the conditions and the classes of functions which satisfy them.

A regulated function is one such that, for every x, $f(x+)$ and $f(x-)$ exist. We usually set $f(x) = \frac{1}{2}[f(x+) + f(x-)]$ since, if $S[f]$ is convergent or summable at x, it will be to that value. Note that, if a function satisfies the GW condition, it is clear that it must be regulated.

It is important to observe that our original result was for *continuous* functions. In Goffman's 1970/1971 paper [13], he asserts that the GW result goes over, without change of proof, to the more general regulated functions. Thus the results of his paper, which depended on the application of the GW result, were shown to be valid only for continuous functions. In all fairness, I must admit that, until I tried to do it, I also believed that it would be simple to extend the GW theorem to regulated functions.

In our papers published in the 1970s, both Cas and I assumed that the GW theorem was valid for regulated functions. In fact, in the paper in which I defined ΛBV and HBV, I repeated Cas's claim and credited him with having proved his result for regulated functions. I proved the sufficiency portion of the GW theorem for regulated functions in 1985 [29] and the complete result, still later, with Pierce [19].

Let me now give a rationale for considering sums of the type that appear in Goffman's results and in my definition of ΛBV.

Let us suppose that f does not satisfy the GW condition at x but has one-sided limits there. Without loss of generality we may assume that there is a $\Delta > 0$ and a right system such that $\sum_{i=1}^{k_n} \frac{1}{i} f(I_{ni}) > \Delta$ for every n. I will assume here that $k_n \to \infty$.

We may choose an n_1 so large that $I_{n_1 k_{n_1}}$ is to the left of I_{11}, $k_{n_1} > k_1$, and $\sum_{i=1}^{k_1} \frac{1}{i} f(I_{n_1 i})$ is so small that $\sum_{i=k_1+1}^{k_{n_1}} \frac{1}{i} f(I_{n_1 i}) > \Delta$. We continue this process to form a sequence of sums

$$\sum_{i=1}^{k_1} \frac{1}{i} f(I_{n_1 i}), \sum_{i=k_1+1}^{k_{n_1}} \frac{1}{i} f(I_{n_1 i}), \sum_{i=k_{n_1}+1}^{k_{n_2}} \frac{1}{i} f(I_{n_2 i}), \ldots,$$

each exceeding Δ, where the nonoverlapping intervals have indices ni, with $n = 1, n_1, n_2, \cdots$ and $i = 1, 2, \cdots$. Re-indexing the intervals by eliminating the n-portion so that each has only the index i and adding the sums together, we have

$$\sum_{i=1}^{\infty} \frac{1}{i} f(I_i) = \infty.$$

It is easy to see that the existence of such a sequence of intervals I_i is equivalent to the existence of a sequence of nonoverlapping intervals J_i such that

$$\sum_{i=1}^{\infty} \frac{1}{i} |f(J_i)| = \infty.$$

Goffman stated his result in terms of Köthe spaces. He showed that if f is such that the set of sequences $\{f(I_1), f(I_2), \cdots, f(I_n), 0, \cdots\}$, where the I_i are disjoint intervals, is bounded in the space X and $\{1/n\} \in X^*$, the associated Köthe space, then $S[f]$ converges everywhere.

I relocated my family at the time that Cas was working on his paper, and we were not in close contact. He sent me a copy of a measure-theoretic result which later appeared in his paper, but I was not aware of the result on convergence. This was fortunate, for if I had read his paper, I might not have taken the approach to the problem that I did .

I noted that, for any nondecreasing sequence of positive numbers, $\Lambda = \{\lambda_n\}$, the class of functions such that

$$\sum_{n=1}^{\infty} \frac{1}{\lambda_n} |f(I_n)| < \infty$$

for every sequence of nonoverlapping intervals was, with a suitable norm, a Banach space ΛBV. The Λ-variation of f is the supremum of such sums, which is finite. Equivalently, one may consider finite sums. If they have a common upper bound, then $f \in \Lambda BV$, and the supremum of such sums is the same as that previously considered. If Λ is a bounded sequence, then $\Lambda BV = BV$.

For a comprehensive account of what is known about ΛBV and outstanding problems, see the paper of Prus-Wiśniowski [21].

An extensive account of the various generalizations of bounded variation, including many original results, is to be found in the paper of Avdispahić [1].

Functions in ΛBV are regulated. The case $\Lambda = \{1/n\}$ I called HBV, harmonic bounded variation. I showed that

If $f \in HBV$, then $S[f]$ converges everywhere and converges uniformly on any closed interval of points of continuity.

This was accomplished by showing that HBV functions satisfy the Lebesgue test [35], and so the result is independent of the GW condition. If $\Lambda BV \supsetneq HBV$, then there is $f \in \Lambda BV$ such that $S[f]$ diverges at a point [26, 27]. Later I gave another proof of the convergence theorem for HBV which is independent of both the GW condition and the Lebesgue test [28].

The validity, for regulated functions, of the result of Salem for ΦBV and of the Garsia–Sawyer theorem on the Banach indicatrix are consequences of the HBV theorem. For regulated functions the definition of the Banach indicatrix must be extended. If you adjoin to the graph of f the vertical line segments connecting $(x, f(x-))$ and $(x, f(x+))$ at every jump, then $N_f(t)$ will be the number of intersections of the line $y = t$ with the resulting set. For regulated functions, the Garsia–Sawyer type theorem yields pointwise convergence. The following results are proved in [26]:

(i) If $f \in \Phi BV$ and $\sum \Psi(1/n) < \infty$, then $f \in HBV$.
(ii) If the range of f is contained in a finite interval I and $L(x)$ is a finite-valued increasing function such that $L(n) \sim \sum_1^n 1/\lambda_k$, then $\int_I L(N_f(y)dy < \infty$ implies that $f \in \Lambda BV$.

It has been shown by Berezhnoi [4] that the HBV theorem on convergence of $S[f]$ is the best possible result of bounded variation type, yielding the Dirichlet–Jordan conclusions.

The notion of ΛBV has had many generalizations and applications. Perhaps the most interesting applications are in the area of Fourier analysis, localization, and convergence theorems for double Fourier series and convergence theorems for partial sums of interpolating polynomials. There have also been applications in summability and number theory.

The Riemann localization principle for functions of one variable asserts that if $f \in L^1$ and $f = 0$ on an open interval, then $S[f]$ converges uniformly to 0 on any compact subset of the interval. For functions of more than one variable, strong additional assumptions are required [36], not even differentiability suffices.

A rectangular partial sum of $S[f] = \sum_{m,n} a_{mn} e^{i(mx+ny)}$ is

$$S_{N_1,N_2}[f; x, y] = \sum_{|k_1| \leq N_1} \sum_{|k_2| \leq N_2} a_{k_1 k_2} e^{i(k_1 x + k_2 y)}.$$

By convergence of rectangular partial sums (or Pringsheim convergence) we mean existence of $\lim S_{N_1,N_2}$ as $\min(N_1,N_2) \to \infty$. We speak of convergence of square partial sums if we set $N_1 = N_2$.

If $w(x,y)$ is defined on an interval in $\mathbb{R}^2$, let $V_x(w,[a,b])$ denote the harmonic variation of w on $[a,b]$ as a function of x with y fixed. $V_y(w,[a,b])$ is similarly defined.

Goffman and I established the following result:

Let $f \in (\mathbb{T}^2)$ and let there exist functions g and h equivalent to f such that$V_x(g,[-\pi,\pi])$ and $V_y(h,[-\pi,\pi])$ are integrable functions of y and x, respectively. Then the localization principle for rectangular partial sums holds for f.

This result is best possible in the sense that it fails if HBV is replaced by ΛBV such that $\Lambda BV \supsetneq HBV$.

This definition of bounded variation for two variables generalizes those of Tonelli and Cesari. It differs from Cesari's in that we replace BV with HBV.

Dyachenko and I [8] investigated the possibility of using Λ-bounded variation to prove a Dirichlet–Jordan type theorem for double Fourier series.

Saakyan [22] and Sablin [23] had given the following definition:
Let f be a measurable function on the rectangle $A = [a,b] \times [c,d]$. Then $f \in \Lambda BV(A)$ if and only if

(i) $f(\cdot,c) \in \Lambda BV$ on $[a,b]$ and $f(a,\cdot) \in \Lambda BV$ on $[c,d]$, and
(ii) if $\mathbb{I}_1$ and $\mathbb{I}_2$ are sets of nonoverlapping intervals $I_k = [\alpha_k,\beta_k]$ and $I_j = [\gamma_j,\delta_j]$, and

$$f(I_k \times I_j) = f(\alpha_k,\gamma_j) - f(\alpha_k,\delta_j) - f(\beta_k,\gamma_j) + f(\beta_k,\delta_j),$$

then $V_\Lambda(f;A) = \sup_{\mathbb{I}_1,\mathbb{I}_2} \sum_{j,k} \frac{|f(I_j \times I_k)|}{\lambda_j \lambda_k} < \infty.$

Note that if $\lambda_n \equiv 1$, or, what is the same, $\lambda_n = O(1)$, then $\Lambda BV(A)$ is the set of functions of Hardy-Krause bounded variation on A.

Dyachenko [7] showed that this definition has significant defects. For example, if $\sum 1/\lambda_n^2 < \infty$, then the requirement of measurability cannot be omitted. Even with measurability, $\Lambda BV(A)$ contains an everywhere discontinuous function. Saakyan proved a convergence theorem for $f \in HBV(A)$, and Sablin extended this to higher dimensions. However, as Dyachenko showed, there are functions in this class for which the result is inapplicable due to their discontinuous nature.

Dyachenko and I gave the following definition:
Let f be a real function on $A = [a,b] \times [c,d]$. We say $f \in \Lambda^* BV(A)$ if

(i) $f(\cdot,c) \in \Lambda BV([a,b])$, $f(a,\cdot) \in \Lambda BV([c,d])$ and if Γ is the set of finite collections of nonoverlapping intervals $A_k = [\alpha_k,\beta_k] \times [\gamma_k,\delta_k] \subset A$ and

$$f(A_k) = f(\alpha_k,\gamma_k) - f(\alpha_k,\delta_k) - f(\beta_k,\gamma_k) + f(\beta_k,\delta_k),$$

then

(ii) $V_{\Lambda^*}(f,A) = \sup_{\Gamma} \sum_k \frac{|f(A_k)|}{\lambda_k} < \infty$.

For $f \in \Lambda^* BV(A)$ we set

(iii) $\|f\|_{\Lambda^*} = \|f\|_{\Lambda^*(A)} = |f(a,c)| + V_\Lambda f(\cdot,c) + V_\Lambda f(a,\cdot) + V_{\Lambda^*}(f,A)$.

This class has good continuity properties. For any f in this class there exist at most countable sets $P \subset [a,b]$ and $Q \subset [c,d]$ such that f is continuous at every $(x,y) \in A$ such that $x \notin P$ and $y \notin Q$. Also, at every point $(x_0,y_0) \in A$, $\lim_{(x,y)\to(x_0,y_0)} f(x,y)$ exists in each coordinate quadrant.

Our principal result was

Let f be a real function on R^2 which is 2π-periodic in each variable and is in $\Lambda^* BV(\mathbb{T}^2)$ with $\Lambda = \left\{\frac{n}{ln(n+1)}\right\}$. Then the rectangular partial sums of $S[f]$ are uniformly bounded and converge at each point to the arithmetic mean of the quadrant limits.

We also showed that, in a certain sense, this result cannot be improved.

Let $\Lambda = \left\{\frac{n}{ln(n+1)}\xi_n\right\}$, where $\xi_n \nearrow \infty$ as $n \to \infty$. Then there exists a function $f \in \Lambda^* BV(\mathbb{T}^2)$ such that the square partial sums of $S[f]$ diverge unboundedly at $(0,0)$.

We turn now to an application to interpolating polynomials. Hualing Xing and I investigated the convergence of the partial sums of trigonometric interpolating polynomials [31].

Let f be a Riemann integrable function of period $2\pi, t_0^{(n)}$ an arbitrary real number, and $t_j^{(n)} = t_0^{(n)} + \frac{2\pi j}{2n+1}, j = 0,\ldots,2n$. Then $I_n(x,f)$ will denote the trigonometric polynomial which coincides with f at the fundamental points $\left\{t_j^{(n)}\right\}$. If $D_n(t)$ denotes the *n-th* Dirichlet kernel, $\sin(n+\frac{1}{2})t/2\sin\frac{1}{2}t$, then

$$I_n(x,f) = \frac{2}{2n+1}\sum_{j=0}^{2n} f(t_j^{(n)})D_n(x-t_j^{(n)}) = \frac{1}{\pi}\int_{-\pi}^{\pi} f(t)D_n(x-t)d\omega_{2n+1}(t),$$

where $\omega_{2n+1}(t)$ is a step function with jumps $2\pi/(2n+1)$ at the points $t_j^{(n)}, j = 0, \pm 1, \pm 2, \ldots$. We can write

$$I_n(x,f) = \frac{1}{2}a_0 + \sum_{\upsilon=0}^{n} (a_\nu^{(n)} \cos \nu x + b_\upsilon^{(n)} \sin \nu x),$$

where the Fourier–Lagrange coefficients $a_\nu^{(n)}$ and $b_\nu^{(n)}$ are given by

$$a_\nu^{(n)} = \frac{1}{\pi}\int_0^{2\pi} f(t)\cos\nu t d\omega_{2n+1}(t) \text{ and } b_\nu^{(n)} = \frac{1}{\pi}\int_0^{2\pi} f(t)\sin\nu t d\omega_{2n+1}(t).$$

The complex Fourier–Lagrange coefficients $c_\nu^{(n)}$ are defined analogously. We shall be interested in the partial sums of the interpolating polynomials,

$$I_{n,\nu}(x,f) = \frac{1}{2}a_0 + \sum_{k=0}^{\nu}(a_k^{(n)}\cos kx + b_k^{(n)}\sin kx) = \frac{1}{\pi}\int_{-\pi}^{\pi} f(t)D_\nu(x-t)d\omega_{2n+1}(t),$$

where $n \geq \nu$.

Zygmund [36, Chap. X, 5.4] has shown

If f is of bounded variation, then $I_{n,\nu}(x,f) \to f(x)$ as $\nu \to \infty, n \geq \nu$, at every point of continuity of f. The convergence is uniform on every closed interval of points of continuity of f.

The proof he gives relies on his estimate of the order of magnitude of the Fourier–Lagrange coefficients and the $(C,1)$-summability of $\{I_{n,\nu}\}$. He then applies a Tauberian theorem of Hardy to obtain the desired result.

We gave an alternate proof of this result based directly on the definition of bounded variation. This method may be applied to analyze the convergence of the sequence $\{I_{n,\nu}(x,f)\}$ for functions of generalized bounded variation. In particular, it may be used for the class HBV. For $f \in HBV$, the conclusion of Zygmund's theorem holds, but if $\Lambda BV \supsetneq HBV$, then there is an $f \in \Lambda BV$ for which the result fails. We also found an estimate of the magnitude of the Fourier–Lagrange coefficients of functions in ΛBV which agrees with the known estimate when $\Lambda BV = BV$.

References

1. Avdispahić, M.: Concepts of generalized bounded variation and the theory of Fourier series. Internat. J. Math. Math. Sci. **9**, 223–244 (1986)
2. Baernstein, A. II: On the Fourier series of functions of bounded Φ-variation. Studia Math. **42**, 243–248 (1972)
3. Baernstein, A. II, and Waterman, D.: Functions whose Fourier series converge uniformly for every change of variable. Indiana Univ. Math. J. **22**, 569–576 (1972/73)
4. Berezhnoi, E.I.: Spaces of functions of generalized bounded variation. II. Problems of the uniform convergence of Fourier series. Sibirsk. Mat. ZH. **42**, 515–532 (2001), Siberian Math. J. **43**, 435–449 (2001)
5. Birnbaum, Z., Orlicz, W.: Über die Verallgemeinerung des Begriffes der zueinander konjugierten Potenzen. Sudia Math. **3**, 1–67 (1931)
6. Chanturiya, Z.A.: On Uniform Convergence of Fourier Series. Math. Sb. **100**, 534–554 (1976), Math. USSR Sb. **29**, 475–495 (1976)
7. Dyachenko, M.I.: Waterman classes and spherical partial sums of double Fourier series. Anal. Math. **21**(1), 321 (1995)
8. Dyachenko, M.I., Waterman, D.: Convergence of double Fourier series and W-classes. Trans. Amer. Math. Soc. **357**, 397–407 (2005)
9. Dyachenko, M.I.: Multidimensional alternative Waterman classes,Vestnik Moskov. Univ. Ser. I Mat. Mekh. **71**, 18–25 (2006); translation in Moscow Univ. Math. Bull. **61**, 18–24 (2006)
10. Garsia, A.M., Sawyer, S.: On some classes of continuous functions with convergent Fourier series. J. Math. Mech. **13**, 589–601 (1964)

11. Goffman, C., Waterman, D.: Functions whose Fourier series converge for every change of variable. Proc. Amer. Math. Soc. **19**, 80–86 (1968)
12. Goffman, C., Waterman, D.: Some aspects of Fourier series. Amer. Math. Monthly **77**, 119–133 (1970)
13. Goffman, C.: Everywhere convergence of Fourier series. Indiana Univ. Math. J. **20**, 107–112 (1970/1971)
14. Goffman, C.: The localization principle for double Fourier series. Studia Math. **69**, 41–57 (1980)
15. Goffman, C., Nishiura, T., Waterman, D.: Homeomorphisms in analysis. In: Mathematical Surveys and Monographs, vol. 54. Amer. Math. Soc., Providence, RI (1997)
16. Love, E.R., Young, L.C.: Sur une classe de functionelles linéaires. Fund. Math. **28**, 243–257 (1937)
17. Musielak, J., Orlicz, W.: On generalized variations (I). Studia Math. **18**, 11–41 (1959)
18. Oskolkov, K.I.: Generalized variation, the Banach indicatrix and the uniform convergence of Fourier series. Mat. Zametki **12**, 313–324 (1972), also, Math. Notes **12**, 619–625 (1972/1973)
19. Pierce, P., Waterman, D.: Regulated functions whose Fourier series converge for every change of variable. J. Math. Anal. Appl. **214**, 264–282 (1997)
20. Pierce, P., Waterman, D.: AΔ_2-equivalent condition. Real Anal. Exchange 26 (2000/01), no. 2
21. Prus-Wiśniowski, F.: Functions of bounded Λ-variation. In: Topics in Classical Analysis and Applications in Honor of Daniel Waterman, pp. 173–190, World Sci. Publ., Hackensack (2008)
22. Saakyan, A.A.: On the convergence of double Fourier series of functions of bounded harmonic variation. Izv. Akad. Nauk Armyan. SSR Ser. Mat. **21**(6), 517–529 (1986)
23. Sablin, A.I.: Λ-variation and Fourier series. Soviet Math. (Iz. VUZ) **31**, 87–90 (1987)
24. Salem, R.: Essais sur les séries trigonométrique, Chap.VI, Actualities Sci. Ind. No.862, Hermann, Paris (1940)
25. Salem, R.: Œuvres mathématiques. Hermann, Paris (1967)
26. Waterman, D.: On convergence of Fourier series of functions of generalized bounded variation. Studia Math. **44**, 107–117 (1972)
27. Waterman, D.: On Λ-bounded variation. Studia Math. **57**, 33–45 (1976)
28. Waterman, D.: Fourier series of functions ofΛ-bounded variation. Proc. Amer. Math. Soc. **74**, 119–123 (1979)
29. Waterman, D.: Change-of-variable invariant classes of functions and convergence of Fourier series. In: Classical Real Analysis (Madison, Wis., 1982), vol. 42, pp. 203–211, Contemp. Math., Amer. Math. Soc., Providence, RI (1985)
30. Waterman, D.: A generalization of the Salem test. Proc. Amer. Math. Soc. **105**, 129–133 (1989)
31. Waterman, D., Xing, H.: The convergence of partial sums of interpolating polynomials. J. Math. Anal. Appl. **333**, 543–555 (2007)
32. Wiener, N.: The quadratic variation of a function and its Fourier coefficients. J. Mass. Inst. Tech. **3**, 73–94 (1924)
33. Young, L.C.: An inequality of the Hölder type connected with Stieltjes integration. Acta Math. **67**, 251–282 (1936)
34. Young, L.C.: Sur une généralization de la notion de variation de puissance p-ième bornée au sens de M. Wiener, et sur la convergence des séries de Fourier, C. R. Acad. Sci. Paris **204**, 470–472 (1937)
35. Young, L.C.: General inequalities for Stieltjes integrals and the convergence of Fourier series. Math. Ann. **115**, 581–612 (1938)
36. Zygmund, A.: Trigonometric Series, 2nd edn. Cambridge Univ. Press, New York (1959)

Stability and Robustness of Weak Orthogonal Matching Pursuits*

Simon Foucart

Abstract A recent result establishing, under restricted isometry conditions, the success of sparse recovery via orthogonal matching pursuit using a number of iterations proportional to the sparsity level is extended to weak orthogonal matching pursuits. The new result also applies to a pure orthogonal matching pursuit, where the index appended to the support at each iteration is chosen to maximize the subsequent decrease of the squared norm of the residual vector.

1 Introduction and Main Results

This note considers the basic compressive sensing problem, which aims at reconstructing s-sparse vectors $\mathbf{x} \in \mathbb{C}^N$ (i.e., vectors with at most s nonzero entries) from the knowledge of measurement vectors $\mathbf{y} = A\mathbf{x} \in \mathbb{C}^m$ with $m \ll N$. By now, it is well established—see [1, 3] and references therein—that this task can be carried out efficiently using the realization of a random matrix as a measurement matrix $A \in \mathbb{C}^{m \times N}$ and by using ℓ_1-minimization as a reconstruction map $\Delta : \mathbb{C}^m \to \mathbb{C}^N$. There exist alternatives to the ℓ_1-minimization, though, including as a precursor the orthogonal matching pursuit introduced by Mallat and Zhang in [7] in the context of sparse approximation. The basic idea of the algorithm is to construct a target support by adding one index at a time and to find the vector with this target support that best fits the measurement. It had become usual, when trying to reconstruct

*This work was triggered by the participation of the author to the NSF-supported Workshop in Linear Analysis and Probability at Texas A&M University.

S. Foucart
Drexel University, Philadelphia, PA, USA
e-mail: foucart@math.drexel.edu

D. Bilyk et al. (eds.), *Recent Advances in Harmonic Analysis and Applications*, Springer Proceedings in Mathematics & Statistics 25, DOI 10.1007/978-1-4614-4565-4_30,

s-sparse vectors, to analyze the convergence of this algorithm after s iterations since the sth iterate is itself an s-sparse vector. For instance, the exact reconstruction of all s-sparse vectors $\mathbf{x} \in \mathbb{C}^N$ from $\mathbf{y} = A\mathbf{x} \in \mathbb{C}^m$ via s iterations of orthogonal matching pursuit can be established under a condition on the coherence of the matrix A or even on its cumulative coherence; see [10] for details. Lately, there has also been some work [2,5,6,8] establishing sparse recovery under some (stronger than desired) conditions on the restricted isometry constants of the measurement matrix. We recall briefly that the sth restricted isometry constant δ_s of A is defined as the smallest constant $\delta \geq 0$ such that

$$(1-\delta)\|\mathbf{z}\|_2^2 \leq \|A\mathbf{z}\|_2^2 \leq (1+\delta)\|\mathbf{z}\|_2^2 \qquad \text{for all } s-\text{sparse } \mathbf{z} \in \mathbb{C}^N.$$

However, it was observed by Rauhut [9] and later by Mo and Shen [8] that s iterations of orthogonal matching pursuit are not enough to guarantee s-sparse recovery under (more natural) restricted isometry conditions. Nonetheless, it was recently proved by Zhang [11] that $30s$ iterations are enough to guarantee s-sparse recovery provided $\delta_{31s} < 1/3$. Zhang's result also covers the important case of vectors $\mathbf{x} \in \mathbb{C}^N$ which are not exactly sparse and which are measured with some error via $\mathbf{y} = A\mathbf{x} + \mathbf{e} \in \mathbb{C}^m$—in fact, it covers the case of measurement maps $A : \mathbb{C}^N \to \mathbb{C}^m$ that are not necessarily linear, too. This note imitates the original arguments of Zhang and extends the results to a wider class of algorithms. Namely, we consider algorithms of the following type:

Generic orthogonal matching pursuit

Input: measurement matrix A, measurement vector $\mathbf{y}$, index set S^0
Initialization: $\mathbf{x}^0 = \operatorname{argmin}\left\{\|\mathbf{y} - A\mathbf{z}\|_2, \operatorname{supp}(\mathbf{z}) \subseteq S^0\right\}$
Iteration: repeat the following steps until a stopping criterion is met at $n = \bar{n}$

$$S^{n+1} = S^n \cup \{j^{n+1}\}, \tag{GOMP$_1$}$$

$$\mathbf{x}^{n+1} = \operatorname{argmin}\left\{\|\mathbf{y} - A\mathbf{z}\|_2, \operatorname{supp}(\mathbf{z}) \subseteq S^{n+1}\right\} \tag{GOMP$_2$}$$

Output: $\bar{n}$-sparse vector $\mathbf{x}^\star = \mathbf{x}^{\bar{n}}$

The recipe for choosing the index j^{n+1} distinguishes between different algorithms, which are customarily initiated with $S^0 = \emptyset$ and $\mathbf{x}^0 = 0$. The classical orthogonal matching pursuit algorithm corresponds to a choice where

$$|(A^*(\mathbf{y} - A\mathbf{x}^n))_{j^{n+1}}| = \max_{1 \leq j \leq N} |(A^*(\mathbf{y} - A\mathbf{x}^n))_j|. \tag{1}$$

The weak orthogonal matching pursuit algorithm with parameter $0 < \rho \leq 1$ corresponds to a choice where

$$|(A^*(\mathbf{y} - A\mathbf{x}^n))_{j^{n+1}}| \geq \rho \max_{1 \leq j \leq N} |(A^*(\mathbf{y} - A\mathbf{x}^n))_j|. \tag{2}$$

We also single out a choice of $j^{n+1} \notin S^n$ where

$$\frac{|(A^*(\mathbf{y}-A\mathbf{x}^n))_{j^{n+1}}|}{\mathrm{dist}(\mathbf{a}_{j^{n+1}}, \mathrm{span}\{\mathbf{a}_i, i \in S^n\})} = \max_{j \notin S^n} \frac{|(A^*(\mathbf{y}-A\mathbf{x}^n))_j|}{\mathrm{dist}(\mathbf{a}_j, \mathrm{span}\{\mathbf{a}_i, i \in S^n\})}, \tag{3}$$

where $\mathbf{a}_1 \ldots, \mathbf{a}_N \in \mathbb{C}^m$ denote the columns of A. We call this (rather unpractical) algorithm pure orthogonal matching pursuit, because it abides by a pure greedy strategy of reducing as much as possible the squared ℓ_2-norm of the residual at each iteration. Indeed, the following observation can be made when $\mathbf{a}_1, \ldots, \mathbf{a}_N$ all have norm one.

Theorem 1. *For any generic orthogonal matching pursuit algorithm applied with a matrix $A \in \mathbb{C}^{m \times N}$ whose columns $\mathbf{a}_1, \ldots, \mathbf{a}_N$ are ℓ_2-normalized, the squared ℓ_2-norm of the residual decreases at each iteration according to*

$$\|\mathbf{y} - A\mathbf{x}^{n+1}\|_2^2 = \|\mathbf{y} - A\mathbf{x}^n\|_2^2 - \Delta_n,$$

where the quantity Δ_n satisfies

$$\Delta_n = \|A(\mathbf{x}^{n+1} - \mathbf{x}^n)\|_2^2 = x_{j^{n+1}}^{n+1} \overline{(A^*(\mathbf{y}-A\mathbf{x}^n))_{j^{n+1}}} = \frac{|(A^*(\mathbf{y}-A\mathbf{x}^n))_{j^{n+1}}|^2}{\mathrm{dist}(\mathbf{a}_{j^{n+1}}, \mathrm{span}\{\mathbf{a}_i, i \in S^n\})^2}$$
$$\geq |(A^*(\mathbf{y}-A\mathbf{x}^n))_{j^{n+1}}|^2.$$

This result reveals that the quantity Δ_n is computable at iteration n. This fact is apparent from its third expression, but it is hidden in its first two expressions. Maximizing Δ_n leads to the pure orthogonal matching pursuit, as hinted above. We informally remark at this point that Δ_n almost equals its lower bound when A has small restricted isometry constants, i.e., when its columns are almost orthonormal. Hence, under the restricted isometry property, the pure orthogonal matching pursuit algorithm almost coincides with the classical orthogonal matching pursuit algorithm. This is made precise below.

Theorem 2. *Given an integer $\bar{n} \geq 1$ for which $\delta_{\bar{n}} < (\sqrt{5}-1)/2$, the pure orthogonal matching pursuit algorithm iterated at most $\bar{n}$ times is a weak orthogonal matching pursuit algorithm with parameter $\rho := \sqrt{1 - \delta_{\bar{n}}^2/(1-\delta_{\bar{n}})}$.*

The proofs of Theorems 1 and 2 are postponed until Sect. 3. We now state the main result of this note, which reduces to Zhang's result for the classical orthogonal matching pursuit algorithm when $\rho = 1$ (with slightly improved constants). Below, the notation $\mathbf{x}_{\overline{S}}$ stands for the vector equal to $\mathbf{x}$ on the complementary set $\overline{S}$ of S and to zero on the set S, and the notation $\sigma_s(\mathbf{x})_1$ stands for the ℓ_1-error of best s-term approximation to $\mathbf{x}$.

Theorem 3. *Let $A \in \mathbb{C}^{m \times N}$ be a matrix with ℓ_2-normalized columns. For all $\mathbf{x} \in \mathbb{C}^N$ and all $\mathbf{e} \in \mathbb{C}^m$, let $(\mathbf{x}^n)$ be the sequence produced from $\mathbf{y} = A\mathbf{x} + \mathbf{e}$ and $S^0 = \emptyset$ by the* *weak orthogonal matching pursuit algorithm with parameter* $0 < \rho \leq 1$. *If* $\delta_{(1+3r)s} \leq 1/6$, $r := \lceil 3/\rho^2 \rceil$, *then there is a constant* $C > 0$ *depending only on* $\delta_{(1+3r)s}$ *and on* ρ *such that for any* $S \subseteq \{1,\ldots,N\}$ *with* $\mathrm{card}(S) = s$,

$$\|\mathbf{y} - A\mathbf{x}^{2rs}\|_2 \leq C\|A\mathbf{x}_{\overline{S}} + \mathbf{e}\|_2. \tag{4}$$

Furthermore, if $\delta_{(2+6r)s} \leq 1/6$, *then, for all* $1 \leq p \leq 2$,

$$\|\mathbf{x} - \mathbf{x}^{4rs}\|_p \leq \frac{C}{s^{1-1/p}} \sigma_s(\mathbf{x})_1 + D s^{1/p-1/2} \|\mathbf{e}\|_2 \tag{5}$$

for some constants $C, D > 0$ *depending only on* $\delta_{(2+6r)s}$ *and on* ρ.

2 Sparse Recovery via Weak Orthogonal Matching Pursuits

This section is dedicated to the proof of Theorem 3. We stress once again the similitude of the arguments presented here with the ones given by Zhang in [11]. Throughout this section and the next one, we will make use of the simple observation that

$$(A^*(\mathbf{y} - A\mathbf{x}^k))_{S^k} = 0 \qquad \text{for all } k \geq 1. \tag{6}$$

Indeed, with $n := k - 1 \geq 0$, the definition of $\mathbf{x}^{n+1}$ via Eq. (GOMP$_2$) implies that the residual $\mathbf{y} - A\mathbf{x}^{n+1}$ is orthogonal to the space $\{A\mathbf{z}, \mathrm{supp}(\mathbf{z}) \subseteq S^{n+1}\}$. This means that for all $\mathbf{z} \in \mathbb{C}^N$ supported on S^{n+1}, we have $0 = \langle \mathbf{y} - A\mathbf{x}^{n+1}, A\mathbf{z}\rangle = \langle A^*(\mathbf{y} - A\mathbf{x}^{n+1}), \mathbf{z}\rangle$, hence the claim. It will also be convenient to isolate the following statement from the flow of the argument.

Lemma 4. *Let* $A \in \mathbb{C}^{m \times N}$ *be a matrix with* ℓ_2*-normalized columns. For* $\mathbf{x} \in \mathbb{C}^N$ *and* $\mathbf{e} \in \mathbb{C}^m$, *let* $(\mathbf{x}^n)$ *be the sequence produced from* $\mathbf{y} = A\mathbf{x} + \mathbf{e}$ *by the weak orthogonal matching pursuit algorithm with parameter* $0 < \rho \leq 1$. *For any* $n \geq 0$, *any index set* U *not included in* S^n, *and any vector* $\mathbf{u} \in \mathbb{C}^N$ *supported on* U,

$$\begin{aligned} \|\mathbf{y} - A\mathbf{x}^{n+1}\|_2^2 &\leq \|\mathbf{y} - A\mathbf{x}^n\|_2^2 - \rho^2 \frac{\|A(\mathbf{u} - \mathbf{x}^n)\|_2^2}{\|\mathbf{u}_{\overline{S^n}}\|_1^2} \max\{0, \|\mathbf{y} - A\mathbf{x}^n\|_2^2 - \|\mathbf{y} - A\mathbf{u}\|_2^2\} \\ &\leq \|\mathbf{y} - A\mathbf{x}^n\|_2^2 - \frac{\rho^2(1-\delta)}{\mathrm{card}(U \setminus S^n)} \max\{0, \|\mathbf{y} - A\mathbf{x}^n\|_2^2 - \|\mathbf{y} - A\mathbf{u}\|_2^2\}, \end{aligned} \tag{7}$$

where $\delta := \delta_{\mathrm{card}(U \cup S^n)}$.

Proof. The second inequality follows from the first one by noticing that

$$\begin{aligned} \|A(\mathbf{u} - \mathbf{x}^n)\|_2^2 &\geq (1-\delta)\|\mathbf{u} - \mathbf{x}^n\|_2^2 \geq (1-\delta)\|(\mathbf{u} - \mathbf{x}^n)_{\overline{S^n}}\|_2^2, \\ \|\mathbf{u}_{\overline{S^n}}\|_1^2 &\leq \mathrm{card}(U \setminus S^n)\|\mathbf{u}_{\overline{S^n}}\|_2^2 = \mathrm{card}(U \setminus S^n)\|(\mathbf{u} - \mathbf{x}^n)_{\overline{S^n}}\|_2^2. \end{aligned}$$

Now, according to Theorem 1, it is enough to prove that

$$|(A^*(\mathbf{y}-A\mathbf{x}^n))_{j^{n+1}}|^2 \geq \rho^2 \frac{\|A(\mathbf{u}-\mathbf{x}^n)\|_2^2}{\|\mathbf{u}_{\overline{S^n}}\|_1^2}\{\|\mathbf{y}-A\mathbf{x}^n\|_2^2 - \|\mathbf{y}-A\mathbf{u}\|_2^2\} \tag{8}$$

when $\|\mathbf{y}-A\mathbf{x}^n\|_2^2 \geq \|\mathbf{y}-A\mathbf{u}\|_2^2$. Making use of Eq. (6), we observe on the one hand that

$$\begin{aligned}\Re\langle A(\mathbf{u}-\mathbf{x}^n), \mathbf{y}-A\mathbf{x}^n\rangle &= \Re\langle \mathbf{u}-\mathbf{x}^n, A^*(\mathbf{y}-A\mathbf{x}^n)\rangle = \Re\langle(\mathbf{u}-\mathbf{x}^n)_{\overline{S^n}}, (A^*(\mathbf{y}-A\mathbf{x}^n))_{\overline{S^n}}\rangle \\ &\leq \|(\mathbf{u}-\mathbf{x}^n)_{\overline{S^n}}\|_1 \|(A^*(\mathbf{y}-A\mathbf{x}^n))_{\overline{S^n}}\|_\infty \\ &\leq \|\mathbf{u}_{\overline{S^n}}\|_1 |(A^*(\mathbf{y}-A\mathbf{x}^n))_{j^{n+1}}|/\rho. \end{aligned} \tag{9}$$

We observe on the other hand that

$$\begin{aligned}2\Re\langle A(\mathbf{u}-\mathbf{x}^n), \mathbf{y}-A\mathbf{x}^n\rangle &= \|A(\mathbf{u}-\mathbf{x}^n)\|_2^2 + \|\mathbf{y}-A\mathbf{x}^n\|_2^2 - \|A(\mathbf{u}-\mathbf{x}^n)-(\mathbf{y}-A\mathbf{x}^n)\|_2^2 \\ &= \|A(\mathbf{u}-\mathbf{x}^n)\|_2^2 + \{\|\mathbf{y}-A\mathbf{x}^n\|_2^2 - \|\mathbf{y}-A\mathbf{u}\|_2^2\} \\ &\geq 2\|A(\mathbf{u}-\mathbf{x}^n)\|_2 \sqrt{\|\mathbf{y}-A\mathbf{x}^n\|_2^2 - \|\mathbf{y}-A\mathbf{u}\|_2^2}. \end{aligned} \tag{10}$$

Combining the squared versions of Eqs. (9) and (10), we arrive at

$$\|A(\mathbf{u}-\mathbf{x}^n)\|_2^2\{\|\mathbf{y}-A\mathbf{x}^n\|_2^2 - \|\mathbf{y}-A\mathbf{u}\|_2^2\} \leq \|\mathbf{u}_{\overline{S^n}}\|_1^2 |(A^*(\mathbf{y}-A\mathbf{x}^n))_{j^{n+1}}|^2/\rho^2.$$

Rearranging the terms leads to the desired inequality (8). □

Theorem 3 would actually hold for any sequence obeying the conclusion of Lemma 4 by virtue of the following proposition.

Proposition 5. *Let $A \in \mathbb{C}^{m\times N}$ be a matrix with ℓ_2-normalized columns. For an s-sparse vector $\mathbf{x} \in \mathbb{C}^N$ and for $\mathbf{e} \in \mathbb{C}^m$, let $\mathbf{y} = A\mathbf{x}+\mathbf{e}$ and let $(\mathbf{x}^n)$ be a sequence satisfying Eq. (7). If $\delta_{(1+3\lceil 3/\rho^2\rceil)s} \leq 1/6$, then*

$$\|\mathbf{y}-A\mathbf{x}^{\bar{n}}\|_2 \leq C\|\mathbf{e}\|_2, \qquad \bar{n} := 2\lceil 3/\rho^2\rceil \operatorname{card}(S\setminus S^0),$$

where $C>0$ is a constant depending only on $\delta_{(1+3\lceil 3/\rho^2\rceil)s}$ and on ρ.

Proof. The proof proceeds by induction on $\operatorname{card}(S\setminus S^0)$, where $S := \operatorname{supp}(\mathbf{x})$. If it is zero, i.e., if $S \subseteq S^0$, then the definition of $\mathbf{x}^0$ implies

$$\|\mathbf{y}-A\mathbf{x}^0\|_2 \leq \|\mathbf{y}-A\mathbf{x}\|_2 = \|\mathbf{e}\|_2,$$

and the result holds with $C=1$. Let us now assume that the result holds up to an integer $s'-1$, $s'\geq 1$, and let us show that it holds when $\operatorname{card}(S\setminus S^0) = s'$. We consider subsets of $S\setminus S^0$ defined by

$$U^0 = \emptyset \text{ and } U^\ell = \{\text{indices of } 2^{\ell-1} \text{ largest entries of } \mathbf{x}_{\overline{S^0}} \text{ in modulus}\} \text{ for } \ell \geq 1,$$

to which we associate the vectors

$$\tilde{\mathbf{x}}^\ell := \mathbf{x}_{\overline{S^0 \cup U^\ell}}, \quad \ell \geq 0.$$

Note that the last U^ℓ, namely, $U^{\lceil \log_2(s') \rceil + 1}$, is taken to be the whole set $S \setminus S^0$ (it may have less than $2^{\ell-1}$ elements), in which case $0 = \|\tilde{\mathbf{x}}^\ell\|_2^2 \leq \|\tilde{\mathbf{x}}^{\ell-1}\|_2^2/\mu$ for a constant μ to be chosen later. We may then consider the smallest integer $1 \leq L \leq \lceil \log_2(s') \rceil + 1$ such that

$$\|\tilde{\mathbf{x}}^{L-1}\|_2^2 \geq \mu \|\tilde{\mathbf{x}}^L\|_2^2.$$

Its definition implies the (possibly empty) list of inequalities:

$$\|\tilde{\mathbf{x}}^0\|_2^2 < \mu \|\tilde{\mathbf{x}}^1\|_2^2, \ldots, \|\tilde{\mathbf{x}}^{L-2}\|_2^2 < \mu \|\tilde{\mathbf{x}}^{L-1}\|_2^2.$$

For each $\ell \in [L]$, we apply inequality (7) to the vector $\mathbf{u} = \mathbf{x} - \tilde{\mathbf{x}}^\ell$ supported on $S^0 \cup U^\ell$ while noticing that $(S^0 \cup U^\ell) \cup S^n \subseteq S \cup S^n$ and $(S^0 \cup U^\ell) \setminus S^n \subseteq (S^0 \cup U^\ell) \setminus S^0 = U^\ell$, and we subtract $\|\mathbf{y} - A\mathbf{u}\|_2^2 = \|A\tilde{\mathbf{x}}^\ell + \mathbf{e}\|_2^2$ from both sides to obtain

$$\begin{aligned}
&\max\{0, \|\mathbf{y} - A\mathbf{x}^{n+1}\|_2^2 - \|A\tilde{\mathbf{x}}^\ell + \mathbf{e}\|_2^2\} \\
&\quad \leq \left(1 - \frac{\rho^2(1 - \delta_{s+n})}{\mathrm{card}(U^\ell)}\right) \max\{0, \|\mathbf{y} - A\mathbf{x}^n\|_2^2 - \|A\tilde{\mathbf{x}}^\ell + \mathbf{e}\|_2^2\} \\
&\quad \leq \exp\left(-\frac{\rho^2(1 - \delta_{s+n})}{\mathrm{card}(U^\ell)}\right) \max\{0, \|\mathbf{y} - A\mathbf{x}^n\|_2^2 - \|A\tilde{\mathbf{x}}^\ell + \mathbf{e}\|_2^2\}.
\end{aligned}$$

For any $K \geq 0$ and any $n, k \geq 0$ satisfying $n + k \leq K$, we derive by immediate induction that

$$\begin{aligned}
&\max\{0, \|\mathbf{y} - A\mathbf{x}^{n+k}\|_2^2 - \|A\tilde{\mathbf{x}}^\ell + \mathbf{e}\|_2^2\} \\
&\quad \leq \exp\left(-\frac{k\rho^2(1 - \delta_{s+K})}{\mathrm{card}(U^\ell)}\right) \max\{0, \|\mathbf{y} - A\mathbf{x}^n\|_2^2 - \|A\tilde{\mathbf{x}}^\ell + \mathbf{e}\|_2^2\}.
\end{aligned}$$

By separating cases in the rightmost maximum, we easily deduce that

$$\|\mathbf{y} - A\mathbf{x}^{n+k}\|_2^2 \leq \exp\left(-\frac{k\rho^2(1 - \delta_{s+K})}{\mathrm{card}(U^\ell)}\right) \|\mathbf{y} - A\mathbf{x}^n\|_2^2 + \|A\tilde{\mathbf{x}}^\ell + \mathbf{e}\|_2^2.$$

For some integer κ to be chosen later, applying this successively with

$$k_1 := \kappa \,\mathrm{card}(U^1), \ldots, k_L := \kappa \,\mathrm{card}(U^L), \text{ and } K := k_1 + \cdots + k_L,$$

yields, with $\nu := \exp(\kappa\rho^2(1-\delta_{s+K}))$,

$$\|\mathbf{y}-A\mathbf{x}^{k_1}\|_2^2 \le \frac{1}{\nu}\|\mathbf{y}-A\mathbf{x}^0\|_2^2 + \|A\tilde{\mathbf{x}}^1+\mathbf{e}\|_2^2,$$

$$\|\mathbf{y}-A\mathbf{x}^{k_1+k_2}\|_2^2 \le \frac{1}{\nu}\|\mathbf{y}-A\mathbf{x}^{k_1}\|_2^2 + \|A\tilde{\mathbf{x}}^2+\mathbf{e}\|_2^2,$$

$$\vdots$$

$$\|\mathbf{y}-A\mathbf{x}^{k_1+\cdots+k_{L-1}+k_L}\|_2^2 \le \frac{1}{\nu}\|\mathbf{y}-A\mathbf{x}^{k_1+\cdots+k_{L-1}}\|_2^2 + \|A\tilde{\mathbf{x}}^L+\mathbf{e}\|_2^2.$$

Dividing the first row by ν^{L-1}, the second row by ν^{L-2}, and so on, then summing everything, we obtain

$$\|\mathbf{y}-A\mathbf{x}^K\|_2^2 \le \frac{\|\mathbf{y}-A\mathbf{x}^0\|_2^2}{\nu^L} + \frac{\|A\tilde{\mathbf{x}}^1+\mathbf{e}\|_2^2}{\nu^{L-1}} + \cdots + \frac{\|A\tilde{\mathbf{x}}^{L-1}+\mathbf{e}\|_2^2}{\nu} + \|A\tilde{\mathbf{x}}^L+\mathbf{e}\|_2^2.$$

Taking into account that $\mathbf{x}-\tilde{\mathbf{x}}^0$ is supported on $S^0\cup U^0 = S^0$, the definition of $\mathbf{x}^0$ implies that $\|\mathbf{y}-A\mathbf{x}^0\|_2^2 \le \|\mathbf{y}-A(\mathbf{x}-\tilde{\mathbf{x}}^0)\|_2^2 = \|A\tilde{\mathbf{x}}^0+\mathbf{e}\|_2^2$; hence

$$\|\mathbf{y}-A\mathbf{x}^K\|_2^2 \le \sum_{\ell=0}^{L}\frac{\|A\tilde{\mathbf{x}}^\ell+\mathbf{e}\|_2^2}{\nu^{L-\ell}} \le \sum_{\ell=0}^{L}\frac{2(\|A\tilde{\mathbf{x}}^\ell\|_2^2+\|\mathbf{e}\|_2^2)}{\nu^{L-\ell}}.$$

Let us remark that, for $\ell \le L-1$ and also for $\ell = L$,

$$\|A\tilde{\mathbf{x}}^\ell\|_2^2 \le (1+\delta_s)\|\tilde{\mathbf{x}}^\ell\|_2^2 \le (1+\delta_s)\mu^{L-1-\ell}\|\tilde{\mathbf{x}}^{L-1}\|_2^2.$$

As a result, we have

$$\begin{aligned}\|\mathbf{y}-A\mathbf{x}^K\|_2^2 &\le \frac{2(1+\delta_s)\|\tilde{\mathbf{x}}^{L-1}\|_2^2}{\mu}\sum_{\ell=0}^{L}\left(\frac{\mu}{\nu}\right)^{L-\ell} + 2\|\mathbf{e}\|_2^2\sum_{\ell=0}^{L}\frac{1}{\nu^{L-\ell}}\\ &\le \frac{2(1+\delta_s)\|\tilde{\mathbf{x}}^{L-1}\|_2^2}{\mu(1-\mu/\nu)} + \frac{2\|\mathbf{e}\|_2^2}{1-\nu}.\end{aligned}$$

We choose $\mu = \nu/2$ so that $\mu(1-\nu/\mu)$ takes its maximal value $\nu/4$. It follows that, with $\alpha := \sqrt{8(1+\delta_s)/\nu}$ and $\beta := \sqrt{2/(1-\nu)}$,

$$\|\mathbf{y}-A\mathbf{x}^K\|_2 \le \alpha\,\|\tilde{\mathbf{x}}^{L-1}\|_2 + \beta\,\|\mathbf{e}\|_2. \tag{11}$$

On the other hand, with $\gamma := \sqrt{1-\delta_{s+K}}$, we have

$$\begin{aligned}\|\mathbf{y}-A\mathbf{x}^K\|_2 &= \|A(\mathbf{x}-\mathbf{x}^K)+\mathbf{e}\|_2 \ge \|A(\mathbf{x}-\mathbf{x}^K)\|_2 - \|\mathbf{e}\|_2\\ &\ge \gamma\|\mathbf{x}-\mathbf{x}^K\|_2 - \|\mathbf{e}\|_2 \ge \gamma\|\mathbf{x}_{\overline{S^K}}\|_2 - \|\mathbf{e}\|_2.\end{aligned}$$

We deduce that

$$\|\mathbf{x}_{\overline{S^K}}\|_2 \le \frac{\alpha}{\gamma}\|\tilde{\mathbf{x}}^{L-1}\|_2 + \frac{\beta+1}{\gamma}\|\mathbf{e}\|_2. \tag{12}$$

Let us now choose $\kappa = \lceil 3/\rho^2 \rceil$, which guarantees that

$$\frac{\alpha}{\gamma} = \sqrt{\frac{8(1+\delta_s)}{(1-\delta_{s+K})\exp(\kappa\rho^2(1-\delta_{s+K}))}} < 1, \qquad \text{since } \delta_{s+K} \le \delta_{(1+3\lceil 3/\rho^2\rceil)s} \le 1/6, \tag{13}$$

where we have used the fact that

$$K = \kappa(1+\cdots+2^{L-2}+\mathrm{card}(U^L)) < \kappa(2^{L-1}+s') \le 3\kappa s' \le 3\lceil 3/\rho^2\rceil s.$$

Thus, in the case $((\beta+1)/\gamma)\|\mathbf{e}\|_2 < (1-\alpha/\gamma)\|\tilde{\mathbf{x}}^{L-1}\|_2$, we derive from Eq. (12) that

$$\|\mathbf{x}_{\overline{S^K}}\|_2 < \|\tilde{\mathbf{x}}^{L-1}\|_2, \quad \text{i.e.,} \quad \|(\mathbf{x}_{\overline{S^0}})_{S\setminus S^K}\|_2 < \|(\mathbf{x}_{\overline{S^0}})_{(S\setminus S^0)\setminus U^{L-1}}\|_2.$$

But according to the definition of U^{L-1}, this yields

$$\mathrm{card}(S\setminus S^K) < \mathrm{card}((S\setminus S^0)\setminus U^{L-1}) = s' - 2^{L-1}.$$

Continuing the algorithm from iteration K now amounts to starting it from iteration 0 with $\mathbf{x}^0$ replaced by $\mathbf{x}^K$; therefore, the induction hypothesis implies that

$$\|\mathbf{y}-A\mathbf{x}^{K+\bar{n}}\|_2 \le C\|\mathbf{e}\|_2, \qquad \bar{n} := 2\lceil 3/\rho^2\rceil(s'-2^{L-1}).$$

Thus, since we also have the bound $K \le \kappa(1+\cdots+2^{L-2}+2^{L-1}) < \lceil 3/\rho^2\rceil \cdot 2^L$, the number of required iterations satisfies $K+\bar{n} \le 2\lceil 3/\rho^2\rceil s'$, as expected. In the alternative case where $((\beta+1)/\gamma)\|\mathbf{e}\|_2 \ge (1-\alpha/\gamma)\|\tilde{\mathbf{x}}^{L-1}\|_2$, the situation is easier since Eq. (11) yields

$$\|\mathbf{y}-A\mathbf{x}^K\|_2 \le \frac{\alpha(\beta+1)}{\gamma-\alpha}\|\mathbf{e}\|_2 + \beta\|\mathbf{e}\|_2 =: C\|\mathbf{e}\|_2,$$

where the constant $C \ge 1$ depends only on $\delta_{(1+3\lceil 3/\rho^2\rceil)s}$ and on ρ. This shows that the induction hypothesis holds when $\mathrm{card}(S\setminus S^0) = s'$. The proof is now complete. □

With Proposition 5 at hand, proving the main theorem is almost immediate.

Proof of Theorem 3. Given $S \subseteq [N]$ for which $\mathrm{card}(S) = s$, we can write $\mathbf{y} = A\mathbf{x}_S + \mathbf{e}'$, where $\mathbf{e}' := A\mathbf{x}_{\overline{S}} + \mathbf{e}$. Applying Proposition 5 to $\mathbf{x}_S$ and $\mathbf{e}'$ with $S^0 = \emptyset$ then gives the desired inequality

$$\|\mathbf{y}-A\mathbf{x}^{2rs}\|_2 \le C\|\mathbf{e}'\|_2 = C\|A\mathbf{x}_{\overline{S}}+\mathbf{e}\|_2$$

for some constant $C > 0$ depending only on $\delta_{(1+3r)s}$ and on ρ. Deducing the second part of the theorem from the first part is rather classical (see [4], for instance), and we therefore omit the justification of this part. □

Remark. For the pure orthogonal matching pursuit algorithm, the parameter ρ depends on the restricted isometry constants as $\rho^2 = 1 - \delta_{\bar{n}}^2/(1-\delta_{\bar{n}})$. In this case, to guarantee that $\alpha/\gamma < 1$ in Eq. (13), we may choose $\kappa = 3$ when $\delta_{s+K} \le \delta_{\bar{n}} \le 1/6$. This would yield the estimate (4) in $6s$ iterations provided $\delta_{10s} \le 1/6$ and estimate (5) in $12s$ iterations provided $\delta_{20s} \le 1/6$.

3 Pure OMP as a Weak OMP Under RIP

This section is dedicated to the proofs of Theorems 1 and 2. The latter justifies that the pure orthogonal matching pursuit algorithm is interpreted as a weak pure orthogonal matching pursuit algorithm with parameter depending on the restricted isometry constants.

Proof of Theorem 1. We start by noticing that the residual $\mathbf{y} - A\mathbf{x}^{n+1}$ is orthogonal to the space $\{A\mathbf{z}, \mathrm{supp}(\mathbf{z}) \subseteq S^{n+1}\}$ and in particular to $A(\mathbf{x}^{n+1} - \mathbf{x}^n)$, so that

$$\|\mathbf{y} - A\mathbf{x}^n\|_2^2 = \|\mathbf{y} - A\mathbf{x}^{n+1} + A(\mathbf{x}^{n+1} - \mathbf{x}^n)\|_2^2 = \|\mathbf{y} - A\mathbf{x}^{n+1}\|_2^2 + \|A(\mathbf{x}^{n+1} - \mathbf{x}^n)\|_2^2.$$

This establishes the first expression for Δ_n. As for the other statements about Δ_n, we separate first the case $j^{n+1} \in S^n$. Here, the ratio $|(A^*(\mathbf{y} - A\mathbf{x}^n))_{j^{n+1}}|/\mathrm{dist}(\mathbf{a}_{j^{n+1}}, \mathrm{span}\{\mathbf{a}_i, i \in S^n\})$ is not given any meaning, but we have $\Delta_n = 0$ in view of $S^{n+1} = S^n$ and $\mathbf{x}^{n+1} = \mathbf{x}^n$, and the awaited equality $(A^*(\mathbf{y} - A\mathbf{x}^n))_{j^{n+1}} = 0$ follows from Eq. (6). We now place ourselves in the case $j^{n+1} \notin S^n$. For the second expression of Δ_n, we keep in mind that $\mathbf{x}^{n+1} - \mathbf{x}^n$ is supported on S^{n+1}, that $(A^*\mathbf{y})_{S^{n+1}} = (A^*A\mathbf{x}^{n+1})_{S^{n+1}}$, and that $(A^*(\mathbf{y} - A\mathbf{x}^n))_{S^n} = 0$ to write

$$\begin{aligned}
\|A(\mathbf{x}^{n+1} - \mathbf{x}^n)\|_2^2 &= \langle \mathbf{x}^{n+1} - \mathbf{x}^n, A^*A(\mathbf{x}^{n+1} - \mathbf{x}^n)\rangle = \langle \mathbf{x}^{n+1} - \mathbf{x}^n, (A^*A(\mathbf{x}^{n+1} - \mathbf{x}^n))_{S^{n+1}}\rangle \\
&= \langle \mathbf{x}^{n+1} - \mathbf{x}^n, (A^*(\mathbf{y} - A\mathbf{x}^n))_{S^{n+1}}\rangle \\
&= \langle (\mathbf{x}^{n+1} - \mathbf{x}^n)_{S^{n+1}\setminus S^n}, (A^*(\mathbf{y} - A\mathbf{x}^n))_{S^{n+1}\setminus S^n}\rangle \\
&= x^{n+1}_{j^{n+1}} \overline{(A^*(\mathbf{y} - A\mathbf{x}^n))_{j^{n+1}}}.
\end{aligned}$$

For the third expression, we first notice that the step (GOMP$_2$) is equivalent to

$$\mathbf{x}^{n+1} = A^{\dagger}_{S^{n+1}}\mathbf{y}, \qquad A^{\dagger}_{S^{n+1}} := (A^*_{S^{n+1}} A_{S^{n+1}})^{-1} A^*_{S^{n+1}}.$$

We decompose $A_{S^{n+1}}$ as $A_{S^{n+1}} = \left[A_{S^n} \,\middle|\, \mathbf{a}_{j^{n+1}} \right]$. We then derive that

$$A^*_{S^{n+1}}\mathbf{y} = \left[\begin{array}{c} A^*_{S^n}\mathbf{y} \\ \hline \mathbf{a}^*_{j^{n+1}}\mathbf{y} \end{array}\right], \qquad A^*_{S^{n+1}}A_{S^{n+1}} = \left[\begin{array}{c|c} A^*_{S^n}A_{S^n} & A^*_{S^n}\mathbf{a}_{j^{n+1}} \\ \hline \mathbf{a}^*_{j^{n+1}}A_{S^n} & 1 \end{array}\right].$$

Among the several ways to express the inverse of a block matrix, we select

$$(A^*_{S^{n+1}}A_{S^{n+1}})^{-1} = \left[\begin{array}{c|c} M_1 & M_2 \\ \hline M_3 & M_4 \end{array}\right],$$

where M_3 and M_4 (note that M_4 is a scalar in this case) are given by

$$M_4 = (1 - \mathbf{a}^*_{j^{n+1}}A_{S^n}(A^*_{S^n}A_{S^n})^{-1}A^*_{S^n}\mathbf{a}_{j^{n+1}})^{-1} = (1 - \mathbf{a}^*_{j^{n+1}}A_{S^n}A^{\dagger}_{S^n}\mathbf{a}_{j^{n+1}})^{-1},$$

$$M_3 = -M_4\mathbf{a}^*_{j^{n+1}}A_{S^n}(A^*_{S^n}A_{S^n})^{-1}.$$

It follows from the block decomposition of $\mathbf{x}^{n+1} = (A^*_{S^{n+1}}A_{S^{n+1}})^{-1}A^*_{S^{n+1}}\mathbf{y}$ that

$$x^{n+1}_j = M_3A^*_{S^n}\mathbf{y} + M_4\mathbf{a}^*_{j^{n+1}}\mathbf{y} = M_4\mathbf{a}^*_{j^{n+1}}(-A_{S^n}A^{\dagger}_{S^n}\mathbf{y} + \mathbf{y}) = M_4\mathbf{a}^*_{j^{n+1}}(\mathbf{y} - A\mathbf{x}^n).$$

We note that $\mathbf{a}^*_{j^{n+1}}(\mathbf{y} - A\mathbf{x}^n)$ is simply $(A^*(\mathbf{y} - A\mathbf{x}^n))_{j^{n+1}}$. As for M_4, we note that $A_{S^n}A^{\dagger}_{S^n}\mathbf{a}_{j^{n+1}}$ is the orthogonal projection $P_n\mathbf{a}_{j^{n+1}}$ of $\mathbf{a}_{j^{n+1}}$ onto the space $\{A\mathbf{z}, \mathrm{supp}(\mathbf{z}) \subseteq S^n\} = \mathrm{span}\{\mathbf{a}_i, i \in S^n\}$ so that

$$M_4 = (1 - \langle P_n\mathbf{a}_{j^{n+1}}, \mathbf{a}_{j^{n+1}}\rangle)^{-1} = (1 - \|P_n\mathbf{a}_{j^{n+1}}\|_2^2)^{-1} = \mathrm{dist}(\mathbf{a}_{j^{n+1}}, \mathrm{span}\{\mathbf{a}_i, i \in S^n\})^{-2}.$$

Note that $\|P_n\mathbf{a}_{j^{n+1}}\|_2 < \|\mathbf{a}_{j^{n+1}}\|_2 = 1$ justifies the existence of M_4 and that $\|P_n\mathbf{a}_{j^{n+1}}\|_2 \geq 0$ shows that $M_4 \geq 1$, which establishes the lower bound on Δ_n. □

Proof of Theorem 2. According to the previous arguments and to the definition of the pure orthogonal matching pursuit algorithm, for any $j \notin S^n$, we have

$$\frac{|(A^*(\mathbf{y} - A\mathbf{x}^n))_{j^{n+1}}|^2}{1 - \|P_n\mathbf{a}_{j^{n+1}}\|_2^2} = \frac{|(A^*(\mathbf{y} - A\mathbf{x}^n))_{j^{n+1}}|^2}{\mathrm{dist}(\mathbf{a}_{j^{n+1}}, \mathrm{span}\{\mathbf{a}_i, i \in S^n\})^2} \geq \frac{|(A^*(\mathbf{y} - A\mathbf{x}^n))_j|^2}{\mathrm{dist}(\mathbf{a}_j, \mathrm{span}\{\mathbf{a}_i, i \in S^n\})^2}$$
$$\geq |(A^*(\mathbf{y} - A\mathbf{x}^n))_j|^2.$$

But, in view of $|\langle A\mathbf{u}, A\mathbf{v}\rangle| \leq \delta_k\|\mathbf{u}\|_2\|\mathbf{v}\|_2$ for all disjointly supported $\mathbf{u}, \mathbf{v} \in \mathbb{C}^N$ satisfying $\mathrm{card}(\mathrm{supp}(\mathbf{u}) \cup \mathrm{supp}(\mathbf{v})) \leq k$, we have

$$\|P_n\mathbf{a}_{j^{n+1}}\|_2^2 = \langle P_n\mathbf{a}_{j^{n+1}}, \mathbf{a}_{j^{n+1}}\rangle = \langle A_{S^n}A^{\dagger}_{S^n}\mathbf{a}_{j^{n+1}}, \mathbf{a}_{j^{n+1}}\rangle = \langle AA^{\dagger}_{S^n}\mathbf{a}_{j^{n+1}}, A\mathbf{e}_{j^{n+1}}\rangle$$
$$\leq \delta_{n+1}\|A^{\dagger}_{S^n}\mathbf{a}_{j^{n+1}}\|_2 \leq \frac{\delta_{n+1}}{\sqrt{1 - \delta_n}}\|AA^{\dagger}_{S^n}\mathbf{a}_{j^{n+1}}\|_2 = \frac{\delta_{n+1}}{\sqrt{1 - \delta_n}}\|P_n\mathbf{a}_{j^{n+1}}\|_2.$$

After simplification, we obtain $\|P_n \mathbf{a}_{j^{n+1}}\|_2 \leq \delta_{n+1}/(\sqrt{1-\delta_n})$. The condition $\delta_{\bar{n}} \leq (\sqrt{5}-1)/2$ ensures that this bound does not exceed one. It then follows that, for $j \notin S^n$,

$$|(A^*(\mathbf{y} - A\mathbf{x}^n))_{j^{n+1}}|^2 \geq \left(1 - \frac{\delta_{n+1}^2}{1-\delta_n}\right)|(A^*(\mathbf{y} - A\mathbf{x}^n))_j|^2.$$

The latter also holds for $j \in S^n$ in view of Eq. (6); hence, Eq. (2) is established when $n+1 \leq \bar{n}$ with $\rho := \sqrt{1 - \delta_{\bar{n}}^2/(1-\delta_{\bar{n}})}$. □

Acknowledgement The author is supported by the NSF grant DMS-1120622.

References

1. Candès, E.J.: Compressive sampling. In: Proceedings of the International Congress of Mathematicians, Madrid, Spain (2006)
2. Davenport, M., Wakin, M.: Analysis of orthogonal matching pursuit using the restricted isometry property. IEEE Trans. Inform. Theory **56**(9), 4395–4401 (2010)
3. Donoho, D.L.: Compressed sensing. IEEE Trans. Inform. Theory **52**(4), 1289–1306 (2006)
4. Foucart, S.: Hard thresholding pursuit: An algorithm for compressive sensing. SIAM J. Numer. Anal. **49**(6), 2543–2563 (2011)
5. Liu, E., Temlyakov, V.: Orthogonal Super Greedy Algorithm and Applications in Compressed Sensing. IEEE Trans. Inform. Theory **58**(4), 2040–2047 (2012)
6. Maleh, R.: Improved RIP Analysis of Orthogonal Matching Pursuit. CoRR abs/1102.4311 (2011)
7. Mallat, S.G., Zhang, Z.: Matching pursuits with time-frequency dictionaries. IEEE Trans. Signal Process. **41**(12), 3397–3415 (1993)
8. Mo, Q., Shen, Y.: A Remark on the Restricted Isometry Property in Orthogonal Matching Pursuit. IEEE Trans. Inform. Theory **58**(6), 3654–3656 (2012)
9. Rauhut, H.: On the impossibility of uniform sparse reconstruction using greedy methods. Sampling Theory Signal Image Process. **7**(2), 197–215 (2008)
10. Tropp, J.: Greed is good: Algorithmic results for sparse approximation. IEEE Trans. Inform. Theory **50**(10), 2231–2242 (2004)
11. Zhang, T.: Sparse recovery with orthogonal matching pursuit under RIP. IEEE Trans. Inform. Theory **57**(9), 6215–6221 (2011)

[illegible] modification, we obtain [illegible]. The condition [illegible] ensures that this bound does not exceed one. It then follows that, for [illegible]

[illegible]

The [illegible] for $j \in S^n$ in view of Eq. (5) [illegible] Eq. (2) is satisfied [illegible] with [illegible]

Acknowledgements The author is supported by the NSF grant DMS-[illegible]

References

1. [illegible] sampling [illegible] Spring [illegible]
2. [illegible] using the restricted [illegible] IEEE [illegible]
3. [illegible] Trans. Inform. Theory 52 [illegible]
4. [illegible]
5. [illegible]
6. [illegible]
7. [illegible]
8. [illegible] Matching [illegible]
9. [illegible]
10. [illegible] Trans. Inform. [illegible]
11. [illegible]